Handbook of Industrial Waste Disposal

Handbook of Industrial Waste Disposal

Richard A. Conway

Development Associate
Union Carbide Corporation
South Charleston, West Virginia

Richard D. Ross

Partner
Read-Ferry Company, Inc.
Haddonfield, New Jersey

Van Nostrand Reinhold Environmental Engineering Series

VNR VAN NOSTRAND REINHOLD COMPANY
NEW YORK CINCINNATI ATLANTA DALLAS SAN FRANCISCO
LONDON TORONTO MELBOURNE

Van Nostrand Reinhold Company Regional Offices:
New York Cincinnati Atlanta Dallas San Francisco

Van Nostrand Reinhold Company International Offices:
London Toronto Melbourne

Library of Congress Catalog Card Number: 79-15218
ISBN: 0-442-27053-4

Manufactured in the United States of America

Published by Van Nostrand Reinhold Company
135 West 50th Street, New York, N.Y. 10020

Published simultaneously in Canada by Van Nostrand Reinhold Ltd.

15 14 13 12 11 10 9 8 7 6 5 4 3 2 1

Library of Congress Cataloging in Publication Data

Conway, Richard A
Handbook of industrial waste disposal.

(Van Nostrand Reinhold environmental engineering series)
Includes index.
1. Factory and trade waste. I. Ross, Richard D., joint author. II. Title.
TD897.C36 628.5'4 79-15218
ISBN 0-442-27053-4

Van Nostrand Reinhold Environmental Engineering Series

THE VAN NOSTRAND REINHOLD ENVIRONMENTAL ENGINEERING SERIES is dedicated to the presentation of current and vital information relative to the engineering aspects of controlling man's physical environment. Systems and subsystems available to exercise control of both the indoor and outdoor environment continue to become more sophisticated and to involve a number of engineering disciplines. The aim of the series is to provide books which, though often concerned with the life cycle—design, installation, and operation and maintenance—of a specific system or subsystem, are complementary when viewed in their relationship to the total environment.

The Van Nostrand Reinhold Environmental Engineering Series includes books concerned with the engineering of mechanical systems designed (1) to control the environment within structures, including those in which manufacturing processes are carried out, and (2) to control the exterior environment through control of waste products expelled by inhabitants of structures and from manufacturing processes. The series includes books on heating, air conditioning and ventilation, control of air and water pollution, control of the acoustic environment, sanitary engineering and waste disposal, illumination, and piping systems for transporting media of all kinds.

Van Nostrand Reinhold Environmental Engineering Series

ADVANCED WASTEWATER TREATMENT, by Russell L. Culp and Gordon L. Culp

ARCHITECTURAL INTERIOR SYSTEMS–Lighting, Air Conditioning, Acoustics, John E. Flynn and Arthur W. Segil

SOLID WASTE MANAGEMENT, by D. Joseph Hagerty, Joseph L. Pavoni and John E. Heer, Jr.

THERMAL INSULATION, by John F. Malloy

AIR POLLUTION AND INDUSTRY, edited by Richard D. Ross

INDUSTRIAL WASTE DISPOSAL, edited by Richard D. Ross

MICROBIAL CONTAMINATION CONTROL FACILITIES, by Robert S. Rurkle and G. Briggs Phillips

SOUND, NOISE, AND VIBRATION CONTROL (Second Edition), by Lyle F. Yerges

NEW CONCEPTS IN WATER PURIFICATION, by Gordon L. Culp and Russell L. Culp

HANDBOOK OF SOLID WASTE DISPOSAL: MATERIALS AND ENERGY RECOVERY, by Joseph L. Pavoni, John E. Heer, Jr., and D. Joseph Hagerty

ENVIRONMENTAL ASSESSMENTS AND STATEMENTS, by John E. Heer, Jr. and D. Joseph Hagerty

ENVIRONMENTAL IMPACT ANALYSIS: A New Dimension in Decision Making, by R. K. Jain, L. V. Urban and G. S. Stacey

CONTROL SYSTEMS FOR HEATING, VENTILATING, AND AIR CONDITIONING (Second Edition), by Roger W. Haines

WATER QUALITY MANAGEMENT PLANNING, edited by Joseph L. Pavoni

HANDBOOK OF ADVANCED WASTEWATER TREATMENT (Second Edition), by Russell L. Culp, George Mack Wesner and Gordon L. Culp.

HANDBOOK OF NOISE ASSESSMENT, edited by Daryl N. May

NOISE CONTROL: HANDBOOK OF PRINCIPLES AND PRACTICES, edited by David M. Lipscomb and Arthur C. Taylor

AIR POLLUTION CONTROL TECHNOLOGY, by Robert M. Bethea

POWER PLANT SITING, by John V. Winter and David A. Conner

DISINFECTION OF WASTEWATER AND WATER FOR REUSE, by Geo. Clifford White

LAND USE PLANNING: Techniques of Implementation, by T. William Patterson

BIOLOGICAL PATHS TO SELF-RELIANCE, by Russell E. Anderson

HANDBOOK OF INDUSTRIAL WASTE DISPOSAL, by Richard A. Conway and Richard D. Ross

HANDBOOK OF ORGANIC WASTE CONVERSION, by Michael W. Bewick

LAND APPLICATIONS OF WASTE (Volume 1), by Raymond C. Loehr, William J. Jewell, Joseph D. Novak, William W. Clarkson and Gerald S. Friedman

LAND APPLICATIONS OF WASTE (Volume 2), by Raymond C. Loehr, William J. Jewell, Joseph D. Novak, William W. Clarkson and Gerald S. Friedman

STRUCTURAL DYNAMICS: Theory and Computation, by Mario Paz

Preface and acknowledgments

Due to both regulatory pressures and internal concerns, environmental protection has grown to play a major role in current industrial operations. All parties are reconciled that much must be done toward reducing various discharges to watercourses, to the atmosphere, and to the land. The points of concern are that the priority problems be properly identified and that the planned solutions be cost effective.

This book was prepared to aid those engineers, scientists, and managers now working on industrial pollution control, as well as those studying in this area and those concerned with regulation, to

1. Identify the real problems
2. Review alternative solutions
3. Ascertain how each solution applies
4. Recognize the key factors controlling each operation or process
5. Evaluate promising control techniques
6. Develop design criteria for each applicable treatment or disposal approach

The technology useful in everyday field practice is provided along with theoretical considerations when appropriate.

The content of this book provides an in-depth coverage of the treatment and disposal of industrial wastewater and solid wastes. With regard to wastewater, concerns of the wet-process industries, especially chemical and petroleum, are emphasized, but the technology presented is widely applicable. Industrial solid waste disposal is covered from a practical viewpoint. Air pollution control considerations are meshed with topics such as incineration; other highlights are presented in a separate chapter. The book is arranged as follows:

- Problem definition, regulations, and disposal alternatives
- Wastewater treatment and disposal
  - Equalization, neutralization, and clarification
  - Biological processes
  - Dewatering excess biomass and other sludges
  - Activated carbon adsorption
  - Special separations and reactions
  - Ultimate disposal
- Solid wastes
  - Preparation, recovery, and reuse
  - Incineration
  - Pyrolysis and wet-air oxidation
  - Land and ocean disposal
  - Contract disposal
  - Disposal of hazardous materials
- Air pollution considerations in industrial operations

Acknowledgments of R. A. Conway. The portion of this handbook concerning disposal of industrial wastewaters and sludges (Chapters 1–7) reflects the experience of the senior author and his coworkers in the Chemicals and Plastics Operations Division of Union Carbide Corporation throughout 22 years of investigations of means of handling organic chemical wastewaters. Key elements of prior technical papers and publications are drawn upon with suitable updating; coauthors of these works include P. R. Bradley, J. A. Fisher, C. W. Ganze, D. M. Guisti, J. C. Hovious, G. W. Kumke, C. T. Lawson, D. C. Macauley, K. S. Price, and G. T. Waggy. Mr. Bradley's contribution of certain equipment descriptions is especially acknowledged. The review of specific sections of this portion of the manuscript by T. W. Siegrist, E. J. McKenna, D. C. Macauley, R. F. Nelson, M. H. Spiegel, W. B. Arbuckle, F. D. Bess, and C. T. Lawson was most helpful. Preparation of the initial manuscript was accomplished by Linda Burford, without whose aid timely completion of this work would have been improbable. Marybeth Hoover and Linda Perkins prepared manuscript revisions.

The foresight of Union Carbide Corporation in sponsoring the long-term technical studies underlying much of the work cited in the wastewater sections of the book deserves recognition. Another thread through the cited studies are the many contributions from among former students, along with the senior author, of the environmental engineering program at the Massachusetts Institute of Technology while under the direction of Dr. C. N. Sawyer, Dr. R. E. McKinney, Mr. W. E. Stanley, and Dr. R. Eliassen. Finally, the patience and understanding of the Conway family (Nancy, David, and Dianne) during the lengthy period of research and writing required in completion of this manuscript are sincerely appreciated.

Acknowledgments of R. D. Ross. A book of this type takes many months to prepare, and it is never the work of the authors, the editors, and the publishers alone which brings it into print. It would be impossible to acknowledge everyone for their part in this book, but specific thanks go to my wife, Pat, and my family, and my partners at Read-Ferry Company for their patience and understanding during the birth of this manuscript, and to Barbara Judah for her excellent typing and collating efforts.

Special Acknowledgment to F. L. Cross, Jr. F. L. Cross, Jr., P.E., P.A., a Consulting Engineer in Orlando, Florida, provided much of the material used as a basis for Chapter 14, "Air Pollution Considerations in Industrial Plant Operations." He also provided material on recovery and reuse of solid waste. His considerable experience and extensive publication in this area were most helpful. This contribution is gratefully acknowledged.

Richard A. Conway
Richard D. Ross

Contents

Handbook of Industrial Waste Disposal

1

Problem definition, regulations, and disposal approaches

Environmental protection has taken its place beside efficient manufacturing, sales, and distribution as a major concern of industry. Pressures to handle wastes properly arise from several sources:

- Legislative and regulatory actions
- Concerns over known or suspected effects of a specific material, which is as-yet unregulated
- Process economics improvement through waste reduction
- Conservation of resources, including water
- Protection of workers' health

Responses by industry to these pressures largely demonstrate a high degree of responsibility. However, of concern to all parties is a comparison of the cost of the environmental-protection measures versus their true benefits to the environment.

The underlying purpose of this book is to provide practical technology useful in selecting and designing various environmental-protection operations and processes which economically and reliably fill identified needs. Theoretical considerations are introduced when required; however, everyday engineering practice and problem solutions are emphasized.

The achievement of cost-effective means of disposing of liquid and solid wastes from manufacturing operations involves the systematic consideration of a range of alternative approaches. The arrangement of this book was developed to facilitate a review of the alternatives, identify where each applies, enumerate the key factors controlling each operation or process, provide approaches for their evaluation, and describe major types of embodiments for each approach. References for further information are cited at the end of each chapter and on many tables and figures.

1-1 DISPOSAL ALTERNATIVES

The schematic in Table 1-1 depicts an overview of the available alternative means of liquid waste disposal. A comprehensive list of treatment alternatives for dilute and concentrated wastes is presented in Table 1-2. A treatment/ disposal schematic for solid wastes is shown in Figure 1-1. The processes and

TABLE 1-1. Overview of General Strategies in Liquid Waste Disposal

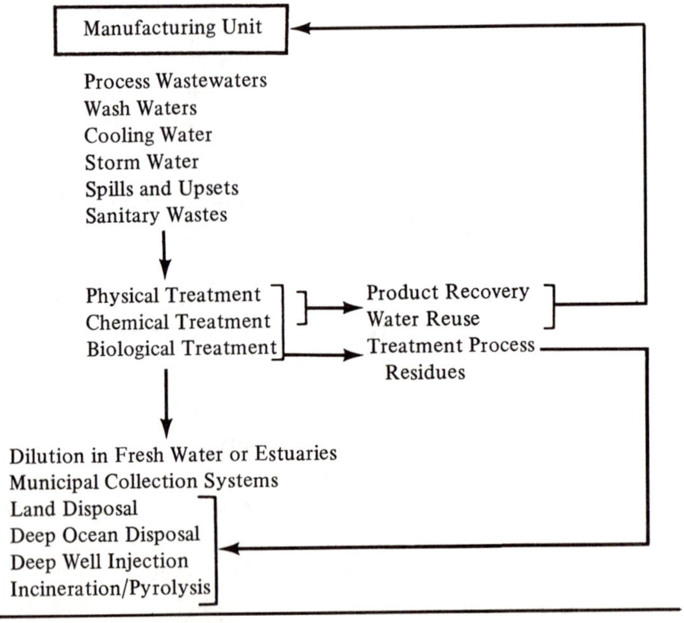

TABLE 1-2. Alternative Treatments for Liquid Wastes

1. Physical
   - Equalization/holding
   - Adsorption
   - Sizing—screening/straining, depth filtration, grinding/comminuting
   - Phase change—stripping, distillation/evaporation, freezing, sublimation
   - Ion exchange
   - Membrane processes—ultrafiltration, reverse osmosis, dialysis, liquid membranes, pervaporization
   - Force field separation—centrifugation, gravity-sedimentation, dissolved-air flotation, dispersed-air flotation
   - Surface methods—foaming, nonfoaming bubble separation, flotation, skimming
   - Extraction/absorption
2. Biological
   - Activated sludge—completely mixed, multistaged/multistepped, contact stabilization, oxygen-based
   - Fixed film processes—trickling filter, rotating disks
   - Aerated stabilization
   - Anaerobic—lagoons, contact stabilization, submerged filter
   - Algae stabilization ponds
   - Balanced ecosystems—aquatic plants, animals
   - Enzyme conversions—immobilized enzymes/cells
3. Chemical
   - Acid-base treatments
   - Chemical precipitation, coagulation, flocculation, polymerization
   - Oxidations—chlorination, ozonation, peroxides, permanganates, wet combustion, incineration, catalytic oxidation
   - Reduction reactions
   - Complexation
   - Photochemical reactions
   - Hydration and clathrates
   - Electromotive displacement
   - Thermal decomposition
4. Electrical and Electromagnetic
   - Ultraviolet irradiation
   - Electrolysis
   - Magnetic separation
   - Electrodialysis
   - Electron beam radiation
5. Acoustical
   - Ultrasonic
6. Nuclear
   - Irradiation

Source: Union Carbide Corp., prepared by E. H. Crum, R. F. Nelson, W. B. Arbuckle, and R. A. Conway.

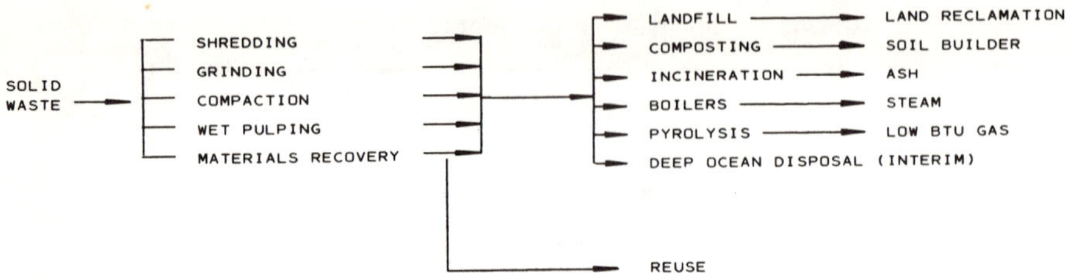

Fig. 1-1. Treatment/disposal sequence for solid wastes.

operations in fairly wide usage with liquids, sludges, and industrial solid wastes sludges are discussed individually in subsequent chapters. Operations specific to municipal-type solid waste disposal are covered in more detail in a handbook by Pavoni, Heer, and Hagerty (Ref. 1-1).

1-1.1 At-Source Modification

When deciding between in-plant modifications and end-of-the-pipe or terminal treatment for liquid wastes, a detailed cost comparison needs to be made. In the case of an existing plant, the costs of terminal treatment can be estimated according to the procedures given in Appendix 1-1. These costs per unit of pollution are also valuable as a control tool in minimizing waste generation, as they can be used to charge each manufacturing source for its true contribution to the total wasteload. A computerized method for doing this, based on production plans and standardized loading factors, has been described by Heath (Ref. 1-2).

Considerations other than cost at times dictate in-plant modifications. A toxic pollutant which cannot be reduced practically to a sufficiently low level due to equilibrium or kinetic considerations when treated after dilution with the total wastestream from a plant needs to be handled by at-source modifications, either by process changes or by an added recovery or destruction step. Recovery of needed energy or materials, including water when in scarce supply, can also motivate at-source modifications. Processes and operations presented in Chapter 6 are especially applicable at source units. Several practical techniques of conserving water and reducing organic discharges have been presented by Grover (1-3).

1-1.2 Terminal Treatment

After all efforts at recovery and reuse of process wastes have been made at the production units, the remaining dilute wastewater is routed through the plant

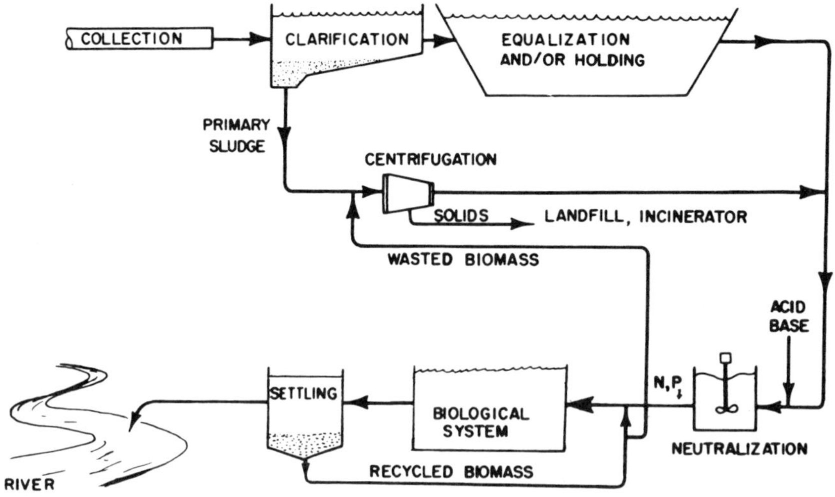

Fig. 1-2. Typical system components for treating dilute process wastes.

collection system to a central treatment facility. These dilute process wastes should be kept segregated from the return cooling water to minimize the required size of the treatment system.

The selection of applicable treatment-disposal approaches depends on the waste composition, the permissible discharge levels, and the amenabilities of the pollutants to treatment by the multitude of available methods. A conceptual design of a typical terminal treatment system for dilute wastewaters from an organic chemical plant is shown in Figure 1-2; Table 1-2 lists alternatives for these and other types of wastewater.

Treatment requirements and applicable techniques are discussed in this book to the extent necessary to provide a firm starting point in a waste disposal investigation. Selected bibliographies are provided to assist in detailed studies.

1-2 WASTE CHARACTERIZATION

The initial step in the rational development of a treatment scheme is to characterize the wastewater or residue in terms of source, generation rate, physical properties, chemical composition, and biological effects. Basic characterization parameters are listed in Table 1-3. The emphasis to be placed on each facet is dictated by the general nature of the waste and the permissible discharge levels.

TABLE 1-3. Basic Parameters in Wastewater Characterization

1. Source Information for the Individual Points of Origin
   - Waste components, individually or at least by classes (see below)
   - Rate of discharge during production run (average and maximum)
   - Periodic discharges due to batch operations
   - Duration and frequency of production runs
   - Susceptibility to emergency discharges or spills
2. Chemical Composition
   - Organic and inorganic components by compounds or classes
   - Gross organics: chemical oxygen demand (COD), total organic carbon (TOC), biochemical oxygen demand (BOD), extractables
   - Specific problem ions (As, Ba, Cd, Cr, CN, Hg, Pb, Se, Ag, NO_3)
   - Specific problem organics, e.g., phenol, certain pesticides, benzidine, polychlorinated biphenyls, certain polynuclear aromatics
   - Total dissolved salts
   - pH, acidity, alkalinity
   - Nitrogen and phosphorus
   - Oils and greases (extractables)
   - Oxidizing or reducing agents (e.g., sulfides)
   - Surfactants
   - Chlorine demand
3. Biological Effects
   - Biochemical oxygen demand
   - Toxicity (aquatic life, bacteria, mammals, plants)
   - Pathogenic bacteria
4. Physical Properties
   - Temperature range and distribution
   - Insoluble components: colloidal, settleable, floatable
   - Color
   - Odor
   - Foamability
   - Corrosiveness
   - Radioactivity
5. Flow Data for Total Discharge
   - Average daily flow rate
   - Duration and level of maximum flow rate
   - Maximum rate of change of flow rate

1-2.1 Environmental Risk

To assess the environmental risk of a given material in an effluent or leachate, a four-step process is necessitated:

1. Determine generation and discharge patterns.
2. Project a concentration in the environment considering transport, stability

(biological, chemical, and photo), and partitioning (water, air, soil, and cells) phenomena.

3. Determine toxic levels to the biological life of concern.
4. Compare projected environmental concentrations with toxic levels to determine the risk. When a decision is not clear from the preliminary assessment, repeat the process after developing more in-depth information.

A set of selected measurements to aid decision-making is presented in Table 1-4. The first-tier measurements usually are to be conducted on most materials known to persist in effluents. Second-tier measurements are used to aid in making border-line decisions. Various committees of the American Society for Testing and Materials are active in the area of environmental risk assessment.

1-2.2 Sampling/Analysis

Meaningful characterization information only can be obtained through proper analysis of representative samples or through the use of on-line water quality monitoring instrumentation.

Acceptable sampling and analytical methods have been compiled in *Standard Methods for the Analysis of Water and Wastewater* (Ref. 1-4); *Annual Book of ASTM Standards, Part 31 Water* (Ref. 1-5); and EPA's *Methods for Chemical Analysis of Water and Wastes* (Ref. 1-6). Common sampling pitfalls include: insoluble components not collected in proper proportion to the sample volume, peak discharges missed by collecting grab samples instead of using compositing equipment, samples not composited in proportion to flow rate, and samples not properly preserved. A state-of-the-art assessment of flow measurement in wastewater collection systems has been made by Shelley and Kirkpatrick (Ref. 1-7).

Emphasis needs to be placed upon obtaining as complete as possible an analysis of the components in the wastes. This was impossible from a practical viewpoint when only wet-chemical analytical methods were available. Atomic absorption and emission spectroscopy now provide means for achieving nearly complete metal analysis. Likewise, chromatographic techniques can be used for organic identifications. After proper sample pretreatment to achieve the desired fractionation, preconcentration, and/or matrix changes, gas chromatography, especially interfaced with mass spectroscopy, is useful for the identification of volatile components and liquid chromatography for the nonvolatiles.

A factor often given insufficient attention is the variability in waste components and strength. This has a major impact on achievable effluent quality, and needs to be defined. Factors to be considered are listed in Table 1-5.

When first faced with a waste disposal problem, the engineer in industry

TABLE 1-4. Selected Measurements to Assess Aquatic Environmental Risk

First-Tier Measurements

Theoretical oxygen demand, chemical oxygen demand, and total organic carbon

To measure the maximum demand on oxygen resources and determine amenability to common monitoring analyses.

Biochemical oxygen demand (5-day, 10-day, 20-day)

To measure the effect on oxygen resources and to gain the first reading of biodegradability.

Biodegradation

Testing with a more ideal environment (activated sludge, soil column) than is the case in the dilution-bottle BOD test.

Cell-water exchange (octanol/water partition)

Estimate cell-water distribution to evaluate the potential for bioconcentration

Atmosphere/water exchange (volatilization)

Loss potential to atmosphere from natural waters, soils, and waste treatment processes.

Soil/water exchange

Adsorptive losses to sediments.

Photolysis and hydrolysis

Nonbiological transformations.

Static acute fish toxicity

96-hr median tolerance level (TL_m) with fathead minnows.

Static acute invertebrate toxicity

96-hr TL_m with *Daphnia* (water flea).

Second-Tier Measurements

Life-cycle tests of aquatic toxicity

Daphnia as first choice to determine MATC (maximum acceptable toxicant concentration); this provides data for an application factor for translating acute-toxicity data to "no effect" level.

Fathead minnows (full or partial life-cycle) as backup choice.

Bioconcentration

Fathead minnows: follow water/organism partitioning with time at various dosage levels; follow depuration ("flush out") rate when dosage is removed.

Salt-water organism

Mysidopsis bahia

Algal growth stimulation

Standard USEPA laboratory test.

Static acute cold-water fish toxicity

Trout.

Microcosm

Balanced ecosystem in aquaria; follow fate and effect of radio-tagged materials.

Genetic acitivity assay

Rat toxicity

Plant toxicity

NOTE: See Ref. 1-4, 1-5, 1-6, and *Federal Register,* **44,** 53. March 16, 1979. Part IV, pp. 16240–16292.

TABLE 1-5. Causes of Variability in Manufacturing Wastes

Production Rate Changes
Variations in Plant Product Mix
Batch-Operated Units
Variations in Production Unit Efficiency
Changes in Raw Materials
Production Process Upsets
Maintenance (Equipment Shutdown and Cleanout)
Miscellaneous Leaks and Spills
Contaminated Drainage from Rainstorms

Source: Conway, R. A., Hovious, J. C., and Macauley, D. C., "Predicting Achievable Effluent Quality and Variability in the Organic Chemicals Industry," presented at International Conference on Effluent Variability from Waste Treatment Processes and Its Control, New Orleans, Louisiana, December 2–4, 1974, *J. Prog. Water Technol.* 7, Part 5, 1975.

is tempted to rely exclusively on the services of a consultant, internal or external. As in most situations, disposal problems are rarely solved merely by passing on responsibilities. Proper waste characterization by the engineers and chemists closest to the problem will greatly enhance the value of consulting services.

1-3 TREATMENT REQUIREMENTS

Once the waste has been adequately characterized, the treatment requirements can be established by comparing the quantity of pollutants to be discharged with the allowable level. The allowable discharge level is based on either effluent quality guidelines or receiving-water quality standards established by regulatory agencies. Since some latitude is provided for individual situational needs, a close examination of effluent quality achievable by conservatively designed, state-of-the-art facilities is worthwhile. A rational approach to be considered in negotiating direct discharge permits is presented in Appendix 1-2.

The assimilative capacity of the receiving body of water or municipal treatment plant needs to be examined, either by review of the background data affecting regulatory agency standards or by studies conducted by involved industries. Typical studies for direct discharges include stream surveys to define oxygen depletion and bioassays for effects on fish and diatoms. When discharges are to a municipal wastewater treatment plant, bench-scale investigations of the effects of the industrial wastes on unit operations at that plant can be conducted.

A list of the common effects of industrial pollutants on receiving systems is presented in Table 1-6. Detailed information on aquatic toxicity and other effects of specific materials has been compiled from several thousand literature

TABLE 1-6. Major Effects of Waste Material Upon Receiving Systems

1. Natural Bodies of Water
    a. Oxygen depletion (microbiological uptake, depressed reaeration)
    b. Interference with aquatic life (toxicity, fish-flesh tainting)
    c. Danger to public health and safety (pathogens, toxic chemicals)
    d. Miscellaneous economic damage (water potability, recreational losses)
2. Municipal Sewerage Systems
    a. Hazards to maintenance personnel (explosions, toxic gases)
    b. Damage to sewers (corrosion, deposition, overload)
    c. Decrease in efficiencies of sewage treatment processes (overload of unit capacities, microbiological toxicity)
3. Underground Strata
    a. Contamination of potable water supplies
    b. Contamination of formation brines and minerals
4. Land
    a. Irrigation: spray, overflow (vegetation damage)
    b. Burial (seepage to ground or surface waters)

references by the California State Water Quality Control Board (Ref. 1-8). A complete coverage of both the technical and judicial bases for the establishment of permissible discharge levels is included in this California publication. More recent data of a similar nature have been compiled by the United States Environmental Protection Agency (Ref. 1-9, 1-10). These detailed data have been reduced to limits in Ref. 1-11, additional quality criteria on priority pollutants separately have been established. The U.S. EPA standards for drinking water also can be of use in considering tolerable pollutant levels in water (Ref. 1-12, 1-13).

1-4 ENVIRONMENTAL REGULATION IN THE UNITED STATES

The scope of environmental regulation in the United States is indicated in Table 1-7. The key federal laws affecting disposal of industrial liquid and solid wastes are the Water Pollution Control Act Amendments and the Resources Conservation and Recovery Act; however, various sections of other legislation relate to specific operations like injection wells, stripping of volatile pollutants, ocean dumping, and discharge of toxic pollutants.

Implementation of the legislation is through regulations drawn up by the U.S. EPA. Considerable opportunity for input to the development of such regulations is provided to conservation organizations, state and local authorities, and industry. Such participation is required to develop regulations that meet these criteria:

TABLE 1-7. Major U.S. Environmental Legislation

Title	Year
Solid Waste Disposal Act	1965
Air Quality Act (PL 90-148)	1967
National Environmental Policy Act	1969
Clean Air Amendments (PL 91-604)	1970
Resource Recovery Act	1970
Occupational Safety and Health Act	1970
Marine Protection, Research, and Sanctuaries Act (PL 92-532)	1972
Water Pollution Control Act Amendments (PL 92-500)	1972
Federal Insecticide, Fungicide, and Rodenticide Act (PL 92-516)	1972
Consumer Product Safety Act	1972
Noise Pollution Control Act	1972
Hazardous Materials Transportation Act (PL 93-633)	1974
Safe Drinking Water Act (PL 93-523)	1974
Energy Supply and Environmental Coordination Act	1974
Energy Supply and Environmental Coordination Act (Amendments)	1975
Toxic Substances Control Act (PL 94-469)	1976
Noise Pollution Control Act Amendments	1976
Resource Conservation and Recovery Act (PL 94-580)	1976
Clean Air Act Amendments (PL 95-95)	1977
Water Pollution Control Act Amendments (PL 95-217)	1977

Source (updated): Daniels, S. L., "The Impact of the Resource Conservation and Recovery Act (PL 94-580) on Hazardous Chemical Waste Management," presented at 70th Annual Meeting of American Institute of Chemical Engineers, New York, November 17, 1977.

1. Improvements sought will have significant positive impact on environmental quality.
2. Anticipated environmental gains are commensurate with associated cost and resource commitments.
3. Goals are achievable with available technology.

Federal legislation and regulations provide for administration of control programs by the individual states when certain criteria have been met. In general, state regulations need to be at least as strict as national standards; at times, more restrictive standards are imposed due to local situations. Also state regulations can cover topics not yet resolved on a national level; examples in the past concerned protection of groundwater and discharge of heavy metals.

As regulations become more restrictive in terms of general pollutional parameters and more specific as to controlling certain hazardous materials, industry's best recourse in staying competitive world-wide on the long-term is through at-source modifications. As a starting point, manufacturing needs to be based on processes which generate very low levels of waste. Then, highly efficient recovery systems are needed at process units which unavoidably generate problem materials. Reliable controls are needed to minimize process upsets and excessive discharges during shut downs and start ups. All environmental control systems, at-source and end-of-pipe, need to be based on sound process technology, have adequate capacity and reliability, and be closely controllable.

1-4.1 Water Pollution Control Act Amendments

Key provisions from the Federal Water Pollution Control Act (FWPCA) Amendments affecting industry are presented in Table 1-8. The original overall goals were to eliminate discharge of pollutants by 1985, to achieve high water quality (suitable for fishing and swimming) by 1983, and eliminate the discharge of toxic pollutants in toxic concentrations (Ref. 1-14).

A National Pollutant Discharge Elimination System (NPDES) encompassing a comprehensive permit program was established. Effluent guidelines and standards used in NPDES are set by the EPA for various industries when discharging into navigable waters; the standards are end-of-pipe limitations generally expressed as milligrams of pollutant per liter of wastewater or pounds (kilograms) of pollutant per 1000 pounds (kilograms) of product. The regulations cover all types of liquid effluents from process wastes to stormwater when discharged through any type of conveyance into virtually any surface water.

Effluent standards and guidelines were issued in three levels of increasing restrictiveness: 1) those to be met by existing plants by 1977, using best practicable technology (BPT); 2) those to be met by new plants upon start up; and 3)

TABLE 1-8. Key Industrial Provisions of the U.S. Water Pollution Control Act Amendments

Establish best practicable treatment (BPT) guidelines for effluent limits for existing sources in various industrial categories.
Establish best available treatment (BAT) guidelines for effluent limits for existing and new sources in various industrial categories.
For new sources, publish list of categories and promulgate standards of performance.
Publish list of toxic pollutants and establish standards.
Promulgate standards for pretreatment of industrial discharges to municipal systems.
Publish guidelines and control methods for nonpoint sources, e.g., mining and agriculture.
Specify changes in water quality standards for interstate waters.
Establish guidelines for state programs, reporting, and test procedures.
Establish penalties for discharges of oil and hazardous substances.
Establish guidelines on ocean discharges.
Establish a national discharge permit system.

Note: The BPT program was to be met by 1977. BAT for toxics is to be met by 1984, as is best conventional control technology (BCT) for conventional pollutants such as BOD, suspended solids, pH, and fecal coliform. Non-conventional, non-toxic pollutants and new-source-performance standards also were addressed in the 1977 and 1978 changes.

those to be met by all plants by 1983 or 1984, using best available technology (BAT). The progressive restriction was aimed at achieving "zero discharge" of contaminants, i.e., effluent streams containing substances harmful to the receiving body of water (Ref. 1-15). A test of "reasonableness" is provided in the 1977 amendments before the EPA further reduces effluent standards for industry that has met the 1977 BPT standards for conventional (e.g., BOD, suspended solids) pollutants.

Control of contaminants by at-source changes or treatment needs to go beyond the prescribed effluent standards and guidelines when quality standards in the receiving watercourse would be exceeded.

The Effluent Guidelines and Standards and the Pretreatment Standards, which cover discharge into municipal systems, are similar for the various industrial classes. The only difference is an exemption for compatible contaminants, i.e., those which can be removed in municipal systems. Biochemical oxygen demand (BOD) and suspended solids are compatible contaminants.

Regulations and penalties for spills of oil and hazardous substances have been established. Spill Prevention, Control, and Countermeasure Plans (SPCC) are required of all facilities engaged in drilling, producing, refining, transferring, distributing, or consuming oil.

The FWPCA and the Marine Protection, Research, and Sanctuaries Act provide for ocean discharge criteria. Applicable rulings cover required permits, control of materials that may be dumped, and the location of dumping grounds.

Under Section 311 of the FWPCA, the EPA can regulate management practices for hazardous materials with regard to leaks, spills, and drainage from storage or plant-site runoff. Lists of hazardous substances are to be maintained by the EPA, removability of these substances from waterways determined, harmful quantities for each substance established, and a schedule of penalties developed.

1-4.2 Integrated Toxics Strategy

As gross pollution of the environment is lessened through concerted efforts of industry, agriculture, and municipalities, the emphasis will shift toward the control of toxic materials. In the United States, the regulatory program uses an integrated strategy involving provisions of the Federal Water Pollution Control Act (FWPCA), the Toxic Substances Control Act (TSCA), the Safe Drinking Water Act, the Resources Conservation and Recovery Act (RCRA), and the Marine Protection, Research, and Sanctuaries Act (ocean dumping act).

Section 307(a) of FWPCA is one tool for controlling discharge of toxic pollutants; a list of pollutants with documented health effects is being added as data are developed. In a related action, a consent decree to a suit brought by the National Resources Defense Council, *et al.* versus the EPA (Ref. 1-16)

(Appendix 1-3) has resulted in a list of priority pollutants. Water quality criteria initially were prepared for 65 of these materials and restrictions written for NPDES permits using Best Available Treatment effluent guidelines, to take effect by 1984. The 1977 amendments to FWPCA incorporate the 65 original Consent-Decree compounds and classes; the additional compounds listed in Appendix 1-3 indicate some potentially troublesome compounds in these classes.

Risk factors affecting toxic-compound guidelines include:

Toxicity data
Environmental fate information
Estimates of human and environmental exposure
Geographical distribution of the pollutant

Risk assessments for each priority pollutant are to be prepared as well as information on economic and technological factors. Environmental benefits are to be balanced against technological practicalities, i.e., reasonable-cost BAT to arrive at environmentally significant, cost-effective regulations.

1-4.3 Resource Conservation and Recovery Act

The Solid Waste Disposal Act of 1965 and the Resource Recovery Act of 1970 are directed toward municipal solid waste problems; the Resource Conservation and Recovery Act (RCRA) of 1976 has major impact on disposal of industrial solid wastes. The overall objective of RCRA is to provide technical and financial assistance to the development of management plans and facilities for the recovery of energy and other resources from discarded materials and for the safe disposal of discarded materials, and to regulate the management of hazardous wastes. The hazardous waste aspect (Sections 3 and 11) has considerable impact on a multitude of industrial operations. Pertinent provisions of RCRA are presented in Table 1-9.

Solid waste management has been divided in three sections for regulatory purposes: 1) generator—produces the waste, 2) transporter—hauls the waste from generator to processer, and 3) processor—stores, treats, and disposes of the waste. In practice, these roles can overlap for a given organization (Figure 1-3). Specific technical practices and record-keeping are required of all functions.

For purposes of this legislation as applied to industrial operations, a *solid waste* is discarded material in any form excluding industrial discharges which are point sources under FWPCA PL92-500. A *sludge* is any solid, semisolid, or liquid waste from an industrial wastewater treatment plant, water supply treatment plant, or air pollution control facility. A *hazardous waste* is a solid waste which, because of quantity, concentration, or characteristics may: 1) cause or contribute to increase in mortality or serious illness, or 2) pose substantial

TABLE 1-9. Hazardous Waste Management Aspects of the Resource Conservation and Recovery Act

Define the term "hazardous waste" and list specific hazardous wastes by criteria, and the quantities, concentrations, and forms in which they are a threat to human health or the environment after disposal.

Establish guidelines on alternative solid waste management practices to aid states and regional authorities in developing solid waste programs in the areas of hazardous wastes and open dumping.

Develop standards applicable to

1. Generation of hazardous wastes
2. Transportation of hazardous wastes
3. Owners and operators of hazardous waste treatment, storage, and disposal facilities

Establish a permit system for treatment, storage, or disposal of hazardous waste.

Provide guidelines to enable states to develop authorized hazardous waste programs.

Develop criteria for classifying waste disposal facilities as sanitary landfills or open dumps.

Inspect facilities, copy records, and obtain samples for enforcement purposes.

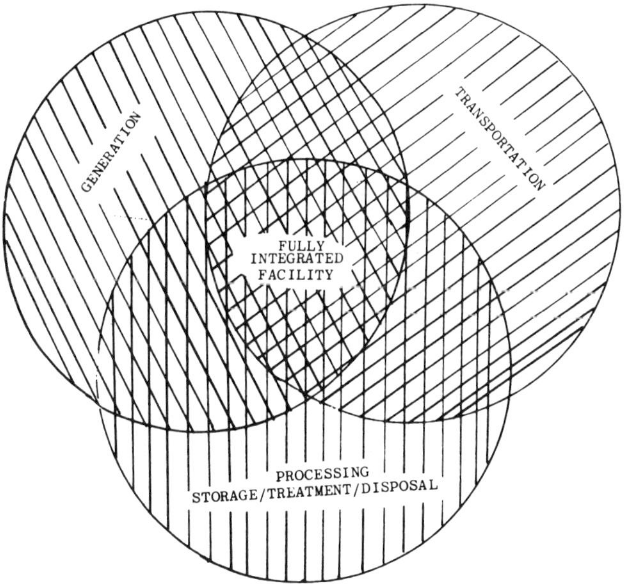

Fig. 1-3. Regulated Roles in Hazardous Solid Waste Management. ***Source:*** **Daniels, Stacy, L., "The Impact of the Resource Conservation and Recovery Act (PL 94-580) on Hazardous Waste Management," presented at 70th Annual Meeting of Am. Inst. of Chem. Engr., New York, November 17, 1977.**

present or potential hazard to human health or the environment when improperly managed.

Typical criteria used to identify characteristics of hazardous wastes include:

1. Flammability (e.g., flash point less than 60°C)
2. Corrosivity (e.g., pH below 3 or above 12)
3. Reactivity (e.g., autopolymerizable materials)
4. Radioactivity
5. Infectivity (e.g., source from laboratory handling pathogens)
6. Toxicity (e.g., acutely toxic, able to harm genetic material, bioaccumulative, persistent)

These criteria are subject to evolving regulations by the EPA; the above examples are cited only as guides. Toxicity criteria provide for either a direct toxicity determination or a comparison of constituents with drinking water standards, oral mammalian LD_{50} (median lethal dose) data, 96 hr LC_{50} (median lethal concentration) data, phytotoxicity data, lists of genetically active material, lists of persistent/bioaccumulative material (or low BOD/COD ratio or high octanol/water partition coefficient). The leachates generated by standardized laboratory techniques are used in toxicity determinations. (Ref. 1-20)

Besides identifying hazardous solid waste, generators need to be concerned with:

1. Preparation of manifests for each shipment, including immediate-response information telling what to do in an emergency.
2. Periodic reports to the EPA on manifest information.
3. Proper containerization and labeling practices.

Currently applicable EPA regulations need to be consulted by generators and transporters.

Storage, treatment, and disposal operations are controllable under a permit system. *Storage* is containment for a definite time before further processing. *Treatment* involves changing the character of the solid waste or eliminating the hazard. *Disposal* concerns incineration or the placing of solid waste into or on land for an indefinite time in a way so as to minimize environmental contamination. The overall philosophy of the hazardous waste program under RCRA is cradle-to-grave control.

Minimum criteria are established in EPA regulations for determining which solid waste landfills will be classified as posing no reasonable probability of harming public health or the environment. All point source discharges of pollutants from disposal facilities, including surface runoff, surface leachate, or leachate treatment effluent must comply with a National Pollutant Discharge Elimination System (NPDES) permit issued under the Water Pollution Control Act of 1972. Nonpoint sources have to be controlled to protect surface and ground waters. Draft criteria also forbid new disposal sites in wetlands under most circumstances and prohibit expansion of existing sites unless an NPDES permit has been obtained. All facilities must control air emissions to comply with applicable federal, state, or local air quality regulations. Open burning of industrial solid waste would be prohibited. (Ref. 1-20)

1-4.4 Other Legislation

National Environmental Policy Act. A draft environmental impact statement is required for projects involving federal funds or approval prior to taking any action which could significantly affect the environment. Locating a new plant in an undeveloped area would require consideration of

1. Environmental impact of the proposal
2. Any unavoidable adverse environmental effects
3. Alternatives
4. Impact of short-term use of the area on long-term productivity
5. Any irreversible and irretrievable commitment of resources

Clean Air Act Amendments of 1977. For hazardous pollutants (typically carcinogens), emission standards to the atmosphere are to be set; these are termed National Emission Standards for Hazardous Air Pollutants (NESHAP). Also, design or operational standards for control equipment are to be specified for over 600 manufacturing processes; these are termed New Source Performance Standards (NSPS). Reasonably Available Control Technology (RACT) and Best Available Control Technology (BACT) are established for such processes.

Obviously, pollutants may not be merely transferred from water and solid wastes to the air as might be the case if these interfaces were not considered. Source performance standards can apply to new or modified wastewater treatment plants; a major initial concern is stripping of organics. For problem organic emissions, at-source removal of the constituent(s) and the use of covered systems are control approaches.

National Primary and Secondary Ambient Air Quality Standards (AAQS) have been promulgated to set maximum ambient air standards for oxidants, CO, NO_2, SO_2, nonmethane hydrocarbons, particulate matter, asbestos, and lead.

Safe Drinking Water Act. Primary drinking water regulations are established for the protection of public health. Maximum contaminant levels are set as close as possible to the water quality level at which no known or anticipated adverse effects on health will occur (Ref. 1-12, 1-13). Interim primary water regulations are summarized in Table 1-10; these are periodically updated. Secondary regulations are set for elements in water related to taste, odor, and appearance (Ref.

TABLE 1-10. National Interim Primary Drinking Water Regulations—Maximum Contaminant Levels (Ref. 1-12)

Turbidity
5 turbidity units (based on an average for two consecutive days)

Coliform Bacteria (based on membrane filter technique)
4 colonies/100 ml in more than one sample when less than 20 are examined per month
4 colonies/100 ml in more than 5% of the samples when 20 or more are examined per month.

Inorganic Chemicals

Chemical	mg/1
Arsenic	0.05
Barium	1
Cadmium	0.010
Chromium	0.05
Lead	0.05
Mercury	0.002
Nitrate (as N)	10
Selenium	0.01
Silver	0.05
Fluoride[a]	1.4–2.4

Organic Chemicals

Chemical	mg/1
Endrin	0.0002
Lindane (γ isomer)	0.004
Methoxyclor	0.1
Toxaphene	0.005
2, 4-D	0.1
2, 4, 5-TP Silvex	0.01

[a]Depends on the annual average of the maximum daily air temperatures.

1-17). Emergency action can be evoked by the EPA when standards are exceeded.

Injection wells are to be regulated under a permit system; effects on groundwater of any subsurface waste disposal and of spills also are considered (Ref. 1-18).

Federal Insecticide, Fungicide and Rodenticide Act. The FIFRA provides the basis for EPA regulations concerning disposal of pesticides. All are strictly controlled; some are banned.

Occupational Safety and Health Act. Standards set for worker safety and health with respect to physical, chemical, and biological hazards, apply to waste treatment and disposal operations.

Toxic Substances Control Act (TSCA). Adverse environmental effects constitute one criterion used to determine chemical substances controllable under this act. Such substances are regulated in manufacturing, distribution, processing, use, and disposal. Of concern are serious long-term toxic effects, in particular those which might cause cancer, birth defects, or genetic damage. New chemicals and existing chemicals with substantial new uses are subject to extensive testing depending, for example, on expected environment persistance and toxicity to aquatic organisms (Ref. 1-19).

Marine Protection, Research, and Sanctuaries Act. Amendments in 1977 allow for no permits for ocean dumping of sewage sludge to be issued after 1981. Extensions can be granted when passing strict tests. Provisions under this act when coupled with FWPCA cover permits, materials that may be dumped, and location of dumping grounds. Ocean dumping is viewed in general as a last resort, as is deep-well injection.

1-5 TREATMENT AND DISPOSAL OPERATIONS

Common means of removing excess pollutants and of ultimately disposing of residues and treated wastewater are discussed in the ensuing chapters. Emphasis is placed upon the methods for developing design criteria and the basic features of the available process modifications and equipment.

Preliminary/Primary Treatment. Chapter 2 covers operations and processes often classed as preliminary and primary treatments. Equalization through back-mixing and hold-up serves to level out loadings to make subsequent operations more effective and economical. Neutralization of excess acidity and alkalinity is needed to allow optimum performance of following processes, especially biological systems. Coagulation and flocculation render insoluble pollutants removable. Clarification by sedimentation, flotation, straining, or filtration is an effective yet relatively inexpensive means of removing pollutants.

Biological Treatment. Microbiological treatment of wastewater to remove dissolved organic pollutants is the subject of Chapter 3. A basic consideration of the involved microbiology is followed by an in-depth discussion of the various activated sludge processes. Aerated stabilization, various fixed-film systems, anaerobic systems, and stabilization ponds are discussed as alternative approaches.

Sludge Dewatering and Stabilization. Dewatering sludges produced in primary clarification and in biological processes is covered in Chapter 4. The involved technology applies to other solid wastes from wet-process industries. Thickening, stabilization, and conditioning prior to the use of a range of types of dewatering equipment are important considerations. Centrifugation, pressure filtration, rotary vacuum filtration, and other dewatering operations are described.

Adsorption. Chapter 5 deals with the use of activated carbon to adsorb dissolved pollutants which are resistant to biodegradation, are toxic, or are economically recoverable. Such materials need to be readily adsorbable and to be subsequently removable from carbon during regeneration, e.g., phenol and color bodies. The amenability of various materials to adsorption, the selection of adsorbents and system conditions, and the physical enbodiments of the system are covered in the text.

Special Separations and Reactions. Separations and reactions which have narrower applicability in terms of materials and situations are described in Chapter 6 and include:

Membrane separation
Ion exchange
Physicochemical oxidation and reduction
Liquid–liquid extraction
Stripping
Evaporation
Other phase-change operations
Precipitation
Magnetic separation
Foam separation
Resin and other adsorptions
Electrochemical
Heat removal

Fundamental considerations and example applications are presented. These processes generally are applied to more concentrated wastestreams at the source manufacturing unit.

Ultimate Disposal of Wastewater and Residues. Chapter 7 covers the ultimate disposal of treated wastewater and by-product sludges and brines from treatment operations. Discussions of alternatives of discharge to surface waters or publicly owned treatment systems, disposal on land or by incineration, injection in deep-wells or barging to ocean dumping areas, and reuse and recovery are included. Disposal of solid-type wastes is an overview; more detail is provided in ensuing chapters. As in all chapters, practical considerations as well as fundamentals are presented.

Solid Waste Preparation, Recovery, and Reuse. Solid wastes generally are difficult and costly to handle as they are not uniform. Chapter 8 covers means (e.g., compaction, separation, and shredding) of preparing solid wastes for various final disposal methods. Recovery and reuse also are discussed.

Solid Waste Incineration. Incineration is an alternative considered for most solid wastes in land-short areas. Basic design considerations are presented in chapter 9 along with detailed descriptions of the most applicable types of equipment for common industrial solid wastes.

Solid Waste Pyrolysis and Wet Oxidation. *Pyrolysis* is defined as the decomposition of a material in the absence or near absence of oxygen under high temperature conditions. Solid wastes by this means can be converted largely into a storeable, transportable fuel, either liquid or gas, as discussed in Chapter 10. An aqueous-phase oxidation system also is covered.

Land and Ocean Disposal or Solid Wastes. Land disposal is generally the most cost-effective alternative for handling solid wastes. Chapter 11 discusses various land-disposal practices, as well as the applicability of ocean disposal.

Contract Disposal. Disposal of waste through contract with an outside organization often is attractive to industrial generators of each materials. Chapter 12 presents the considerations involved with governmental agencies and private disposal contractors.

Hazardous Material Disposal. Often little is known concerning the properties of materials requiring disposal. In such cases and for materials having identified hazards, special precautions are needed in disposal processes as outlined in Chapter 13.

Air Pollution Considerations. An overview of problems and practices in the control of air pollution from industrial operations is presented in the final chapter of this book. Control techniques for particulates and gaseous emissions are described.

REFERENCES

1-1. Pavoni, J. L., Heer, J. E., Jr., and Hagerty, D. J., *Handbook of Solid Waste Disposal; Materials and Energy Recovery*, Van Nostrand Reinhold Co., New York, 1975.

1-2. Heath, H. W., "Control of Wastewater Composition from a Large Chemical Plant," presented at Mid-Atlantic Industrial Waste Conference, January 12, 1976.

1-3. Grover, P., "Aqueous Waste Management," presented at Symposium New Techniques in Pilot Plants, American Institute of Chemical Engineers Meeting, Houston, Texas, March 20, 1977.

1-4. *Standard Methods for the Examination of Water and Wastewater,* 14th Ed. American Public Health Association, New York, 1976.

1-5. *Annual Book of ASTM Standards, Part 31 Water,* American Society for Testing and Materials, Philadelphia, 1977.

1-6. *Methods for Chemical Analysis of Water and Wastes,* U. S. Environmental Protection Agency, Washington, D. C., 1978.

1-7. Shelley, P. E. and Kirkpatrick, G. A., *Sewer Flow Measurement–A State-Of-The-Art Assessment,* EPA-600/2-75-027, U. S. Environmental Protection Agency, Cincinnati, Ohio 45268, 1975.

1-8. *Water Quality Criteria,* 2nd Ed., California State Water Quality Board, Sacramento, 1963.

1-9. *Water Quality Criteria Book, Vol. 3–Effects of Chemicals on Aquatic Life,* U. S. Environmental Protection Agency, May 1971.

1-10. Ibid, Vol. 5, U. S. EPA, September 1973.

1-11. *Quality Criteria for Water,* U. S. Environmental Protection Agency, Washington, D. C. 20460, 1976.

1-12. "Interim Primary Drinking Water Standards," *Fed. Regist.* **40**, 51, pp. 11990-119908, March 13, 1975.

1-13. "Drinking Water and Health, Recommendations of the National Academy of Sciences," *Fed. Regis.* **42**, 132, pp. 35761-35779, July 11, 1977.

1-14. Zarytkiewicz, E. D., "Federal Environmental Laws and Regulations," *Chem. Eng.*, 9-17, October 6, 1975.

1-15. "Water Bill–A Reasonable Compromise," *Chem Week,* 15-16, November 23, 1977.

1-16. *Natural Resources Defense Council, et al. v. Train,* 8 ERC 2120 (D. D. C. 1976).

1-17. "Secondary Regulations for Drinking Water," *Fed. Regis.* **42**, 62, p. 17143, March 31, 1977.

1-18. Hernandez, J. W., "Underground Injection Program Set Under Safe Drinking Water Act," *Water Sewage Works,* 67-69, October 1977.

1-19. Librizzi, W., "TSCA–An EPA View," *Chem. Tech.*, 405–411, July 1977.

1-20. "Hazardous Waste, Proposed Guidelines and Regulations and Proposal on Identification and Listing", *Federal Register,* **43**, 243, 58945–59028, December 16, 1978.

APPENDIX 1-1

COST ALLOCATION FOR GUIDING AT-SOURCE REDUCTION

Step 1

Calculate the total operating cost of the plant's water pollution control effort on an annual basis, consider:

Chemicals
Power
Assigned labor
Supervision
Maintenance
Laboratory
Travel
R/D and Engineering support
Fixed costs (overhead)

Step 2

Select the important operations and processes in the collection-treatment-disposal sequence at the plant; consider:

Collection system
Equalization
Clarification
Neutralization
Biological treatment
Filtration
Adsorption
Sludge dewatering and disposal

Step 3

Allocate the total cost from Step 1 among the important operations and processes identified in Step 2.

Step 4

Select the pollutional parameters critical in the plant's situation, consider:

Flow
Organic content (BOD, COD, TOC, slowly degrading or nondegradable organics, adsorbable organics)
Suspended solids (SS: total, settleable, floatable)
Acidity/alkalinity
Dissolved solids
Foam generation
Heat
Toxic components
Variability

Step 5

Allocate the cost of each selected operation and process among the critical pollutional parameters selected in Step 4.

Step 6

Sum the costs associated with each critical pollutional parameter.

Step 7

Divide the sums from Step 6 by the total number of units of each parameter (e.g., 1000 gal. of flow, pounds of BOD) projected for the year of concern for the plant. These "pers" (cost per pollutional unit) can be used to:

1. Charge generating facilities a realistic fee for waste treatment for the number of each type of pollution unit generated each month.
2. Project the value of load changes due to at-source reductions under consideration.

Investment or depreciation costs may be handled in a similar manner.

Example

1. Total annual operating costs of the plant's water pollution control effort is estimated to be $1,000,000 from Step 1.
2. Projected distribution of costs (Steps 2–6).

Key Treatment Steps	Total (M$/yr)[a]	Allocation to Parameters M$/yr) Flow	BOD	SS	Acidity
Primary					
Collection system	150[b]	100	30	10	10
Equalization	30	10	10		10
Clarification	100			100	
pH Adjustment	100				100
Secondary					
1st Step	570	70	500[c]		
2nd Step	None				
Sludge Disposal					
Sludge holding lagoon	50		40	10	
Sludge dewatering	None yet				
Ultimate disposal	None yet				
Tertiary					
Filtration	None yet[d]				
Carbon treatment	None yet[e]				
	1000	180	580	120	120

[a]M = 1000.
[b]Includes at-source monitoring.
[c]Could divide into two parameters, i.e., average BOD and maximum BOD to account for variability. Also, removable BOD might apply if a second biological step were involved.
[d]Costs might be allocated to BOD and/or total dissolved solids (causative agent).
[e]Costs might be allocated to a nondegradable COD parameter.

3. Projected plant discharges and dollar rates (Steps 6–7).

	Flow (M gal.)	BOD (lb)	SS (lb)	Acidity (lb as $CaCO_3$)
Estimated Annual Discharge Rate	730,000	7,300,000	3,650,000	3,650,000
Cost Rate ($/unit)[a]	0.247	0.079	0.033	0.033

[a]Divide dollar sum allocated to that parameter by the discharge rate.

4. Charges made to a typical source unit (e.g., monthly).

Measured or Estimated Discharge	Cost Rate ($/Unit)	Charges ($)
6083 M gal.	0.247	1503
60,833 lb BOD	0.079	4806
30,417 lb SS	0.033	1104
30,417 lb acidity	0.033	1104
		8517

APPENDIX 1-2

RATIONAL APPROACH FOR SETTING EFFLUENT QUALITY GUIDELINES

I. Establish Raw Waste Load for the Plant

A. Establish average and capacity production levels. Categorize materials into each of the product categories for which a standard waste discharge has been set by regulatory agencies.

B. Determine the variability of the production schedule, especially the percentile of capacity production the plant reaches on a high month and the day-to-day variability of the product mix in terms of both types and volumes.

C. Apply any available standard raw waste load factors (e.g., lb BOD_5/1000 lb product) to each product category and total up the plant pollutional load.

D. Compare this standard load with historical raw waste load data; consider the age of the processes, prior waste reduction programs and nonprocess load sources (i.e., laboratory, shipment vessel cleaning, and maintenance) when judging appropriateness of any discrepancy.

E. Review the volume of flow in which raw waste load is diluted; consider average flow rate factors for each product category and prior water conservation programs.

II. Calculate effluent quality for a well-designed treatment system based on raw waste load (Use pilot plant results and literature reports). The following example relates to the activated sludge treatment of organic chemical manufacturing wastes.

A. Use the raw waste load data to establish a long-term average concentration of BOD_5, COD, and TOC in the wastewater before treatment.

B. Project the removal expected in primary treatment (usually 8–15% for organic chemical manufacturing wastes depending on the oil content).

C. Project an expected effluent quality for the high month (S_{em}) using this model.

$$S_{em} = S_{ea}(S_o/1000)^c(\Theta_a)^{30-T}(V) + (B)(X_{eh}) \qquad \text{(Eq. 1-1)}$$

S_{ea} = Long-term average soluble 5-day BOD in mg/liter achievable with a single biological system in the field for a typical petrochemical feed of 1000 mg BOD_5/liter (S_{ea} = 40 mg/liter suggested in absence of other data).

S_o = Wastewater BOD_5 to biosystem, mg/liter.

c = Exponent for concentration effect (0.9 suggested).

Θ_a = Temperature coefficient for long-term effluent quality adjustment (1.06 suggested).

T = Projected biosystem temperature in °C for coldest month in the particular plant location.

V = Adjustment factor for effluent variability to convert from long-term average to maximum month, ranges from 2.0 for plant with frequently changing product mix to about 1.3 for plant with relatively steady manufacturing operations.

B = Conversion factor, mg BOD_5/mg suspended solids (0.3 suggested).

X_{eh} = Effluent suspended solids in high month (50 to 150 mg/liter suggested for well operated clarifier with petrochemical wastes; value depends on system stress factors such as temperature variation, feed strength, composition, and variability).

D. Project a highest daily pollutional discharge level; this is in the order to 2.0 to 3.0 times the high monthly average for plants having batch-type operations and frequently changing product mixes.

E. The effluent COD level from a biosystem approximates 25% of the influent level for a range of petrochemical wastes. An independent estimate can be made from totaling contributions of highly bioresistant components (can run about 10% of feed level for petrochemical wastes), residual slowly degradable soluble organic feed components and biointermediates, and effluent suspended solids.

F. When the dissolved oxygen level in the receiving body of water is a critical criterion, the effluent BOD level for warm-weather conditions should be calculated.

G. When laboratory studies are used to determine achievable effluent levels, a factor for inherent scale-up problems (e.g., reactor mixing and clarifier performance) needs to be applied to determine the parameter S_{ea} in the cited equation, i.e., multiply minimum effluent values obtained in the laboratory by a factor of at least 2 (possibly 3–4).

H. Other load sources for the plant, such as those in once-through cooling water flows and in surface water run-off, need to be considered in arriving at the total poundage discharge for the plant.

I. Prior discharge reductions, incineration, and other control approaches need to be considered along with the biological system reductions. Factors such as land cost and availability, age of facility, and energy requirements need to be considered when evaluating treatment alternatives available to the plant.

Source: Conway, R. A., Hovious, J. C., Macauley, D. C., "Predicting Achievable Effluent Quality and Variability in the Organic Chemicals Industry," presented at the International Conference on Effluent Variability from Waste Treatment Processes and Its Control, New Orleans, Louisiana December 2–4, 1974, *J. of Progress in Water Technology,* 8, 1, 165–182 (1976).

APPENDIX 1-3

PRIORITY WATER POLLUTANTS FROM NRDC VERSUS TRAIN CONSENT DEGREE

Compound Name

1. *acenaphthene
2. *acrolein
3. *acrylonitrile
4. *benzene
5. *benzidine
6. *carbon tetrachloride (tetrachloromethane)

**Chlorinated benezenes (other than dichlorobenzenes)*

7. chlorobenezene
8. 1, 2, 4-trichlorobenzene
9. hexachlorobenzene

**Chlorinated ethanes (including 1, 2-dichloroethane, 1, 1, 1-trichloroethane, and hexachloroethane)*

10. 1, 2-dichloroethane
11. 1, 1, 1-trichloroethane
12. hexachloroethane
13. 1,1-dichloroethane
14. 1, 1, 2-trichloroethane
15. 1, 1, 2, 2-tetrachloroethane
16. chloroethane

**Chloroalkyl ethers (chloromethyl, chloroethyl, and mixed ethers)*

17. bis(chloromethyl) ether
18. bis(2-chloroethyl) ether
19. 2-chloroethyl vinyl ether (mixed)

**Chlorinated naphthalene*

20. 2-chloronaphthalene

**Chlorinated phenols (other than those listed elsewhere; includes trichlorophenols and chlorinated cresols)*

21. 2, 4, 6-trichlorophenol
22. parachlorometa cresol

23. *chloroform (trichloromethane)
24. *2-chlorophenol

Dichlorobenzenes

25. 1, 2-dichlorobenzene
26. 1, 3-dichlorobenzene
27. 1, 4-dichlorobenzene

Dichlorobenzidine

28. 3, 3′-dichlorobenzidine

Dichloroethylenes (1, 1-dichloroethylene and 1, 2-dichloroethylene)

29. 1, 1-dichloroethylene
30. 1, 2-trans-dichloroethylene
31. *2, 4-dichlorophenol

Dichloropropane and dichloropropene

32. 1, 2-dichloropropane
33. 1, 2-dichloropropylene (1, 3-dichloropropene)
34. *2, 4-dimethylphenol

Dinitrotoluene

35. 2, 4-dinitrotoluene
36. 2, 6-dinitrotoluene
37. *1, 2-diphenylhydrazine
38. *ethylbenzene
39. *fluoranthene

Haloethers (other than those listed elsewhere)

40. 4-chlorophenyl phenyl ether
41. 4-bromophenyl phenyl ether
42. bis(2-chloroisopropyl) ether
43. bis(2-chloroethoxy) methane

Halomethanes (other than those listed elsewhere)

44. methylene chloride (dichloromethane)
45. methyl chloride (chloromethane)
46. methyl bromide (bromomethane)
47. bromoform (tribromomethane)
48. dichlorobromomethane
49. trichlorofluoromethane
50. dichlorodifluoromethane
51. chlorodibromomethane
52. *hexachlorobutadiene
53. *hexachlorocyclopentadiene
54. *isophorone

55. *naphthalene
56. *nitrobenzene

**Nitrophenols (including 2, 4-dinitrophenol and dinitrocresol)*

57. 2-nitrophenol
58. 4-nitrophenol
59. *2, 4-dinitrophenol
60. 4, 6-dinitro-o-cresol

**Nitrosamines*

61. N-nitrosodimethylamine
62. N-nitrosodiphenylamine
63. N-nitrosodi-n-propylamine
64. *pentachlorophenol
65. *phenol

**Phthalate esters*

66. bis(2-ethylhexyl) phthalate
67. butyl benzyl phthalate
68. di-n-butyl phthalate
69. di-n-octyl phthalate
70. diethyl phthalate
71. dimethyl phthalate

**Polynuclear aromatic hydracarbons*

72. benzo(a)anthracene (1, 2-benzanthracene)
73. benzo (a) pyrene (3, 4-benzopyrene)
74. 3, 4-benzofluoranthene
75. benzo(k)fluoranthane (11, 12-benzofluoranthene)
76. chrysene
77. acenaphthylene
78. anthracene
79. benzo(ghi)perylene (1, 12-benzoperylene)
80. fluroene
81. phenathrene
82. dibenzo (a, h) anthracene (1, 2, 5, 6-dibenzanthracene)
83. indeno (1, 2, 3-cd) pyrene (2, 3-o-phenylenepyrene)
84. pyrene
85. *tetrachloroethylene
86. *toluene
87. *trichloroethylene
88. *vinyl chloride (chloroethylene)

Pesticides and metabolites

89. *aldrin
90. *dieldrin

91. *chlordane (technical mixture and metabolites)

DDT and metabolites

92. 4, 4′-DDT
93. 4, 4′-DDE (p, p′-DDX)
94. 4, 4′-DDD (p, p′-TDE)

Endosulfan and metabolites

95. a-endosulfan-Alpha
96. b-endosulfan-Beta
97. endosulfan sulfate

Endrin and metabolites

98. endrin
99. endrin aldehyde

Heptachlor and metabolites

100. heptachlor
101. heptachlor epoxide

Hexachlorocyclohexane (all isomers)

102. a-BHC-Alpha
103. b-BHC-Beta
104. r-BHC (lindane)-Gamma
105. g-BHC-Delta

Polychlorinated biphenyls (PCB's)

106. PCB-1242 (Arochlor 1242)
107. PCB-1254 (Arochlor 1254)
108. PCB-1221 (Arochlor 1221)
109. PCB-1232 (Arochlor 1232)
110. PCB-1248 (Arochlor 1248)
111. PCB-1260 (Arochlor 1260)
112. PCB-1016 (Arochlor 1016)
113. *Toxaphene
114. *Antimony (Total)
115. *Arsenic (Total)
116. *Asbestos (Fibrous)
117. *Beryllium (Total)
118. *Cadmium (Total)
119. *Chromium (Total)
120. *Copper (Total)
121. *Cyanide (Total)
122. *Lead (Total)
123. *Mercury (Total)
124. *Nickel (Total)

125. *Selenium (Total)
126. *Silver (Total)
127. *Thallium (Total)
128. *Zinc (Total)
129. **2, 3, 7, 8, -tetrachlorodibenzo-p-dioxin (TCDD)

*Specific compounds and chemical classes as listed in the consent degree.
**This compound was specifically listed in the consent degree. Because of the extreme toxicity (TCDD), acquisition by laboratories of analytical standard for this compound is not recommended.

Source: Chemical Regulation Reporter, p. 1301, 1977.

2

Wastewater equalization, neutralization, and clarification

An initial consideration in treating industrial wastewaters is to achieve as high a degree of stability in quality and quantity of the feed stream as is practical. This "equalization" step both increases the effectiveness of subsequent processes and reduces their associated investment and operating costs. Neutralization of excess acidity or alkalinity is the next step often used to prepare wastes for subsequent treatments; although, when other contaminants are not present, it can serve as a sole treatment step. Clarification to remove suspended solids and immiscible liquids is the third step in a typical sequence before soluble constituents are addressed. Principles, design, and applications of equalization, neutralization, and clarification follow.

2-1 EQUALIZATION

Most approaches to design and operation of treatment facilities are based upon a steady feed stream in terms of composition and flow rate. However, this condition is seldom experienced in actual practice. Equalization techniques are desirable both to lessen facility costs by allowing design based on conditions nearer to the average and to improve operation by decreasing the variability of the feed. Of particular importance are:

- Leveling of contaminant concentration
- Self-neutralization
- Flow equalization
- Spill containment
- Holding for steady feed to a biological system during the manufacturing unit downtime
- Concurrent treatments, e.g., settling and oxidation

Several case histories of equalization applications have been compiled by Nemerow (Ref. 2-1).

Fundamental to the rational selection of a design basis or retention time is the acquisition both of wastewater flow data from charts showing instantaneous flow rates and of feed concentration data based on grab samples. Statistical methods for design using these data have been described by Adams and Eckenfelder (Ref. 2-2) and by Wallace (Ref. 2-3). An alternative approach is to use the duration of cyclic plant operations as a basis for determining retention time. An equalization basin capacity of 24 hr retention of the average daily flow and a surge basin capacity of 12 hr retention are suggested. Mixing energy is about 0.01 hp/1000 gal. (2×10^{-6} kW/liter). Equalization at individual process units, as well as at terminal facilities, should be considered.

In order to be effective in leveling out problem contaminants, an in-line basin needs to be mixed. Floating flow inducers, low-speed gate-type agitators, and aerators have been employed. The natural forces of wind and diffusion are important in larger basins. Multiple inlets, step feeding along the tank, and baffling to promote vertical mixing and prevent thermal stratification are other aids.

In addition to a flow-through basin in the main flow path, a parallel holding basin often is desirable. Spills or unusual discharges can be diverted to such a basin and later fed back into the main flow at a controlled rate. When discharging to a municipal sewer, such a holding basin also allows for discharge either in proportion to flow or during periods of underloading of the municipal treatment facility. For hydraulic equalization, a varying volume equalization basin is needed.

2-2 NEUTRALIZATION

2-2.1 Applications of pH Adjustment

Neutralization of excessively acidic or basic wastewaters is required prior to treatment and/or discharge. Bacteria and aquatic life are sensitive both to rapid pH variations and to pH levels outside the range of about 6 to 9. Consequently, surface waters and biological treatment units must be protected against unneutralized wastewater. A pH adjustment also is dictated in several other situations, including protection of construction materials, breaking of emulsions, insolubilization of certain organic materials, and control of chemical reaction rates (e.g., chlorination).

Many reagents can be used in neutralization reactions. The reaction systems themselves range from simple limestone pits to complex in-line mixing units. System design is a three-step procedure: 1) determination of neutralization requirements, 2) selection of reagents, and 3) design of the reaction and control system.

2-2.2 Neutralization Requirements

Accurate measurement of pH can be achieved readily with presently available equipment if frequent cleaning of electrodes is practiced to prevent fouling. The distribution of pH level of the wastewater usually is measured using an on-stream pH instrument with automatic recording of the read out. Composite samples should be used only when the neutralization facility is to be preceded by an equalization system. Also, an in-line holding tank with a detention time equal to that of the planned equalization basin or clarification unit can be used to smooth out pH fluctuations of a sample side stream prior to measurement.

A probability plot of the pH data should be made to show the frequency of occurrence of the various pH levels. Particular attention should be paid to the extreme levels since these control the required maximum rate of reagent feed. The maximum rate of change of pH level is also important in design of the neutralization control system.

In addition to pH level, buffer capacity and flow rate are factors in determining reagent needs. Composite samples are titrated to obtain the concentration of reagent, in terms of milliequivalents per liter, required to adjust the pH of the wastewater to the desired level. The titration data also are useful in reagent selection since weak acids are more difficult to neutralize with lime bases than with sodium basis. The flow rate data are used to convert the concentration of excess acidity or alkalinity to a weight rate of acid or base to be neutralized. The range of flows and rate of change of flow are also important in cases where waste equalization prior to the neutralization system is not practical due to space limitations.

2-2.3 Reagent Selection

Acids. The choice of an acidic reagent for neutralization of an alkaline wastewater is generally between sulfuric acid and hydrochloric acid. Sulfuric acid usually is used due to its lower cost, although if subsequent processes involve anaerobic biological treatment, the production of hydrogen sulfide can be a problem. Hydrochloric acid has the advantage of soluble reaction end products; however, the discharge of chlorides into fresh waters can be controlled by regulation, as can sulfates. Carbon dioxide from flue gas or submerged combustion also has been utilized in a few instances; this has the advantage of adding less to the total level of dissolved salts. Nitric acid is used when available as a waste material; it can provide a source of oxygen to a subsequent biological treatment system, but, if not removed, can contribute to eutrophication.

Bases. The selection of a caustic reagent usually is between sodium hydroxide and various limes; although, ammonium hydroxide is occasionally used. The availability of waste streams, such as acetylene sludge, frequently simplifies the rather involved study required to select the most suitable alkaline reagent. The factors to be considered in choosing the most suitable reagent include: purchase cost, neutralization capacity, reaction rate, storage and feeding, and neutralization products.

The cost of the various reagents is summarized in Table 2-1. Although sodium hydroxide costs much more than the other materials, it is frequently used due to uniformity, ease of storage and feeding, rapid reaction rate, and solubility of end products. Sodium carbonate is not as reactive as sodium hydroxide and can produce frothing problems due to carbon dioxide release. The lime materials have the obvious advantage of relatively low cost. This low material cost at least partially is offset by increased capital and operating costs for the rather complex feeding and reaction system which is required.

High calcium hydrate is fed as a 15% contained CaO slurry. High calcium quicklime usually is slaked prior to slurry feeding. When sulfuric acid is in the wastewater being neutralized, sludge and scale problems can be expected due to the insoluble calcium sulfate formed. These high calcium products have high reactivity.

Dolomitic hydrate and dolomitic quicklime have moderate reactivity and usually do not form insoluble end products due to the increased solubility of magnesium sulfate. The reactivity of these materials decreases at high pH levels and low temperatures. The inert content of all limes needs to be specified in purchases as it increases sludges and causes increased maintenance through wear.

Although high-calcium limestone is more reactive than dolomitic limestone, the reaction rate is still quite slow. Pulverizing can improve reactivity, but this can more than double the cost. The carbon dioxide which is produced must be removed by aeration if the pH is to be increased above 4.3, unless acid buffers are absent. Fouling of the surfaces of the limestone due to silt and end products can reduce neutralization capacity.

While the rate of reaction between the completely ionized sodium hydroxide

TABLE 2-1. Comparison of Commercial Alkaline Agents

Agent	Chemical Formula	Price[d] ($/ton)	Price[d] Relative	Basicity[e] Factor	Relative Cost per Basicity Unit
Sodium hydroxide[a]	NaOH	250	1.00	0.69	1.00
Sodium carbonate[b]	Na_2CO_3	105	0.42	0.51	0.57
High calcium hydrate	$Ca(OH)_2$	45	0.18	0.71	0.17
Dolomitic hydrate	$Ca(OH)_2 \cdot MgO$	47	0.19	0.91	0.14
High calcium quicklime	CaO	43	0.17	0.94	0.12
Dolomitic quicklime	$CaO \cdot MgO$	45	0.18	1.11	0.11
High calcium limestone[c]	$CaCO_3$	11	0.044	0.49	0.06
Dolomitic limestone[c]	$CaCO_3 \cdot MgCO_3$	13	0.052	0.56	0.06

[a]Solid, 75% Na_2O.
[b]Powder, 58% Na_2O.
[c]Pulverized–basicity somewhat dependent upon application.
[d]Representative price quotation for carlot quantities in bags or drums, 1975.
[e]Representative values for total basicity in terms of grams CaO per gram, as cited in Ross, R. D., ed., *Industrial Waste Disposal,* Van Nostrand Reinhold Co., New York, 1968.

and a strong acid waste is virtually instantaneous, the reactions of lime bases require considerable time for completion. Even with high calcium hydrate as much as 10 min can be required. Reaction time can be minimized by several approaches: a relatively high end point pH level, efficient mixing, and slurry feeding as opposed to dry feeding.

2-2.4 Reaction Systems

Reactors. The most common system for neutralizing acidic or basic waste streams utilizes a multiple compartmented basin usually constructed of concrete. This basin is lined with acid brick or coated with a material resistant to the expected environment. Where organics are expected in the waste stream, care must be taken to ensure that the coating is resistant to the organics as well as to corrosion produced by a widely varying pH level. In some cases the neutralization may be accomplished in the discharge sewer.

In order to reduce the required volume of the neutralization basin, mixers are installed in each compartment to provide more intimate contact between the waste and neutralizing reagents, thus speeding up reaction time. Stainless steel plates mounted on the floor of the pit and directly below the mixers will reduce erosion damage to the structure. Basin inlets are baffled to provide for flow distribution, while effluent baffles can help to prevent foam from being carried over into the receiving stream. Reaction times can be set on a rational basis (Ref. 2-4) depending upon:

1. maximum rate of acidity change, from records
2. overall process lag time (sum of reaction, measurement, and transport lags)
3. minimum normal residence time of a perfectly agitated basin to give process controllability

The reaction rate with certain limes can control reactor size; otherwise, 5 to 10 min in each stage of a 2-stage reaction system, based on maximum flow, is adequate when a feed forward control system is used.

Series operated in-line mixers and blenders are used in one large industrial waste treatment process (Ref. 2-5) to adjust stepwise a rapidly varying influent pH. A pH variation from 1 to 13, changing one pH unit/min, was satisfactorily controlled by this system. This process is computerized to control the addition of either acid or caustic to the basins; each basin is operated by an independent control loop. Both feed-forward and feed-back control systems are utilized.

Another chemical plant dissolved waste anhydrous hydrogen chloride gas in large quantities of water and neutralized the resultant material in limestone pits. Limestone is fed from hopper cars located directly above the pits. As the limestone is consumed in the pit, it is replaced from the hopper cars. Unless the effluent is aerated to remove the contained carbon dioxide, the pH of the discharge is about 4.3. Aeration will increase the pH to about 7.0. Inerts in the limestone and in the dilution water cause solids to build up in the pits, necessitating frequent shutdowns for cleaning.

In a columnar operation, an upflow expanded bed is preferable due to the carryout of some of the problem precipitates. Acids can be applied at 0.3 to 0.6%. Recycle can be used to dilute high strength feeds. Flow rates in the order of 100 gpm/sq ft (6000 m^3/m^2-day can be used when neutralizing hydrochloric acid.

Control. The operation of a neutralization process generally is controlled by measuring the influent and effluent pH, and adjusting the reagent feed rate accordingly. Where the incoming waste flows and pH levels are relatively stable, the influent measurements can be fed into a controller for adjustment of chemical feed rates to the reaction system. Where substantial fluctuations in flow and pH occur, the pH of the effluent from the reaction system is measured and appropriate changes in chemical addition are made.

Multiple units, complete with control systems, are generally used where widely varying pH levels are to be handled. For example, if the reagent feed needs to vary over a range of 300:1, from 0.3 to 0.001 gpm, use two valves (0.300 to 0.015 gpm and 0.020 to 0.001 gpm) in a feed forward system and, if necessary, trim with a final valve of low capacity in a second-stage feedback system. In order to avoid overcompensation, an adjustable signal dampener may be installed between the pH sensing element and chemical addition system. The pH is measured continuously, but changes in chemical feed rates are made only in given time intervals such as every 10 min.

The feeding of liquid chemicals to neutralization facilities is most easily accomplished by pumps. Slurries can be fed through gravity piping or pumps, but fewer problems occur when open flumes are utilized. Liquid caustic is satisfactorily stored in stress-relieved, bare steel or coated tanks, while rubber lined tanks are used for dilute acid storage. Limestone may be stored in the open due to its low solubility in water. Lime, on the other hand, must be protected from the weather; and storage in waterproof silos, hoppers, or bags is the normal practice.

2-3 CLARIFICATION

2-3.1 Clarification Applications

Removal of suspended solids and immiscible liquids from a wastewater discharge, along with neutralization, usually is the minimum requirement of pollution control regulations. These insoluble materials not only constitute a visible indication of pollution but also can have significant harmful effects in the receiving streams. Proper measurement of the quantity and character of these materials is the first step toward selection of the most economical removal process. The range of available separation equipment is sufficiently broad to allow tailoring of each facet of the separation process to the unique character of the suspension problem.

When used prior to biological treatment units, liquid-solids separation equipment functions both to collect rapidly settling organic matter in a cost-effective manner and to protect the biological unit from a buildup of inorganic and nondegradable organic material in the biological floc. Clarification is used after most biological treatment systems to remove settleable growth from the effluent and to thicken the biomass for recycle. A portion of the settled biomass is dewatered by filtration or centrifugation prior to its ultimate disposal.

2-3.2 Gravity Separation Studies

Concentration. The weight concentration of suspended solids can be suitably measured by standard gravimetric techniques, usually involving glass-fiber filters (Ref. 1-4). Immiscible liquids can be measured conveniently by allowing the wastewater to stand in a separatory funnel until separation is complete. After draining off the settled sludge and water layers, the volume and weight of the floatable material can be determined. A sufficient number of samples should be tested to define the frequency distribution of the insoluble components' concentrations.

Heavy Suspensions. The simplest method of determining settling rates of fairly heavy suspensions is to measure the subsidence of the sludge-liquid interface in a graduated cylinder. The data in Figure 2-1 illustrate this technique for estimating the clarifier size required to accomplish the desired separation (Ref. 2-6, 2-7). A stirrer is used to agitate the compacted sludge gently to simulate the raking action of the sludge collector in the prototype units (Ref. 2-8). The clarifier area predicted by these laboratory tests, conducted under ideal conditions, is increased by a hydraulic factor of about 1.5 to compensate for the turbulence and short circuiting occurring in the full-scale unit. When clarifying mixed liquor from an activated sludge system, thickening of sludge in the zones after free settling often controls design as discussed in a later section.

In some cylinder tests, a sharp interface does not develop if the initial solids

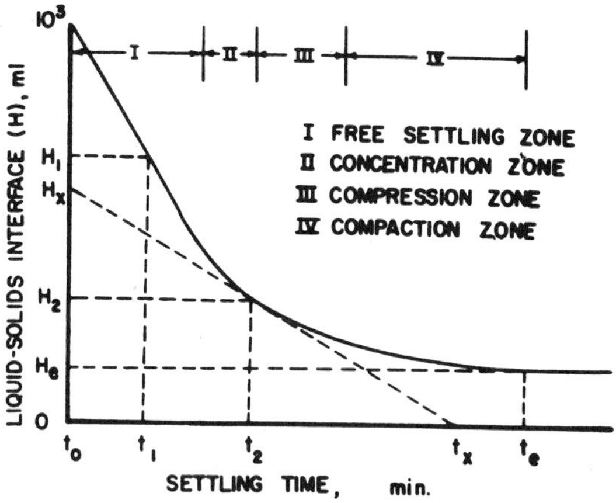

Area (sq ft) based on free-settling rate (ft/hr).

$$A = \frac{\text{Overflow}}{\text{Settling Rate}}$$

$$= \frac{\text{Feed Rate cu ft/hr - Underflow Rate cu ft/hr}}{(1000 \text{ ml} - H_1 \text{ ml}) \left(\frac{X \text{ in.}}{\text{liter}}\right) \left(\frac{1 \text{ liter}}{1000 \text{ ml}}\right) \left(\frac{1 \text{ ft}}{12 \text{ in.}}\right) \left(\frac{1}{t_1 \text{ min}}\right) \left(\frac{60 \text{ min}}{\text{hr}}\right)}$$

Area based on settling rate at start of compression zone (for thick pulps).

$$A = \frac{\text{Overflow}}{\text{Settling Rate}} = \frac{\text{Area (sq ft) required to settle a slurry feed}}{\text{rate of 1 ton of dry solids per 24 hr}}$$

$$\text{Overflow Rate} = (1.33)\,(H_x - H_e)/(\text{solids concn. gm/liter})$$

$$\text{Settling Rate} = (H_x - H_2) \left(\frac{X \text{ in.}}{\text{liter}}\right) \left(\frac{1 \text{ liter}}{1000 \text{ ml}}\right) \left(\frac{1 \text{ ft}}{12 \text{ in.}}\right) \left(\frac{1}{t_2 \text{ min}}\right) \left(\frac{60 \text{ min}}{\text{hr}}\right)$$

Note: t_2 is located by discontinuity in plot of log $(H–E)$ vs. t.
Line $H_x t_x$ is drawn tangent to settling curve at t_2.

Fig. 2-1. Graduated cylinder test for primary clarifier sizing.

concentration is too low. Sizing is then based upon both turbidity observations and the rate of build-up of the settled sludge deposit (Ref. 2-6). At a suitable turbidity level and constant sludge volume, the settling rate is determined by dividing the supernatant height by the settling time.

Light Suspensions. The settling rate of light suspensions often can be determined using plastic test cylinders about 9 ft (2.7 m) deep and 6 in. (15 cm) in diameter (Ref. 2-8, 2-9). Sampling taps are installed at 1-ft intervals. The temperature of the freshly collected wastewater is first adjusted to the ambient temperature of the test area to minimize thermal currents. When a constant-temperature jacketed tube is available, the settling test should be made at the original temperature of the wastewater. Any required temperature adjustment can be carried out quickly in a vat equipped with mixing facilities and coils through which hot or cold water is passed. Also at this time, any desired modifications of the waste, such as pH adjustment or flocculant addition, can be made.

The waste is then pumped into the test cylinder, and samples are taken periodically at various depths until settling is nearly complete. The suspended solids concentration of each sample is determined. Plots of these concentrations versus sampling depths are made for various detention times in order to provide the format required to predict the amount of suspended solids removed for various clarifier sizes. A simplified graphical means of data interpretation has been described by Conway and Edwards (Ref. 2-9). See Appendix 2-1

A simplified approach for interpreting settling tube data is to calculate the overflow rate required to obtain a measured solids concentration. This is done by dividing the depth of the particular sampling point by the settling time. The plot of settling rate versus effluent suspended solids shown in Figure 2-2 then

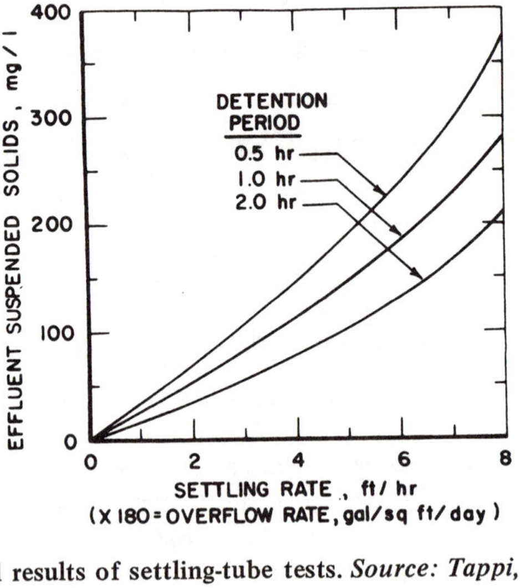

Fig. 2-2. Typical results of settling-tube tests. *Source: Tappi,* 46, 5, pp. 155A-160A, 1963.

can be obtained for each detention period. The overflow rate, which is the rate at which the liquid can be allowed to rise through the clarifier, is obtained directly from the rate at which the solids are settling. Overflow rates are usually expressed either as feet per hour (meters/hour) or gallons per square foot per day.

Floatables. Generally, the techniques for determining settling rates also apply in cases in which the controlling feature of clarifier design is the removal of various floatable liquids and solids. A separator design for removing oil globules from wastewater has been developed by the American Petroleum Institute (Ref. 2-10). The rise rate of a 0.015-cm diameter globule is used for design calculations. Separator depths are specified at 3 to 8 ft (0.9 to 2.4 m), widths at 6 to 20 ft (1.8 to 6.1 m), and depth to width ratio at 0.3 to 0.5. The maximum horizontal velocity is specified at 3 ft/min (0.9 m/min) or 15 times the rise rate of the standard globule. Clarifier design depends upon waste flow rate (process wastes and pad runoff), presence or absence of emulsions, wastewater temperature, wastewater viscosity, wastewater specific gravity, settleable solids, and the specific gravity of the contained oil. Although some wastes from sources other than petroleum refineries may be amenable to the empirical relationships developed by API, experimental determinations of rise rates are usually necessary. Design parameters of an overflow rate of 1000 gal./day/sq ft and a maximum horizontal velocity of 3 ft/min are common.

Pilot Studies. A reliable means of experimentally obtaining data for clarifier design is to conduct pilot-scale studies. Pilot models of each type of clarifier are available from equipment manufacturers. Properly designed models can produce reliable design data (Ref. 2-11). In addition to data from experimental studies, considerable design information can be gathered from the performance of full-scale units handling similar wastewater. The equipment manufacturers can supply installation lists and operating data, which are useful in this respect.

2-3.3 Dissolved-Air Flotation

For insoluble components which have densities near that of water and, consequently, neither settle nor rise at significant rates, dissolved-air flotation can be used to advantage. The suspension is pressurized at 40 to 60 psig (280 to 410 kN/m^2) in the presence of sufficient air to approach saturation. When the pressurized slurry is discharged to nearly atmospheric pressure in the flotation unit, minute air bubbles are released which attach to the solids. The air-sludge mixture quickly rises to the surface and is skimmed off as illustrated in Figures 2-3 and 2-4.

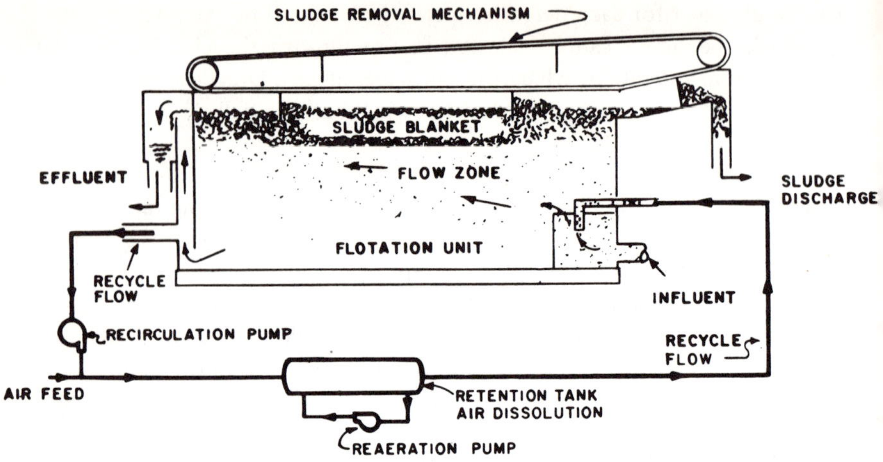

Fig. 2-3. Dissolved-air flotation unit. ***Source: Process Design Manual for Suspended Solids Removal,*** **EPA 625/1-75-003a, U.S. Environmental Protection Agency, Technology Transfer, Cincinnati, Ohio, 1975.**

The amenability of the suspension to clarification by the dissolved-air flotation process can be determined using a laboratory flotation unit, consisting of a pressure chamber and a large graduated cylinder (Ref. 2-12). Water or clarified effluent is charged to the pressure chamber, and the graduated cylinder is partially filled with the suspension to be clarified. After applying compressed air at 40 psig (280 kN/m^2) or greater at an appropriate ratio of air to contained solids (about 0.02 lb/lb), the pressure chamber is shaken for 1 min and allowed to stand for an additional 2 min to effect saturation. The air-saturated water is then introduced through a nozzle into the bottom of the graduated cylinder. The position of the sludge-liquid interface is periodically recorded to provide the data needed for rise-rate calculations.

The key variables in this process are air to solids ratio (dependent upon air pressure and the relative volumes involved), hydraulic loading, and detention time in the flotation unit. For wastes with moderate to low quantities of oil, the recycle ratio is about 30 to 50%; a total hydraulic loading of 1500 to 6000 gal./day/sq ft (61 to 240 m^3/m^2-day) at a retention of 20 to 40 min generally is used. Partial feed-flow pressurization, without recycle, also can be used. For wastes with large amounts of oil, full feed-flow pressurization has application.

Dispersed air flotation is an alternative process in which the air is introduced directly by mechanical means as small bubbles. Retention time in the order of 4 min is used. It has been used extensively in the ore industry; some waste disposal applications include recovery of latex, carbon black, and wood fibers.

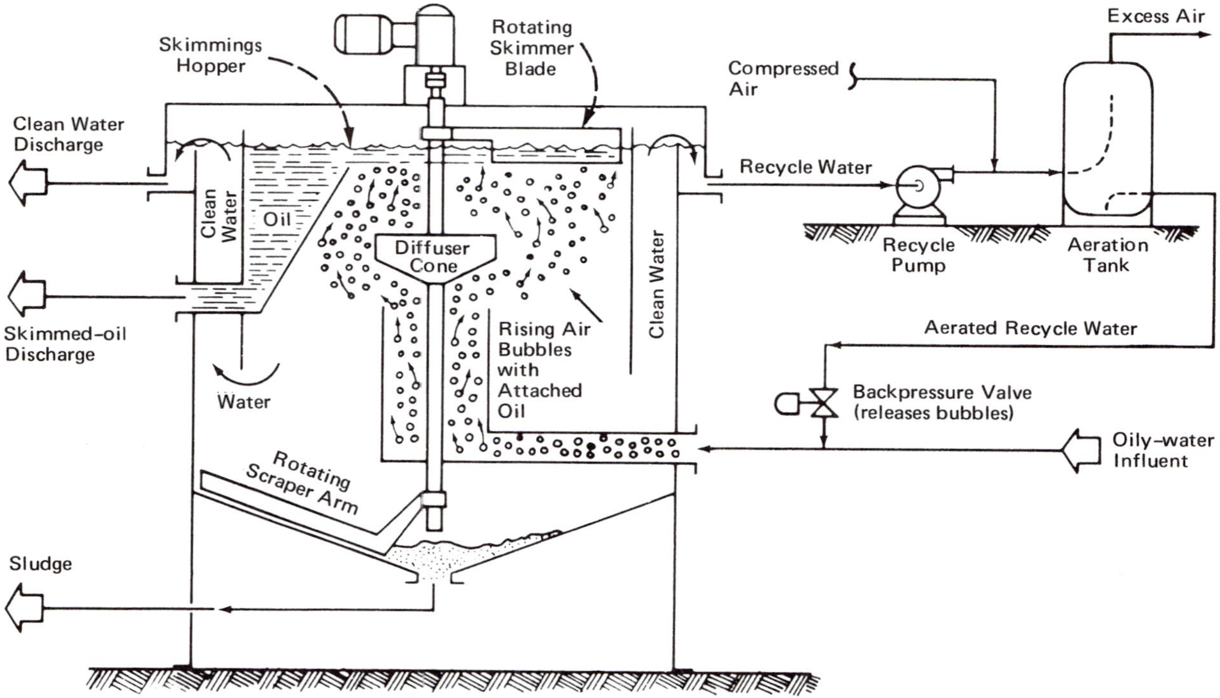

Fig. 2-4. Circular dissolved-air flotation system with recycle-flow pressurization to separate solids-laden or oily water influents. *Source:* Lash, L. D. and Kominek, E. G., "Primary-Waste Treatment Methods," *Chem. Eng.*, Deskbook Issue, pp. 49–61, October 6, 1975.

2-3.4 Precipitation/Coagulation/Flocculation

Removal of metals as their hydroxides is the most common precipitation application in wastewater treatment operations. The amount of lime or other alkali added is dependent upon the required effluent concentration; an amount of precipitant about twofold in excess of the stoichiometric amount is added to reduce the solubility due to the common-ion effect. Polyelectrolytes are added at about a 1 mg/liter dosage. Precipitation carried out in the presence of previously formed sludge reduces the problem of supersaturation and serves to develop larger particles. Either a solids contact unit involving internal recirculation or an external recycle of underflow solids may be used. Removal of metals to 1.5 times their solubility in water is the level of effectiveness to be expected. (See also Section 6-8.)

Coagulation by added chemicals serves to chemically destabilize the particles so that agglomeration may occur. Mechanisms include charge neutralization, compaction of the double-ionic layer, polymer bridging, and enmeshment in hydroxide floc. Flocculation is the process of bringing the particles into contact to allow agglomeration. Metal hydroxides of iron or aluminum serve both to reduce the particle charge and to adsorb the colloids. Organic polymers serve to bond the particles together. A typical polymer feed system is shown in Figure 2-5.

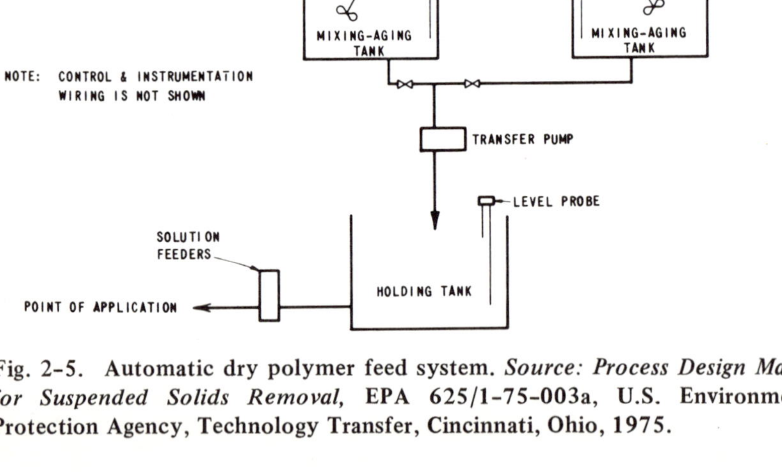

Fig. 2-5. Automatic dry polymer feed system. ***Source: Process Design Manual for Suspended Solids Removal,*** **EPA 625/1-75-003a, U.S. Environmental Protection Agency, Technology Transfer, Cincinnati, Ohio, 1975.**

Liquid-solids separation rates usually can be greatly enhanced through the application of the appropriate inorganic flocculants and/or organic polyelectrolytes. Organic flocculants are added as dilute solutions (0.2 to 1.0%) to effect rapid dispersion. Coagulants are added to a rapid mix tank having a retention time of about 2 min. More gentle agitation in a following flocculation section of about 20 min retention is used to build up the size of the floc particles (Figure 2-6). The flocculation operation also can be combined with clarification in a single unit as discussed in later sections. Of prime importance in flocculator design is the velocity gradient (*G* value) which is related to power input divided by the dynamic viscosity.

The effects of candidate flocculants can be predicted through jar tests simultaneously conducted using a multiple mixer apparatus. Varying the pH and at times other solute properties, such as surface tension, can have a marked effect on separation rates and should be considered in developing reliable design criteria.

2-3.5 Application of Experimental Data in Equipment Selection

The liquid-solids separation data obtained through laboratory and pilot-scale studies serve both to define the utility of each separation approach and to allow the system to be properly sized. Equipment suppliers can predict, through correlation, the probable performance of their units based on the supplied experimental data. A participative review of the underlying basis for such predictions usually results in a more reliable design, which is to the betterment of all concerned.

Closer bids and firmer performance guarantees can be developed as a result of the waste characterization studies. In the case of gravity settling, for example, performance specifications can be related to a series of settling tests run in a tube with controlled stirring during the same period as are the performance tests of the full-scale unit. Clarifier performance could be required to be at a level equal to 90% of the settling-tube results. Dye dispersion studies also can be run to define dead space, short circuiting, and severe turbulence due to inlet and outlet conditions (Ref. 2-23). The equipment supplier is more likely to agree to firm performance tests if the unit design is based upon sound experimentation. The cost and time required for exhaustive testing, however, needs to be balanced against the potential savings in each situation.

In a similar manner, carefully developed experimental data can serve to rationalize the design of flotation and sludge-dewatering equipment and to check that the manufacturer's claims are proven out. The commonly used "rules of thumb" too often result in either overdesigned or inefficient operations, whereas a design based on prior experimentation can be tailored closely to the individual

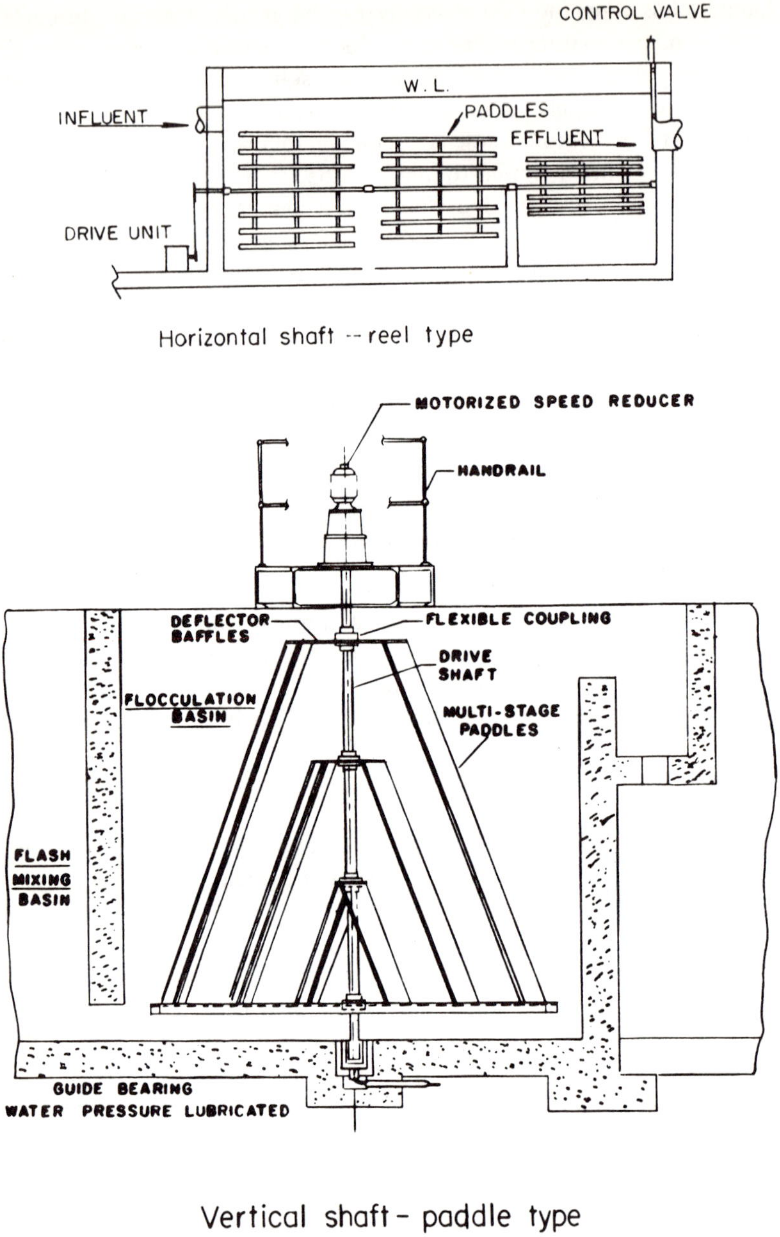

Fig. 2-6. Mechanical flocculation basins. *Source: Process Design Manual for Suspended Solids Removal,* EPA 625/1-75-003a, U.S. Environmental Protection Agency, Technology Transfer, Cincinnati, Ohio, 1975.

circumstances. Many sources of comprehensive information on types and applicability of wastewater treatment equipment are available (Ref. 2-14, 2-15).

2-3.6 Circular Clarifiers

Circular clarifiers are constructed of steel or concrete, and range from 8 to 25 ft (2.4 to 7.6 m) in depth and from about 10 ft (3 m) to over 300 ft (90 m) in diameter. The two basic types of circular clarifiers are center-fed and peripheral-fed units. These are shown in Figure 2-7. Flow is introduced to center-fed units by influent piping extending into a feed well located in the center of the clarifier; the feed well should be at least 20% of the tank diameter and about 60% of the tank depth (Ref. 2-15). Inflow velocity is reduced to 1 to 2 ft/min (0.3 to 0.6 m/min) by dividing the waste flow into many separate streams; vertical and horizontal flow distribution is accomplished by circular baffling at the tank's center. Waste flow is generally horizontal, moving outward from the center of the tank. The periphery of the clarifier is equipped with a baffle extending a few inches above and from 1 to 2 ft (0.3 to 0.6 m) below the liquid surface. Scums and oils are confined to the center portion of the tank, while the clarified wastewater flows under the baffle and over a weir into a peripheral collection channel; weir loading rates of 10,000 to 20,000 gal./day/ft (120 to 240 m^3/m-day) are used depending on the size of the floc (Ref. 2-15).

Scum is collected in the center-fed clarifier by means of a sweeping arm extending from the center of the clarifier to the peripheral baffle. The arm is supported both at the periphery and at the center of the clarifier, and power to rotate the arm is supplied at either point depending upon the equipment manufacturer's design. Scum is pushed in front of the arm as the arm rotates, and a trough extending radially into the settling zone and located approximately at water level collects the scum as the skimmer arm passes over the trough.

In the case of peripheral-fed clarifiers, the incoming flow is distributed either by means of a channel of constant width and decreasing depth or by a skirt extending to within a few feet of the clarifier bottom. The peripheral channel distribution system is equipped with orifices sized and spaced in the bottom of the channel, so that head loss through the orifices is large as compared to head loss in the channel. Flow is thereby distributed in a nearly uniform fashion. Flow is directed downward by a baffle to near the bottom and then moves toward the tank's center in a radial flow pattern. Effluent is collected either in the center of the clarifier by an overflow weir or at the rim by a weir attached to the inner side of the peripheral baffle. Flow distribution is obtained in the skirted clarifier by varying the distance between the skirt and the outer wall of the clarifier. Flow entering the clarification zone is a combination of sink and forced-vortex flow moving toward the center of the tank in a spiral flow pattern.

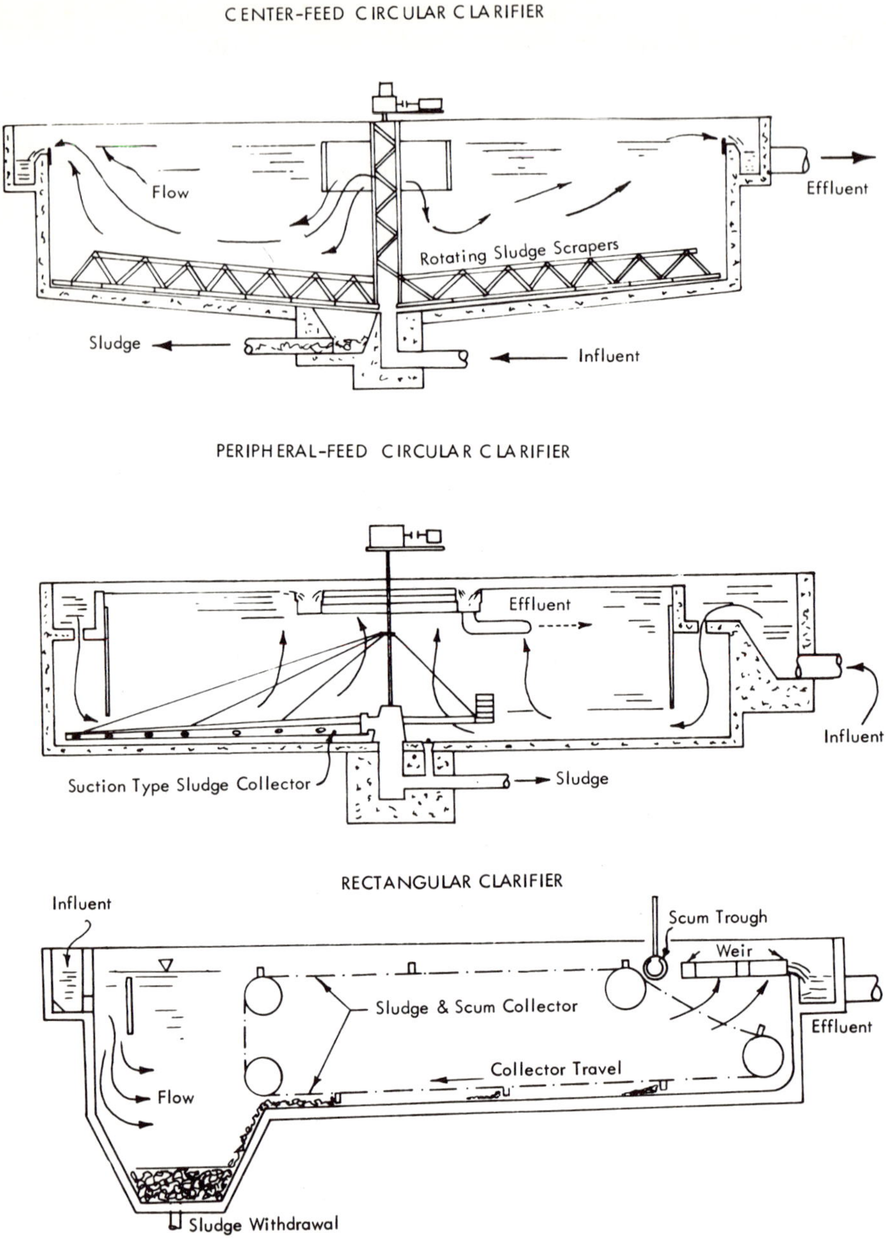

Fig. 2-7. Typical clarifier sections.

Effluent is collected by a combination radial-circular overflow weir trough in the center of the tank. The space between the clarifier wall and the skirt is used for scum collection; however, scums or foams of a rigid nature tend to bridge and are difficult to remove from this type equipment.

Influent flow velocities of center-fed clarifiers are highest at the point of introduction, the centrally located feed well, and decrease as the periphery of the tank is approached. In opposition, peripheral-fed clarifiers introduce the wastewater at a low velocity, and as the flow moves toward the center of the tank, the velocities increase. There are pro and con arguments for the efficacy of each design. Examination of operating results with wastes similar to those under study would seem to be the best means for judging relative performance. Also, well-defined tests to measure fulfillment of the specified performance levels are a necessary part of any purchase specifications.

Removal of sludge from the floor of a circular clarifier is accomplished in a variety of ways. The most common device employed is a rotating arm that extends from the center to the outside wall of the clarifier. The arm is equipped with a series of blades, set at an angle to the radial arm which is rotated by means of an electric motor and speed reducer. Sludge is displaced by each rake as the arm rotates and is thereby directed down the sloped (1 to 12) floor toward the center of the clarifier to a sludge sump. The velocities of the scrapers are kept below 1 ft/min (0.3 m/min) to lessen disturbance of the settled sludge. From the sump, the sludge is withdrawn hydraulically for final disposal. Another means of sludge removal involves the use of two spiral scraper blades which extend from the outside circular wall to the centerwell.

An alternative method of sludge removal utilizes a tapered rectangular header arm or pipe equipped with orifices or nozzles. When rotated through the sludge blanket, the difference in head between the clarifier and the interconnecting sludge piping forces the sludge into the header and out of the clarifier. This type of sludge collection tends to remove the solids from the clarifier faster than does the rotating rake.

Although surface area controls liquid rise rate and consequently is the prime factor affecting solids removal efficiency, clarifier depth also is important due to impact on detention time and flow patterns. A minimum depth of 11 ft (3.4 m) is recommended for clarifiers up to 65 ft (20 m) in diameter (Ref. 2-15). The suggested minimum depth increases to 15 ft (4.8 m) for clarifiers 200 ft (61 m) in diameter or larger.

Many modifications of standard circular clarifiers are available. Combined flocculation-clarification units have found considerable utility (Figure 2-8). Also, equipment which provides preaeration in the center of the unit, thus, reducing the immediate oxygen demand of the incoming waste while providing a degree of flotation for solids and immiscible liquid removal, is available. Equipment can be obtained which has multiple effluent weirs, thereby providing

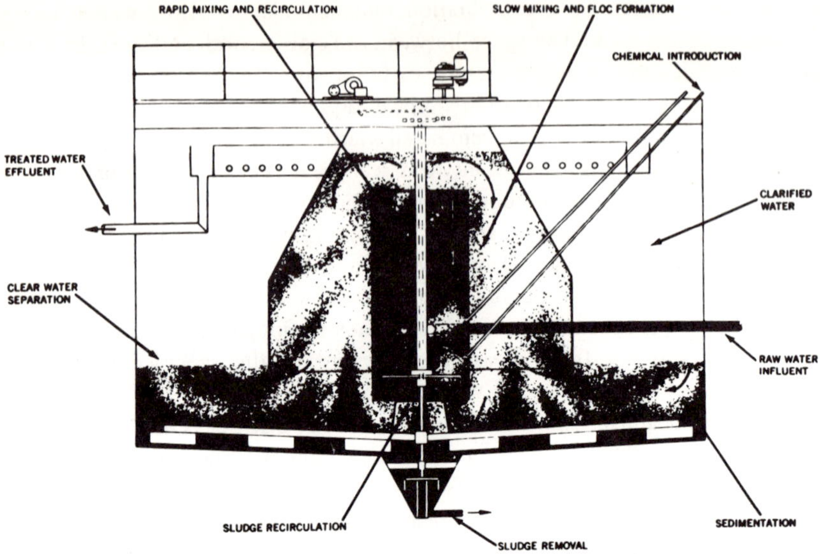

Fig. 2-8. Solids-contact unit combining mixing, coagulation and flocculation, liquid-solids separation. *Source:* **Lash, L. D. and Kominek, E. G., "Primary-Waste Treatment Methods,"** ***Chem. Eng.*****, Deskbook Issue, pp. 49-61, October 6, 1975.**

lower weir overflow rates and reduced solids carryover in the effluent. Some equipment is designed primarily for handling slurries, while other units treat dilute wastewater only.

2-3.7 Rectangular Clarifiers

Rectangular clarifiers generally are about 40 to 150 ft (12 to 46 m) long and 10 to 40 ft (3 to 12 m) wide. Flow is introduced at the head end of the clarifier through multiple-ported influent channels. Various baffling arrangements are used to distribute the incoming flow across the width and depth of the unit, thereby producing uniform flow, low velocities and minimum turbulence throughout the tank length. Overflow weirs and troughs are used to collect and remove the treated wastewater from the clarifier, as shown in Figure 2-7.

Floors of rectangular clarifiers are generally sloped downward from the effluent end of the unit to facilitate the movement of collected sludge to hoppers located near the inlet of the tank. Sludge collection is accomplished by a number of blades or flights suspended between endless chains on either side of the tank. As the chains are moved by a sprocket arrangement, the sludge is pushed by the flights into the sludge hoppers. Where multiple parallel units or very wide clarifiers are used, cross collectors are often used to move the sludge to a central hopper. A cross collector, similar to the endless chain collector, moves the sludge perpendicular to the length of the basin.

If removal of oils and other floating materials is required, the sludge collecting system is installed so that as the flights at the bottom move sludge to one end of the basin, the returning flights move along the surface pushing scum in the opposite direction. Removal of scum is accomplished when the returning flights force the floating material into a trough or split pipe where it flows by gravity to a sump. Rectangular clarifiers generally are preferable to circular clarifiers when oil removal is the prime consideration. An API-type design as described in a previous section is used when levels of oil are high.

A second type of sludge and scum collector manufactured for rectangular tanks consists of a moving bridge spanning the clarifier and moving parallel to the wastewater flow. A retractable blade moves sludge to one end of the clarifier as the bridge traverses the clarifier's length. The blade is then withdrawn, and as the bridge returns to its starting point, a skimmer blade is lowered to remove scum. Another method of sludge removal involves a series of arms moving along the floor of the clarifier in a reciprocating fashion. The arms consist of tapered tubes with orifice holes spaced along the length and are supported on a chain drive mechanism. Sludge is withdrawn through flexible hoses connecting the arms to sludge piping; the difference in hydrostatic head between the clarifier liquid and the sludge piping provides the force or suction necessary to remove sludge. This type installation requires no sludge hopper.

Another variation in rectangular clarifier design is a floor sloped from both ends to a centrally located sludge withdrawal point. Square clarifiers are available having circular sludge collectors with hinged arms that extend into the corners of the tank to remove sludge.

Where flocculation enhances the settling characteristics of solids, rectangular clarifiers are available with flocculating mechanisms built into the influent end of the tank. These consist of a number of paddles attached to slowly revolving shafts. In some instances, floc development is carried out in a separate tank.

2-3.8 Tube/Plate Clarifiers

The conventional circular or rectangular clarifier can be modified to increase capacity by the installation of modules of tubes, each about 2 in. by 6 in. (5 X 15 cm) in cross section and 4 ft (1.2 meters) in length, inclined at 60°. As the wastewater flows up through the tubes, the solids settle along the inclined walls and drop to the bottom of the basin. Tubes also are available with a V-grooved base. Inclined plates are used in the treatment of refinery wastewaters as shown in Figure 2-9 (Ref. 2-16).

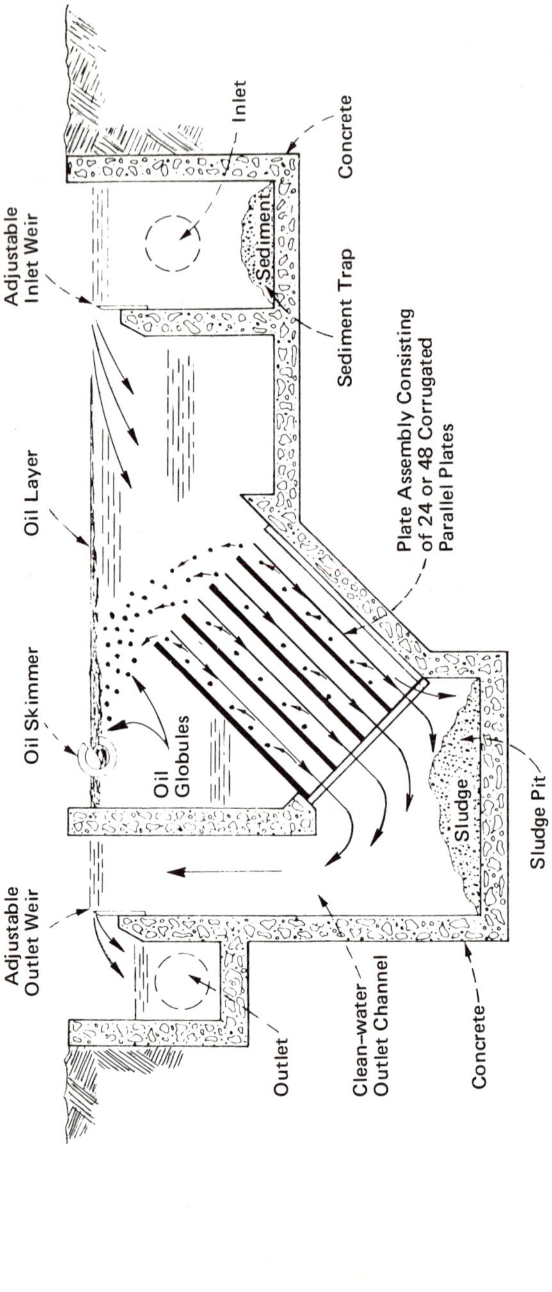

Fig. 2-9. Corrugated-plate interceptor for oil wastewater. *Source:* Lash, L. D. and Kominek, E. G., "Primary-Waste Treatment Methods," *Chem. Eng.*, Deskbook Issue, pp. 49–61, October 6, 1975.

2-3.9 Earthen Basins

Where sufficient land area is available, earthen basins sometimes are used to remove settleable solids from dilute wastewater. When small basins are used, multiple units are constructed to provide for waste treatment when one basin is out of service for cleaning or repair. With larger basins the solids can be removed by hydraulic dredging or with draglines.

Inlet baffling of earthen basins is provided to prevent gross short-circuiting. Baffles are provided on the effluent end of basins to collect oils and scums. The basin detention time usually is based on sludge storage requirements, as dictated by solids concentration and removal efficiency, compaction, rate of biological degradation, and the length of time sludge is to be stored. Consideration should be given to subsurface conditions to ensure against dam failures due to hydrostatic pressure and leakage of impounded waters.

2-3.10 Straining

Straining processes remove solids on a thin medium by virtue of physical restrictions. Screens generally precede the other waste treatment processes and are used either to protect the treatment equipment from damage, to reduce the strength of the incoming waste by the removal of organic solids, or to treat storm waters (Ref. 2-17). A second application for finer strainers is the removal of suspended solids after secondary or biological treatment. A summary of the physical straining operation is presented in Table 2-2.

Bars. Bar screens used in the removal of coarse materials are constructed of parallel steel rails spaced about 1 in. (2.5 cm) or greater apart. The screen is installed at a 30° to 45° angle to the horizontal and sloped away from the waste flow. Cleaning of the screens is usually accomplished with a series of moving combs which deposit the screenings in a receptacle located at the top of the unit.

Inclined. Inclined screens operate by gravity and function as a drainage board with a wedge-wire construction; openings are traverse to the flow (Figure 2-10). A rotating configuration of the wedge wire screen also has been developed.

Vibrating/Centrifugal. Vibrating and centrifugal screens with openings ranging down to 100 μm (microns) are applicable to a variety of solids removal problems. One type consists of a circular fine mesh screen, which is vibrated both vertically and horizontally at a very high frequency as wastewater is admitted at the center. Solids trapped by the screen are moved to the periphery and discharged through a collection nozzle while the liquids pass through the

TABLE 2-2. Physical Straining Processes

Principal Applications	Device	Straining Surface	Hydraulic Capacity	Performance[a]	
				Retained Solids (% by weight)	Suspended Solids (Removals, %)
Pretreatment and primary treatment	Inclined wedge–wire stainless steel screens	Coarse 0.01–0.06 in. (250–1500 μ)	High flow rates 4–16 gpm/in. of screen width	10–15	5–25
Pretreatment and primary treatment	Rotary stainless steel wedge wire screens	Coarse 0.01–0.06 in. (250–1500 μ)	16–112 gpm/sq ft	16–25	5–25
Pretreatment and primary treatment	Centrifugal screens	"Medium 105"	40–100 gpm/sq ft	0.05–0.1	60–70
Secondary and tertiary SS removal	Micro-screens	Medium[b] (15–60 μ)	Medium flow rates, 3–10 gpm/sq ft	0.03–0.07	50–80
Secondary and tertiary SS removal	Diatomite filters	N/A[b]	Medium flow rates, 0.5–1.0 gpm/sq ft	–	up to 90
Secondary and tertiary SS removal	Ultra-filters	Fine[c] (10^{-3}–15 μ)	Low flow rates, 5–50 gpd/sq ft	1.5	>99

[a]Data derived from municipal wastewater applications.
[b]Straining occurs through particulate mat of solids on screening surface.
[c]These values typify the range of solids filtered by the media.
Source: Process Design Manual for Suspended Solids Removal, EPA 625/1-75-003a, U.S. Environmental Protection Agency, Technology Transfer, Cincinnati, Ohio, 1975.

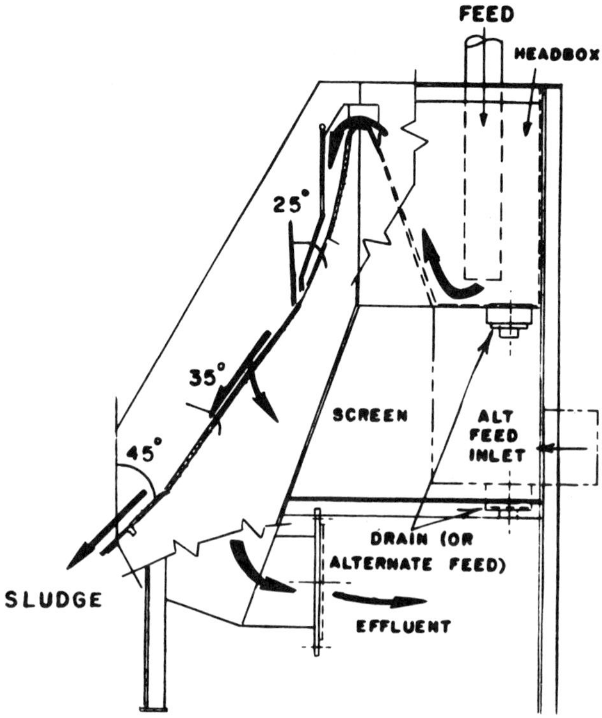

Fig. 2-10. Hydrasieve screening unit schematic. ***Source: Process Design Manual for Suspended Solids Removal,*** **EPA 625/1-75-003a, U.S. Environmental Protection Agency, Technology Transfer, Cincinnati, Ohio, 1975.**

screen. The process is continuous, and relatively high waste flows can be handled. Rectangular vibrating screens are generally installed in an inclined position so that the waste flows from top to bottom. The solids are entrapped on the screen and move downward to a collection hopper or conveyor belt while the liquid passes through the screen to collection wells. In a centrifugal screen the influent is directed against the inside of a rotating cylinder case. The relatively clear effluent is taken from the periphery, and the suspended solids and reject portion of the liquid flow are removed from the bottom.

Microstrainers. Microstraining, which can remove all particles down to 23 μm in diameter and some particles between 5 and 23 μm, has been used to advantage by some industries for final reduction in effluent suspended solids or to reuse the carrying liquid directly and utilize the recovered solids. Typical loading rates range from 2.5 to 10 gpm/sq ft (100 to 400 liters/m^2-min) depending upon the drum fabric. About 80% of the solids from a well-operated activated-sludge-process effluent can be removed by microstraining. Wastewater is introduced to the center of a slowly revolving drum that is covered with a fine mesh cloth. As the liquid passes through the cloth, the solids are retained on the surface. A water spray is employed to remove the accumulated solids at the top of the rotation. Cloths with openings less than 23 μm now are available.

2-3.11 Granular-Media Filtration

The advent of tighter restrictions on discharge of suspended solids along with the development of multimedia systems has resulted in a broader application of granular-media filtration for wastewater treatment. The typical applications of granular media filters include:

- Final treatment of effluents from biological systems prior to discharge.
- Clarification of influents to activated carbon columns.
- Removal of residual oils from API separator effluents.
- Removal of fines from pulp and paper wastes.
- Final clean up of heavy metal precipitates after a solids contact clarifier.
- Clarification of steel mill wastes.

Beds of 4 ft (1.2 m) or less composed of 0.5 mm sand and 0.9 mm anthracite frequently are used in most of the above operations (Figure 2-11), except for handling steel mill wastes, in which case deeper filters having a depth of about 7 ft (2.1 m) of a single, coarse-sand medium of 1 to 2 mm is used. Coal, 3 to 6 mm in diameter and about 2 ft deep (0.6 m), can be used at the top of deep-bed filters to increase the duration of filter runs. Typical designs are shown in Table 2-3.

Filters find economic application in handling streams containing less than 100 to 200 mg/liter suspended solids or oils, depending on the required effluent level. Prior chemical flocculation is needed for dispersed solids. Gravity feed at rates of 2 to 6 gpm/ft^2 (80 to 240 liters/m^2-min) for the shallow, dual-bed media and 10 to 15 gpm/ft^2 for the coarse deep-bed system are common (Ref. 2-15). With oily wastes of low solids, flow rates to multimedia filters can be 10 to 15 gpm/sq ft to achieve 50% solids reduction and 75% oil reduction (Ref. 2-18).

Gravity filters are operated to a terminal head loss of 5 to 10 ft (1.5 to 3 m) of water before being backwashed. An air scour at 4 std cu ft/min/sq ft can help to separate the solids and oils from the media before backwashing, initially at about 25 gpm/sq ft. The backwash rate depends on size of media and degree of expansion required. This rate is slackened to allow restratification of the coal and the denser sand to provide the coarse-to-fine path desired. The duration of the backwash is about 20 min per cycle. Backwash water, which amounts to 1 to 5% of the total flow, can be routed to the primary clarifier often via a storage

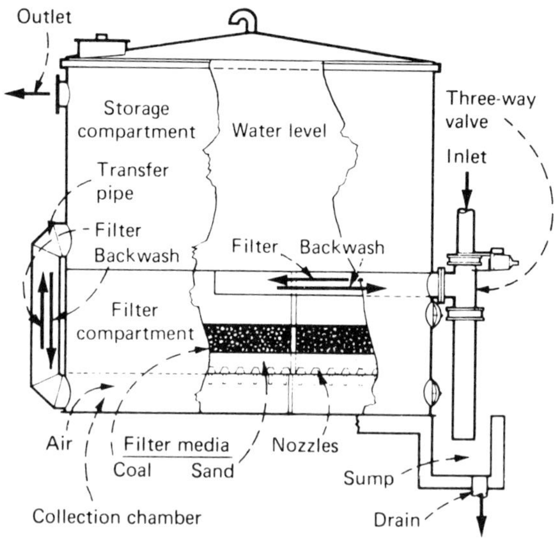

Fig. 2-11. Granular-media filter operating on a controlled cycle. *Source:* **Lash, L. D. and Kominek, E. G., "Primary-Waste Treatment Methods,"** *Chem. Eng.*, **Deskbook Issue, pp. 49-61, October 6, 1975.**

vessel to allow flow equalization. Several filters are used in parallel to allow continuous processing during backwashing; the backwash cycle usually is automated. Other processes must be sized to handle this recycle flow.

Some vendors discourage the use of an air scour when backwashing multimedia filters as it can unnecessarily extend the downtime of the filter and result in a greater chance for media loss. A surface wash of 0.5 to 1 gpm/sq ft during backwashing can be used to help prevent the formation of balls of mud.

Pressure filters can be operated to 15 to 20 ft (4.6 to 6.1 m water) headloss, which usually makes them less expensive than gravity filters in small and medium-sized plants. Pressure operation is an advantage when the solids loading is variable. Also, it results in a smaller percent backwash requirement; and the effluent can be sent to the downstream treatment step without pumping after the filter.

Multimedia filters usually remove 30 to 70% of the applied solids and oil concentration, unless flocculants are used. Due to the coarser medium, bleeding of solids into the effluent in deep-bed filters is likely unless flow surges are avoided. Recycle of the initial filtrate after backwash is practiced until filter flow patterns are established.

The use of shredded polyurethane foam as a filter medium in pilot studies

TABLE 2-3. Typical Media Designs for Filters

		Coal			Sand			Garnet		
Media Design	Typical Application Conditions[a]	Dia. (mm)	Depth (in.)	Unif. Coeff.	Dia. (mm)	Depth (in.)	Unif. Coeff.	Dia. (mm)	Depth (in.)	Unif. Coeff.
Single	A	–	–	–	1–2	60	1.2	–	–	–
Single	A	–	–	–	2–3	72	1.11	–	–	–
Dual	B	0.9	36	<1.6	0.35	12	<1.85	–	–	–
Dual	A	1.84	15	<1.1	0.55	15	<1.1	–	–	–
Tri	B	1.0–1.1	17	1.6–1.8	0.42–0.48	9	1.3–1.5	0.21–0.23	4	1.5–1.8
Tri	C	1.2–1.3	30	–	0.8–0.9	12	–	0.4–0.8	6	–

[a]Code: A = Heavy loadings, strong floc
B = Moderate loadings, weaker floc
C = Moderate loadings, strong floc

Source: Process Design Manual for Suspended Solids Removal, EPA 625/1-75-003a, U.S. Environmental Protection Agency, Technology Transfer, Cincinnati, Ohio, 1975.

has resulted in long run times, high filtrate quality, high media loadings, and very low head losses (Ref. 2-19).

REFERENCES

2-1. Nemerow, N. L., *Theories and Practices of Industrial Waste Treatment,* Addison-Wesley Publishing Co., Reading, Massachusetts, 1963.

2-2. Adams, C. E., Jr. and Eckenfelder, W. W., Jr., *Process Design Techniques for Industrial Waste Treatment,* Environmental Press, Inc., Nashville, Tennessee, 1974.

2-3. Wallace, A. T., "Analysis of Equalization Basins," *J. Environ. Eng. Div., Proc. Am. Soc. Civ. Eng.* **94**, SA6, p. 1161, 1968.

2-4. Greer, W. N., *The Measurement and Control of pH,* Leeds and Northrup Co., Philadelphia, 1966.

2-5. Field, W. B., "Design of a pH Control System by Analog Simulation," *J. Inst. Soc. Am.* **6**, p. 1, 1959.

2-6. Anderson, A. A. and Sparkman, J. E., "Review of Sedimentation Theory," *Chem. Eng.* **59**, p. 75, Nov. 2, 1959.

2-7. Foster, J. H., "Fundamentals of Primary Treatment," *Tappi,* **46, 5**, p. 155A, 1963.

2-8. Mancini, J. L., "Gravity Clarifier and Thickener Design," *Proc. 17th Ind. Waste Conf., Purdue University Eng. Bull. Ext. Ser.* **112**, p. 267, 1963.

2-9. Conway, R. A. and Edwards, V. H., "How to Design Sedimentation Systems from Laboratory Data," *Chem. Eng.* **68**, p. 167, September 18, 1961.

2-10. *Manual on Disposal of Refinery Wastes,* American Petroleum Institute, Washington, 1969.

2-11. Baumann, E. R., Cleasby, J. L., and Morgan, P. E., "Performance of Sedimentation Tanks Evaluated in Model Tests," presented at the National Symposium on Sanitary Engineering Research, Development, and Design, Pennsylvania State University, July 29, 1965.

2-12. Rohlich, G. A., "Application of Air Flotation to Refinery Waste Waters," *Ind. Eng. Chem.*, **46**, p. 304, 1954.

2-13. Brumo, G. W., "Interpretation of Tracer Tests on Sedimentation Tanks," *TAPPI,* **55**, 7, pp. 1097-1102, 1972.

2-14. *Environmental Wastes Control Manual and Catalog File,* Public Works Journal Corp., Ridgewood, New Jersey, 1975.

2-15. "Environmental Engineering Deskbook on Pollution-Control Equipment," *Chem. Eng.* 82, p. 21, October 6, 1975.

2-16. Rubin, E. and Zahave, E., "Enhanced Settling Rates of Solid Suspensions in the Presence of Inclined Plates," presented at American Institute of Chemical Engineers Meeting, Houston, March 16–20, 1975.

2-17. Matthews, C. W., "Screening," *Chem. Eng.*, pp. 99–104, February 15, 1971.

2-18. Wallace, J. T., Jr. and Brown, J. S., "Deep-Bed Filtration of Oilfield Produced Water," 27th Annual Purdue University Industrial Waste Conference, Lafayette, Indiana, May 4, 1972.

2-19. Romagnoli, R. J. and Berglund, R. L., "Polyurethane Foam Filtration", *Proc. 33rd Industrial Waste Conf., May 9–11, 1978, Purdue Univ.*, p. 343, Ann Arbor Science, Ann Arbor, Mich., 1979.

APPENDIX 2-1

A PROCEDURE FOR DESIGNING SEDIMENTATION SYSTEMS FROM LABORATORY DATA WITH LIGHT SUSPENSIONS

I. Scope

This method applies to lighter, flocculent suspensions (0.01–1% by weight) when an interface does not form in a settling-column test. For heavy suspensions forming interfaces or for discrete, free-settling particles much less complex approaches apply.

II. Equipment

Nine-foot test cylinder, 6-in. diameter, made from rigid plastic tubing, supported by a small bench and securely clamped to a laboratory grid. Joint between tube sections reinforced with 2-in. flat plastic strip, heated and bent to fit. Ethylene dichloride solution (keep and handle in hood) of plastic chips used for adhesion.

III. Data-Collection Procedure

Adjust waste water temperature to ambient. Adjust pH, and add coagulants as desired. Pump into clarification cylinder through bottom drain to depth of 8 ft. Collect four 100-ml samples from evenly spaced depths from 0.25-in. sampling ports to define initial solids concentration. Resample at 0.5-hr intervals for 2-hr period. Determine suspended solids concentration in samples by glass-fiber filter method.

IV. Assemble data, as per the following example:

Settling Period, Min.	Depth Sampled, Ft.*	Suspended Solids Content: Concentration, Mg./L	Suspended Solids Content: Remaining %†
0	1.0	325 } 322	100
0	3.0	326 }	
0	5.0	319 }	
0	7.0	319	
30	0.9	224	70
30	2.9	265	82

(continued)

Settling -Period, Min.	Depth Sampled, Ft.*	Suspended Solids Content Concentration, Mg/L	Remaining %†
30	4.9	266	83
30	6.9	303	94
60	0.8	103	32
60	2.8	121	38
60	4.8	134	42
60	6.8	194	60
90	0.7	38	12
90	2.7	42	13
90	4.7	58	18
90	6.7	63	20
120	0.6	27	8
120	2.6	26	8
120	4.6	29	9
120	6.6	33	10

* Each set of samples reduced the water depth about 0.1 ft.
† Calculated from: (final conc/initial conc.) X 100

V. Plot data from Part IV at different settling times:

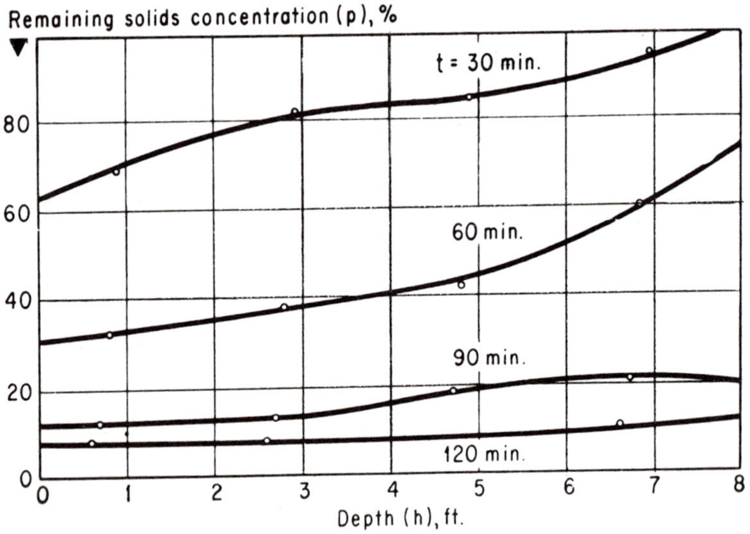

VI. Integrate graphically Plot in Section IV

A. Graphical integration method ("p" = remaining solids concentration as % of original, "h" = height in feet)

Use this graphical integration method

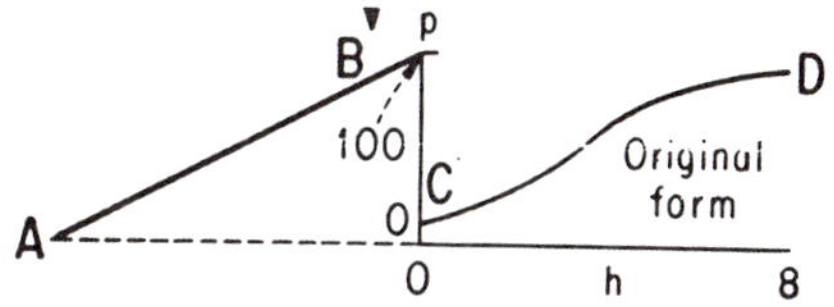

1. Locate reference point *A* for integration of curve *CD* by constructing line *AB* at the same slope as line *EF*.

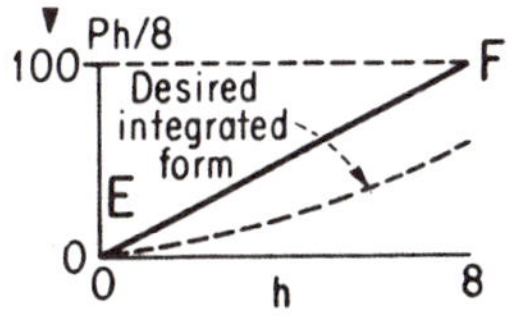

2. Divide curve *CD* vertically into segments of about equal slope. Bisect each segment horizontally, keeping the areas of the two "triangles" equal.

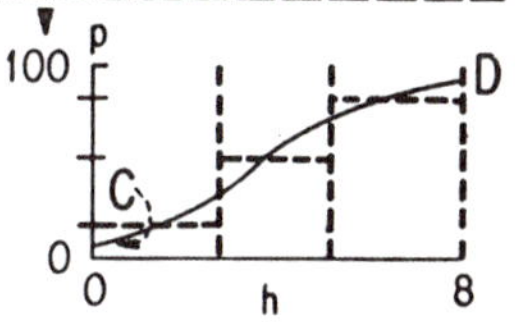

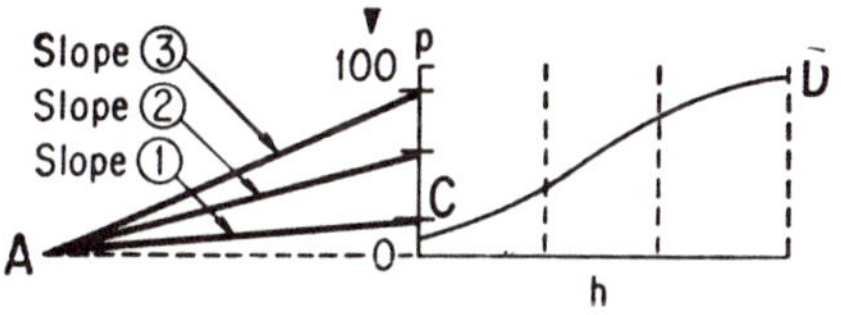

3. Draw reference lines by connecting *A* with intersections of above horizontal lines with the vertical axis. Construct the integrated form of CD by keeping the slope within each segment equal to the slope of its reference line.

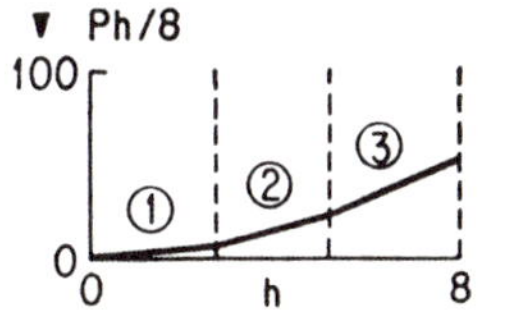

B. Results

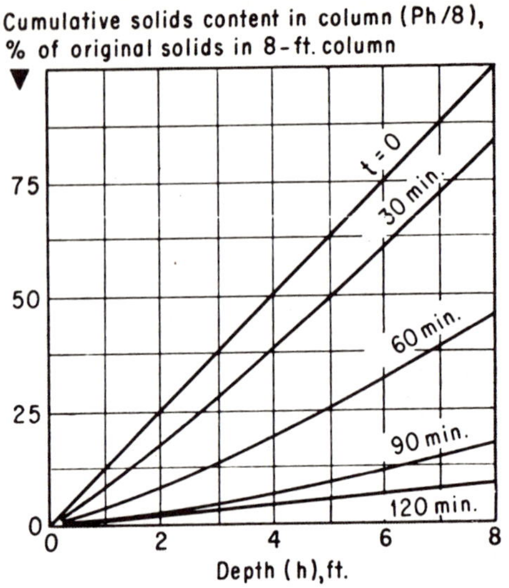

VII. Translate data from Section VI-B to Tabular form:

Depth, Ft.	Settling Period, Min.	Overflow Rate, Gal./Day/ Sq. Ft.*	Remaining Solids Content, *Ph*/8**	Solids Removal, %†
4	30	1,440	39	22
4	60	720	19	62
4	90	480	7.5	85
4	120	360	4.3	91
6	30	2,160	61	19
6	60	1,080	31	59
6	90	720	12	84
6	120	540	6.8	91
8	30	2,880	85	15
8	60	1,440	45	55
8	90	960	18	82
8	120	720	9.0	91

* Calculated from: 10,800 (depth, ft.)/(settling period, min.).

**Data from IV–B

† Calculated from: 100–(*Ph*/8) × (8/depth).

VIII. Construct correlations from Section VII for design purposes: apply appropriate scale-up factors.

A. Solids Removal versus Overflow Rate

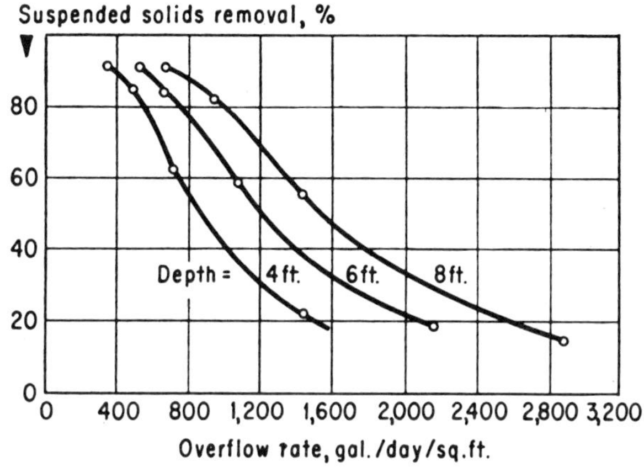

B. Solids Removal versus Depth

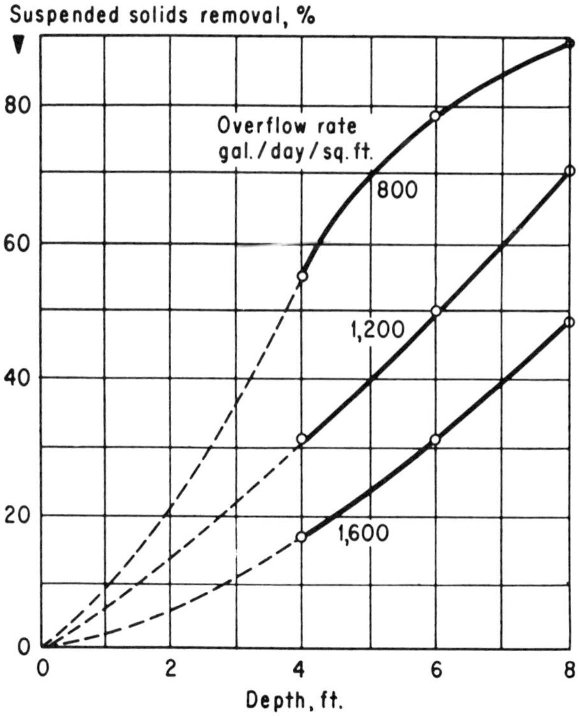

NOTE: For theoretical considerations and additional procedural details, see: Conway, Richard A. and Edwards, V. H., "How to Design Sedimentation Systems from Laboratory Data," *Chem. Eng.*, pp. 167–170, September 18, 1961.

3

Biological treatment of wastewater

The function of biological treatment processes is to remove organic matter and at times ammonia from wastewater through the metabolic means of oxidation and cell synthesis. Although several physical and chemical approaches have been proposed as substitutes, biological treatment remains the only proven means of economically abating the oxygen depletion caused by the discharge of wastewater contaminated with low concentrations of a mixture of biodegradable organic compounds. If this organic matter were to reach the receiving body of water, bacteria would oxidize it and thereby reduce the oxygen resources upon which aquatic life is dependent. Biological treatment processes are basically an accelerated form of these natural water purification mechanisms. Organics causing problems in addition to oxygen depletion also are economically removed, e.g., materials toxic to fish and odorous in water supplies. The fit of biological processes into overall wastewater treatment sequences is shown in Figure 3-1.

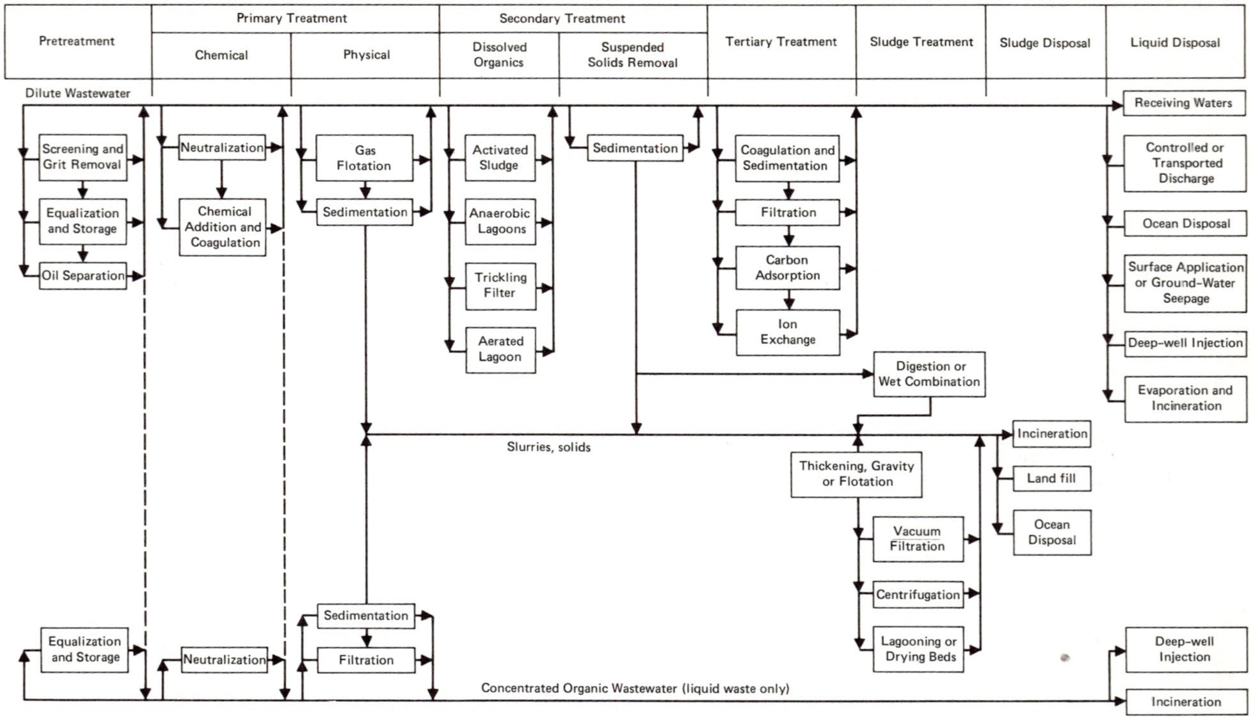

Fig. 3-1. Fit of biological treatment into overall wastewater treatment sequences. *Source:* Eckenfelder, W. W., Jr. and Ford, D. L., "Economics of Wastewater Treatment," *Chem. Eng.* pp. 109-118, August 25, 1969.

3-1 ENVIRONMENTAL REQUIREMENTS

In order that the bacteria and higher microbial forms in the biological treatment unit may stabilize the waste organics at a suitably high rate, key environmental conditions must be maintained at near optimum levels. The conditions which are basic to all aerobic treatment processes include:

- Oxygen availability
- pH near neutral
- Available nitrogen and phosphorus
- Absence of toxic materials
- Adequate mixing

Since anaerobic processes utilize materials other than molecular oxygen as acceptors for the hydrogen atoms removed from the organics during oxidation, oxygen availability can be omitted from the above list for these processes. Most industrial biological treatment processes are aerobic, however, since aerobic processes can stabilize organics faster at lower temperatures and have less potential for odor problems. Some of the pollutants that make pretreatment necessary before the wastes can be handled in the biological unit are listed in Table 3-1.

In addition to the basic environmental factors, biological processes are specifically designed to operate efficiently by proper manipulation of these rate-controlling variables:

- Microorganism concentration
- Bacterial acclimation or adaptation
- Temperature level
- Contact duration and mode
- Organic feed concentration

The rate at which organic matter is stabilized, and to some degree the extent of degradation, is dependent upon the concentration of active bacteria in the system. Unless the organic matter is of natural origin, waste stabilization often is delayed until the bacteria in the system have been allowed either to develop adaptive enzyme systems or to undergo shifts in species mix in order to oxidize the organic material at a high rate. The oxidation rate increases with increasing temperature until a maximum level is attained either based on metabolic or solids-retention considerations; the optimum temperature varies for each type of system. The minimum allowable contact time is dictated by the rate of conversion of the organic matter to carbon dioxide, water, new cells, and stored or adsorbed products. The conversion obtained in a given time also is dependent upon whether the system is completely mixed or has reactors in series (staged). The maximum time is governed by the extent to which the stored organics and

TABLE 3-1. Concentration of Pollutants that Make Prebiological or Primary Treatment Desirable

Pollutant or System Condition	Limiting Concentration	Kind of Treatment
Suspended solids	>50–125 mg/1	Sedimentation, flotation, lagooning
Oil or grease	>35–50 mg/1	Skimming tank or separator
Toxic ions		Precipitation or ion exchange
Pb	≤0.1 mg/1	
Cu + Ni + CN	≤1 mg/1	
Cr^{+6} + Zn	≤3 mg/1	
Cr^{+3}	≤10 mg/1	
pH	6–9	Neutralization
Alkalinity	0.5 lb alkalinity as $CaCO_3$/lb BOD removed	Neutralization for excessive alkalinity
Acidity	Free mineral acidity	Neutralization
Organic-load variation	>2:1–4:1[a]	Equalization
Sulfides	>100 mg/1	Precipitation or stripping with recovery
Phenols	>70–300 mg/1[b]	Extraction, adsorption, internal dilution
Ammonia	>1.6 g/1	Dilution, ion exchange, pH adjustment and stripping
Dissolved salts	>10–16 g/1	Dilution, ion exchange
Temperature	13–38°C in reactor	Cooling, steam addition

[a]Higher level if constituents remain in same ratio.
[b]Up to at least 1500 mg/1 if bacterial acclimation is maintained.
Based on: Eckenfelder, W. W., Jr. and Ford, D. L., "Economics of Wastewater Treatment," *Chem. Eng.*, pp. 109–118, August 25, 1969 (with modifications).

synthesized cells are allowed to be oxidized in turn, since excess oxidation can cause problems in retention of viable cells. The organic feed concentration governs the rate at which food can be supplied to the biological system while still maintaining the minimum desirable contact with the bacteria; it also affects the equilibrium level achievable. The function of the various types of biological units is to utilize these rate-controlling factors to the best advantage. These qualitative statements will be expressed as mathematical relationships in subsequent sections to aid in their being understood and used.

3-2 MICROBIAL METABOLISM AND BIODEGRADABILITY

Familiarity with a few selected terms and concepts regarding microbial life should allow the engineer or chemist to realize better the full potential of biological systems for treating industrial wastewaters.

3-2.1 Protista

The microbial life of interest in treatment systems belongs to the Protista Kingdom; the members of which, as opposed to plants and animals, do not have similar cells arranged in a tissue structure. Among the protista, the classes of microorganisms described in Table 3-2 are of most importance.

Bacteria are the primary organisms which serve to stabilize organic and certain inorganic constituents in wastewaters. The structural components of a hypothetical bacterial cell are shown in Figure 3-2. The functions of each component also are indicated. The wide range of chemical transformations capable of being conducted at a fast rate by these organisms will be discussed in later sections.

Fungi occur in liquid waste treatment systems, usually when nitrogen is deficient or the pH is low. Except when they are needed to degrade certain large molecules, fungi are undesirable as they cause problems during separation of the biomass from the effluent liquor. Fungi are important in composting.

Algae are important in lagoon-type waste treatment systems where they photosynthetically convert carbon dioxide and water to oxygen and cellular material. Their growth often is stimulated in receiving waters rich in nitrogen and phosphorous.

Protozoa and rotifers play an important role in biological waste treatment systems by feeding on bacteria. A clearer effluent is produced, and additional organic material is converted to carbon dioxide and water. The maturity or state of the system can be determined by the rotifers and types of protozoa present; as the system reaches a higher degree of conversion, growth of the protozoan classes and rotifers progresses as follows:

Sarcodina (amoebae)
Phytoflagellates
Zooflagellates
Free-swimming ciliates
Suctoria (tentacled)
Stalked ciliates
Rotifers

TABLE 3-2. Characteristics of Microbial Classes Important in Waste Treatment Systems

1. Bacteria
   - Unicellular
   - Procaryotic (no distinct nucleus)
   - Absorb soluble nutrients
   - Primarily binary-fission reproduction
   - Cylindrical, spherical, or helical
   - 0.5 to 3.0 μm (cylinders or rods)
2. Fungi (true fungi or molds)
   - Multicellular
   - Generally reproduce by spore formation
   - Nonphotosynthetic
   - Low nitrogen requirement
   - Survive at low pH
   - Most are strict aerobes
   - Heterotrophic (organic carbon source)
3. Algae
   - Photosynthetic
   - Unicellular or multicellular
   - Autotrophic (inorganic carbon source)
   - Types:
     - green algae (*Chlorella*)–chloroplasts, eucaryotic (nuclear membrane)
     - motile green (*Eugelena*)–flagellated
     - yellow green or golden brown–diatoms
     - blue green–unicellular, dispersed chlorophyll, procaryotic
4. Protozoa
   - Unicellular
   - Eucaryotic
   - Majority aerobic
   - Types in treatment systems:
     - Amoebae–pseudopods for movement and particulate food engulfment
     - Flagellates–phytoflaggelates (chloroplasts) and zooflagelates (absorb food)
     - Ciliates–cilia for motion and capturing solid food, free-swimming (e.g., *Paramecium*), and stalked (e.g., *Vorticella*)
     - Tentacles–capture other protozoa and draw out protoplasm
5. Rotifers
   - Multicellular
   - Aerobic
   - Heterotrophic
   - Two sets of rotating cilia for mobility and capturing food
6. Other
   - Yeasts
     - Unicellular
     - Eucaryotic
     - Reproduce by budding
   - Viruses
     - Obligate parasites, i.e., redirect invaded cells to produce viral cells

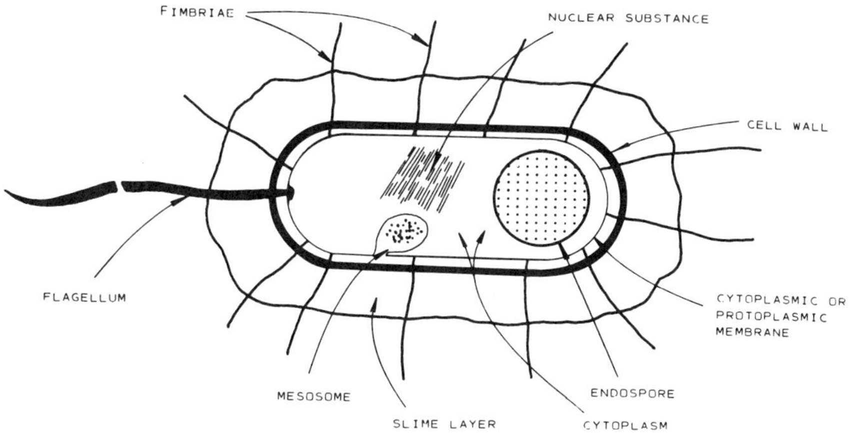

Fig. 3–2. Hypothetical bacterial cell.

Cell Wall—10–40% of cell weight, provides rigidity to protect against osmotic pressure changes.

Cytoplasm—colloidal suspension of proteins, lipids, carbohydrates; granular in appearance; contains ribonucleoprotein structures (ribosomes) which are the site of protein synthesis. Inclusions found in various bacteria under various conditions include glycogen, lipid, and volutin granules, and sulfur droplets.

Glycogen Granule—polysaccharide granules formed from excess organic matter if the medium is N deficient.

Lipid Granule—natural fats or poly-β-hydroxybutyric acid (an ester polymer formed from excess organic material if N is deficient in the medium).

Volutin Granule—polyphosphate accumulation, also termed metachromatic granule, often due to nutrient deficiency impeding nucleic acid synthesis.

Sulfur Globule—associated with sulfur as an end product.

Cytoplasmic or Protoplasmic Membrane—a semipermeable lipoprotein complex which controls material passage in and out of the cell by passive (equilibrium) and active transport.

Endospore—some bacteria have the ability to transpose into small spheres which are highly resistant to environmental factors such as heat, etc., and can germinate to vegetative cells.

Fimbriae (Pili)—employed by the bacterium for attachment; implicated in conjugation, DNA (deoxyribonucleic acid) transfer.

Flagellum—responsible for motility, visible with special stain. Flagella may be disposed in a polar or peritrichous (many all around) fashion, and are macromolecular protein structures appearing as rope-like spirals.

Mesosomes—involved in cross-wall formation during cell division.

Nuclear Substance (chromatinic area)—not enclosed by nuclear membrane; contains the genetic information of the cell and is composed of a circular, double helical strand of DNA.

Slime Layer—generally polysaccharide, called a capsule if of a regular thickness and too viscous to diffuse away.

Microscopic examination of the system under low power magnification is an effective tool to visualize the succession of these forms during treatment.

3-2.2 Growth

In order for the bacteria and other protista to accomplish the desired stabilization (energy-reduction) reactions at optimum rates, proper growth conditions need to be maintained. These conditions are summarized in Table 3-3; important terms are cited to facilitate reading of the specialized literature.

The time interval required for the bacterial cell to divide is termed the *generation time;* it ranges from 10 to 60 min for some heterotrophs, about 20 hr for one nitrite oxidizer, and 3 days for some methanogenic bacteria. As the cells divide, the population doubles (1, 2, 4, 8, 16, etc.) until nutrients become limiting or end-product inhibition occurs. A typical growth progression for a batch-fed system starting with a low level of seed microorganisms is shown in Figure 3-3. In a steady-state treatment system, an overall growth state in the declining growth phase is desirable. In actuality, various species are likely in different growth states due to the inherent variability of the system.

The complex chemical transformations resulting in the growth of microorganisms can occur at ambient temperatures and pressure levels due to the catalytic action of enzymes produced by the microorganisms. A given species will have a set of enzymes always available to conduct both energy-yielding (catabolic) reactions and cell-synthesis (anabolic) reactions, while others are produced only upon the external stimulation of a new substrate. The former enzymes are termed constitutive and the latter adaptive. The transfer of energy released in the catabolic reactions to the energy-consuming anabolic reactions is accomplished through the linkage of the adenosine-triphosphate/adenosine disphosphate (ATP/ADP) couple (Figure 3-4).

Hydrolysis of large molecules to facilitate transport across the cellular membrane is accomplished by extracellular enzymes; while intracellular enzymes are responsible for oxidative and synthesis reactions. The transformation of an organic molecule to carbon dioxide, water, and cellular constituents involves a sequence or chain of single steps of the removal or addition of atoms or electrons. Each step is catalyzed by an enzyme and has its own rate of reaction. One of the steps becomes rate limiting. The enzymes have three components:

Apoenzyme—high molecular weight organic compound, a protein, determines where reaction occurs on substrate molecule.

Coenzyme—smaller organic molecule, determines what reaction occurs, not always required.

Metal activator—helps to line up enzyme and substrate, not always required.

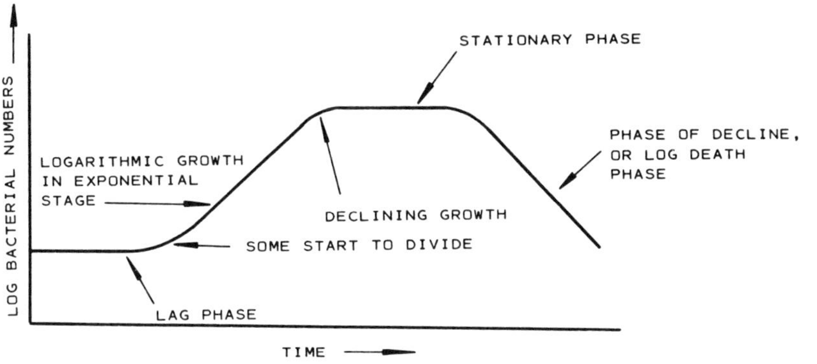

Lag phase–enzyme development (catalysts), new protoplasm (elongation), adjustment to physical factors.
Log growth–divide at rate according to generation time, excess food: stored, adsorbed, soluble.
Declining growth–food supply becoming limited.
Stationary–nutrients limiting, toxicants produced, division balanced with death, endogenous growth (utilizes stored and lysed nutrients).
Log death–lack of nutrients, end products toxic, rapid die-off, auto oxidation.

Fig. 3-3. Typical bacterial growth curve.

Familiarization of engineers and chemists with these concepts and terms should aid in a deeper understanding of the biosystem involved.

3-2.3 Energy Metabolism

The basic features of bacterial metabolism are summarized in Figure 3-4. The inputs are the cellular building blocks, chemical or light energy, and a final acceptor of the hydrogen atoms or electrons. The catabolic reactions yield the energy used for synthesis, as well as the indicated soluble products. Typical reactions are listed in Table 3-4. The reactions shown for organic matter oxidation are of an overall nature. The actual process involves transforming the waste constituent into a form which upon entering the terminal respiratory cycle (tricarboxylic acid or TCA cycle) yields carbon dioxide, water, cellular building blocks, and energy.

Conversion of organic waste constituents to simple acids, mainly acetic, generally is needed as a first step to enter the TCA cycle. The preliminary reactions for a range of organic classes are shown in Figure 3-5. Familiarity with these steps should allow judgments to be made concerning the amenability to biodegradation of some of the waste constituents encountered.

TABLE 3-3. Factors Influencing Bacterial Growth

1. Nutritional Requirements
   - a. Energy source
     - Oxidize chemical compounds (chemotrophs)
       - organics–organotrophic (heterotrophs)
       - inorganics–lithotrophic (autotrophs)
     - Employ radiant energy (phototrophs)
       - bacterial photosynthesis–sulfur bacteria
   - b. Carbon source (~50% of cell on dry weight basis)
     - Organics (saphrophytic heterotrophs)
     - Carbon dioxide as sole source (autotrophs)–usually oxidize reduced, S, N, Fe forms
   - c. Nitrogen source (~12% of dry cells)
     - Inorganic-N suitable for algae, most bacteria
   - d. Phosphorus and sulfur (2.6 and 0.7%)
     - Phosphorus–usually supplied as PO_4
     - Sulfur–some bacteria require inorganic, some organic, some elemental
   - e. Trace metals
     - Na, K, Ca, Mg, Fe, Mn, Zn, Cu, Co
     - (Na, K, Ca, Mg, Fe:) 2.5%
   - f. Vitamins and vitamin-like compounds
     - Some bacteria can manufacture all; others need one or more in the growth media
   - g. Water–nutrients must be in solution.
   - h. Hydrogen acceptor (Oxidant)
     - Respiration
       - molecular oxygen
       - nitrate, nitrite, sulfate ions–in the absence of oxygen
       - Carbon dioxide
     - Fermentation
       - Oxidized organics

2. Physical Conditions
   - a. Temperature

	Range (°C)	Best (°C)
Psychrophiles	7–35	20–30
Mesophiles	25–40	30
Thermophiles	25–75	40–60

Growth rate: generally doubles per 10°C rise (up to denaturing temp.)

   - b. Gaseous requirements
     - Aerobic–grows in presence of free oxygen
     - Anaerobic–grows in absence of free oxygen
     - Facultatively anaerobic–grows in either case
     - Microaerophilic–obligate aerobes which grow best under low partial pressures of oxygen
   - c. pH
     - 6.5–7.5 optimum for most (6–8 at outside)
     - 5–9 minimum to maximum for growth–most systems

TABLE 3-3. (Continued)

- d. Light (phototrophs)
- e. Salt (obligate halophiles)
- f. Oxidation–reduction potential (ORP)–influences anaerobic reactions

3. Lack of Toxicants
   - a. Organic (concentration, ratio to enzyme level, and organism acclimation state influence degree of toxicity)
   - b. Heavy metals
   - c. High cation, ammonia, and sulfide levels

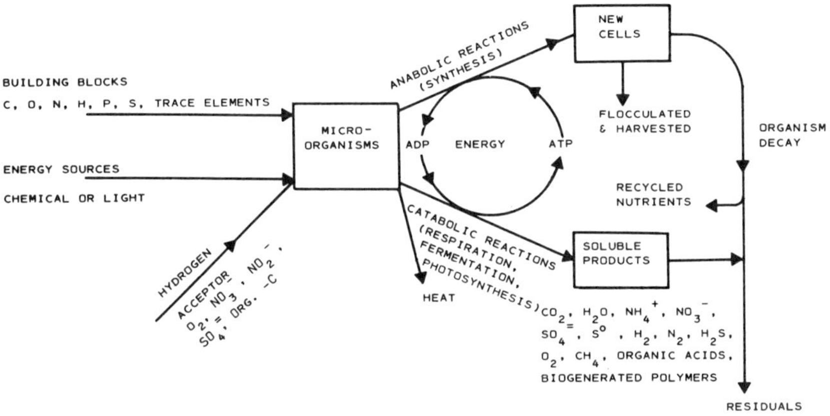

Fig. 3-4. Biological transformations. ***Source:*** **Andrews, J. F. University of Houston, Houston, Texas.**

3-2.4 Biodegradability

The resistance of certain organic materials to rapid degradation in a given situation can be related to several mechanisms (Ref. 3-1):

Nonexistence of an active organism or enzyme
Violation of comparative biochemical sequences
Violation of enzyme specificity
Lack of sufficient energy or carbon for growth
Lack of an essential nutrient
Exceeding microbial tolerance to environmental factors
Toxicity of substrate or products of its metabolism
Inhibition or inactivation of an extracellular enzyme
Failure of a chemical to penetrate a cell
Concentration of substrate in aqueous solution being too low

TABLE 3-4. Typical Biological Treatment Reactions

1. Anaerobic nonphotosynthetic reactions (molecular oxygen absent)
   A. Nitrate reduction (denitrification)
   $5\ CH_3COOH + 8\ NO_3^- \longrightarrow 10\ CO_2 + 4\ N_2 + 6\ H_2O + 8\ OH^-$
   $5\ S + 6\ NO_3^- + 2\ H_2O \longrightarrow 5\ SO_4^= + 3\ N_2 + 4\ H^+$
   B. Sulfate reduction
   $2\ CH_3CHOHCOOH + SO_4^= \longrightarrow 2\ CH_3COOH + H_2S + 2\ OH^-$
   $4\ H_2 + SO_4^= \longrightarrow 2\ H_2O + H_2S + 2\ OH^-$
   C. Organic carbon reduction (fermentation)
   $CH_3COOH \longrightarrow CH_4 + CO_2$
   $4\ CH_3OH \longrightarrow 3\ CH_4 + CO_2 + 2\ H_2O$
   $C_6H_{12}O_6 \xrightarrow{\text{bacteria}} 3\ CH_3COOH$
   $C_6H_{12}O_6 \xrightarrow{\text{yeast}} 2\ CH_3CH_2OH + 2\ CO_2$
   D. Carbon dioxide reduction
   $2\ CH_3CH_2OH + CO_2 \longrightarrow 2\ CH_3COOH + CH_4$
   $4\ H_2 + CO_2 \longrightarrow CH_4 + 2\ H_2O$
   $4\ H_2 + 2\ CO_2 \longrightarrow CH_3COOH + 2\ H_2O$

II. Aerobic nonphotosynthetic bacterial reactions
   A. Oxygen limited systems
   $CH_3CH_2OH + O_2 \longrightarrow CH_3COOH + H_2O$
   $2\ CH_3CHO + O_2 \longrightarrow 2\ CH_3COOH$
   $2\ CH_3CHOHCH_3 + O_2 \longrightarrow 2\ CH_3COCH_3 + 2\ H_2O$
   B. Complete oxidation
   $CH_3COOH + 2\ O_2 \longrightarrow 2\ CO_2 + 2\ H_2O$
   $2\ H_2 + O_2 \longrightarrow 2\ H_2O$
   $CH_4 + 2\ O_2 \longrightarrow CO_2 + 2\ H_2O$
   C. Nitrification
   $2\ NH_3 + 3\ O_2 \longrightarrow 2\ NO_2^- + 2\ H^+ + 2\ H_2O$
   $2\ NO_2^- + O_2 \longrightarrow 2\ NO_3^-$
   D. Sulfur oxidation
   $2\ H_2S + O_2 \longrightarrow 2\ S + 2\ H_2O$
   $2\ S + 2\ H_2O + 3\ O_2 \longrightarrow 2\ SO_4^= + 4\ H^+$
   $S_2O_3^= + H_2O + 2\ O_2 \longrightarrow 2\ SO_4^= + 2\ H^+$
   E. Nitrogen fixation
   $N_2 \longrightarrow$ Nitrogenous Organics

III. Photosynthetic reactions
   $CO_2 + 2\ H_2S \xrightarrow{\text{light}} (CH_2O) + H_2O + 2\ S$
   $3\ CO_2 + 2\ S + 5\ H_2O \xrightarrow{\text{light}} 3\ (CH_2O) + 4\ H^+ + 2\ SO_4^=$
   $CO_2 + 2\ H_2O \xrightarrow{\text{light, algae}} (CH_2O) + H_2O + O_2$
   $9\ CH_3COOH \xrightarrow{\text{light}} 2\ CO_2 + 4\ (C_4H_6O_2) + 6\ H_2O$
   $CO_2 + 2\ H_2 \xrightarrow{\text{light}} (CH_2O) + H_2O$
   $2\ CH_3COOH + H_2 \xrightarrow{\text{light}} (C_4H_6O_2) + 2\ H_2O$

Source: Fisher, J. A., Hovious, J. C., Kumke, G. W., and Conway, R. A., "Pilot Demonstrations of Basic Designs for Anaerobic Treatment of Petrochemical Wastes," *Chem. Eng. Sym. Ser., Water–1970* 67, 107, pp. 485–494, 1971.

1. Carbohydrates

   a. Hydrolyze to at least disaccharide stage before transfer through cell wall.

      Starch (alpha linkage) to glucose
      Cellulose (beta linkage) anaerobically to glucose
      Hemicellulose to glucose and xylose (pentose)
      Sucrose to glucose and fructose
      Lactose to glucose and galactose
      Maltose to glucose

   b. Convert to pyruvic acid

$$C_6H_{12}O_6 \longrightarrow 2\,CH_3\overset{\overset{\displaystyle O}{\|}}{C}COOH + 4\,H$$

   c. Convert to CO_2, H_2O, organic acids, ethanol

      Aerobic

$$2\,CH_3\overset{\overset{\displaystyle O}{\|}}{C}COOH + 4\,H + 6\,O_2 \longrightarrow 6\,CO_2 + 6\,H_2O$$

      Anaerobic

$$2\,CH_3\overset{\overset{\displaystyle O}{\|}}{C}COOH + 4\,H \longrightarrow \begin{cases} 2\,CH_3\overset{\overset{\displaystyle OH}{|}}{C}HCOOH \text{ (lactic)} \\ CH_3CH_2COOH + CH_3COOH + HCOOH \\ 2\,CH_3CH_2OH + 2\,CO_2 \end{cases}$$

(continued)

Fig. 3-5. Preliminary biochemical reactions. ***Sources:*** **Rose, A. H., *Chemical Microbiology*, Plenum Press, New York, 1968. McCarty, P. L. and Sawyer, C. N., *Chemistry for Sanitary Engineers*, McGraw-Hill Book Co., New York, 1967. Stanier, R. Y., Doudoroff, M., and Adelberg, E. A., *The Microbial World*, 3rd Ed., Prentice-Hall, Inc., Englewood Cliffs, New Jersey, 1970.**

2. Proteins

a. Hydrolysis

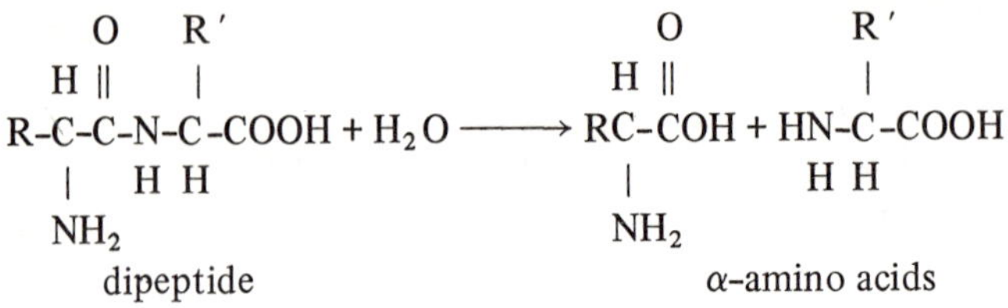

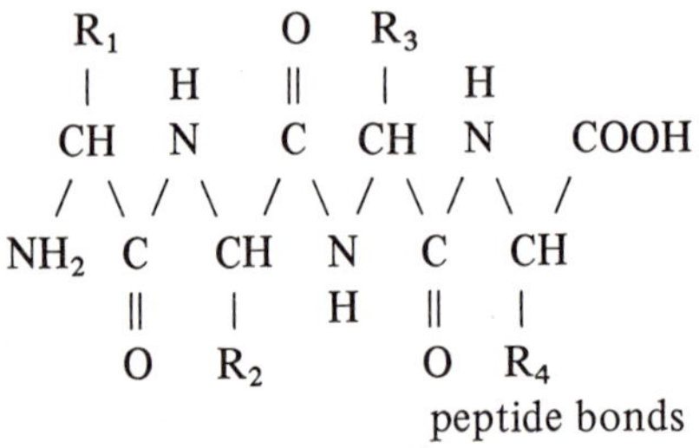

b. Deamination

Aerobic

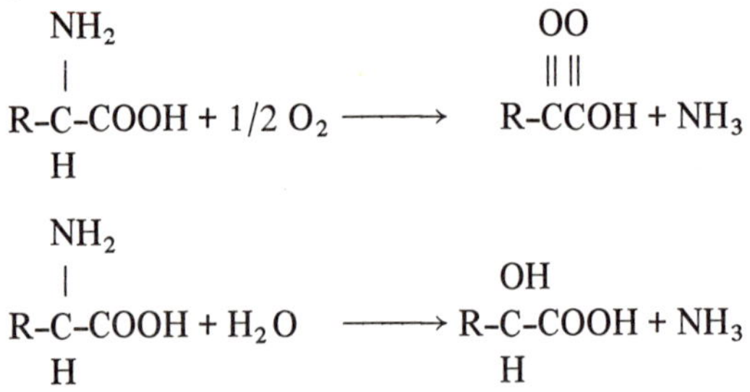

Anaerobic

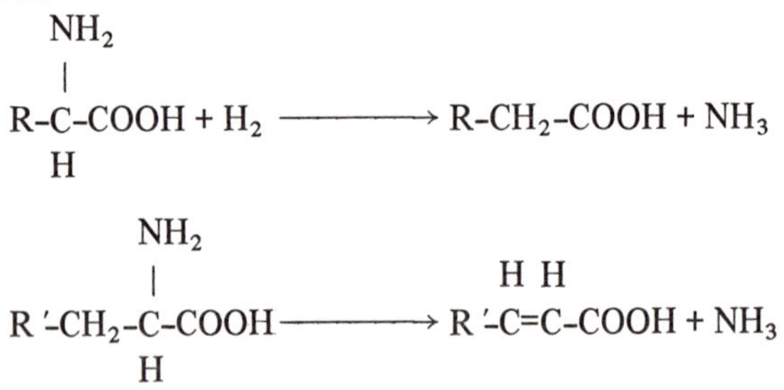

Fig. 3-5. (cont.)

3. Fats and Oils

Hydrolysis

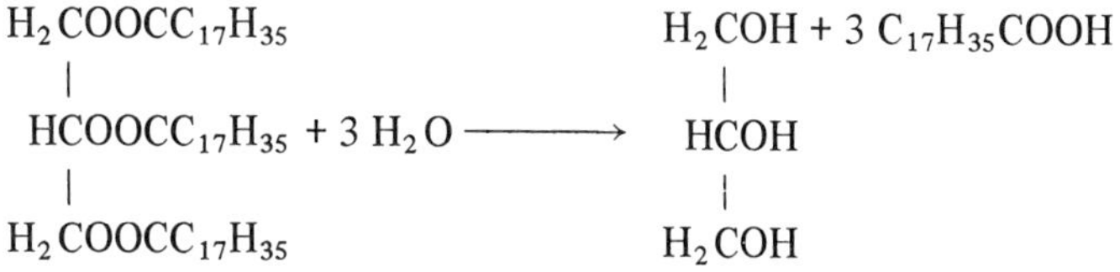

Stearin Glycerol Stearic Acid

4. Acids (β Oxidation with Coenzyme A)

Aerobic

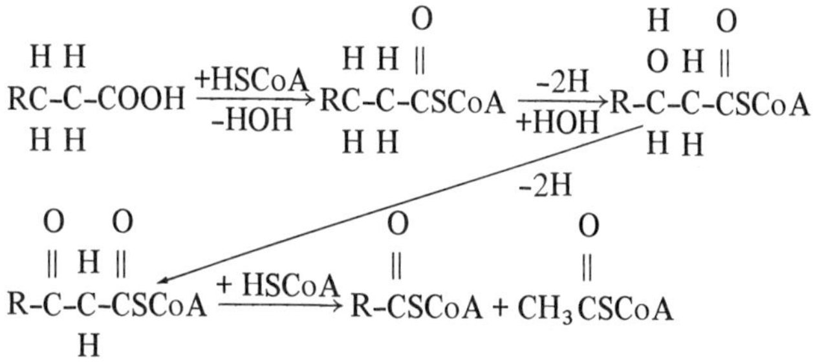

$$4\,H + O_2 \longrightarrow 2\,H_2O$$

$$CH_3COSCoA + 2\,O_2 \longrightarrow 2\,CO_2 + 2\,H_2O \quad \text{(via TCA cycle)}$$

Anaerobic

$$CH_3COOH + 8/5\,NO_3^- \longrightarrow 8/5\,HCO_3^- + 4/5\,N_2 + 2/5\,CO_2 + 6/5\,H_2O$$

$$CH_3COOH + SO_4^= \longrightarrow 2\,CO_2 + 2\,H_2O + S^=$$

$$CH_3COOH \longrightarrow CO_2 + CH_4$$

$$\text{also, } 4\,H + 1/2\,CO_2 \longrightarrow 1/2\,CH_4 + H_2O$$

5. Other Aliphatics

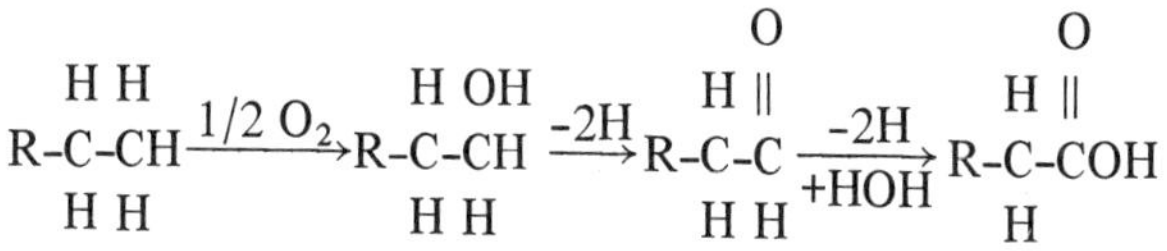

(continued)

Fig. 3-5. (cont.)

$$\underset{\text{H H}}{\text{R-C=C-R}} \xrightarrow{1/2\,O_2} \underset{\text{H H}}{\overset{\overset{\text{O}}{/\backslash}}{\text{R-C-C-R}'}} \xrightarrow{HOH} \underset{\text{H H}}{\overset{\overset{\text{H H}}{\text{O O}}}{\text{RC-CR}'}}$$

$$\underset{\text{H}}{\overset{\overset{\text{Cl}}{|}}{\text{R-C-R}}} + \text{HOH} \xrightarrow{\text{(resistant)}} \underset{\text{H}}{\overset{\overset{\text{H}}{\text{O}}}{\text{R-C-R}}} + \text{HCl}$$

$$\overset{\overset{\text{O}}{\|}}{\text{R-C-O-R}'} + \text{HOH} \longrightarrow \overset{\overset{\text{O}}{\|}}{\text{R-COH}} + \text{R}'\text{OH}$$

$$\text{R-O-R}' + \text{HOH} \xrightarrow{\text{(resistant)}} \text{ROH} + \text{R}'\text{OH}$$

$$\overset{\overset{\text{O}}{\|}}{\text{R-CNH}_2} + \text{HOH} \longrightarrow \overset{\overset{\text{O}}{\|}}{\text{R-COH}} + \text{NH}_3$$

$$\text{NH}_2\text{-}\overset{\overset{\text{O}}{\|}}{\text{C}}\text{-NH}_2 + \text{H}_2\text{O} \longrightarrow \text{CO}_2 + 2\,\text{NH}_3 \xrightarrow{H_2O} (\text{NH}_4)_2\text{CO}_3$$

6. Aromatics

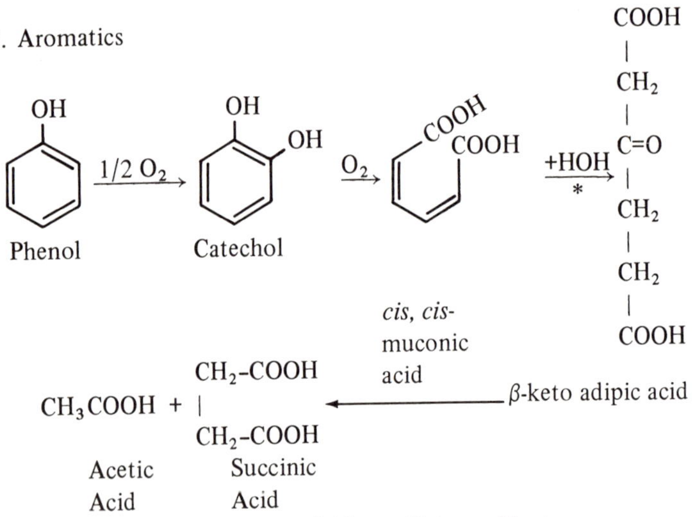

*Via (+) - muconolactone and β-ketoadipic enol lactone

Fig. 3-5. (cont.)

Lack of induction of a requisite enzyme
Need for extracellular enzymes from different organisms
Inaccessibility of a substrate to a microorganism
Complexing of a substrate with a polyaromatic or with other resistant organic compounds
Inaccessibility of the site on substrate to be acted on enzymatically

Consideration of these factors can lead to improvements ranging from slight modifications in problem molecules for increased biodegradability to the addition of a trace nutrient required by specific bacterial species.

While only a few naturally occurring chemicals are resistant to rapid biodegradation, e.g., polyaromatics such as lignin, problems are encountered in degrading many synthetic organic chemicals. Foremost examples of resistant synthetic materials are certain compounds within the classes of chlorinated hydrocarbon pesticides, polychlorinated biphenyls, synthetic polymers, and surfactants. Other compounds which need special consideration when planning treatment approaches are listed in Table 3-5.

Many techniques available for the experimental determination of biodegradability have been described by Waggy, Price, Conway, and others (Ref. 3-2, 3-3, 3-4). Test methods include:

Biochemical oxygen demand (BOD) by dilution bottle method
Oxygen uptake by respirometry or oxygen electrode
Die-away under river water conditions
Shake culture tests
Techniques simulating biological treatment systems
Tests in soil environment
Techniques for photochemical and chemical alteration

Results from BOD tests on a range of synthetic organic chemicals are presented in Table 3-6.

Biological treatment processes are categorized according to the growth-phase of the bacteria, the method of contacting the organic food with the bacteria, and the presence or absence of molecular oxygen. The common types of systems are:

Activated sludge
Aerated stabilization
Fixed film systems
Anaerobic systems
Stabilization ponds

The basic features of each system, along with experimental design approaches and process equipment descriptions will be covered in the remainder of this section.

TABLE 3-5. Example Organic Compounds Potentially Resistant to Rapid Degradation

Acetylethanolamine	2, 2-Dimethyl–1, 3-propanediol
Acetyl morpholine	2, 2-Dimethylsuccinic acid
Aminobenzenesulfonic acids	Dinitrobenzenes
tert-Amyl alcohol	Dioxane
Anthraquinone sulfonic acid	*bis*-2-Ethoxy ethyl ether
tert-Butanol	Ethylenediamine tetraacetic acid
Chloroanilines	Methoxyanilines
Chloronitrobenzenes	Morpholine
Chlorophenols, *meta*-substituted	Nitroanilines
Diaminobenzenes	Nitroanisoles
2, 6-Dibromohexanoic acid	Nitrobenzenesulfonic acids
Dichloroacetic acid	Nitrotoluenes
2, 3-Dichloropropionic acid	Pentaerythritol
Diethylaniline	*tert*-Pentanol
Diethylene glycol	Phenoxyalkyl carboxylic acids, α-substituted
Diethyl ether	2-Phenylbutyric acid
3, 3-Diethylglutaric acid	Poly(ethylene glycol)400
3, 3-Dimethyl-1-butanol	Tetraethylene glycol
2, 2-Dimethylglutaric acid	Toluene sulfonic acid
1, 1-Dimethylhexanol	Triethanolamine
Dimethylmalonic acid	Triethylene glycol
2, 2-Dimethyloctanoic acid	Trimethylacetic acid
1, 1-Dimethyl–1-pentanol	
2, 2-Dimethyl-1-pentanol	

Note: Omitted from list are certain members of the chlorinated-hydrocarbon pesticide, polychlorinated-biphenyl, synthetic-polymer, and surfactant classes.
Source: Alexander, M., "Nonbiodegradable and Other Recalcitrant Molecules," *Biotechnology and Bioengineering* **XV**, pp. 611–647, 1973.

3-3 ACTIVATED SLUDGE REACTION SYSTEMS

3-3.1 Flow Schemes and Reaction Kinetics

The activated-sludge process can stabilize soluble organic material to a low level in a short period because a high concentration of flocculant biological solids is maintained in the aerated or oxygenated tank in which the waste and bacteria are contacted. This concentration is maintained by routing the mixed liquor flowing from the aeration tank through a secondary clarifier, and recycling most of the settled solids back to the aeration tank. This concept is illustrated in Figure 3–6 for three basic types of systems: staged or plug, completely mixed, and contact-stabilization. The staged or plug-flow system has inherent kinetic advantages when the reaction rate is dependent upon the residual organic con-

TABLE 3-6. BOD Determinations on Typical Synthetic Organic Chemicals

				Bio-oxidation, percent							
				Non-Acclimated				Acclimated			
				Days				Days			
Product Tested	Water Solubility, g/100 ml of Solution	Theoretical Oxygen Demand, mg/mg	Measured COD mg/mg	5	10	15	20	5	10	15	20
Acetic acid (neutralized)	C	1.07	1.09	76	82	85	96	–	–	–	–
Acetone	C	2.20	2.00	56	76	83	84	–	–	–	–
Benzene	0.17	3.10	1.40	24	27	24	29	58	67	76	80
n-Butanol	7.7	2.59	2.45	68	87	92	92	–	–	–	–
n-Butyl acetate	0.68	2.20	2.32	58	68	70	83	–	–	–	–
Cumene	0.05*	3.50	1.13	40	62	63	70	–	–	–	–
Di-decylphthalate	0.2	2.69	2.50	1	1	1	0	7	10	3	7
Diethanolamine	95	1.52	1.47	17	72	81	88	–	–	–	–
Diethylene glycol	C	1.50	1.49	5	8	13	30	43	55	61	67
Diethylene glycol monoethyl ether	C	1.90	1.74	17	71	75	87	–	–	–	–
Diethylenetriamine	C	1.55	1.57	4	4	4	0	23	46	51	70
Di-2-ethyl-hexyl phthalate	0.2	2.58	2.74	0	0	–	0	13	0	6	23
Diisobutyl ketone	0.05	2.93	2.88	4	39	57	88	–	–	–	–
Epoxidized Soybean Oil	0.03*	–	2.24	1	1	0	0	4	–	13	24
Ethanol	C	2.10	1.99	74	74	95	84	–	–	–	–
Ethyl acetate	8.5	1.82	1.69	62	62	69	69	–	–	–	–
Ethyl acrylate	2	1.92	1.71	28	32	32	35	66	74	76	79
Ethylenediamine	C	1.33	1.30	24	44	55	47	36	45	56	70
Ethylene dichloride	0.81	0.97	0.025	0	18	–	–	–	–	–	–
Ethylene glycol	C	1.30	1.29	34	86	92	100	–	–	–	–
Ethylene glycol monobutyl ether	C	2.30	2.25	26	74	82	88	–	–	–	–

(continued)

TABLE 3-6. (Continued)

				Bio-oxidation, percent							
				Non-Acclimated				Acclimated			
				Days				Days			
Product Tested	Water Solubility, g/100 ml of Solution	Theoretical Oxygen Demand, mg/mg	Measured COD mg/mg	5	10	15	20	5	10	15	20
Ethylene glycol monoethyl ether acetate	23	1.81	1.76	36	79	82	80	–	–	–	–
Ethylene glycol monoethyl ether	C	1.80	1.98	36	88	92	100	–	–	–	–
Ethylene glycol monomethyl ether	C	1.68	1.64	30	62	74	88	–	–	–	–
Ethyl hexanol	0.07	2.95	2.79	26	75	78	86	–	–	–	–
Ethyl hexyl acrylate	<0.01	2.60	2.48	0	24	29	30	9	15	23	40
Ethoxy triglycol	C	1.89	1.78	8	47	63	71	–	–	–	–
Formic acid (neutralized)	C	0.40	0.36	48	45	60	60	45	41	34	53
Hexylene glycol	C	2.30	2.11	2	29	47	48	55	85	88	90
Isobutanol	8.5	2.59	2.39	64	73	76	72	46	71	75	81
Isobutyl acetate	0.63	2.20	1.43	60	74	79	81	–	–	–	–
Isodecanol (mixed isomers)	0.014*	3.03	2.87	24	44	42	39	14	26	45	32
Isophorone	1.2	2.78	2.64	0	13	47	42	–	–	–	–
Isopropanol	C	2.40	2.30	28	77	80	78	–	–	–	–
Isopropyl acetate	2.9	2.04	1.67	61	72	74	76	47	53	54	62
Linear alkyl benzene	0.04*	3.18	2.31	–	–	–	–	16	5	5	11
Methanol	C	1.50	1.50	76	88	91	95	–	–	–	–
Methyl amyl acetate	0.13	2.44	2.36	20	87	62	69	–	–	–	–
Methyl amyl alcohol	0.62*	2.82	2.64	50	72	90	94	–	–	–	–
Methyl ethyl ketone	26.8	2.44	2.23	76	82	84	89	–	–	–	–
Methyl isobutyl ketone	1.9	2.70	2.40	56	66	69	69	–	–	–	–
Octyl epoxytallate	0.06*	–	2.25	–	–	–	–	1	1	3	6
Phenol	8.4	2.40	2.30	90	89	87	96	–	–	–	–

Primary amyl acetate	0.2	2.33	2.22	64	76	67	72	–	–	–	–
n-Propanol	C	2.40	2.18	64	76	81	75	–	–	–	–
n-Propyl acetate	1.03*	2.04	2.04	62	80	75	72	–	–	–	–
Propylene glycol	C	1.68	1.63	62	68	75	79	–	–	–	–
Styrene	0.03	3.07	2.88	65	65	78	87	–	–	–	–
Tetrahydronaphthalene	0.035*	3.15	2.50	–	–	–	–	–	–	–	–
Tetraethylene glycol	C	1.65	1.64	4	10	15	22	9	58	71	88
Toluene	0.08*	3.13	1.40	53	62	70	80	73	74	80	86
Triethanolamine	C	1.61	1.52	8	9	45	66	–	–	–	–
Triethylene glycol	C	1.60	1.60	4	10	17	24	32	64	77	80
Vinyl acetate	2	1.67	1.39	34	34	31	32	62	70	66	72

*Solubility measured at 25°C in synthetic seawater. Other values are for distilled water. C = completely miscible. BOD tests run in standard dilution water, not seawater.

Source: Price, K. S., Waggy, G. T., and Conway, R. A., "Brine Shrimp Bioassay and Seawater BOD of Petrochemicals," *J. Water Pollut. Control Assoc.* **46**, 1, pp. 63–77, 1974.

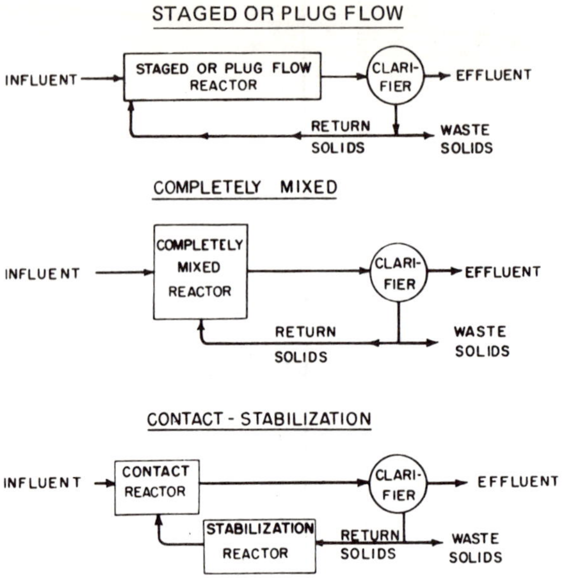

Fig. 3-6. Major activated sludge process variations.

centration; however, the effect of transient loads at the influent end of the basin may be more pronounced. The completely mixed system operates under more steady reactor conditions, but a lower driving force for the reactions is present. In a contact-stabilization system the waste is contacted with the bacteria for only a brief period before flowing to the settling tank. The return biological solids are aerated for a relatively long period to allow oxidation of the adsorbed and stored organic matter to occur prior to reentry of the cells into the contact chamber.

Evaluation of the activated sludge system for a particular application can be achieved best by first representing the system as a mathematical model, and then determining the necessary coefficients by running laboratory or pilot reactors at various loading rates while measuring organic reduction, biomass production, oxygen utilization, and biomass settleability. Results from these determinations need to be tempered by practical engineering judgements before their application. All notations and symbols used are compiled in Appendix 3-1.

The rate of the biological reactions governs the selection of the key design parameters of the biological system, i.e., the detention time and the biomass concentration, or more basically, as will be discussed later, the cell residence time and food-to-microorganism ratio. In order for the microorganisms to grow and metabolize the waste constituents, they obviously must be left in the system long enough to reproduce. The kinetics of biological growth have been studied in great depth and many models are available. Only the key elements of two approaches are summarized herein without extensive derivation (Ref. 2-2, 3-5 to 3-10).

The effluent quality in terms of soluble BOD from a completely mixed activated sludge system may be predicted under many conditions by using the relationship of liquid detention time and concentration of biological solids as developed by Adams, *et al.* (Ref. 2-2, 3-6).

$$\frac{S_e}{S_o} = \frac{1}{1 + KX_v t} \qquad \text{(Eq. 3-1)}$$

where

S_e = effluent dissolved BOD, mg/liter
S_o = influent dissolved BOD, mg/liter
K = reaction rate coefficient (independent of S_o in this model, liter/mg-hr)
X_v = mixed liquor volatile suspended solids (MLVSS), mg/liter
t = average liquid retention time, hr

While the relationship deals with the usually more important soluble BOD, organics in the form of large suspended solids in the wastewater are enmeshed in the floc; also, colloidal organics are flocculated and adsorbed. Their subsequent dissolution and assimilation needs to be considered in these models when concentrations are significant relative to soluble constituents. Also, refinements can be made to the above and subsequent relationships by expressing the MLVSS level as active biomass when the nonviable fraction is high due to contained nonliving volatile matter. Another limitation is that it applies best when treating to relatively low effluent levels.

The relationship is Eq. 3-1 assumes an approximate first-order reaction rate. When considering the total amount of soluble substrate remaining at any time, however, the actual kinetic relationship is likely the simulation of multiple zero order reactions down to low substrate levels. This is not true when sequential reactions occur.

Although the Eckenfelder "first-order" model has considerable utility, the trend seems to be toward modeling biological growth kinetics (growth and substrate removal) by procedures developed by Monod, based on the Michaelis-Menten hypothesis for rates of enzyme-catalyzed reactions. Monod kinetics were related to waste treatment processes by Lawrence and McCarty (Ref. 3-8) and later a multiprocess biological treatment model was published by Christenson and McCarty (Ref. 3-10). Their contribution was to incorporate removal and growth kinetics into a single approach. The Monod equations can be written as

$$\mu = \frac{dX_a}{dt} \cdot \frac{1}{X_a} \qquad \text{(Eq. 3-2)}$$

where

μ = specific (per unit mass) growth rate in the sytem, hr^{-1} . . . mg growth/mg biomass-hr

X_a = active biomass concentration usually approximated by MLVSS, mg/liter

t = time, hr

also,

$$\mu = \mu_{max}\left(\frac{S}{K_s + S}\right) \qquad \text{(Eq. 3-3)}$$

where

μ_{max} = maximum specific growth rate under defined conditions or rate that a plot of μ vs S asymptotically reaches (see Fig. 3-7), hr^{-1}. Note that specific growth rate is directly proportional to specific rate of substrate utilization.

S = substrate concentration in the reactor, mg/liter

K_s = half velocity constant or substrate concentration resulting in μ = 0.5 μ_{max} (see Fig. 3-7), mg/liter

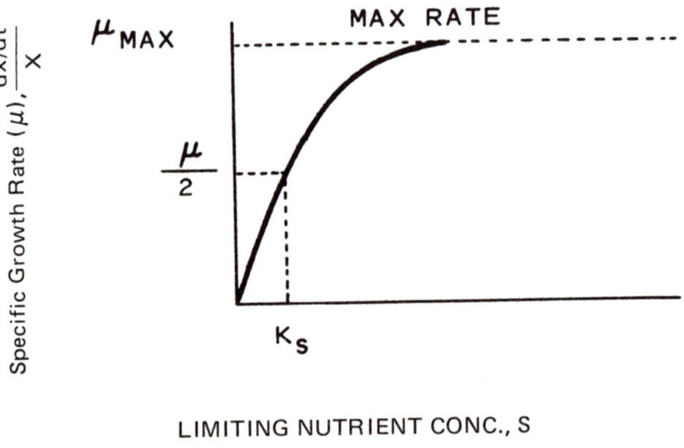

*Or, specific rate of substrate utilization, $\frac{ds/dt}{X}$

Fig. 3-7. Monod kinetic relationships.

If one assumes that cell yield can be defined as mass of cells produced per unit mass of substrate consumed, then the rate of substrate metabolism can be related to the specific growth rate via the constant of proportionality, Y

$$Y = \frac{dX}{dS} \qquad \text{(Eq. 3-4)}$$

where

Y = maximum cell yield coefficient, g bacteria synthesized/g BOD_L consumed

therefore

$$\frac{dS}{dt} = -\frac{\mu_{max} X_a}{Y}\left(\frac{S}{K_s + S}\right) \qquad \text{(Eq. 3-5)}$$

Division of μ_{max} by Y changes the expression to contain a substrate removal rate constant

$$\frac{dS}{dt} = \frac{K'' X_a S}{K_s + S} \qquad \text{(Eq. 3-6)}$$

The net growth of active organisms is related to the rate of substrate utilization

$$\frac{dX_a}{dt} = Y\frac{dS}{dt} - bX_a \qquad \text{(Eq. 3-7)}$$

where b is the cell autooxidation rate coefficient, day^{-1}. Equations 3-6 and 3-7 and a mass balance for a complete mix reactor can be used to yield the expression

$$\frac{1}{\theta_c} = \frac{YK''S}{K_s + S} - b \qquad \text{(Eq.3-8)}$$

as

$$\mu = \frac{1}{\theta_c} \qquad \text{(Eq. 3-9)}$$

where

θ_c = solids or cell retention time, days

Solving for S, which is the same as S_e for a completely mixed reactor, in Eq. 3-8, yields

$$S = \frac{K_s(1 + b\theta_c)}{\theta_c(YK'' - b) - 1} \qquad \text{(Eq. 3-10)}$$

BOD levels preferably are expressed as those after long-term incubation. Monod relationships also can be used to predict sludge yield and oxygen requirements (Ref. 3-8, 3-10).

In cases where $K_s >> S$, i.e., low levels of substrate in a reactor (same as effluent in completely mixed system), Eq. 3-6 becomes first order with respect to substrate level, i.e.,

$$\frac{dS}{dt} = -kX_aS \qquad \text{(Eq. 3-11)}$$

Integration yields the first order relationship for a continuously fed, complete mixed system.

$$\frac{S_e}{S_o} = \frac{1}{1 + KX_at} \qquad \text{(Eq. 3-12)}$$

which is identical to Eq. 3-1. In cases where $S >> K_s$, Eq. 3-6 becomes zero order

$$\frac{dS}{dt} = -kX_a \qquad \text{(Eq. 3-13)}$$

or

$$\frac{S_o - S_e}{X_at} = k_{\max} \qquad \text{(Eq. 3-14)}$$

as shown in Figure 3-7, which means that substrate removal proceeds at a constant, linear rate. In cases where S_e is neither very high nor about an order of magnitude smaller than K_s (i.e., the central curved portion of Figure 3-7) other techniques need to be used. This can be the case when feed BOD levels are high and/or removal efficiencies are low; a typical value of K_s for a mixed waste is 200 mg/liter (Ref. 3-10). Equation 3-10 covers the whole range of S levels in a continuous manner. The application of these relationships along with calculative means of reactor design, solids production, and oxygen requirements have been published by Christenson and McCarty (Ref. 3-10). Reaction kinetics for a plug flow system are included. A simplified mathematical model of the complete mix activated sludge system has been developed by McKinney (Ref. 3-11). Field data have verified its applicability to design and operational evaluation.

3-3.2 Temperature Effects

The removal rates for BOD assume an adequacy of other nutrients, such as nitrogen and phosphorus, and optimum levels of pH, oxygen, and mixing. The reaction rate constant is temperature dependent according to:

$$K_2 = K_1 \theta_t^{(T_2 - T_1)} \qquad \text{(Eq. 3-15)}$$

where

K_2 = organic removal rate coefficient at T_2, day^{-1}
K_1 = organic removal rate coefficient at T_1, day^{-1}
θ_t = temperature coefficient
T_2, T_1 = temperature, °C

The value for θ_t typically ranges from 1.03 to 1.15. This coefficient appears to increase with increasing waste strength and with the content of synthetic organic compounds that are more difficult to degrade (Ref. 3-8, 3-12). At a high S_o of 4500 mg/liter with a synthetic organic chemical waste, a θ_t of 1.25 was measured. Experimental determination is made by operating parallel systems at ambient and cold-weather conditions, and inserting the rate coefficients obtained into Eq. 3-15. At high θ_t values and low temperatures, the effects on effluent quality are drastic.

The temperature of an open reactor using surface-entrainment aerators can be predicted based on flow rate, influent wastewater temperature, ambient air temperature, and installed horsepower (Ref. 3-9). With closed systems, such as those using pure oxygen, heat losses are less.

Temperature level has effects other than upon reaction rates. Protozoan and rotifier populations are less above a temperature of about 36°C and usually are very low above 40°C. Decreased water viscosity tends to enhance flocculation and settling at a higher temperature. However, at about 35°C deflocculation can start to occur. At temperatures below about 18°C effluent solids level often increases. Some evidence suggests that less sludge is produced at higher temperatures as the rates of the oxidation reactions increase more than that of the synthesis reactions. Values for K_s generally increase with decreasing temperature, unlike rate constants.

3-3.3 Feed Concentration Effects

The substrate removal rate also can be affected by the initial strength of the wastewater according to the Grau model as extended by Adams, Eckenfelder, and Hovious (Ref. 3-9). For a completely mixed reactor the relationship is:

$$\frac{S_o(S_o - S_e)}{X_v t} = K' S_e \qquad \text{(Eq. 3-16)}$$

where

K' = reaction rate coefficient dependent on S_o for completely mixed system, liter/mg-hr

while for batch or plug-flow systems, the relationship is:

$$\frac{S_e}{S_o} = e^{-kX_v t/S_o} \qquad \text{(Eq. 3-17)}$$

where

k = batch-fed reaction coefficient

These models, which are new and require additional evaluation, imply that as the concentration of organics remaining in the reactor decreases, the rate of removal also decreases since the organics remaining are progressively becoming more difficult to remove. The completely mixed model holds particularly when the influent level is varying with step increases in incoming strength (Ref. 3-9).

In cases where either COD or TOC is used in the model and nonbiodegradable organics are present, Eq. 3-16 can be modified:

$$\frac{(S_o - Z)(S_o - S_e)}{X_v t} = K'(S_e - Z) \qquad \text{(Eq. 3-18)}$$

where

Z = concentration of nondegradable organics in mg/liter

This relationship would apply where data are taken in terms of COD or TOC and the effect of influent concentration needs to be considered.

3-3.4 Reaction Rate Determination

The value of K is determined by operating reactors at different loading levels (varying feed concentration, mixed liquor suspended solids, and retention time as appropriate) and measuring the achieved effluent level. Equation 3-1 is rearranged:

$$\frac{S_o - S_e}{X_v t} = K(S_e) \qquad \text{(Eq. 3-19)}$$

A plot on the ordinate (Y axis) and abscissa (X axis) respectively of $S_o - S_e/X_v t$ vs. S_e with slope = K (S_o independent) yields the desired rate coefficient. Some typical values are listed in Table 3-7. An example at the top end of the typical range cited is a high strength waste of 4000 mg BOD/liter, containing mostly volatile acids in which K was measured to be about 10×10^{-4} liter/mg-hr; however, a value of 2.5×10^{-4} liter/mg-hr does seem typical for many chemical wastes.

Likewise, after rearrangement of Eq. 3-16 and appropriate plotting $S_o(S_o - S_e)/X_v t$ vs. S_e with slope = K' (S_o dependent) yields the rate coefficient when it is dependent on influent concentration.

TABLE 3-7. Typical Coefficients and Constants for Activated Sludge Treatment of Wastes from the Chemical Industry Using the Eckenfelder Model

Systems	Range of Values of One Standard Deviation	Avg. Value
K, liter/mg-hr	0.00012–0.00076	0.00024
a	0.31–0.72	0.52
a'	0.31–0.76	0.53
b, day^{-1}	0.02–0.18	0.07
b', day^{-1}	0.10–0.24	0.17
$F{:}M$, lb BOD/(lb MLVSS) (day)	0.4–0.8	0.5

Source: Eckenfelder, W. W., Jr. and Ford, D. L., "Economics of Wastewater Treatment," *Chem. Eng.*, pp. 109–118, August 25, 1969.

A unique rate constant does not exist for a particular waste. In reality, rate constants vary considerably from day to day, primarily due to the feed variability. The extent of feed variability is indicated by the scatter in the above plots; a range encompassing a desired degree of confidence can be used. This is an alternative means of arriving at the variability portion of the approach used in Eq. 1-1 (Appendix 1-2) for defining achievable effluent quality.

Similar calculative means are available using Monod kinetics in cases where first-order reaction rates do not apply (Ref. 3-8, 3-11).

3-3.5 System Limitations

One further point concerning the application of kinetic data needs to be kept in mind. In addition to materials which are nondegradable in a structural sense, the concept of a "residual BOD" level should be considered. Biological systems treating relatively high strength wastewaters generally do not decrease the BOD to the low levels predictable from the kinetic models. This can be attributed to several factors, all of which are related to the fact that the sample collected under conditions in the biological treatment system reactor is tested in a BOD dilution bottle under simulated stream conditions:

- Equilibrium established with the high level of reactor solids is shifted under the low level of seed in the BOD bottle.
- Reseeding in the BOD bottle allows a culture to develop specific to the residual structures present.
- Essential micronutrients (for specific species) depleted in the reactor are replenished in the BOD bottle.

- Surface adsorption in the BOD bottle increases the effectiveness of the conversions.
- Dilution of inhibitory end products allows reinitiation or acceleration of the reactions.
- Growth of nitrifiers and higher microbial forms would be enhanced due to the dilution of inhibitory constituents.
- Degradation of the residual materials requires the longer incubation periods afforded in the BOD test.

For these reasons, as well as the relationships developed by Grau, the models should not be extrapolated to achieve very low levels of BOD starting at a high strength, especially with synthetic organic wastewaters. Even with low strength industrial wastes, the lowest achievable levels seem to be about 10 to 15 mg/liter. Some long-term effluent values measured under warm-temperature conditions for reasonably designed treatment systems at three petrochemical plants are plotted in Figure 3-8. Data are also plotted for a single laboratory study of municipal wastes which had been concentrated by vacuum distillation. Practical achievable BOD removal efficiency in a single step system is on the order of 90 to 95%.

The effect of feed variability on achievable effluent quality also needs to be emphasized. The data in Table 3-8 show that not only is the average effluent level increased by a factor of 8, but the range of effluent quality is increased by a factor of 15 under highly variable conditions.

3-3.6 Biomass Production

The production of biomass is the net effect of growth due to substrate conversion and cellular autooxidation

$$\Delta X_{vt} = aS_{rt} - bf_b X_{vt} \qquad \text{(Eq. 3-20)}$$

where

ΔX_{vt} = change in MLVSS, lb/day (kg/day)
a = fraction of substrate removed converted to cells
S_{rt} = substrate removed in terms of BOD_5, lb/day (kg/day)
b = cell autooxidation rate coefficient, day^{-1}
X_{vt} = total weight of volatile suspended solids in aeration tank, lb (kg)
f_b = fraction of generated biomass that is degradable, decreases with sludge age, experimentally determinable or computable by (Ref. 3-9):

$$f_b = \frac{aS_{rt} + bX_{vt} \sqrt{(aS_{rt} + bX_{vt})^2 - (4\, bX_{vt})(0.77\, aS_{rt})}}{2\, bX_{vt}} \qquad \text{(Eq. 3-21)}$$

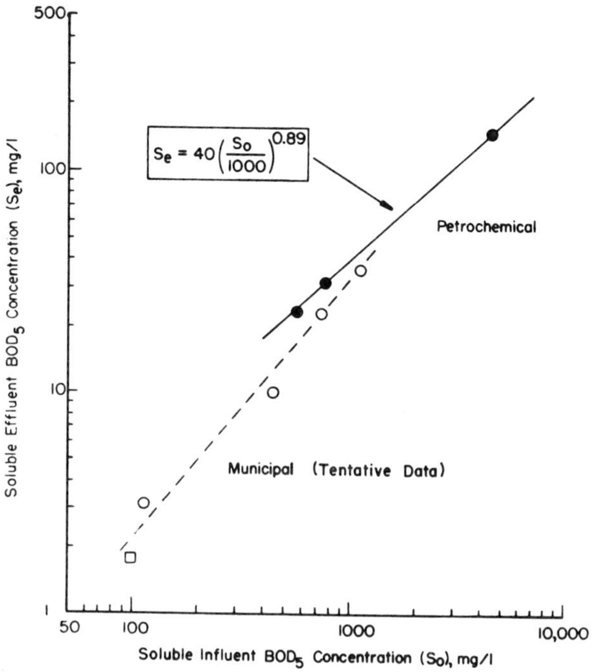

Fig. 3-8. Soluble effluent BOD values for biological treatment systems as a function of influent BOD. ***Source:*** **Conway, R. A., Hovious, J. C., Macauley, D. C., "Predicting Achievable Effluent Quality and Variability in the Organic Chemicals Industry," presented at International Conference on Effluent Variability from Waste Treatment Processes and Its Control, New Orleans, Louisiana, December 2-4, 1974,** ***J. Prog. Water Technol.*** **8, 1, 165-182 (1976).**

The accumulation of solids, both organic and inorganic, from non-biological sources also needs to be considered when calculating excess solids which require disposal. Wood pulping fines and calcium carbonate are examples of materials accumulating in biological flocs.

$$\Delta X_t = rX_{ot} + \frac{\Delta X_v}{f_v} - X_{et} \qquad \text{(Eq. 3-22)}$$

where

ΔX_t = total excess sludge, lb/day (kg/day)
X_{ot} = total influent suspended solids, lb/day (kg/day)
r = nondegradable fraction of influent suspended solids
f_v = ratio of MLVSS/MLSS
X_{et} = total effluent suspended solids, lb/day

TABLE 3-8. Summarized Data from Laboratory Study of Feed Variability Effects

	Constant Feed[b]	Highly Variable Feed[b]
Influent BOD, mg/1[(a)]		
average	1400	1400
standard deviation	120	850
Effluent BOD, mg/1		
average	22	180
standard deviation	16	240
BOD removal (average), percent	98	87
Variability reduction, percent of initial std. deviation on absolute basis	87	72

[a]All BOD data are on a soluble basis. Results shown are the average from duplicated reactors. Average *F/M* and MLVSS concentration were 0.72 d^{-1} and 2200 mg/1, respectively.
[b]Constant feed consisted of a mixture of 14 1-hr composite samples of petrochemical wastewater collected at random over a 3-day period. The units receiving this feed were subject only to a small degree of quantitative variability. Variable feed units were fed the hourly composite samples individually on a 12-hr cycle, i.e., the feed was changed between composites once every 12-hr. The variable-feed units, therefore, were subjected to a high degree of quantitative and qualitative variability, although on the average they received the same feed as the constant-feed units.
Source: Conway, R. A., Hovious, J. C., and Macauley, D. C., "Predicting Achievable Effluent Quality and Variability in the Organic Chemicals Industry," presented at International Conference on Effluent Variability from Waste Treatment Processess and Its Control, New Orleans, Louisiana, December 2–4, 1974, *J. Prog. Water Technol.* 8, 1, 163 (1976).

A common value for f_v is 0.8.

When running the experimental units to determine substrate removal efficiency, biosolids production also is measured by a daily solids balance. Equation 3-20 is rearranged:

$$\frac{\Delta X_{vt}}{f_b X_{vt}} = a\left(\frac{S_{rt}}{f_b X_{vt}}\right) - b \qquad \text{(Eq. 3-23)}$$

Plot on the Y and X axes, respectively:

$$\frac{\Delta X_{vt}}{f_b X_{vt}} \text{ vs. } \frac{S_{rt}}{f_b X_{vt}}$$

$$\frac{\text{lb VSS produced}}{\text{lb biodegradable VSS-day}} \text{ vs. } \frac{\text{lb organics removed}}{\text{lb biodegradable VSS-day}}$$

Slope = a; Y intercept = $-b$.

The fraction of biodegradable solids can be determined by aerating without feeding a sample of mixed liquor and plotting the solids concentration versus time. Extrapolation of the plot will approach the nondegradable level. Published correlations indicate fractions about 0.74 at F/M of 0.8 day^{-1}, 0.36 at F/M of 0.4 day^{-1}, and 0.31 at F/M of 0.1 day^{-1} (Ref. 2-2).

Typical values for other coefficients in the model are presented in Table 3-7; values for use when Monod relationships are used also are available (Ref. 3-10).

Factors which influence cell yield have been reviewed by Ramanathan and Gaudy (Ref. 3-13). Random variations in predominant species in the treatment system had the major effect. Various species have different partitioning of the synthesis and respiration pathways. Although the free energy of a given substrate can be used to predict available energy for synthesis, this concept was overshadowed by species variations.

3-3.7 Oxygen Requirements

Requirements for oxygen to be supplied through aeration or pure oxygenation may be predicted from the additive effects of substrate removal and biomass autooxidation.

$$O_t = a'S_{rt} + b'f_bX_{vt} \qquad \text{(Eq. 3-24)}$$

where

O_t = total oxygen required, lb/day (kg/day)
a' = oxygen utilization coefficient for substrate conversion, lb O_2 required/lb BOD_5 removed (kg/kg)
b' = oxygen utilization rate for endogenous respiration, lb O_2/day-lb MLVSS (kg/day-kg)

When a chemical oxygen demand, such as for sulfite oxidation, must be supplied, that requirement needs to be added. Likewise, when the system converts ammonia to nitrate (nitrification), a requirement of up to 4.6 lb O_2/lb NH_3 oxidized is added. Nitrification requires long mean cell residence times for the slowly growing responsible species. The sensitivity of nitrifiers to pH, temperature, and chemical inhibitors reduces nitrification processes to a negligible rate in many industrial wastes.

Oxygen utilization is followed in experimental units by measuring oxygen uptake rate under a range of conditions. Rearrangement of Equation 3-24 yields:

$$\frac{O_t}{f_bX_{vt}} = a'\frac{S_{rt}}{f_bX_{vt}} - b' \qquad \text{(Eq. 3-25)}$$

Plot on the Y and X axes, respectively:

$$\frac{O_t}{f_b X_{vt}} \text{ vs. } \frac{S_{rt}}{f_b X_{vt}}$$

$$\frac{\text{lb } O_2 \text{ required}}{\text{lb biodegradable VSS-day}} \text{ vs. } \frac{\text{lb organics removed}}{\text{lb biodegradable VSS-day}}$$

Slope = a'; Y intercept = $-b'$.

The sum of the biomass produced and oxygen utilized should be checked, on an oxygen equivalent basis, versus substrate removed according to the relationship:

$$O_t = (y)(S_{rt}) - 1.42\,(X_{wt} + X_{et}) \qquad \text{(Eq. 3-26)}$$

where

y = lb BOD_L/lb BOD_5
BOD_L = BOD measurable through long-term incubation (or removable COD)
X_{wt} = total biological solids wasted, lb/day
X_{et} = total effluent biological solids, lb/day
1.42 is lb O_2 required to oxidize 1 lb of biomass of formulation $C_5H_7O_2N$

3-3.8 Loading Factor

One important parameter affecting design and operation is the food-to-microorganism ratio,

$$F/M = \frac{S_a}{(X_{vt})} \qquad \text{(Eq. 3-27)}$$

where

F/M = food-to-microorganism ratio, day^{-1}
S_a = substrate (BOD_5) applied, lb/day (kg/day)
X_{vt} = average MLVSS in aeration tank, lb (kg)

The previously cited relationships infer that as F/M increases, the extent of soluble BOD removal decreases, the oxygen requirement decreases, and the biomass production increases for a given amount of waste treated.

An additional important effect of the F/M ratio is on the settleability of the biomass. Optimum settleability usually is achieved between values of about 0.1 to 0.8/day; presumably because at lower levels the polymeric materials enhancing cell flocculation are excessively degraded and at higher levels polymer production decreases due to high growth rates. Filamentous growths also are stimulated at high loadings.

When the experimental units are run to determine substrate reduction, oxygen requirements, and solids production, effluent solids levels should be measured and plotted versus F/M to indicate the levels achievable. Results from larger pilot units can be used with more confidence. Suspended solids levels in both the effluent from the test unit and the supernatant from settling tests of the mixed liquor in a slowly stirred graduated cylinder (simulates idealized settler) should be measured. Determinations of soluble and total BOD levels are made on effluent or supernatant samples collected at various loadings. A plot of mg BOD/mg VSS (*B* factor in Eq. 1-1) versus *F/M* is useful in calculating total effluent BOD levels when soluble effluent BOD and effluent suspended solids levels are known. A value of 0.3 mg BOD_5/mg VSS is common.

The initial settling velocity in the graduated cylinder test also should be plotted versus *F/M* to indicate the rates achievable. Sludge thickening rates may need to be noted for use in clarifier design by one method, as discussed in a later section. Levels of MLVSS usually are maintained between 1500 and 8000 mg/liter, depending on an optimized balance between clarifier and reactor costs as well as upon operating and performance considerations.

3-3.9 Cell Residence Time or Sludge Age

Many kinetic models are expressed in terms of mean cell residence time (θ_c); this can be related to *F/M* by:

$$1/\theta_c = (a)(F/M) - b \qquad \text{(Eq. 3-28)}$$

Mean cell residence time is calculatable by this equation:

$$\theta_c = \frac{X_{at}}{X_{wt} + X_{et}} \qquad \text{(Eq. 3-29)}$$

where

X_{at} = average weight of biological solids maintained in reactor, lb
X_{wt} = weight of biological solids wasted, lb/day
X_{et} = weight of biological solids in the effluent, lb/day

or, by this relationship,

$$\theta_c = \frac{X_{vt}}{\Delta X_{vt}} \qquad \text{(Eq. 3-30)}$$

The mean cell residence time or sludge age is a basic control parameter as it can be related to microorganism generation time. When transformations involving slowly growing microorganisms are involved (e.g., those involved in nitrogen oxidation and certain polymeric degradations) longer cell residence times must be provided. As with *F/M*, the settleability of the biological solids can be related to θ_c for various types of wastewaters (Table 3-9). Incidentally, adverse effects of high salt levels on settleability can over shadow θ_c effects.

TABLE 3-9. Example Sludge Settling Properties for Various Modifications of the Activated Sludge Process with a Synthetic, Degradable Feed

Modification	Solids Retenion Time (θ_c), days	Sludge Settling Properties
High rate (rapid assimilation)	0.25–2	Predominantly dispersed growth; practically no zone settling; SVI > 250
Conventional activated sludge	2–6	Well-formed floc; zone settling velocity between 2 and 6 ft hr^{-1}; SVI ranges from 580 (at 2 days) to 125 (at 6 days)
Extended aeration	6–30	Pinpoint floc; deflocculated particles; zone settling velocity averages 15 ft hr^{-1}; SVI about 100

Source: Bisogni, J. J., Jr. and Lawrence, A. W., "Relationships between Biological Solids Retention Time and Settling Characteristics of Activated Sludge," *Water Res.* 5, pp. 753–763, 1971.

3-3.10 Nutrient Requirements

In order to support cell synthesis, nitrogen and phosphorous must be added if deficient in the wastestream. These are usually added at a level in proportion to the BOD removed.

$$\text{Nitrogen required in lb/day} = \frac{S_{rt}}{20} \quad \text{(Eq. 3-31)}$$

$$\text{Phosphorous required in lb/day} = \frac{S_{rt}}{100} \quad \text{(Eq. 3-32)}$$

Nitrogen usually is added as ammonia, and phosphorous as phosphoric acid. Organic sources in the wastes, such as degradable amines, need to be considered when calculating addition requirements.

Instead of using the average ratios given above, the consideration that while new cells average 12.4% nitrogen and 2.6% phosphorous, these levels will decrease after endogenous oxidation to 7.0% and 1.0%, can be taken into account (Ref. 2-2).

$$\text{lb N/day} = \frac{0.124\, f \Delta X_{vt}}{0.77} + 0.07\,\frac{(0.77\text{-}f)\Delta X_{vt}}{0.77} \quad \text{(Eq. 3-33)}$$

$$\text{lb P/day} = \frac{0.026\, f \Delta X_{vt}}{0.77} + 0.01\,\frac{(0.77\text{-}f)\Delta X_{vt}}{0.77} \quad \text{(Eq. 3-34)}$$

In addition to nitrogen and phosphorous, the adequacy of other elements, such as iron, sulfur, magnesium, and manganese, needs to be established, especially for wastewater coming from a single manufacturing operation where contamination by a range of materials is less likely to have occurred. Trace, but not toxic, copper and zinc levels also should be checked. Sodium, potassium, calcium, and chloride are required in substantial quantities, but usually are present. Certain protists, especially higher forms, require boron, molybdenum, vanadium, cobalt, iodine, and/or selenium.

Nutrient deficiencies affect not only substrate conversion but also biomass settleability. Proper floc formation seems to be hampered; and excessive filamentous organisms, which have a lesser nutrient requirement, can develop.

3-4 AERATION/MIXING

When air is used, the oxygen requirement (O_t) for activated sludge systems, calculated using Eq. 3-24, is supplied by one of several means:

Surface-entrainment aerators
Diffused aeration
Sparged turbine aeration
Venturi (jet) aeration
Static-mixer aeration
Brush aerators

The system selected also needs to provide adequate macro-turbulence to keep the biological floc in suspension and micro-turbulence to enhance substrate-bacterial contact.

3-4.1 Transfer Efficiency

The oxygen transfer efficiency of the device can be specified as

$$N_a = N_s \frac{\beta C_{sw} - C_L}{C_s} \alpha \theta_o^{T_w - 20} \qquad \text{(Eq. 3-35)}$$

where

N_a = O_2 transfer efficiency at application conditions in the field, lb O_2/hp-hr (kg/joule × 5.92 × 10^6)

N_s = oxygen transfer efficiency under standard conditions (tap water at zero dissolved oxygen, 20°C, and 760 mm Hg), lb/hp-hr (kg/joule × 5.92 × 10^6)

β = ratio of dissolved oxygen level at saturation in reactor liquor to that in pure tap water

C_{sw} = dissolved oxygen saturation level calculated for reactor liquor at design pressure and liquor temperature, mg/liter
C_L = design operating dissolved oxygen concentration, mg/liter
C_s = dissolved-oxygen saturation concentration at standard conditions, mg/liter
α = ratio of K_La of reactor liquor to that of pure water
K_La = overall oxygen transfer coefficient, mg O_2/mg O_2 deficit-hr
θ_c = temperature adjustment coefficient for oxygen transfer
T_w = design reactor temperature, °C

The standard transfer efficiencies usually are determined for the various types and sizes of aerators by the equipment manufacturers through reoxygenation rate tests either in large test tanks or in field installations as described by Conway and Kumke (Ref. 3-14) and by Nogaj (Ref. 3-15).

The β factor is determined by aerating a sample of filtered mixed liquor (or effluent from a completely mixed reactor) and dividing the equilibrium level by that obtained with tap water at the same pressure and temperature. Salt content depresses the saturation level (0.1 mg/liter by each 100 mg/liter of chloride ion at 15°C) as do other materials. Typical values for β are 0.90 to 0.98. For surface-entrainment aeration, C_{sw} is obtained by multiplying the saturation value (Ref. 1-4) at the design temperature by $p/760$, where p is the design atmospheric pressure in mm mercury (kN/cm^2 × 7.50). Usually, C_L is selected at 2 mg/liter for well-mixed, air-based systems. And, C_s equals 9.2 mg/liter at the surface at 20°C. In the case of submerged aerators, the effect of increased pressure creates an average C_{sw} according to:

$$C_{sw} = C_{sa}\left(\frac{p_b}{29.4} + \frac{O_v}{42}\right) \qquad \text{(Eq. 3-36)}$$

where

C_{sa} = oxygen saturation at one atmosphere pressure, mg/liter
p_b = absolute pressure at depth of air release, psia (kN/m^2 × 0.145 or kgf/cm^2 × 14.2)
O_v = volume concentration of oxygen in gas leaving the test unit, %

An assumed value for O_v can be used for initial calculations; the correctness of C_s will be evident from the straightness of the plot of log oxygen-saturation deficit versus time.

The α factor usually is determined by following the reoxygenation rate in a test vessel aerated under conditions simulating those expected in the field, i.e., similar mechanism of transfer in terms of physical appearance of contact interface and similar mixing-power levels (same K_La). Typical values for α are 0.70 to 0.90. The slope of a plot of $C_s - C_t$ on a log scale versus time in hours, on an arithmetic scale is K_La:

$$K_L a = \frac{2.3 \log_{10}\left(\dfrac{C_s - C_{t_1}}{C_s - C_{t_2}}\right)}{t_2 - t_1} \qquad \text{(Eq. 3-37)}$$

where C_t = dissolved oxygen concentration at times t_1 and t_2, mg/liter

which was derived from the first order relationship:

$$\frac{dD_o}{dt} = -K_3 D_o \qquad \text{(Eq. 3-38)}$$

where

D = dissolved-oxygen deficit
K_3 = proportionality constant

This relationship is used in determining performance of aeration equipment.

$$N_a = \frac{(K_L a)(D)(W)}{P} \qquad \text{(Eq. 3-39)}$$

where

W = weight of aerated water, millions of pounds (kg $\times$ 2.21 $\times$ 10^6)
P = power, horsepower (kW $\times$ 1.34)

The power needs to be specified either as line horsepower to the aeration device or input (brake) horsepower to the rotor after motor and gear efficiencies are considered. In the case of compressed air, motor, gear, and fan, efficiencies are involved. Due to the effects on diffusivity and viscosity, the oxygen transfer coefficient is influenced by temperature; a value for θ_v of 1.024 is typical (Ref. 3-7). The value for θ_o can range from about 1.01 to 1.04 depending upon the type of system.

An in-depth coverage of all aspects of oxygen transfer testing is found in a report by Boyle and Brenner on a 1978 ASCE-EPA Workshop Toward an Oxygen Transfer Standard (Ref. 3-43).

3-4.2 Power and Mixing

The total horsepower required is calculated by:

$$hp = \frac{O_t}{(N_a)(24 \text{ hr/day})} \qquad \text{(Eq. 3-40)}$$

This value needs to be checked for adequacy of mixing; at least 0.1 hp/1000 gal. (0.02 kW/m^3) is required. Surface aerator size selection and placement are critical as interactions occur which adversely affect mixing (Fig. 3-9), as

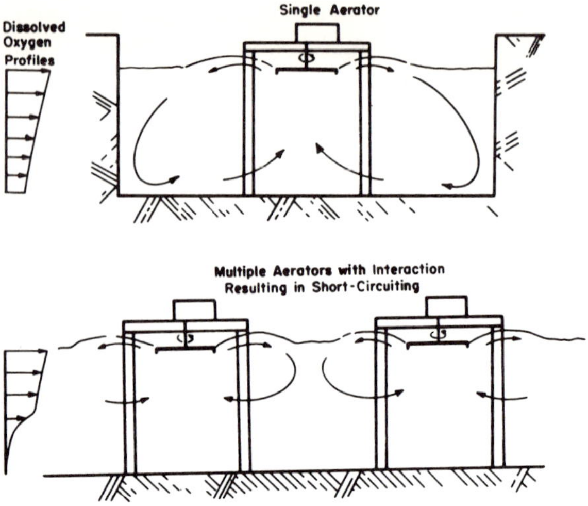

Fig. 3-9. Conceptual representation of aerator mixing patterns. ***Source:*** **Price, K. S., Conway, R. A., and Cheely, A. H., "Surface-Aerator Interactions,"** ***J. Env. Engr. Div., ASCE*** **99, EE 3, 283-300 Proc. Paper 9797, 1973.**

described by Price, Conway, and Cheely. (Ref. 3-16). A rather strict limitation on minimum spacing exists for a given combination of aerator type, horsepower, and basin depth to avoid adverse interactions as regards both oxygen transfer and biosolids suspension. For one type of 100-hp (75 kW) stationary aerator, an aerator spacing of 70 ft (21 m) provided adequate mixing to a depth of 15 ft (4.6 m); at closer spacing, the level of oxygen transfer and macromixing would be reduced. A horizontal velocity of 0.4 to 0.5 ft/sec (0.12 to 0.15 m/sec) measured under viscosity conditions in a functioning biological system is a reasonable minimum value for allowing up to a biosolids level of about 5000 mg/liter to be maintained in suspension. Deep tanks require bottom mixers or draft tubes on the aerators.

3-4.3 Aeration Equipment

The basic types of surface-mechanical aerators which have been developed include (Ref. 2-14, 3-17):

Submerged propeller driving water upwards against a diffuser cone at the top of the unit;
Impeller rotating at the surface of the water;
Combination of a surface impeller with a submerged impeller on the same shaft;

Spiral-vaned rotating cone with a draft tube or scooped impeller;
Downward-pumping impeller in a draft tube with radial inlet troughs at the surface;
Plate-type surface aerators with horizontally vaned impellers;
Combination aerators employing surface or subsurface impellers, or both, with the addition of compressed air;
Submerged, sparged turbine for satisfying high oxygen demand rates;
Paddle wheel, brush, or disks on a horizontal axis revolving with partial submergence.

Diffused aeration equipment utilizing compressed air includes:

Porous ceramic plates or tubes;
Ceramic tubes with integral cast-iron caps and orifices;
Plastic-cord-wrapped metal cores with integral end-caps and orifices;
Flexible plastic cloth on a metal frame with integral end-cap and orifice;
Porous plastic media with a plastic pan holder;
Perforated, weighted hosing for lagoon installation;
Stainless steel or plastic nozzles with check valves;
Tubular grid located 1.5 to 3 ft (0.5 to 0.9 m) below the surface
Shear type nozzle with control orifice;
Adjustable nozzle with 4 to 12 openings;
Bronze, aluminum-magnesium, or plastic multiple orifice
Metal valve with a moving plastic ball.

The jet aeration system consists of a submersible pump for circulating the mixed liquor, air compressors, and a cluster of air-liquid jets operating on a Venturi principle. In the static mixer system air is conveyed in a tubing arrangement to the base of a vertical mixing element.

In the case of surface aerators an evaluation of two additional options needs to be made: Floating versus fixed installation and Direct drive versus gear reduced. Factors to be considered include the trade off between the decreased capital costs of floating and/or direct-drive units and the considerations of oxygen transfer capacity, mixing capabilities, maintenance costs, and available space. Typical oxygen transfer efficiencies in terms of power requirements are presented in Table 3-10.

The surface aerators and diffusers with air are capable of satisfying demands of up to about 100 to 140 mg O_2/liter-hr due to limitations on minimum aerator spacing and depth. Sparged turbines can supply at least 250 mg O_2/liter-hr. In diffused air systems the efficiency of oxygen transfer is reduced by two major components of the system: 1) air compression is an inefficient process, and 2) horsepower loss due to head loss in the piping system cannot be used in the aeration process. In addition, many diffusers require filtered air to reduce

TABLE 3-10. Aerator Oxygen Transfer Properties

	Typical Standard Transfer Capacity (N_s) with Air, lb O_2/hp-hr
Surface Aerators	
Low speed	
<<30 hp rating	5.0
>30 hp rating	3.5
Floating, high-speed	
<30 hp rating	3.0–3.5
>30 hp rating	2.5–3.0
Diffused Aeration	
Porous diffusers	3.7[a]
Nonporous diffusers	1.2[b]
Sparged Turbine	2.0–3.0

[a]At 12% oxygen transfer efficiency.
[b]At 4% oxygen transfer efficiency.
Conversion, hp/1.34 = kW, lb/hp-hr ÷ 5.92 × 10^6 = kg/joule
Source: Nogaj, R. J., "Selecting Aeration Equipment," *Chem. Eng.*, April 17, pp. 92–102, 1972.

problems of internal clogging. External or surface clogging is also a problem when using many types of diffusers, and periodic removal for cleaning is considered normal maintenance.

3-5 CLARIFICATION/THICKENING

3-5.1 Thickener Analysis

In order for the activated sludge system to function, the desired biological transformations must take place in the reactor. Equally important is that the produced biomass be separated from the mixed liquor in the following clarifier, be thickened, and then be either returned to the reactor or wasted. An analysis of the thickening performance of final settling tanks has been performed by Dick and Young (Ref. 3-18). Activated sludge settling rate data except at very low solids levels were found to fit the relationship:

$$v_i = a_s C_i^{-n} \qquad \text{(Eq. 3-41)}$$

where

v_i = settling velocity of sludge, ft/hr (m/hr × 3.28)
C_i = suspended solids concentration, mg/mg (mg/liter × 10^{-6})
a_s, n = constants depending on sludge properties; values in the order of $a_s = 4 \times 10^{-5}$, $n = 2.23$ are typical

Taking the natural logarithum of both sides

$$\ln v_i = -n \ln C_i + \ln a_s \qquad \text{(Eq. 3-42)}$$

Plot on the Y and X axes, respectively (Figure 3-10).

$$\ln v_i \text{ vs. } \ln C_i$$

$$\text{Slope} = -n = -\frac{\ln (v_1/v_2)}{\ln (C_1/C_2)}$$

theoretical intercept (a_s) is calculated from $v_1/C_1{}^n$.

where

v_1 = velocity at any concentration C_1 as related by the straight line
v_2 = velocity at any concentration C_2 as related by the straight line

The settling properties of activated sludge at various initial concentrations are determined by observing the subsidence of the liquid-solids interface in transparent columns, as a function of time. Sludge samples for the lower concentrations are obtained by diluting mixed liquor with clarifier effluent. Intermediate concentrations are obtained by blending varying fractions of mixed liquor and clarifier underflow. The highest concentrations are obtained directly from the clarifier underflow. All tests used for developing a given plot of v_i versus C_i should be made on the same day.

Either a 2-liter graduated cylinder or larger settling column, such as a 6-in. diameter plexiglass column with 9-ft sludge depth, may be used. Either should be equipped with slow speed stirring devices to minimize the artificial conditions encountered in laboratory settling tests. In the larger column, solids should be redistributed after the column is full by plunging the stirring device up and down.

The interrelationship between these process parameters and settler performance, as derived from flux analysis, assuming negligible loss in the effluent is:

$$X_r = k_s/[(R/Q)(OR)]^{1/n} \qquad \text{(Eq. 3-43)}$$

$$k_s = \frac{n}{n-1}\,[180\,(n-1)\,a_s]^{1/n} \times 10^6 \qquad \text{(Eq. 3-44)}$$

where

k_s = constant dependent upon a_s and n
X_r = recycle sludge solids concentration, mg/liter
R/Q = recycle flow fraction
OR = overflow rate or flow volume/clarifier area, gal./sq ft-day (m^3/m^2-day × 24.5)

For existing plants, where the overflow rate is known, the predicted level of return sludge concentration can be calculated for various recycle ratios. For

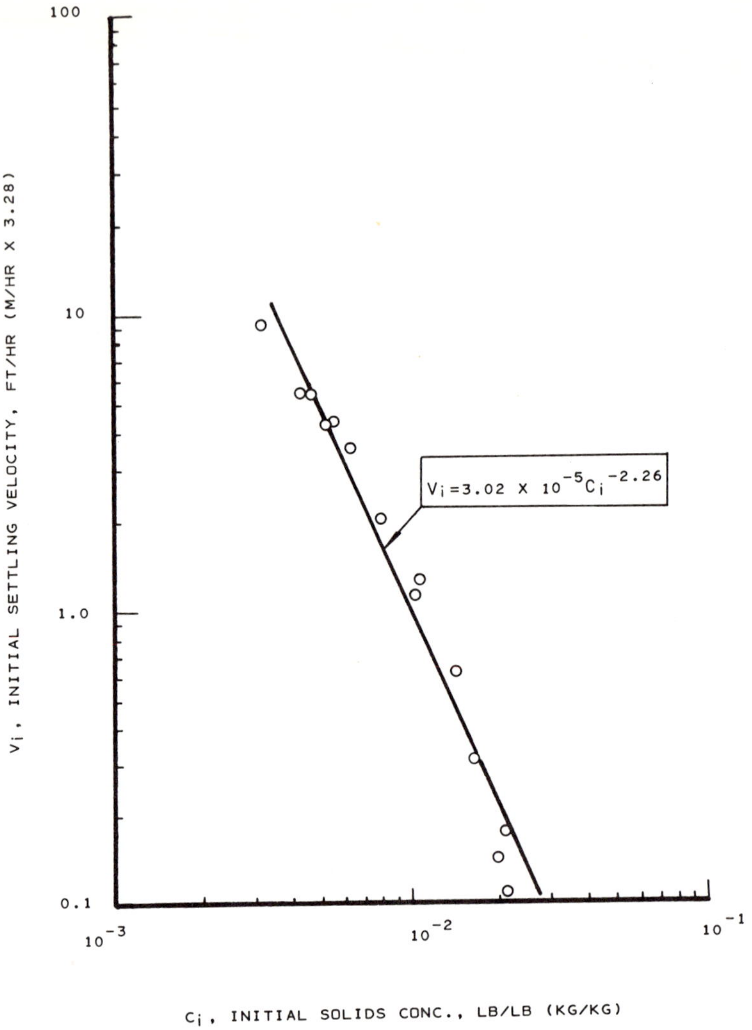

Fig. 3-10. Typical correlation of biological solids settling data. ***Source:*** **Dick, R. I. and Young, K. W., "Analysis of Thickening Performance of Final Settling Tanks," 27th Purdue University Industrial Waste Conference, Lafayette, Indiana, 1972,** ***Proc. Eng. Ext. Series*** **141 (*****Courtesy:*** **Purdue University).**

planned installations, an optimum balance between recycle solids (affects reactor size) and overflow rate (controls-clarifier size) can be established, and the required recycle fraction calculated. Appropriate scale-up factors can be applied, depending either on the variability expected as defined by Nyer and Hovious (Ref. 3-19) or on the factor of safety desired. The observation that larger-scale performance at times exceeds predictions from cylinder tests also needs to be considered. An alternative approach for separately relating initial settling rates and thickening rates to clarifier sizing has been described by Adams and Eckenfelder (Ref. 2-2). Either the clarification or thickening function can control clarifier sizing.

The reactor solids can be calculated directly from X_r and R/Q, ignoring solids entering the basin with the influent flow and solids synthesized in the basin:

$$MLSS = \left(\frac{R/Q}{1 + R/Q}\right) X_r \tag{3-45}$$

This demonstrates the dependence of MLSS on settling tank underflow concentration, an obvious factor often overlooked due to over attention to reactor design. On the other hand, the dependence of clarifier design on reactor loading conditions is illustrated in Table 3-9. Some typical hydraulic design levels for secondary clarifiers are presented in Table 3-11. A common underflow solids level for air activated sludge is 0.7 to 1%.

3-5.2 Settling Problems

The models for clarifier design are aimed at removing and thickening the bulk of the biological solids for recycle and wasting. The phenomena of colloidal growth, pinpoint floc, rising sludge, and bulking sludge also need to be considered. Colloidal biomass can be attributed to high salt levels, shock loadings, lack of protozoa and rotifiers, and too high a loading rate. Pinpoint, or nonsettling floc, usually is due to too high or too low a loading rate or sludge age (Table

TABLE 3-11. Typical Design Parameters for Secondary Clarifiers

	Overflow Rate (OR), (gal./sq ft-day)	
Type of Treatment	Average	Peak
Settling following activated sludge (excluding extended aeration)	400–800	1000–1200
Settling following extended aeration	200–400	800
Settling following trickling filtration	400–600	1000–1200

Conversion: m^3/m^2-day × 24.5 = gal./ft^2-day

Source: Process Design Manual for Suspended Solids Removal, EPA 625/1-75-003a, U.S. Environmental Protection Agency, Technology Transfer, Cincinnati, Ohio, 1975.

3-9). Less frequent causes of nonsettling floc include adsorbed oils and other immiscible liquids, floc breakup due to high shear rate in the reactor, and flotation due to adsorption of gas bubbles in the reactor. Rising of once-settled sludge usually is due to the biological reduction of nitrate ions (denitrification) to nitrogen gas, which buoys up the sludge; faster sludge removal from the clarifier is indicated. Bulking sludge or poorly settling and compacting sludge, is attributable to either filamentous organisms or bound water which swells the floc and reduces its density.

The control of filamentous organisms involves many interrelated factors. Influent wastewater characteristics, such as fluctuating strength, low pH, extreme temperatures, high carbohydrate level, high sulfide levels, and low nitrogen or phosphorous levels can contribute. Operating parameters such as oxygenation level, solids retention time in the clarifier, and high organic loading or low cell residence time also can be involved. The best means to handle this problem is to detect it early and search out the cause before the ratio of filamentous organisms to floc-forming organisms in the biomass becomes large. However, treatment with chlorine or hydrogen peroxide to kill the filamentous organisms is effective at times; addition of inert materials also can help.

3-5.3 Clarification Equipment

Secondary clarifiers can either be circular or rectangular, although the former seem to have wider application. Sludge take-off is either by the suction type or the plow type, as with primary clarifiers. The suction type reportedly can remove the sludge in a denser and more oxidized condition (Ref. 3-20). Weir placement and loading is critical to prevent upwelling. Scum collection is desirable. See Section 2-3.

3-6 OXYGENATED ACTIVATED SLUDGE AND OTHER MODIFICATIONS

3-6.1 Direct Oxygenation

Activated sludge treatment can use direct oxygenation in contrast to air oxygenation by means of the Unox system (Union Carbide Corp.) and other proprietary processes. The Unox system reduces area requirements for treatment of municipal and industrial wastes through a high-rate, high-solids approach in a staged, covered reactor (Fig. 3-11, 3-12). Either surface aerators or sparged turbines can be used. Design of the Unox system is keyed to maintaining a higher biomass concentration in the reactor, and at times loading the system at a higher rate. Most air-operated plants use *F:M* loading ratios of 0.3 to 0.5

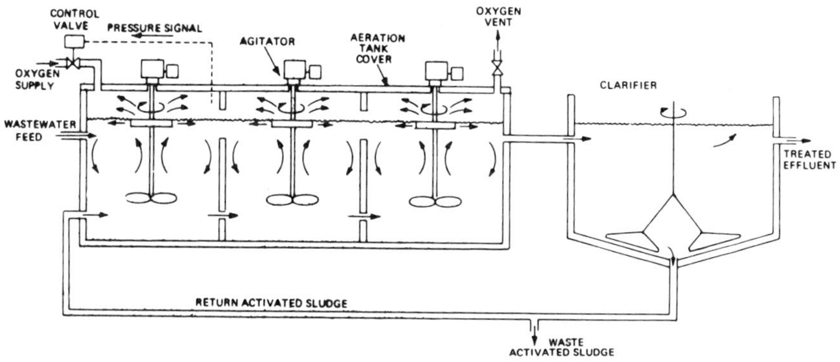

Fig. 3-11. Three-stage Unox pure oxygenation system. ***Source: Oxygen Activated Sludge Wastewater Treatment Systems,*** **EPA 625/4-73-003, Environmental Protection Agency Technology Transfer Seminar Publication, U.S. Environmental Protection Agency, Cincinnati, Ohio, 1973.**

day^{-1}, while operations where oxygen is not limiting can use ratios up to 0.5 to 0.8 day^{-1} as in the Unox System, perhaps because a larger fraction of the volatile solids are aerobically active. BOD removals range generally over 87 to 97%. Mixed-liquor direct-oxygenation levels vary over 4 to 8 mg/liter, as compared to 1 to 2 mg/liter (at best) for conventional air systems (Ref. 3-17, 3-20, 3-21).

The mixed liquor suspended-solids (MLSS) concentrations for the Unox system range from 4500 to 8000 mg/liter, compared to 1500 to 4000 mg/liter common for air systems. Recycle-sludge ratios (R:Q) are typically 0.2 to 0.5 for the Unox system, compared to 0.3 to 1.0 for air systems. Sludge production per pound of BOD removed ranges from 0.4 to 0.55 for direct oxygenation, compared to 0.5 to 0.75 for air systems (Ref. 3-17, 3-21).

The basic features of the Unox pure-oxygen system rest in these considerations:

- The high dissolved oxygen levels that can be practicably achieved can satisfy the high oxygen demand rates by large bacterial flocs without imposing mass-transfer limitations.
- The concurrent staging of both gas and liquid allows high efficiency of oxygen utilization and has kinetic advantages with noninhibitory substrates when reactions are first order.
- The covered system prevents discharge of problem vapors and aerosols; odor control on the low volume of effluent gas is possible.
- The compact biomass that often develops results in a higher MLSS level upon recycle.

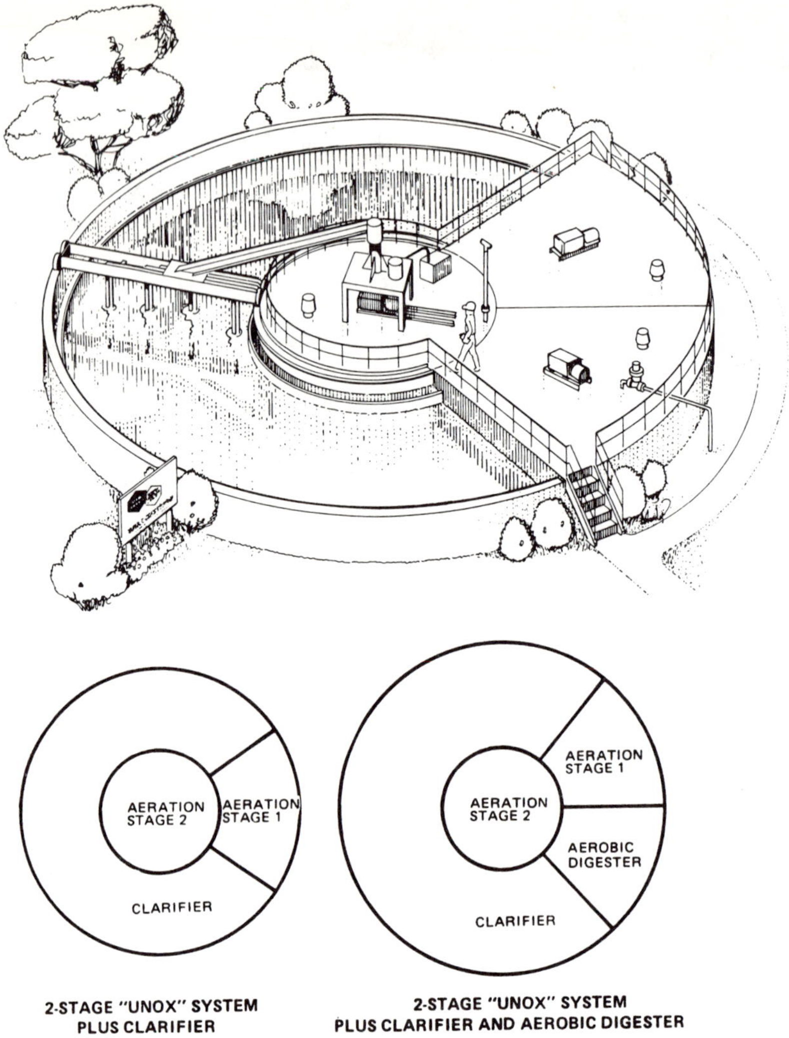

Fig. 3-12. Unox two-stage modular system with integral clarifier (*Courtesy*: Union Carbide Corporation).

- A statistically significant lesser level of net biosolids generation exists, which is believed to be due to increased endogenous respiration.
- More oxygen can be supplied per unit reactor volume at reasonable shear levels due to the higher differential between saturation level and operating level of dissolved oxygen.

These factors of higher biochemical oxidation rates per unit aerated volume, low gas feed rate, and better flocculated biomass have many ramifications:

- Lowered aeration-basin volume; less space needed.
- Recycle capacity can be lower, and wasted solids can be of lower volume and easier to thicken and dewater.
- A covered, compact system has aesthetic (and perhaps health) advantages in congested locations.
- Less sludge is wasted when net synthesis is less, which saves on disposal costs but increases oxygen needs.
- More compressible biomass tends to lower clarifier area, but higher recycle solids tends to increase it.
- If short cell-residence times are used, the development of microorganisms having relatively long generation times could be hampered, especially if environmental conditions such as pH and temperature were suboptimal.
- For some types of inhibition (i.e., dependent on concentration only and not on ratio of inhibitor to enzyme level), low volume and staging can enhance susceptibility to shock loading. However, maintenance of dissolved-oxygen "buffer" level in all parts of the aeration tank can lessen the variability of effluent quality as adverse effects of shock loadings, nutrient shortages, pH variations, and temperature changes, are heightened at marginal DO levels.
- Power requirements, including air separation and oxygen transfer, are at least competitive and many times are lower.
- Oxygen supply can be tailored to demand by a simple pressure control.
- Higher carbon dioxide content in the gas phase can internally neutralize alkaline wastes, but otherwise can result in a pH less than neutral.
- Heat loss is less, which is advantageous at least in cold weather.
- Monitors are needed to detect any flammability hazards and to initiate air purging of the gas space; however, problems are normally precluded by pretreatments for immiscible liquid removal and by the elimination of ignition sources.

As with other alternatives, an evaluation of all cost-performance factors is needed for system selection and design. Considerable pilot plant and plant-scale data in a range of applications are available for such an evaluation.

Other pure-oxygen-based systems more recently have been commercialized. These include the Oases system (Air Products), Marox diffuser units (FMC Corp.), and the Forced-Free Fall oxygen system (Airco), (Ref. 3-20).

The Oases system is a gas-tight closed reactor that provides positive vent control. The Marox system involves fine-bubble diffusion in deep tanks, usually with active fluid flow against the diffuser or with rotation of the diffuser. The underlying principle is to achieve high utilization of oxygen in a single pass of the process gas through the mixed liquor in an uncovered unit; produced carbon dioxide is stripped out. In the Forced Free Fall Oxygen (F^3O) system, oxygen is mixed in a turbulent wastewater fall zone inside a reinforced concrete module. The mixed liquor is forced down a duct by an impeller, flows up an annular space and over a weir, from which it falls into the oxygen chamber. The modules may be installed in open tanks.

Several oxygen supply options are available: liquid oxygen transport, pipeline transport from a central air separation plant, on-site cryogenic processing plant, or on-site generation by a pressure-swing-absorption process using molecular sieves which preferably absorb nitrogen. The choice depends on the oxygen usage rate and facility location.

3-6.2 Other Process Modifications

The activated sludge process is most versatile. Discussed in prior sections have been the concepts or operating modes of:

- Completely mixed reaction systems.
- Staged or plug-flow reaction systems.
- Contact stabilization.
- Extended aeration.
- Pure oxygen systems.

Many other modifications have been developed:

- Addition of powdered activated carbon to reactor primarily for the purpose of increasing removal of bioresistent materials (e.g., Pact process, Du Pont Company).
- Phosphorous removal process involving inducement of high uptakes and subsequent anaerobic stripping of phosphorous from the recycle sludge (Phostrip Process, Union Carbide Corp.).
- Nitrification in a second step system with separate clarifier and sludge recycle.
- Denitrification in a third step using methanol as a carbon source.
- Two step system for carbonaceous BOD removal (e.g., Zurn Attisholz process).
- Oxidation ditch in which the liquor is forced around an oval path using an axially mounted cage rotor.
- Deep-shaft concept in which introduced air reaches a high pressure during its flow down the shaft (ICI).
- Integral aerator-clarifier in package units.

3-7 AERATED STABILIZATION

Aerated stabilization is accomplished using a dispersed-growth system without biological-solids recycle. This process sometimes is termed "aerated lagooning" since the simplest prototype consists of one or more aerators installed in a large pond with the accomplishment of only partial mixing. However, full advantage of this approach can be realized by installing sufficient aeration capacity to keep the system completely mixed, and to satisfy the oxygen requirements of the system. Some typical operating data from a facility which treated a waste flow of 3700 gal./min (230 liters/sec) of fairly concentrated wastewater (2100 mg BOD/liter) are presented in Table 3-12. Due to the already dispersed nature of the culture, the aerated-stabilization approach does not require as close control over variations in temperature, pH, and organic loading as does the activated sludge process. A disadvantage of the aerated-stabilization process is a practical limitation in maximum BOD removal of about 75%, primarily due to the dispersed bacteria in the effluent and inhibition of the biological reactions at cold temperatures.

3-7.1 Aerated Stabilization Applications

Due to the limitations of this type system and the increasing strictness of effluent quality standards, the main applications for aerated stabilization are–

- pretreatment of wastewater prior to another biological system to gain the overall removal efficiency possible with systems in series,
- pretreatment prior to another biological system to level out fluctuating loads and/or reduce problem constituents,

TABLE 3-12. Aerated Stabilization of a Synthetic Organic Chemical Waste

	Summer	Winter
Temperature, °C	35	15
Dissolved oxygen, mg/liter	nil	2.0
Power, bhp/1000 gal.[a]	0.077	0.062
Detention time, days	2.9	2.3
BOD removal, %	54	48
COD removal, %	41	39

[a]hp/1000 gal. × 0.197 = kW/m^3
bhp is brake horsepower (power input to motor)
Source: Bess, F. D. and Conway, R. A., "Aerated Stabilization of Synthetic Organic Chemical Wastes," *J. Water Pollut. Control Fed.*, 38, 6, pp. 939–956, 1966.

- pretreatment to achieve levels permitting discharge into municipal facilities,
- interim treatment until upgrading becomes necessary to meet anticipated future standards,
- treatment of dilute wastes or activated-sludge effluents where suitable final effluent quality can be achieved directly or through a subsequent un-aerated stabilization pond.

Process stability and relatively low cost make this process attractive, despite its efficiency limitations.

3-7.2 System Design

One model for the aerated-stabilization process is similar to the relationship developed for the completely mixed activated-sludge process, as derived from a simulated first-order reaction, except that the level of biomass maintained is not considered:

$$\frac{S_e}{S_o} = \frac{1}{1 + K_a t_d} \qquad \text{(Eq. 3-46)}$$

where

S_e = effluent soluble BOD, mg/liter
S_o = influent soluble BOD, mg/liter
K_a = reaction rate constant, day^{-1}
t_d = liquid retention time, days

A typical range of K_a at 20°C, according to Eckenfelder and Ford (Ref. 3-6, 3-22), is 0.5 to 1.0; although values of up to 1.8 have been experienced with high strength chemical wastes. The lowest level of BOD concentration that this process can reach is subject to the same limitations as discussed for the activated-sludge process, i.e., the conditions in the treatment system are such that the reactions slow as low BOD levels are reached, even though this near-equilibrium level is reflected as degradable organics when measured under more favorable conditions in the BOD test procedure.

The reaction rate constant is adjusted to the temperature levels expected in the system under winter and summer conditions using Eq. 3-15. A typical θ_t value is 1.04; however, measurement of θ_t for the particular waste being treated is desirable. Relationships for predicting basin temperature levels have been developed (Ref. 2-2). Basins in series have the advantage of maintaining a higher temperature in the initial basin. Reaction rates may be of zero order (constant), i.e., independent of reactor substrate level, with wastes of high initial concentration. An alternative kinetic model has been developed using

Monod kinetics (Ref. 3-8, 3-10); Eq. 3-10 applies for the calculation of the effluent substrate level.

In addition to soluble organic removal, the level of biosolids in the basin is important as it also appears in the effluent. These can be predicted using

$$X_v = X_{ov} + a\,S_r - bX_v t_d \qquad \text{(Eq. 3-47)}$$

where

X_v = volatile suspended solids (VSS) in reactor and effluent, mg/liter
X_{ov} = influent degradable VSS, mg/liter
a = fraction of S_r converted to cells
S_r = BOD removed (S_o-S_e), mg/liter
b = cell autooxidation rate, day^{-1}
t_d = liquid retention time, day

Upon rearrangement,

$$X_v = \frac{X_{ov} + a\,S_r}{1 + bt} \qquad \text{(Eq. 3-48)}$$

Application of Eq. 3-46 to 3-48, including typical values for coefficients and experimental means to determine constants, is discussed in the section covering the activated sludge process. Monod-based expressions for biosolids level and oxygen requirements also are available (Ref. 3-8, 3-10).

Another alternative means of modeling the basin logically takes into account the level of biological solids, although it is in one sense an artifact of the system and cannot be controlled. Adams and Eckenfelder (Ref. 2-2) show that

$$\frac{S_o}{S_e} = \frac{1}{1 + K_{ax} X_v t_d} \qquad \text{(Eq. 3-49)}$$

After rearrangement and combination with Eq. 3-47 (assuming X_{ov} is nil), the following relationship for S_e is achieved.

$$S_e = \frac{1}{aK_{ax} t_d} + \frac{b}{aK_{ax}} \qquad \text{(Eq. 3-50)}$$

The slope of a plot of S_e (Y axis) versus $1/t_d$ (X axis) is $1/aK_{ax}$, and the Y intercept is b/aK_{ax}. Substitution of values for a and b yields K_{ax}. Some indications are that K values derived from activated sludge results can be used in this relationship, after converting from hr^{-1} to day^{-1}.

In cases in which the power used for aeration is insufficient to keep the solids in suspension throughout the whole basin, solublization of BOD from the deposited solids through anaerobic decomposition can be significant, especially in the summer when the contribution can reach 20 to 40% of that calculated from Eq. 3-46. Any feedback also needs to be considered when calculating oxygen requirements. A minimum value of 0.01 hp/1000 gal. (0.002 kW/m^3) typically is used when evaluating adequacy of mixing, although the level required is dependent upon the solids level and liquid properties. Procedures for calculating oxygen requirements are given in Section 3-3.7; aeration equipment alternatives and capacity determinations are outlined in Section 3-4.

Nutrient requirements are as discussed in Section 3-3.10. The basic consideration is that cells are 12.6% nitrogen and 2.6% phosphorous and that the net increase in cell weight needs to be provided with the proper nutrient quantities.

In cases in which the effluent from the aerated-stabilization process needs to be settled before being discharged, either a partially mixed second basin in series with a completely mixed first basin or a large quiescent settling basin can be used. In the latter case the development of algae and odors from the stored sludge can be minimized by maintaining liquid retention times from 1 to 2 days.

3-8 FIXED FILM SYSTEMS

3-8.1 Applications

While the activated-sludge and aerated-stabilization systems involve biomass suspended in an aerated or oxygenated reactor, fixed-growth systems employ biomass attached to surfaces which are contacted with the wastewater. The main attractions of this type system are:

- Process stability not closely tied to settleability of the biomass; able to pass shock loads.
- Relatively simple operation.
- Relatively low power requirements.

The disadvantages include:

- Lower practically achievable efficiency than activated sludge.
- Performance markedly decreased in cold weather (covering lessens this problem).
- Scale-up using available evaluation and design procedures is less rational.
- Potential for nuisance conditions—odors, insects.
- Relatively high capital costs and large space requirements.

These considerations result in many of the applications for fixed film systems being either to pretreat wastewater before a final biological treatment step or to handle low-volume wastestreams.

3-8.2 Trickling Filters

The trickling filtration process utilizes biological growth on a contact medium over which the wastewater flows. Sufficient void space in the "filter" is provided to maintain aerobic conditions. The microorganisms are cultivated usually either in a packed bed of crushed stone or on a system of vertical plastic sheeting. Other packings are available, such as flexirings. The waste is fed at such a hydraulic rate that a portion of the attached growth is continuously sloughed from the filter. A clarifier is provided after the filter; and a portion of the supernatant can be recycled for additional treatment, to maintain adequate hydraulic loading, and to equalize the constituents in the influent stream. Figure 3-13 depicts the mass transfer operations involved. Nitrification can be achieved at low loadings.

Filter Construction. Trickling filter construction is quite diverse. Design details are governed by available land area, volume and strength of the wastestream, extreme temperatures, and local construction costs. Periodic dosage

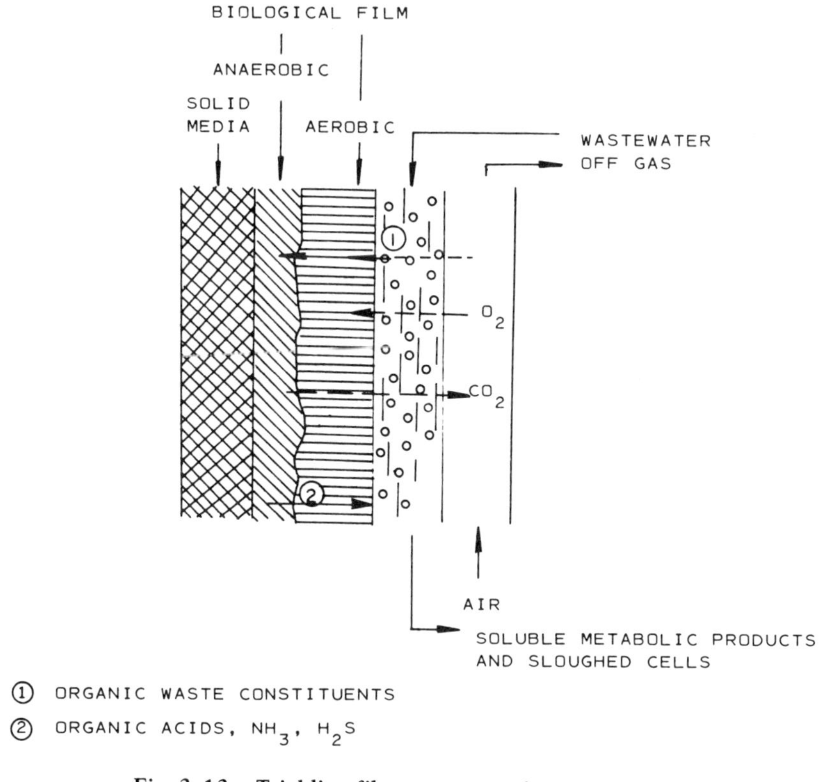

Fig. 3-13. Trickling filter mass transfer operations.

of the influent flow over the filter surface without flooding the system, which would prevent oxygenation of the film, usually is accomplished by using commercially available rotating distribution arms. The arms consist of tapered pipes connected to a rotating stilling well. Orifice holes are drilled along the arm's length, and deflector plates are attached to the arm below each orifice. As an alternative special nozzles are available which fasten directly to the arm, thus eliminating the orifices and deflection plates. The reaction of the escaping wastewater causes the arm to rotate. Fixed nozzle flow distribution is sometimes used to supply waste to rectangular or irregular filters. The nozzles are connected by piping to a dosing siphon which periodically discharges the waste to the filter. Waste may also be supplied to the nozzles directly by the use of pumps.

Stone ranging from 2 to 4 in. (5 to 10 cm) in diameter installed in beds from 6 to 8 ft (1.8 to 2.4 m) deep commonly is used as a filter packing, although other coarse materials such as slag or coal can be employed in addition to the plastic sheeting alternatives. Filters up to 30 to 40 ft (9 to 12 m) in height have been constructed by using commercially available synthetic media. In general, synthetic filter media consist of flat or corrugated sheets of rigid material assembled in self-supporting modules having the appearance of a honeycomb; void space is about 94%, as compared to about 45% for stone filters. Contact between the biomass and wastewater is maximized, and space is provided for unimpeded ventilation and liquid flow. Specific surface areas of synthetic media range from 20 to 35 ft^2/ft^3 (66 to 115 m^2/m^3), while rock packing ranges from 10 to 25 ft^2/ft^3 (33 to 82 m^2/m^3).

The floor of a filter needs to be designed to support the filter media, drain the treated wastewater from the filter, and provide air circulation through the filter. Vitrified clay filter block is available for floor construction. These are rectangular in cross section and hollow in the center. The upper face of the block, the surface upon which the filter media rest, is perforated. Wastewater and air flow in opposite directions through the tile; the wastewater passes through the perforations and is directed by the hollow blocks to centralized liquid collection galleries.

Where temperatures drop below freezing for prolonged periods of time, filters are sometimes covered. Feeds also can be preheated. Filters have been constructed using forced or induced fans for air circulation, although natural ventilation induced by density differences usually is sufficient.

Filter Design. Some typical operating conditions are shown in Table 3-13. Several models for trickling filters have been developed; one described by Thackston and Eckenfelder that seems to apply best to industrial wastes (Ref. 3-23) is:

TABLE 3-13. Typical Trickling Filter Operating Conditions

	Nominal Filter Rate		
	Low	High	"Super"
Hydraulic loading			
gal./ft^2-day	25–100	200–1000	1400–8500
Organic loading			
lb BOD/1000 ft^3-day	5–25	25–300	130–1500
BOD removal efficiency, %			
Single-stage	60–70	40–50	50
Two-stage	90+	85	–

Conversion: gal./ft^2-day × 0.0408 = m^3/m^2-day
lb/1000 ft^3-day × 6.0 = g/m^3-day

Source: A Literature Search and Critical Analysis of Biological Trickling Filter Performance, EPA Report No. 17050-004, 12/71, U.S. Environmental Protection Agency, Washington, D.C., 1971.

$$\frac{S_e}{S_a} = e^{-K_f \, (D_f/Q_h^m)} \qquad \text{(Eq. 3-51)}$$

where

S_e = effluent BOD concentration, mg/liter
S_a = BOD concentration applied to the filter, mg/liter
e = constant (2.72)
K_f = product of the reaction rate constant (gal./ft^2-min) and the specific surface area (ft^2/ft^3) of the packing; reportedly 0.09 is typical, min^{-1}
D_f = media depth, ft (m × 3.28)
Q_h = hydraulic loading (including recycle), gpm/ft^2 (m^3/m^2-min × 24.5)
m = constant (reportedly 0.5 is common, can range from 0.4 for some random packings to 1.0 for some vertical sheets)

This equation is similar to that developed for a batch or plug-flow activated sludge system:

$$\frac{S_e}{S_o} = e^{-kX_v t} \qquad \text{(Eq. 3-52)}$$

The retention time is proportional to the ratio of the depth to the hydraulic loading, and the biomass level is proportional to the specific surface of the filter medium.

Calculation of S_a is by mass balance.

$$S_a = \frac{S_o + (R/Q)S_e}{1 + R/Q} \qquad \text{(Eq. 3-53)}$$

where R/Q = recycle ratio, recycle flow/influent flow

The variables may be evaluated by collection of pilot-scale performance data at several surface loading rates and media depths (Ref. 3-23). Data are reduced by plotting ln $(100\ S_e/S_o)$ versus D at various hydraulic loadings; the slope is K/Q_h^m. The slope of a ln-ln plot of KQ_h^m versus Q_h is b. Then the slope of a plot of ln (S_e/S_o) versus D/Q_h^m (m is known) is K; K is adjusted for expected temperatures using Eq. 3-15. A typical θ_t value is 1.035. By multiplying by the ratio of the specific surface areas, K can also be adjusted to similar packings. Perhaps the broader utility of these equations is for the estimation of allowable surface loadings with and without recirculation in order to achieve desired efficiencies at a given temperature. In this case, a reasonable value of the coefficients is assumed.

Oxygen transfer capabilities limit the waste concentration that can be applied to the filter without recirculation; 500 mg BOD/liter is a limiting guide for readily degradable wastes.

3-8.3 Rotating Biological Contactors and Other Modifications

As opposed to passing the wastewater over a fixed medium, a newer development is a rotating biological contactor. The biological growth is developed on a set of partially submerged, rotating disks. During rotation, the growth is alternately exposed to the wastewater and to air to provide a means of both contacting the growth with the organic impurities and aerating the culture. Rotational speed is adjusted according to waste strength. Successive units can be used to upgrade efficiencies. Shearing forces cause excess biomass to slough into the mixed liquor, which is kept mixed by the disks. Treated wastes flow to a clarifier, where the solids are separated for ultimate disposal without recycle.

Low power requirements are a particularly cited advantage from among the list of considerations previously presented for fixed-film systems. Housing can be provided for protection from inclement weather. In addition to the rotating disks (e.g., Bio-surf Process), another approach is to use a rotating cage-like drum (Bio-drum) filled with hollow plastic balls.

Another development, termed Activated Bio-Filter Process, is a two-stage system employing both fixed film and activated sludge. The fixed film unit consists of stacked racks made of wooden laths attached to supporting rails. The laths and rails are sized to permit free transfer to liquid and air in all directions. The incoming wastewater flows to the fixed film reactor along with recycled biological solids from a final clarifier. A portion of the effluent from the fixed-film reactor is recycled and a portion flows to an aeration tank and then to the final clarifier.

3.9 ANAEROBIC SYSTEMS

3-9.1 Anaerobic Treatment Basics

Although the stepwise chemical reactions for the metabolism of organic chemicals by anaerobic and aerobic bacteria are complicated, more simple, overall equations familiar to chemists and engineers may be written that describe the initial reactants and the final products (Table 3-4). The anaerobic reactions, known as fermentation and anaerobic respiration, occur in closed reactors or in the bottom of open lagoons, and typically involve the reduction of sulfates, nitrates, and organic molecules. Biological cycles of interest in anaerobic systems are shown in Figure 3-14.

Several important economic and technical advantages inherent in the anaerobic treatment process along with potential problem areas are as follows:

Advantages

No aeration equipment is required for organic reduction. Associated capital, power, and maintenance costs are avoided. System loading is not limited by oxygen transfer.

Cellular material is produced in lower quantity and more stable form. Savings in nutrients and in flocculants, equipment, and labor costs for dewatering and final disposal can be realized.

Somc problem organic chemicals difficult to degrade aerobically will degrade anaerobically.

The oxygen in sulfate (and nitrate) ions can be utilized for organic oxidation.

Methane in off-gas potentially can be utilized.

Potential Problems

High temperatures are needed for maximum rates.

High biomass concentration is required for reasonable rates at short retention times.

Generation time for methane bacteria is long (2 to 11 days at 37°C), thereby requiring long solids retention and acclimation times.

Methane bacteria are very sensitive to shock loads, toxic materials, and environmental conditions.

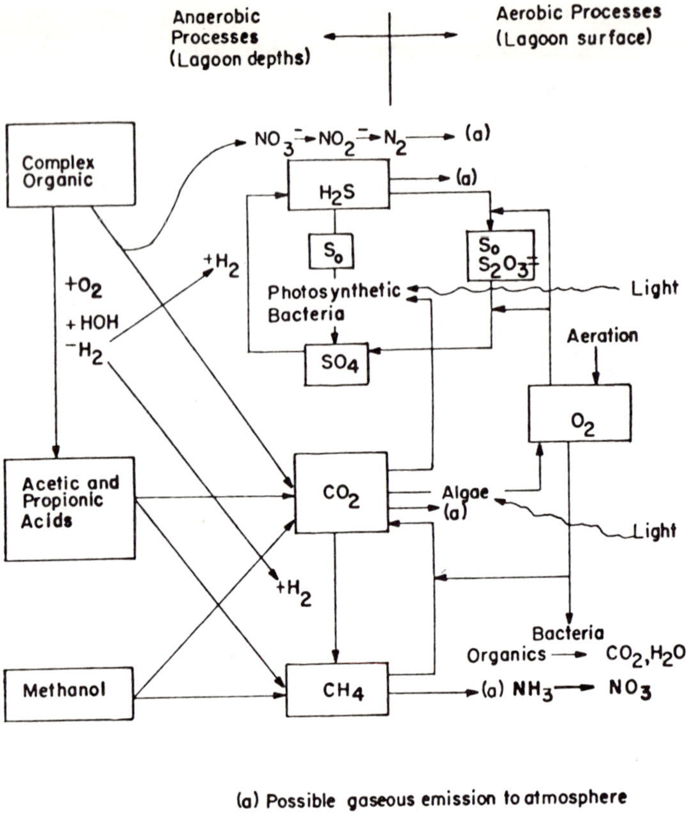

Fig. 3-14. Anaerobic reactions coupled with cycles at aerobic surfaces. ***Source*****: Hovious, J. C., Conway, R. A., and Ganze, C. W., "Anaerobic Lagoon Pretreatment of Petro-chemical Wastes,"** ***J. Water Pollut. Control Fed.*****, 45, 1, pp. 71-84, 1973.**

Effluents low in BOD (<50 mg/liter) with good aesthetic properties are difficult to produce.

Odor emission potential is high.

In some instances, significant savings can result if the anaerobic and aerobic processes are joined to utilize the best features of both to compensate for some of the problems of each. This could involve utilizing a roughing anaerobic treatment followed by an aerobic treatment step for polishing of the wastewater.

Toxicity. As shown in Figure 3-14, for most substrates the anaerobic process proceeds in two steps, i.e., conversion of the organic material to volatile organic acids by one set of bacteria, and then conversion of the acids to methane and carbon dioxide by the methanogenic bacteria. The methane bacteria are sensitive to high levels of alkali and alkaline earth cations (1000 to 10,000 mg/liter), ammonia (1500 to 3000 mg/liter), sulfide ion (100 to 200 mg/liter), ionized heavy metals, dissociated volatile acids (~2000 mg/liter), and various synthetic organic chemicals (Ref. 3-24, 3-25). Toxic effects can be lessened through acclimation in the case of organics and sulfides, through addition of antagonistic ions in the case of cation toxicity, through pH adjustment in the case of ammonia, and through precipitation of heavy metals by controlled sulfide generation.

Using a Warburg respirometric procedure, Hovious, Waggy, and Conway identified the following organic materials (and concentrations which produced a 50% decrease in the activity of an unacclimated biomass) as inhibitory: acrolein (20 to 50 mg/liter), formaldehyde (50 to 100 mg/liter), 2-ethyl-1-hexanol (500 to 1000 mg/liter), methyl isobutyl ketone (100 to 300 mg/liter), diethylamine (300 to 1000 mg/liter), acrylonitrile (100 mg/liter), 2-methyl-5-ethylpyridine (100 mg/liter), ethylene dichloride (150 to 500 mg/liter), ethyl acrylate (300 to 600 mg/liter), and phenol (300 to 1000 mg/liter) (Ref. 3-25). A typical test run is shown in Figure 3-15. Inhibitory effects were more severe at high volatile acid concentrations.

Acclimated cultures were developed in mixed digesters for crotonaldehyde, phenol, and to some degree for ethyl acrylate. No acclimation was observed for sodium acrylate. Cultures acclimated to crotonaldehyde and ethyl acrylate were able to degrade the material, while phenol was not degraded with acclimation but was no longer inhibitory.

Additional acclimation studies were made in continuously fed anaerobic filters. A filter was acclimated to a crotonaldehyde concentration of 600 mg/liter as compared to the 50 to 100 mg/liter level found inhibitory in Warburg studies. Treatment of formaldehyde, ethyl acrylate, phenol, and acrylonitrile indicated synergistic inhibitory effects. These mixed inhibitors were treated satisfactorily at low concentrations in two series anaerobic filters; however, increasing inhibitor concentrations resulted in failure of both filters. The amenability of organic materials to treatment in anaerobic systems is highly dependent upon the nature of the system to be employed.

3-9.2 Anaerobic Reactors

The basic types of anaerobic systems used in treating industrial wastewaters and residues are described in Table 3-14.

Contact Reactor. The contact reactor employs solids separation and recycle

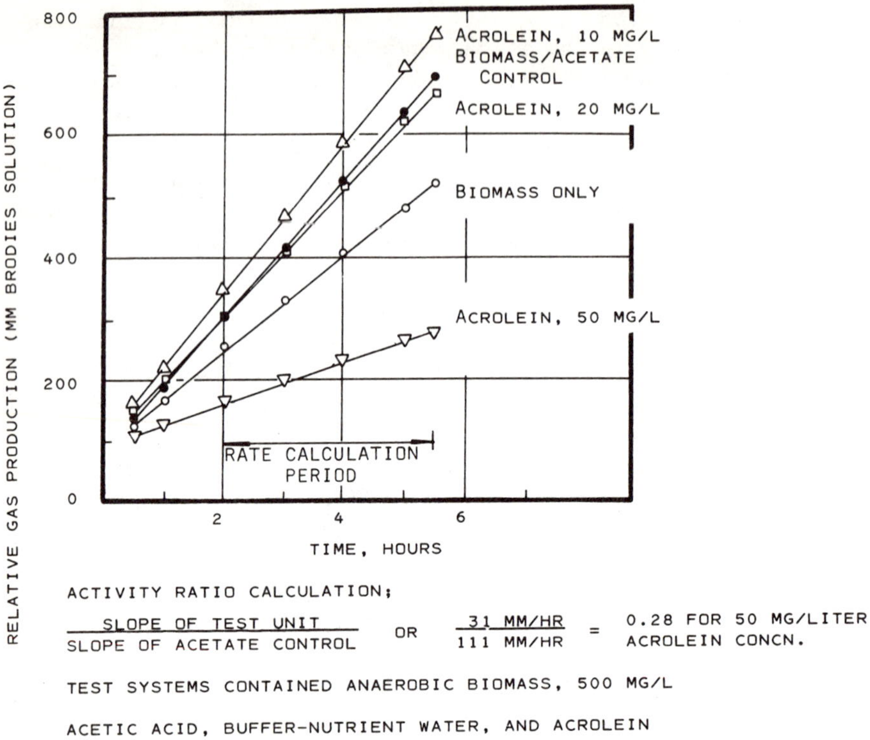

Fig. 3-15. Typical Warburg respirometer data measuring methanogenesis in anaerobic systems. *Source*: Hovious, J. C., Waggy, G. T., and Conway, R. A., "Identification and Control of Petrochemical Pollutants Inhibitory to Anaerobic Processes," Environmental Protection Technology Series EPA-R2-73-194, Office of Research and Monitoring, U.S. Environmental Protection Agency, Washington, D.C. 20460, April 1973.

to increase the organism concentration and the biomass retention time within the digestion system. The increased solids retention time (θ_c) compensates for the long generation time of the methanogenic bacteria. Solids separation usually is accomplished by gravity settling, but if carried out in open clarifiers it would have the problem of attached gas bubbles associated with the decreased partial pressures; this may be alleviated by a vacuum degassing step. The digester gas also may be used in flotation separation of solids. Other means which can be used to increase organism separation efficiency are addition of flocculants or weighting agents.

Filter. Anaerobic filters utilize both the tendency of methane bacteria to grow

TABLE 3-14. Basic Anaerobic Systems

System Type	Submerged Filter	Contact Digester[a]	Open Lagoon
Reactor	Packed column	Completely mixed vessel	Basin with considerable stratification
Flow pattern	Plug flow	Backmixed	Some wind and wave mixing, thermal turnovers
Biosolids level	High biomass through attached growths	High biomass through settling and return	Low suspended solids, bottom sludge layer
Metabolic pathways	Fermentation and anaerobic respiration	Fermentation and anaerobic respiration	Fermentation, anaerobic respiration, sulfur oxidation, photosynthesis, some aerobic respiration
Retention times	1 to 3 days	1 to 10 days	10 to 100 days
Gas collection	Normally collected	Collected	Gas is released, although a plastic covering with peripheral collection tiles is possible
Temperature control	Not normally practiced	Usually practiced	Unfeasible

[a]As opposed to a standard digester, which is only partially mixed and has biomass return only if two digesters are operated as stages with the second unmixed.

Source: Fisher, J. A., Hovious, J. C., Kumke, G. W., and Conway, R. A., "Pilot Demonstration of Basic Designs for Anaerobic Treatment of Petrochemical Wastes," *Water, 1970,* AIChE's Chemical Engineering Progress Symposium Series, **67**, 107, 1971.

on surfaces and the filtering capacity of a flooded, packed bed to overcome the long generation time of the methanogenic culture. The low solids production of the anaerobic system slows filter clogging if a low solids feed is used. High recycle is the key to stable operation in a system developed by Celanese Chemical Co. (Ref. 3-40).

Lagoon. Anaerobic lagoons operate at low suspended microorganism concentrations and, therefore, require a relatively long retention time for adequate reduction of waste constituents. The retention time required in a system is primarily a function of temperature and the amenability of the waste material to transformation by the various types of organisms involved. The key pathways are illustrated in Figure 3-14.

The anaerobic lagoon has the potential to utilize additional organic removal pathways not found in the other anaerobic processes. For example, an uncovered lagoon which is not excessively loaded will develop an algal culture, which may result in a shallow aerobic surface layer. This aerobic layer can account for a significant aerobic oxidation of wastes.

Another process which may function in a lagoon is re-oxidation of reduced sulfur forms. In the conventional digester, sulfates are reduced by *Desulfovibrio* to sulfides with a conversion of complex substrates to volatile acids. If the sulfides remain in the system, no net reduction in COD accompanies this reaction. In an enclosed type of anaerobic system, the sulfides may be lost in part as hydrogen sulfide in the produced gas or may be precipitated with heavy metals. In the anaerobic lagoon, any sulfides lost in the produced gas usually constitute an air pollution problem. Fortunately, additional biological processes may exist in a lagoon to oxidize the produced sulfide. The anaerobic, photosynthetic sulfur bacteria, *Chlorobacteriaceae* and *Thiorhodaceae*, utilize light as an energy source to oxidize sulfides. Aerobic bacteria such as *Beggiatoa* and *Thiobacillus*, which oxidize sulfides, can live in the upper layers of the lagoon. Direct oxidation of sulfides by molecular oxygen in the aerobic upper lagoon layers is also a possibility.

Experimental work on anaerobic lagoons is most meaningful when performed with actual plant wastewater, outdoors under the full spectrum of expected operating conditions and removal processes. Besides organic removal efficiency, the key aspects needing evaluation are inhibition of methanogenesis and emission of sulfide and other odors.

Anaerobic digesters can be designed in a manner analogous to that previously described for aerobic systems. In an approach described by Eckenfelder and Ford (Ref. 3-22), experimental systems are operated at various retention times and data are taken regarding biomass levels and organic matter (COD) reduction. Graphical reduction of the data using Eq. 3-19 is used to determine the reaction rate constant and the nondegradable fraction. Other design constants are defined using relationships of specific removal versus retention time and methane production versus COD removal (Ref. 3-22). Alternative procedures have been described by Christenson and McCarty using Monod-type kinetics (Ref. 3-10). Constants for synthesis and other relationships (Eq. 3-20) have been developed for various substrates by Speece, Lawrence, and McCarty (Ref. 3-8, 3-24, 3-26).

A design procedure which takes into account feed organic concentration, hydraulic loading, and temperature has been developed by Hovious, Conway, and Ganze for anaerobic lagoon pretreatment of petrochemical wastes relatively low in sulfates (Ref. 3-27). Also for lagoons a relationship similar to that in Eq. 3-46 can be used with a reaction rate constant of 0.030 to 0.055 (Ref. 3-6).

3-10 STABILIZATION PONDS

3-10.1 Pond Applications

Stabilization ponds rely upon surface reaeration and algal photosynthesis for the oxygen used in the biological processes. However, most industrial waste ponds are anaerobic in their lower levels and aerobic near the surfaces. In this type pond, termed facultative, methane formation in the lower levels also is a means for organic removal, as can be sulfate reduction. Depths range from 3 to 8 ft (0.9 to 2.4 m); however, 4 to 6 ft (1.2 to 1.8 m) is more common. Often in series operations of ponds, the first one will be virtually all anaerobic, the next facultative, and the last aerobic. The allowable loadings for facultative ponds range from 15 to 80 lb BOD/acre-day (170 to 900 kg/hectare-day) (Ref. 3-28).

The factors that affect the capacity of the pond, as described by Gloyna and Hermann, include: air and waste temperature, wind velocity, orientation of pond with respect to prevailing winds, rainfall, horizontal and vertical movement of the water in the pond, infiltration, evaporation, seepage, depth of the pond, length of daylight, hours of overcast and sunshine, sulfate level, and types of predominating microorganisms (Ref. 3-29). Pond systems are sometimes designed to provide for recirculation to increase treatment efficiencies. Stabilization ponds have primary application in situations when adequate land area is available at reasonable cost and the waste discharge is relatively small (Ref. 3-30). The potential reactions involved are depicted in Figure 3-14; design is set to allow the aerobic reactions to predominate to a considerable depth.

The conditions which favor the use of stabilization ponds are:

Low land costs
Low flow rate to be treated
Turbid effluent allowable

Warm climate
Low latitude and minimum cloud cover (for light)
Steady winds (for reaeration)
Need for impoundment or equalization
Low sulfate level in waste
Low percolation rate
High power costs
Minimum operating capabilities

Often lagoons are used in series with other systems to provide polishing treatment for further removal of soluble organics, nutrients, or suspended solids; however, either algal growth needs to be controlled or a final solids-removal step provided to meet effluent quality standards. In reality, however, the effect of a small amount of algal discharge can be beneficial to fish life in the receiving water.

3-10.2 Design of Ponds

Design of ponds often is based on allowable surface loading of BOD, based on experience with similar wastewaters and climatic conditions. An alternative approach is to apply the first-order kinetic relationships previously described for anaerobic lagoon systems to stabilization ponds using procedures described by Eckenfelder and Ford (Ref. 3-22). For single ponds the relationship is:

$$\frac{S_e}{S_o} = \frac{1}{1 + K_p t_d} \qquad \text{(Eq. 3-54)}$$

where K_p = the rate constant coefficient for ponds, day^{-1}, which for multiple ponds is:

$$\frac{S_e}{S_o} = \frac{1}{(1 + K_p t_{d1})(1 + K_p t_{d2})(1 + K_p t_{d3})} \qquad \text{(Eq. 3-55)}$$

where t_d = hydraulic retention time in days for Ponds 1, 2, and 3. Design data can be collected in large field pilot units, or with some risk, in laboratory units equipped with suitable baffling, lighting, and simulated wind action (Ref. 3-22). Data reduction proceeds as described for the other biological systems.

Adjustment of the reaction rate coefficient needs to be made for temperature effects using Eq. 3-15 with a constant (θ_t) of 1.06 to 1.09; 1.085 is typical. The design approach needs to consider the depth of light penetration, as this affects photosynthesis. Higher loadings create turbid conditions, which limit the depth of aerobic operation. Many pond operations in the southwestern

United States are based on a design loading of 50 to 60 lb BOD_5/acre-day (560 to 670 kg/ha-day), while in northern areas of the USA loadings down to 10–15 lb BOD_5/acre-day (110–170 kg/ha-day) have been used.

Rubber, plastic, or asphaltic liners are available to prevent seepage; the use of soil sealants is an alternative approach. Various aeration devices have been developed especially for usage in this type of lightly loaded biological system (Ref. 2-14).

3-10.3 Pond Upgrading

Due either to increased loads or to more strict effluent quality standards, upgrading of stabilization ponds frequently needs to be considered. The general problems are excessive algae, offensive odors, and seasonal variations. Alternatives which should be considered would be to (Ref. 3-28, 3-29):

Deepen ponds (or final portion of pond system) to increase detention time, and more importantly, to allow sedimentation of algae; wind scour is lessened.

Pretreat wastes by aerated stabilization or other means to decrease loading.

Add another pond in series; utilize interpond and/or intrapond recirculation.

Consider multiple-entry and multiple-exit approaches for fuller utilization of volume.

Install mechanical or compressed-air aeration.

Remove algae by staged 1) chemical coagulation, flocculation, and gravity sedimentation, followed by 2) multimedia, rapid-sand filtration (pressure operation).

Remove algae by staged 1) coagulation and dissolved-air flotation, followed by 2) polishing multimedia filtration.

Remove algae by a dike-type, submerged-rock filter.

Remove algae by intermittent, slow sand filters (Figure 3-16).

Use effluent for spray irrigation.

Use microstrainers (Section 2-3.10).

Use parallel ponds, make into one and discharge from another which is quiescent. Benemann terms this the pond-isolation process (Ref. 3-41).

In the intermittent sand-filtration approach, filters operated by Middlebrooks. *et al.* at 600,000 gal./acre-day (5600 m^3/ha-day) reduced the effluent suspended solids from an average of 26 mg/liter (10 to 70 mg/liter) to less than 7 mg/liter during a 26-day run before plugging required them to be cleaned; some nitrification and BOD reduction also occurred (Ref. 3-31).

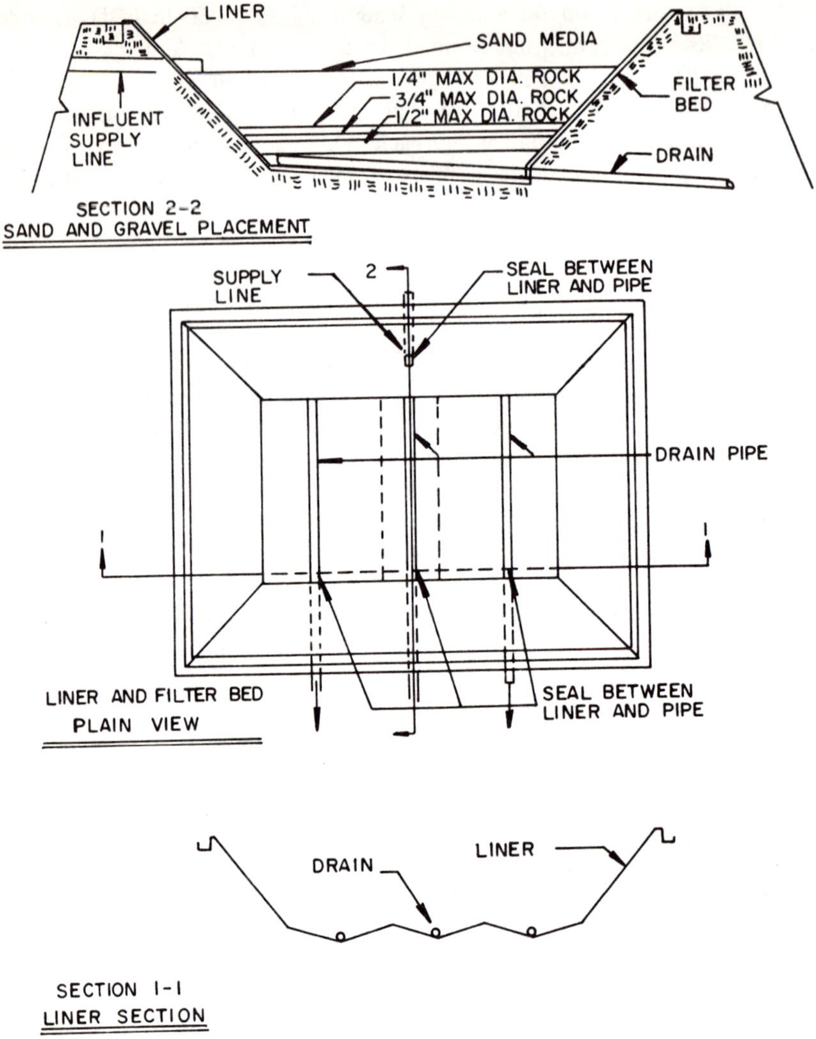

Fig. 3-16. Intermittent sand filter for upgrading stabilization ponds. *Source:* **Middlebrooks, E. J.,** *et. al., Proc. Symp. Upgrading Wastewater Stabilization Ponds to Meet New Discharge Standards,* **Utah State University, Logan, Utah, August 21-23, 1974.**

3-10.4 Natural Systems

Passing waste through a balanced ecosystem, which includes plants and higher animal forms, has been indicated to be an effective, low-cost treatment technique (Ref. 3-32, 3-33). Reeds and bullrushes (or bacteria acting in symbiosis with them), as well as *Daphnia,* shellfish, and brine shrimp in conjunction with algae are involved in various types of these "natural" systems.

3-11 BIOLOGICAL EXPERIMENTAL TECHNIQUES

3-11.1 General Approach

Experimental programs to develop design data for biological treatment processes must be developed individually according to wastewater characteristics, feasible treatment processes, and effluent-quality requirements. Typical approaches are presented in this section; references to comprehensive coverages of the details of experimentation and process-design theory are provided. Considerable information as to collection of process feasibility and design information has been presented along with the discussions of the individual treatment processes in prior sections.

Basic factors in the design of an activated sludge system, as an illustration, are shown below:

Independent Variables	*Dependent Variables*
Hydraulic detention time	Organic removal efficiency
F/M ratio or θ_c	Oxygen uptake rate
Biological solids concentration	Solids synthesis rate
Temperature level	Sludge settling rate
pH level	Oxygen transfer characteristics
Oxygen residual	Foaming and other nuisances
Nutrient supplementation	

The complexity of the experimental work can be minimized by holding several independent variables at near steady state during the initial test period, e.g., temperature at about 16 or 35°C depending upon predicted extreme conditions, pH at 6 to 8, oxygen residual at 0.5 to 2 mg/liter or 5 to 8 mg/liter depending on whether air or oxygen is to be used, and the ratio of available nitrogen and phosphorus to BOD removed at 1:20 and 1:100, respectively. The system then is operated at two or three detention times and biological-solids concentrations while measuring the dependent variables.

The mathematical relationships previously cited are used to relate organic removal efficiency to detention time and biomass concentration through development of a reaction rate constant. Oxygen utilization and solids synthesis rates are similarly determined from the experimental data. The influence of loading factor (F/M ratio) or cell retention time (θ_c) on these rates and upon effluent suspended solids levels and biomass settling rates is measured. Oxygen transfer characteristics, excessive foaming, stripping of volatiles, and other factors of concern are examined.

3-11.2 Equipment

Laboratory studies of activated sludge processes are conducted with either semi-continuous (batch) feed or continuous feed. Batch-fed systems are somewhat easier to operate, provide clearer mass balances, and generate meaningful feasibility data. Their use for design purposes is possible only if shock feeding of waste does not inhibit the system (Ref. 3-34). The laboratory units shown in Figure 3-17 can be used in screening the responses of completely mixed systems to various loading and environmental stresses. For more refined studies, external clarification and pumped sludge return are desirable. Frequently fabrication of a small pilot unit, such as that shown in Figure 3-18, is warranted. Operation of units with on-line feed experiencing any variability of the wastewater under study is most desirable toward achieving meaningful results, as demonstrated by Conway, Hovious, and Macauley (Ref. 3-35).

Detailed procedures for the design and operation of laboratory and pilot units have been published along with techniques for developing design criteria from the experimental data (Ref. 3-22, 3-36, 3-37, 3-38). These references contain apparatus and procedural descriptions for evaluating fixed-film systems, aerated stabilization, and anaerobic systems, as well as many ways of evaluating the activated sludge approach. Techniques for evaluating auxiliary physical-chemical systems are also presented. Relative data on fixed film processes can be achieved using a rotating tube biological filter (Figure 3-19); design studies need to be made in fairly large systems simulating the actual configurations being considered.

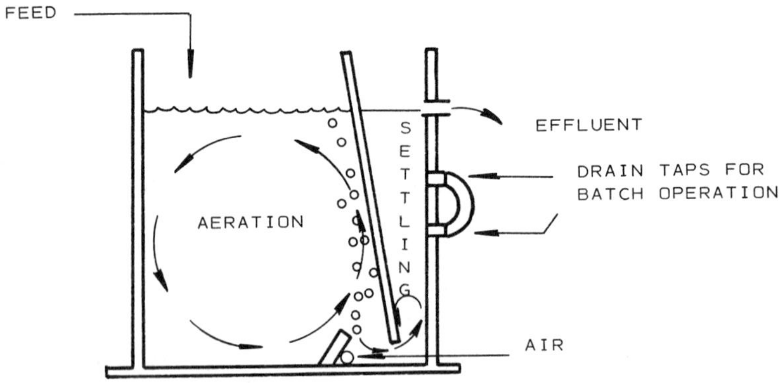

Fig. 3-17. Laboratory activated sludge unit. ***Source:*** **Renn, C. E., The Johns Hopkins University, Baltimore, Maryland.**

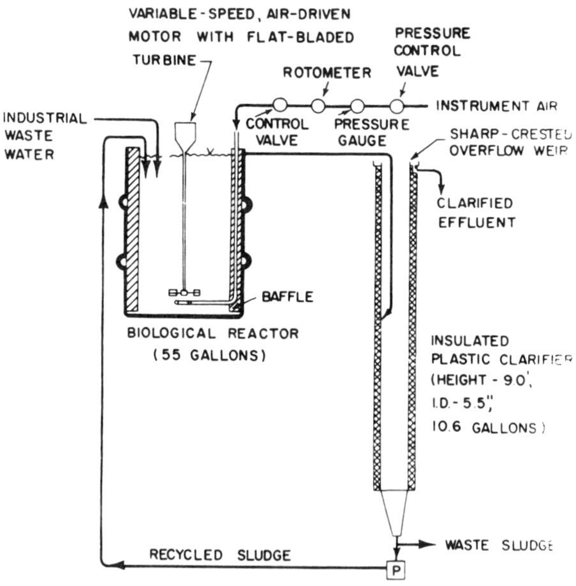

Fig. 3-18. Pilot activated sludge system.

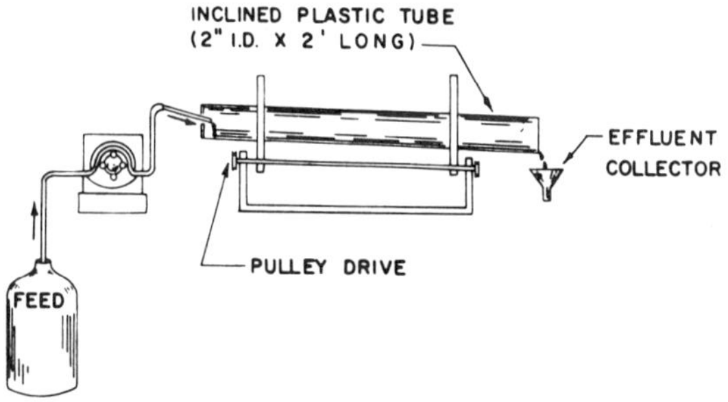

Fig. 3-19. Rotary-tube trickling filter. ***Source:*** **Conway, R. A. and Waggy, G. T., "Biodegradation Testing of Typical Surfactants in Industrial Usage,"** ***Am. Dyest. Rep.,*** **55, 12, pp. 607-15, 1966.**

An apparatus of general utility in evaluating metabolic rates of microorganisms under a range of feed and environmental conditions is the Warburg respirometer. Figure 3-20 depicts this investigative tool as used for aerobic standards; anaerobic systems can also be followed by gas evolution (Ref. 3-35). Microbiological response to a range of conditions can be examined in a single test as the Warburg apparatus accommodates 18 to 20 respirometers. A simpli-

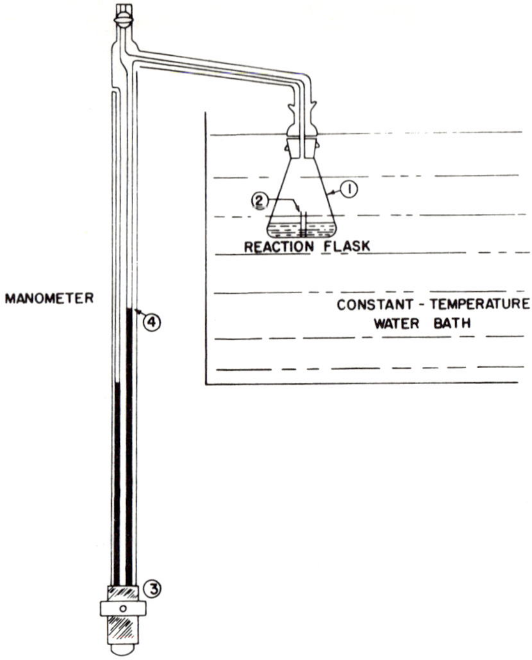

Fig. 3-20. Warburg constant-volume respirometer. (1) Main compartment containing wastewater and microorganisms. (2) Centerwell containing KOH to absorb evolved CO_2. (3) Fluid reservoir with screw clamp for volume adjustment as O_2 is used by microorganisms. (4) Reference level of closed leg of manometer.

fied means of using sealed serum bottles to follow anaerobic degradation (and inhibition) has been described by McCarty (Ref. 3-42.)

3-11.3 Analysis

Thorough analysis of the performance of the biological system is essential. A first step usually is to measure the BOD_5 removal across the system. Since this measurement reflects only a portion of the biodegradable material, longer term incubations in analyzing the BOD level in feed and effluent samples shed more insight into the system, as do tests of COD and TOC removal efficiency. Frequently, measurement of the removal of specific organic compounds or classes of compounds also is essential in understanding the involved transformations. The fate of heavy metals should be checked. The modes of removal of materials by metabolic processes, by volatilization, or by incorporation in the waste biomass should be examined.

3-11.4 Engineering Judgment

Results of experimental studies should be applied to design using considerable engineering judgment. With virtual certainty, the assumption can be made that the volume and composition of the industrial wastewater studied in laboratory or in field pilot units is going to change, even before startup of the treatment facility, as manufacturing plant operations are modified. Effluent goals also will change according to societal needs. Consequently, the experimental information should be viewed as a guide in making decisions, not as the dictator of the decision. A considerable effort should be spent on conceptual engineering work before proceeding to the process and detailed design phases. Process design work should include a least-cost analysis; one such means was described by Middleton and Lawrence (Ref. 3-39).

REFERENCES

3-1. Alexander, M., "Nonbiodegradable and Other Recalcitrant Molecules," *Biotechnol. Bioeng.*, **XV**, pp. 611-647, 1973.

3-2. *Review and Evaluation of Available Techniques for Determining Persistence and Routes of Degradation of Chemical Substances in the Environment,* EPA-560/5-75-006, U.S. Environmental Protection Agency, Washington, D. C., 1975.

3-3. Conway, R. A. and Waggy, G. T., "Biodegradation Testing of Typical Surfactants in Industrial Usage," *Am. Dyes. Rep.*, **55**, 12, pp. 607-615, 1966.

3-4. Price, K. S., Waggy, G. T., and Conway, R. A., "Brine Shrimp Bioassay and Seawater BOD of Petrochemicals," *J. Water Pollu. Control Fed.*, **46**, 1, pp. 63-77, 1974.

3-5. Metcalf and Eddy, Inc., *Wastewater Engineering: Collection, Treatment, Disposal,* McGraw-Hill Book Co., New York, 1972.

3-6. Eckenfelder, W. W., Jr. and Ford, D. L., "Economics of Wastewater Treatment," *Chem. Eng.* pp. 109-118, August 25, 1969.

3-7. Williamson, K. J., and McCarty, P. L., "Rapid Measurement of Monod Half-Velocity Coefficients for Bacterial Kinetics," Stanford University, Palo Alto, California, May 1974.

3-8. Lawrence, A. W., and McCarty, P. L., "Unified Basis for Biological Treatment Design and Operation," *J. Environ. Eng. Div., Proc. ASCE,* **96**, pp. 757-778, 1970.

3-9. Adams, C. E., Eckenfelder, W. W., and Hovious, J. C., "A Kinetic Model for Design of Completely-Mixed Activated Sludge Treating Variable-Strength Industrial Wastewaters," *Water Res.*, **9**, pp. 37–42, 1975.

3-10. Christenson, D. R. and McCarty, P. L., "Multi-process Biological Treatment Model," *J. Water Pollu. Control Fed.*, **47**, 11, pp. 2652–2664, 1975.

3-11. McKinney, R. E., "Design and Operational Model for Complete Mixing Activated Sludge System," *Biotechnol. Bioeng.*, **XVI**, pp. 703–722, 1974.

3-12. Novak, J. T., "Temperature-Substrate Interactions in Biological Treatment," *J. Water Pollu. Control Fed.*, **46**, 8, pp. 1984–1994, 1974.

3-13. Ramanathan, M. and Gaudy, A. F., Jr., "Studies on Sludge Yield in Aerobic Systems," 26th Purdue University Industrial Waste Conference, Lafayette, Indiana, 1972.

3-14. Conway, R. A. and Kumke, G. W., "Field Evaluation of Aeration Equipment," *J. Env. Eng. Div., Proc. Am. Soc. Civil Eng.*, **92**, SA2, pp. 21–42, 1966.

3-15. Nogaj, R. J., "Selecting Aeration Equipment," *Chem. Eng.*, pp. 92–102, April 17, 1972.

3-16. Price, K. S., Conway, R. A., and Cheely, A. H., "Surface-Aerator Interactions," *J. Env. Engr. Div., ASCE*, **99**, EE 3, Proc. Paper 9797, pp. 283–300, 1973.

3-17. Tyer, B. R. and Hayes, C. H., "Oxygenation Equipment–Aerators and Pure Oxygen Systems," *Chem. Eng.*, pp. 75–77, October 6, 1975.

3-18. Dick, R. I. and Young, K. W., "Analysis of Thickening Performance of Final Settling Tanks," 27th Purdue University Industrial Waste Conference, Lafayette, Indiana, 1972.

3-19. Nyer, E. K. and Hovious, J. C., "A Statistical Method of Design for Secondary Settlers," Annual Meeting of Water Pollution Control Federation, Miami, Florida, 1975.

3-20. "Oxygenation and Wastewater Treatment," *Environ. Science Technol.*, **9**, 10, pp. 910–915, 1975.

3-21. *Oxygen Activated Sludge Wastewater Treatment Systems,* EPA Technology Transfer Seminar Publication, U.S. Environmental Protection Agency, Washington, D. C., 1973.

3-22. Eckenfelder, W. W., Jr. and Ford, D. L., *Water Pollution Control, Experimental Procedures for Process Design,* Jenkins Publishing Co., Austin and New York, 1970.

3-23. *Process Design in Water Quality Engineering,* E. L. Thackston and W. W. Eckenfelder, (ed.), Jenkins Publishing Co., Austin and New York, 1972.

3-24. McCarty, P. L., "Anaerobic Waste Treatment Fundamentals," *Public Works,* pp. 107-112, September 1964; pp. 123-126, October 1974, pp. 91-94, November 1974, pp. 95-99 December 1974.

3-25. Hovious, J. C., Waggy, G. T., and Conway, R. A., "Identification and Control of Petrochemical Pollutants Inhibitory to Anaerobic Processes," Series EPA-R2-73-194, Office of Research and Monitoring, U.S. Environmental Protection Agency, Washington, D. C. 20460, April 1973.

3-26. Speece, R. E. and McCarty, P. L., "Nutrient Requirements and Biological Solids Accumulation in Anaerobic Digestion," *Proc. Int. Conf. on Water Pollu. Res.,* London, 1962.

3-27. Hovious, J. C., Conway, R. A., and Ganze, C. W., "Anaerobic Lagoon Pretreatment of Petrochemical Wastes," *J. Water Pollu. Control Fed.,* **45**, 1, pp. 71-84, 1973.

3-28. *Upgrading Lagoons,* Technology Transfer Seminar Publication, U.S. Environmental Protection Agency, Washington, D. C., August 1973.

3-29. Gloyna, E. F. and Hermann, E. R., "Some Design Considerations for Oxidation Ponds," *J. Environ. Eng. Div., Proc. Am. Soc. Civil Eng.,* **82**, SA4, Paper 1047, 1956.

3-30. Porges, R., "Industrial Waste Stabilization Ponds in the United States," *J. Water Pollu. Control Fed.,* **35**, 4, pp. 456-468, 1963.

3-31. Middlebrooks, E. J., *et. al., Proceedings of Symposium on Upgrading Wastewater Stabilization Ponds to Meet New Discharge Standards,* Utah State University, Logan, Utah, August 21-23, 1974.

3-32. "Green Systems for Wastewater Treatment," *Environ. Science Technol.* **9**, 5, pp. 408-409, 1975.

3-33. Walroth, D. and Natter, A. S., "Aquaculture–New Broom Cleans Up Wastewater," *Water Wastes Eng.,* pp. 38-41, February 1976.

3-34. Weston, R. F. and Stack, V. T., Jr., "Fundamentals of Biological Treatment," *Tappi,* **46**, 5, 166A, 1963.

3-35. Conway, R. A., Hovious, J. C., and Macauley, D. C., "Predicting Achievable Effluent Quality and Variability in the Organic Chemicals Industry," *J. Prog. Water Technol.,* **8**, 1, pp. 165-182, 1976.

3-36. Howard, P. H., Saxena, J., Durkin, P. R., and Ou, L.-T., *Review and Evaluation of Available Techniques for Determining Persistence and Routes of Degradation of Chemical Substances in the Environment,* EPA 560/5-75-006, U.S. Environmental Protection Agency, Washington, D. C., May 1975.

3-37. *Environmental Engineering Unit Operations and Unit Processes Labora-*

tory Manual, Association of Environmental Engineering Professors, Dept. of Environmental Health Eng., University of Texas, Austin, 1975.

3-38. Eckenfelder, W. W., Jr. and Ford, D. L., *Laboratory and Design Procedures for Waste Treatment Processes,* The Center for Research in Water Resources, University of Texas, Austin, 1968.

3-39. Middleton, A. C. and Lawrence, A. W., "Least Cost Design of Activated Sludge Systems," *J. Water Pollu. Control Fed.*, **48**, 5, pp. 889-905, 1976.

3-40. Roberts, T. E., Witt, E. R. and Humphreys, W. J., "Full-Scale Anaerobic Filter Treats High Strength Wastes", 34th Purdue Industrial Waste Conference, Lafayette, Indiana, 1979.

3-41. *News Quarterly,* Sanitary Eng. Research Laboratory, Richmond Field Station, Univ. of California, Berkeley, California, April, 1979.

3-42. Stuckey, D. C., Parkin, G. F., Owen, W. F. and McCarty, P. L., "Comparative Evaluation of Anaerobic Toxicity by Batch and Semi-Continuous Assays", 51st Conf. of Water Pollution Control Federation, Anaheim, California, October, 1978.

3-43. *Proc.: Workshop Toward an Oxygen Transfer Standard,* Pacific Grove, California, April 11-14, 1978, W. C. Boyle, ed., EPA-600/9-78-021, MERL, ORD, U.S. Environmental Protection Agency, Cincinnati, Ohio 45268, April 1979.

APPENDIX 3–1

BIOLOGICAL TERMS AND NOTATION

a = fraction of S_r converted to cells

a' = oxygen utilization coefficient for substrate conversion, lb O_2 required/lb BOD_5 removed (kg/kg)

α(alpha) = ratio of $K_L a$ of reactor liquor to that of tap water

a_s = biological solids settling rate constant

B = conversion factor, mg BOD_5/mg effluent suspended solids

b = cell autooxidation rate coefficient, day^{-1}

b' = oxygen utilization rate for endogenous metabolism, lb O_2/day-lb MLVSS (kg/day-kg)

β(beta) = ratio of dissolved oxygen level at saturation in reactor liquor to that in pure tap water

BOD_L = long-term biochemical oxygen demand (~ 20-day incubation period), mg/liter

BOD_5 = 5-day biochemical oxygen demand (5-day incubation period), mg/liter

c = exponent for feed concentration effect

C_i = suspended solids concentration to be clarified, mg/liter

C_L = design DO level, mg/liter

COD = chemical oxygen demand, mg/liter

C_s = DO saturation at standard conditions, mg/liter

C_{sa} = DO saturation at one atmosphere pressure, mg/liter

C_{sw} = DO saturation level calculated for reactor liquor at design pressure and liquor temperature, mg/liter

C_t = DO at times t_1, t_2, etc., mg/liter

C_1 = suspended solids concentration at settling velocity v_1, lb/lb or mg/mg or mg/liter $\times 10^{-6}$

C_2 = suspended solids concentration at settling velocity v_2, lb/lb or mg/mg or mg/liter $\times 10^{-6}$

D_f = biological filter depth, ft (m $\times$ 3.28)

DO = dissolved oxygen concentration, mg/liter

D_o = DO deficit, mg/liter

f_b = fraction of generated biomass that is degradable

F/M = food-to-microorganism ratio, lb BOD_5/lb MLVSS-day

f_v = ratio of MLVSS/MLSS

K = reaction rate coefficient independent of S_o, liter/mg-hr

K' = reaction rate coefficient dependent on S_o for complete mixed system, liter/mg-hr

K'' = maximum rate of substrate utilization per unit weight of microorganisms, g BOD_L/day-g bacteria
k = batch-fed reaction rate coefficient, liter/mg-hr
K_a = reaction rate constant for aerated stabilization basins when level of biosolids is not considered, day^{-1}
K_{ax} = reaction rate constant for aerated stabilization basins when level of biosolids is considered, liter/mg-day
K_f = overall reaction rate constant for a biological filter taking into account actual reaction rate constant (gal./ft^2-min) and specific surface area (ft^2/ft^3), min^{-1}
$K_L a$ = overall oxygen transfer coefficient, mg/liter O_2 per mg/liter DO deficit-hr (or hr^{-1})
K_p = reaction rate coefficient for unaerated stabilization ponds, day^{-1}
K_s = half velocity constant or substrate level resulting in $\mu = 0.5\mu_{max}$, mg/liter
K_1 = reaction rate coefficient at T_1, liter/mg-hr
K_2 = reaction rate coefficient at T_2, liter/mg-hr
K_3 = oxygen transfer proportionality constant
k_1 = thickening constant dependent on a_s and n
m = packing constant for biofilter
MLSS = mixed liquor suspended solids in aeration tank, mg/liter
MLVSS = mixed liquor volatile suspended solids, mg/liter
μ(mu) = specific growth rate, mg growth/mg biomass-hr (or hr^{-1})
μ_{max} = maximum specific growth rate attainable with a given type substrate, hr^{-1}
n = biological solids settling rate exponent
N_a = O_2 transfer efficiency at field conditions, lb/hp-hr (kg/joule $\times$ 5.92 $\times$ 10^6)
N_s = O_2 transfer efficiency at standard conditions (20°C, 760 mm, pure tap water, nil DO), lb/hp-hr (kg/joule $\times$ 5.92 $\times$ 10^6)
OR = overflow rate (flow volume/clarifier area), gal./sq ft-day ($\times$ 24.5 = m^3/m^2-day)
O_t = total oxygen required, lb/day (kg/day)
O_v = volume concentration of oxygen in gas leaving the test unit, %
P = power hp (KW $\times$ 1.34)
p = design atmospheric pressure mm Hg (atm $\times$ 760, psia $\times$ 51.7, kN/cm^2 $\times$ 7.50, Kgf/cm^2 $\times$ 735)
p_b = absolute pressure at depth of air release, psia (kN/m^2 $\times$ 0.145)
Q = influent flow rate not including recycle, gal./day (or liters/day)
Q_h = hydraulic loading to biofilter, gpm/ft^2 (m^3/m^2–min $\times$ 4.5)
R = recycle flow rate, gal./day (or liters/day)
r = nondegradable fraction of influent suspended solids
R/Q = recycle flow fraction
S = substrate level, mg/liter
S_a = BOD concentration applied to the biosystem (after dilution by recycle), mg/liter

S_{at} = total substrate applied, lb BOD_5/day (kg/day)
S_e = effluent soluble BOD_5, mg/liter
S_{ea} = long-term average soluble BOD_5 achievable with a single biological system in the field for a typical petrochemical S_o of 1000 mg/liter, mg/liter
S_{em} = predicted effluent BOD level for high month, mg/liter
S_o = influent soluble BOD_5, mg/liter
S_r = BOD concentration reduction, mg/liter
S_{rt} = total BOD_5 removed, lb/day (kg/day)
SVI = sludge volume index, ml/gm
T = temperature, °C
t = average liquid retention time, hr
θ_a(theta) = temperature coefficient for long-term effluent quality adjustment
θ_c = mean cell residence time, days
θ_t = coefficient for temperature adjustment of reaction rate for BOD removal
TOC = total organic carbon, mg/liter
t_d = retention time, days
T_w = design reactor temperature, °C
t_1, t_2 = time, hr
V = adjustment factor for effluent variability to convert long-term average S_{ea} to maximum month S_{em}
v_i = initial biosolids settling rate at c_i, ft/hr (m/hr $\times$ 3.28)
v_1 = settling velocity at concentration c_1, ft/hr (m/hr $\times$ 3.28)
v_2 = settling velocity at concentration c_2, ft/hr (m/hr $\times$ 3.28)
W = weight of aerated water, millions of pounds (kg $\times$ 2.21 $\times$ 10^6)
X_a = active biological solids concentration in the reactor, mg/liter
X_{at} = active biological solids weight in the reactor, lb (kg $\times$ 2.21)
X_{eh} = effluent suspended solids in high month, mg/liter
X_{et} = total effluent biological solids, lb/day (kg/day)
X_{ot} = total influent suspended solids, lb/day (kg/day)
X_{ov} = influent degradable VSS, mg/liter
X_r = recycled sludge solids concentration, mg/liter
ΔX_t = total excess sludge, lb/day (kg/day)
X_v = MLVSS, mg/liter
X_{vt} = total weight of VSS in aeration tank, lb (kg)
ΔX_{vt} = change in MLVSS, lb/day (kg/day)
X_{wt} = total biological solids wasted, lb/day (kg/day)
Y = maximum cell yield coefficient, g bacteria synthesized/g BOD_L consumed
y = conversion coefficient for BOD_5 to BOD_L, lb BOD_5/lb BOD_L (kg/kg)
Z = concentration of nonbiodegradable organics, mg/liter

4

Dewatering of excess biomass and other sludges

4-1 SOLIDS HANDLING ALTERNATIVES

Slurry side streams requiring disposal are generated both in the primary clarification step and as excess biomass in secondary treatment systems. Handling of these streams warrants careful study as associated costs can reach 50% of that for the primary and secondary treatment operations. Because of high water retentiveness and potential environmental incompatibility, various treatment steps generally are appropriate before final disposal. The alternative steps and their functions are presented in Table 4-1. The operations and processes which may be used in each step are shown in Figure 4-1.

Decisions regarding which steps to include and which operation or process is best for that step in order to develop a cost-effective solution involve considerations such as:

- quantity of sludge and its solids content;

TABLE 4-1. Sludge Treatment Sequence and Process Functions

Unit Processes	Functions
Thickening (Blending)	– Water Removal – Volume Reduction – Post Process Efficiencies – Blending
Stabilization (Reduction)	– Pathogen Destruction – Volume and Weight Reduction – Odor Control – Putrescibility Control – Gas Production
Conditioning (Stabilization)	– Improve Dewatering or Thickening Rate – Improve Solids Capture – Improve Compactability – Stabilization
Dewatering	– Water Removal – Volume and Weight Reduction – Change to Damp Cake – Reduces Fuel Requirements for Incineration/ Drying
Heat Drying	– Water Removal – Sterilization – Utilization
Reduction (Stabilization)	– Destruction of Solids – Water Removal – Conversion – Sterilization
Final Disposal	– Utilization (Cropland) – Utilization (Energy) – Utilization (Land Reclamation) – Disposal (Landfill) – Disposal (Ocean)

Source: Process Design Manual for Sludge Treatment and Disposal, EPA 625/1-74-006, U.S. Environmental Protection Agency, Technology Transfer, Cincinnati, Ohio, 1974.

- condition of the biomass or chemical solids in terms of water retentiveness and extent of stabilization;
- other slurry streams more amenable to dewatering which are available for blending with the problem material;
- local costs and availability of land, energy, and labor;

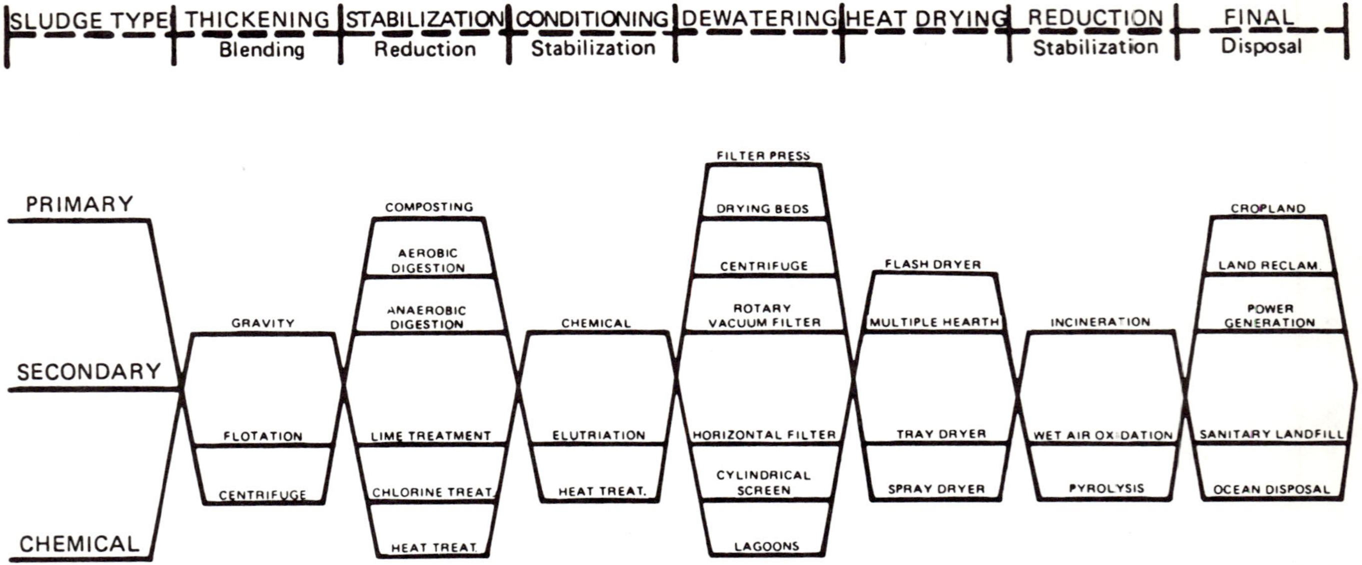

Fig. 4-1. Sludge processing alternatives. *Source: Process Design Manual for Sludge Treatment and Disposal,* EPA 625/1-74-006, U.S. Environmental Protection Agency, Technology Transfer, Cincinnati, Ohio, 1974. (patterned after Stanley Consultants, "Sludge Handling and Disposal, Phase I - State of the Art," Report to Metropolitan Sewer Board of the Twin Cities Area, November 15, 1972).

- option of proceeding step-wise, with time for interim evaluations, versus proceeding with ultimate sequence at the start.

A comprehensive text on the subject of sludge handling and disposal has been prepared by Vesilind (Ref. 4-1); also, the U.S. Environmental Protection Agency has prepared a useful manual on this subject (Ref. 4-2). Many reported studies are with sludges from municipal operations, and their use must be tempered by the observation that waste biomass generated from soluble industrial wastes generally is much more difficult to dewater than is the municipal sludge higher in inorganic particulates.

The more commonly used alternatives for thickening, stabilization, conditioning, and dewatering are described in this section. Ultimate disposal is covered in Chapter 7.

4-2 THICKENING, STABILIZATION, AND CONDITIONING

The interrelationships and purposes of the alternative treatments of sludges prior to the step of dewatering to a handleable cake are depicted in Table 4-1 and Figure 4-1. Additional comments as to the nature and applicability of the more widely applicable systems follow along with references as to design techniques to supplement the techniques and general experience cited in the basic references (Ref. 4-1, 4-2).

4-2.1 Thickening

Gravity-type thickener units commonly employ rakes to aid in the concentration of the sludge, and to prevent stratification of the settling solids. It is designed on the basis of solids loading in terms of lb/ft^2-day (kg/m^2-day) (Ref. 3-22). A typical value for activated sludge is 4.5 lb/ft^2-day (0.9 kg/m^2-hr); rates for oxygenated activated sludge can be two-fold higher (Ref. 4-1). Thickened solids levels of 2 to 3 times the initial levels frequently are achieved. A level of 2.5% may be achieved through thickening of waste activated sludge from air systems. A pilot study with activated sludge at a chemical plant by Kumke, Hall, and Oeben, showed that a 1% feed was concentrated to 3.2% with a solids capture of 97% at an unusually high loading of 33 lb/ft^2-day (6.9 kg/m^2-hr) (Ref. 4-3).

Design values may be determined by thickening tests either in the laboratory with the apparatus shown in Figure 4-2 stirred at 10-30 rph or preferably through pilot studies. With regard to evaluation procedure and data handling, a comprehensive analysis has been made by Vesilind (Ref. 4-1) and by Eckenfelder and Ford (Ref. 3-22). Thickening also was discussed in Section 3-5 along with

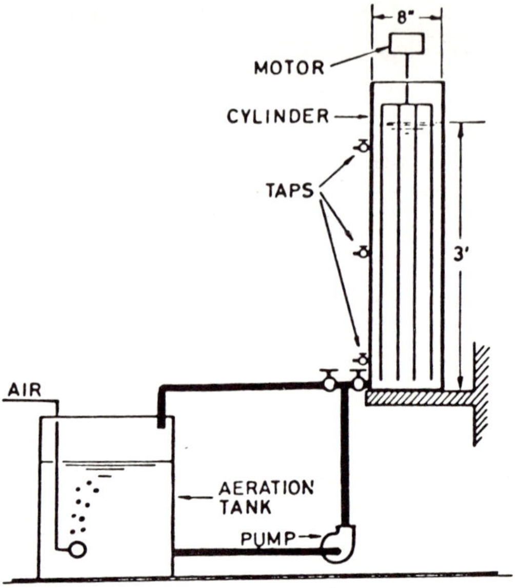

Fig. 4-2. Laboratory apparatus for batch thickening tests. ***Source:*** **Vesilind, P. A.,** ***Treatment and Disposal of Wastewater Sludges,*** **Ann Arbor Science Publishers, Inc., Ann Arbor, Michigan, 1974.**

procedures of Dick and Young (Ref. 3-18). Thickening of a sludge from 0.75 to 2.5% not only reduces the volume to be handled by 70%,

$$\text{Volume Reduction, \%} = 100\left(1 - \frac{\text{original solids conc., \%}}{\text{final solids conc., \%}}\right) \qquad \text{(Eq. 4-1)}$$

but also makes subsequent dewatering operations more effective.

Another thickening alternative is dissolved-air flotation, the basics of which were discussed in a prior section. This approach can be used to preconcentrate secondary-clarifier underflow from a level of about 0.7 to 1.5% to about 4 to 6.0% solids content. Typical design factors are surface loadings of 1 to 4 gal./ft^2-min (0.2 to 0.9 m^3/m^2-hr), solids loading of 0.5 to 5 lb/ft^2-hr (2.5 to 25 kg/m^2-hr), and a detention time of 30 min (Ref. 4-1). The use of gases high in carbon dioxide, e.g., UNOX system and combuster off-gases, to lessen costs was described by Conway, Nelson, and Young (Ref. 4-15).

A nozzle disk centrifuge (Figure 4-3) also can be applied to waste activated sludge to achieve thickened concentrations of 3 to 5%. Clogging is a potential problem unless the sludge is prescreened and degritted. Special designs have been developed to lessen maintenance needs.

Lagooning provides a means of thickening if adverse environmental impacts

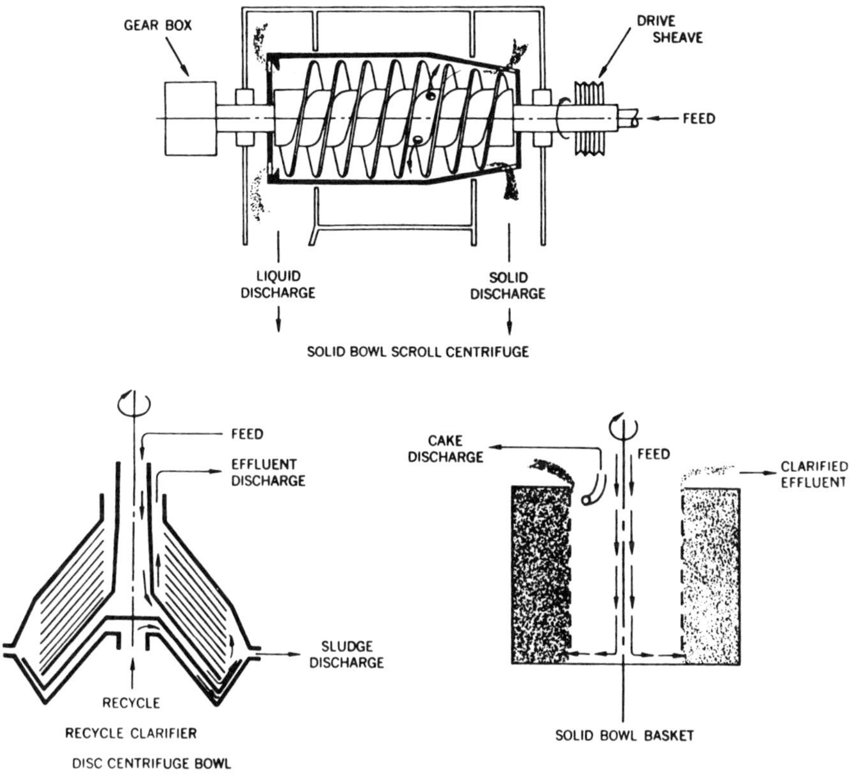

Fig. 4-3. Centrifuges for biomass dewatering. ***Source:*** **Moll, R. T., "Centrifugal Dewatering of Secondary Activated Sludge and Secondary Fiber Streams,"** ***TAPPI,*** **1st Secondary Fiber Pulping Conference, Dayton, Ohio, October 22-25, 1968.**

do not preclude its use. At one chemical plant, waste activated sludge fed at 1% to a lagoon and accumulated to a 9-ft (2.7 m) depth was thickened to 11% at the bottom of the sludge layer and 5% at the top over a period of several months. No significant loss of its volatile content occurred at this northern location.

4-2.2 Stabilization

Aerobic digestion is one means of stabilizing waste biological solids; it serves to reduce the organic content of the mixture and thereby decreases the odor potential if it is to be disposed of on land. Specific resistance to filtration improves, but compressibility can increase (Ref. 4-1). Filterability was found to be best after 4 to 8 days under aeration and worsened with the additional treatment normally needed for stabilization (Ref. 4-4). The microorganisms in the process, conducted in an aerobic vessel in the absence of an external substrate, deplete their stored organic material. Some organisms undergo lysis; the released cellular constituents are utilized by other organisms.

At a hydraulic retention time of 10 to 15 days, a 40 to 60% reduction in volatile suspended solids may be achieved in aerobic digestion. The higher reductions occurred when the *F/M* loading ratio was high (0.5 to 0.6 day^{-1}). Laboratory studies in batch aerobic digesters have been devised (Ref. 2-2) to determine the first-order reaction rate constant; the rate of decrease of volatile solids is proportional to the level of volatile solids remaining. Perhaps of more importance is to follow decrease in odor potential as described by Eikum (Ref. 4-16).

Anaerobic digestion, which is more common to municipal operations, involves the stepwise conversion of the cellular constituents to soluble materials, to volatile acids, and then to methane and carbon dioxide. Since a more limited number of bacterial species is involved (especially in methanogenesis), anaerobic digestion is more susceptible to upset than is aerobic digestion. Digestion is accomplished in a heated, mixed tank in the absence of oxygen (Figure 4-4). The digested sludge is more amenable to land disposal and dewaters better on drying beds. At 35°C the minimum solids retention time (θ_c) is about 10 days to achieve adequate digestion.

Laboratory anaerobic-digestion experiments can be conducted using the apparatus shown in Figure 4-5 to determine the specific growth rate, waste utilization rate, half-velocity coeffieient, yield coefficient, and microorganism decay coefficient, needed to determine the solids retention time required for a particular reduction level (Ref. 4-1). This approach is as discussed in the section on activated sludge kinetics. In regard to full-scale operations, a variety of equipment has been developed for mixing, heating, gas collection, and the other involved operations (Ref. 2-14).

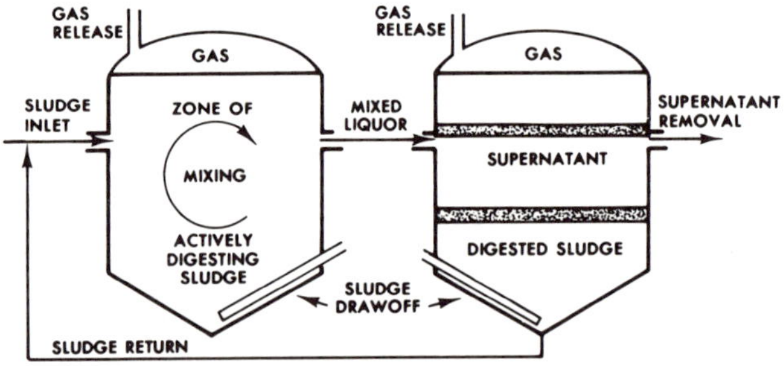

Fig. 4-4. Anearobic digestion process schematic. ***Source: Process Design Manual for Sludge Treatment and Disposal,*** **EPA 625/1-74-006, U.S. Environmental Protection Agency, Technology Transfer, Cincinnati, Ohio 1974.**

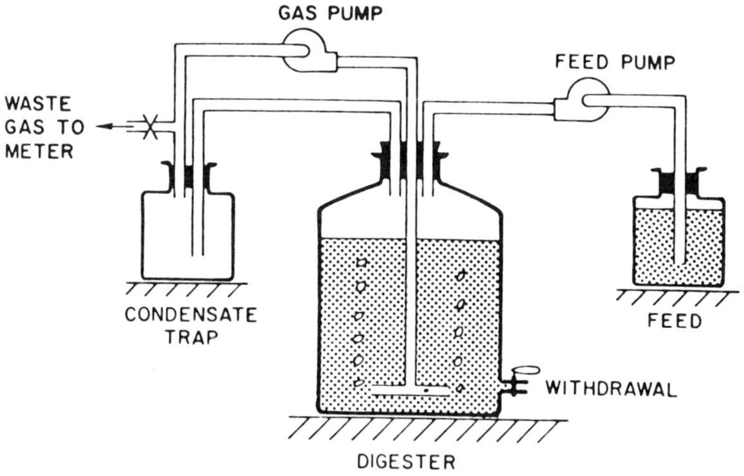

Fig. 4-5. Laboratory apparatus for conducting anaerobic digestion studies. *Source:* Vesilind, P. A., *Treatment and Disposal of Wastewater Sludges,* Ann Arbor Science Publishers, Inc., Ann Arbor, Michigan, 1974.

Low pressure oxidation for about 25 min at about 190°C and 270 psig (1.86×10^6 N/m^2) has been used in some instances to condition sludge for improved dewaterability and to kill pathogens. This destroys the cells and releases the bound water; thus, it is potentially more effective than is mechanical dewatering, which is limited to removing only the water not incorporated in or bound to the cells. However, some organics are solubilized, which increases the load to the treatment system; and corrosion problems have been encountered.

Dosage of the sludge with chlorine at about 2000 mg/liter is another stabilization approach. It can be used to achieve long-term stability; increased filterability sometimes can result. Any soluble chlorinated by-products can be a concern due to their persistence.

4-2.3 Conditioning

Chemical conditioning with ferric chloride, lime, and/or an organic polymeric coagulant, usually cationic, is virtually always required to achieve reasonable yields and solids capture with subsequent dewatering equipment. As opposed to inorganic materials, the use of organic coagulants have the advantages of not adding a significant particulate weight to the sludge, of an operation with less pH sensitivity, and of not increasing the ash of the sludge if incinerated. However, polymers have been associated with clinker problems. In the case of filtration, chemicals are added to a flocculation tank involving staged rapid

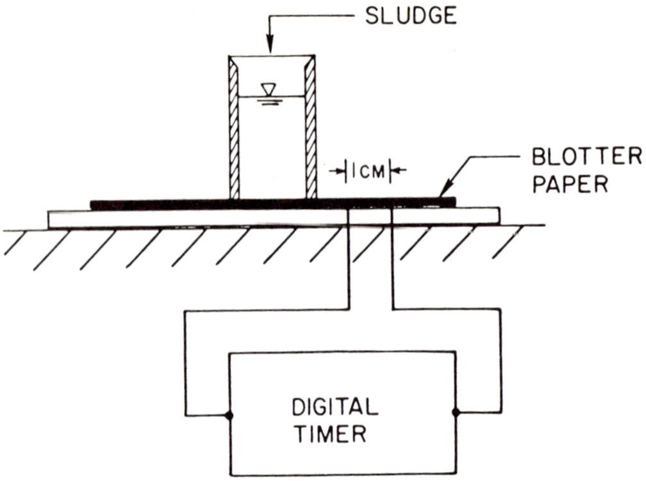

Fig. 4-6. Capillary suction time apparatus. *Source:* **Vesilind, P. A.,** ***Treatment and Disposal of Wastewater Sludges,*** **Ann Arbor Science Publishers, Inc., Ann Arbor, Michigan, 1974.**

mixing and floc development. Coagulants generally are added internally to centrifuges.

Chemical selection and effective dosage levels are determined through laboratory tests. Capillary suction time measurements on conditioned sludges indicate the obtained degree of water release, which especially is pertinent to filtration operations. Using the apparatus shown in Figure 4-6, the time for the free water to migrate across the blotting paper between two electrical connections is determined. This time is correlated with tests more specific to the planned mode of separation as described in subsequent sections. Times of 10 sec or less using a 1.8 cm diameter tube usually are desirable.

The conditioning of sludges with incinerator ash or flyash from power plants can be effective. These ashes also can be used as pressure filter precoats. Smith, *et al.* (Ref. 4-5) found that sludge-incinerator ash was more effective than was flyash; 1.3 lb ash/lb activated sludge solids doubled the filter yield on an ash-free basis. Low pressure oxidation, as discussed in Section 4-2.2 is another means of sludge conditioning.

4-3 DEWATERING EQUIPMENT

Based on reports, the most commonly used mechanical dewatering approach for industrial sludges seems to be centrifugation. Presumably, this is due to: lower capital costs, less required operator attention, enclosed operation, compactness, ease of cleaning, and familiarity of industrial people with this type of operation.

However, high maintenance costs can be encountered unless sludges are degritted and screened. Also, solids recovery efficiencies generally are lower when compared with filtration operations. Polymeric coagulant costs can be high.

As an alternative to centrifugation as energy costs increase, the improved cake dryness that can be achieved with filter presses increases the attractiveness of that approach. Rotary vacuum filters, as well as top-feed horizontal vacuum filters, also have application in some instances. Figure 4-7 shows alternative centrifugation and filtration approaches as described by Moyers. Lagooning is very often used as an interim measure to achieve a degree of stabilization and dewatering while more environmentally compatible measures are being evaluated. For biological sludges belt filter presses are an attractive compromise considering cost, cake dryness, and operability.

4-3.1 Centrifugation

Considering applications with all types of sludges, both municipal and industrial, the continuous, solid-bowl centrifuge seems to be the most widely used. This type centrifuge is illustrated in Figure 4-3. As with all centrifuges, retention time of the solids and centrifugal acceleration are the major variables. Centrifugal accelerations about 3000 times gravity are typical. Performance is dependent upon the degree of clarification achieved (solids recovery), the extent of compaction (cake solids level), and the ability to move the cake out of the centrifuge.

Laboratory tests can indicate the amenability of the waste biomass to centrifugal separation. Tests with bench-top centrifuges are run over different time periods and at various rotation rates or centrifugal acceleration (CA) levels.

$$CA = \frac{(r_1 + r_2)}{2} \omega^2 \frac{1}{g} \qquad \text{(Eq. 4-2)}$$

where

CA = centrifugal acceleration, expressed as number of times greater than the acceleration of gravity

r_1 = radial distance from centrifuge center to top of sludge, in. (cm × 0.39)

r_2 = radial distance from centrifuge center to bottom of the sludge, in. (cm × 0.39)

ω = radial velocity, radians/sec (rpm × 1 min/60 sec × 2π radians/rev)

g = gravitational acceleration, 384 in./sec^2

With activated sludge, the concentration of cake solids generally levels out at 8 to 10% at centrifugal accelerations above 1000 g, as opposed to municipal primary sludge which can reach 25%. However, results collected by Kumke, Hall,

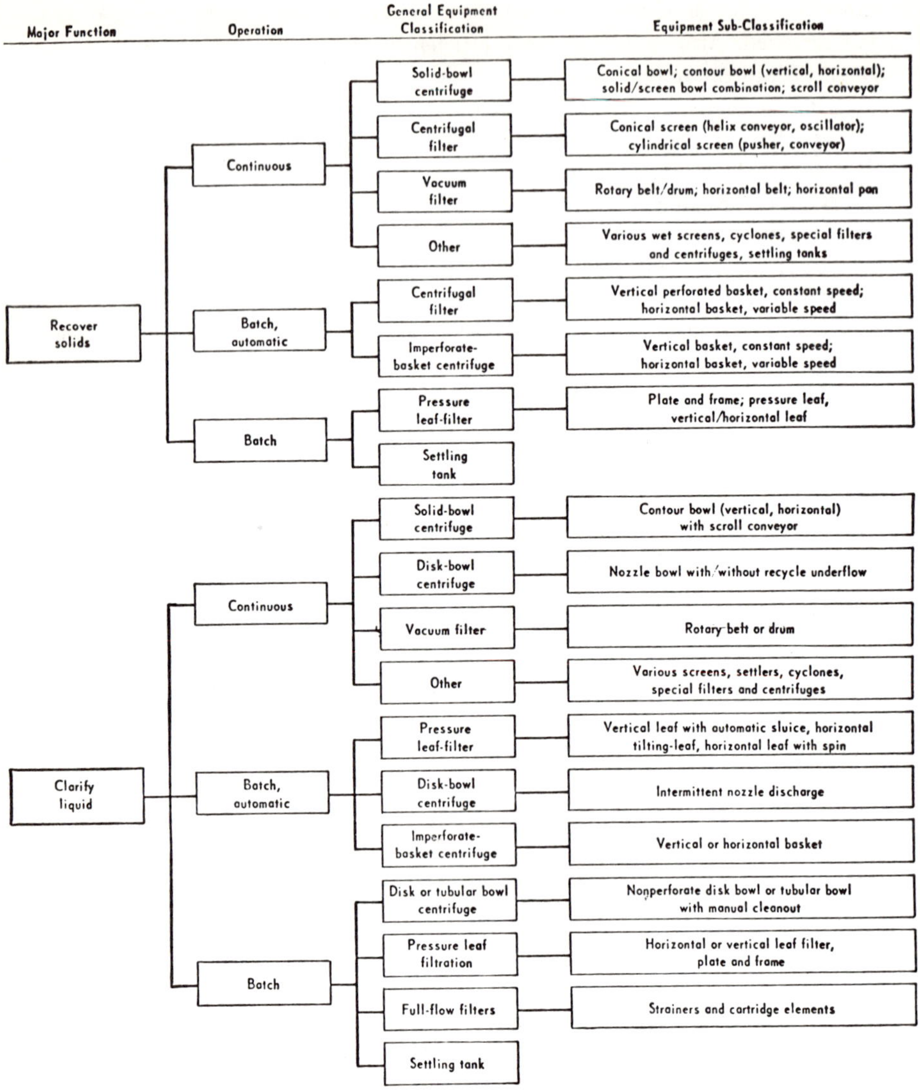

Fig. 4-7. Selection of centrifugation and filtration equipment. *Source:* Moyers, C. G., Jr., "How to Approach a Centrifugation Problem," *Chem. Eng.* pp. 182-189, June 20, 1966.

and Oeben using a small pilot-scale machine with activated sludge at one chemical plant indicate that a cake solids level of about 19 to 22% at an 80 to 90% solids capture was possible at a 3% feed concentration using coagulants (Ref. 4-3).

The degree of solids recovery is calculated in laboratory tests after measuring centrate (supernatant) suspended solids. The cake solids are calculated knowing feed solids, centrate solids, and cake volume. The consistency of the cake, as it affects its amenability to being conveyed out the centrifuge, is indicated using a penetrometer rod (Ref. 4-6). An extremely qualitative test may be made using a glass rod; if the concentrated solids support the rod, conveying problems are not expected; while if the rod sinks to the bottom, conveying generally is ineffective.

Tests with pilot-scale prototypes provide information which can be scaled up with more confidence than can laboratory results. Scale-up is based upon grouping the involved factors into machine variables and process or sludge-dependent variables. The Sigma concept incorporates the variables of ω, r_1, and r_2. A Beta factor also can be used to incorporate the factors of the differential radial velocity of the pool and conveyor, the pitch of the blades and number of blades, the bowl diameter, and the pool depth; these are important when dealing with flocculent solids under hindered settling and cake compression (Ref. 4-1). More detailed techniques for estimating centrifuge performance have been presented by Vesilind in a separate article (Ref. 4-6). To supplement laboratory and pilot-scale studies, available experience with existing installations handling similar sludges should be sought. Also, vendors have developed useful empirical relationships for scale-up based on experience.

Basket-type centrifuges also have been used for dewatering industrial biomasses. This centrifuge is an automated, semi-continuous device consisting basically of a metal basket with a lip that rotates around a vertical axis (Figure 4-3). Units may be equipped with a perforated bowl for centrifugal filtration or, more commonly in the case of biological sludges, with an imperforate bowl for centrifugal sedimentation. A mechanical plow can be run into the basket periodically to remove the concentrated solids. During this period feed is stopped and the basket speed is reduced. One study by Romagnoli with a 2% starting biomass from an oxygenated activated sludge system showed a final solids concentration of about 14% with a 97% solids capture using polyelectrolytes (Ref. 4-7); tests at another location showed 10% cake solids with a 98% capture. Basket centrifuges are used when the solids are not sufficiently firm for the conveyor system in a continuous, horizontal centrifuge to be effective.

Disk centrifuges have utility for thickening thin or oily sludges which do not present clogging problems (Figure 4-3). A flow sheet for utilizing a disk centrifuge after degritting to reduce maintenance requirements is shown in

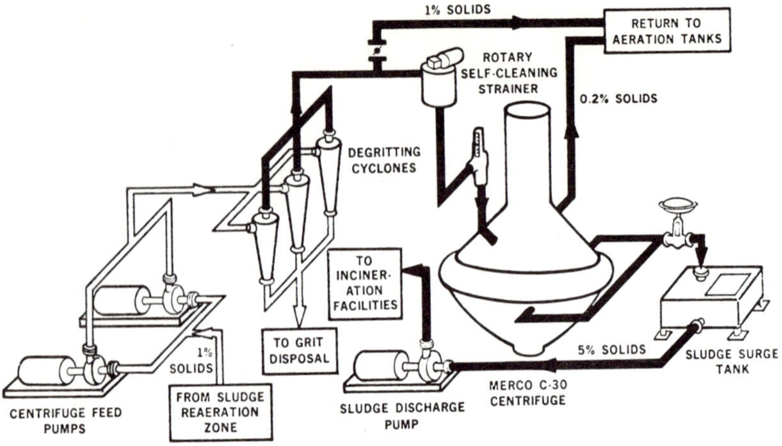

Fig. 4-8. Flow sheet for solids dewatering using a disk centrifuge. ***Source:*** **Stumpf, M. R. and Harper, W. H., "Continuous Feed Centrifuge Replaces Flotation for Removal of Excess Activated Sludge from a Pharmaceutical Wastewater Treatment Plant," Abbott Laboratories, North Chicago, Illinois.**

Figure 4-8. A 1% activated-sludge stream from a pharmaceutical plant was concentrated to 5% with an 80% solids recovery; cleaning was required once per week. This system was found to be preferable to a basket centrifuge from a cost-capacity viewpoint; a continuous solid-bowl centrifuge produced a 16% solids cake at 90% solids recovery, but it was very viscous and coagulant requirements were high (Ref. 4-8). Typical data with activated sludge from paper mills for three commonly used centrifuges are presented in Table 4-2.

TABLE 4-2. Centrifugal Dewatering of Activated Sludge from Paper Mills

Type Centrifuge	Polyelectrolytes (lb/ton)	Solids Recovery (%)	Cake Solids (%)
Solid bowl scroll	10–15	75–85	13–19
Disk	None	80–90	5–7
Solid bowl basket	None	80–90	8–12
Disk in series with solid bowl scroll	10–15	75–85	20–22

Source: Moll, R. T., "Centrifugal Dewatering of Secondary Activated Sludge and Secondary Fiber Streams," *TAPPI,* 1st Secondary Fiber Pulping Conference, Dayton, Ohio, October 22–25, 1968.

4-3.2 Pressure Filtration

Pressure filters can be used to approach or reach the solids levels required for incineration without supplemental fuel. The recessed-plate filter press, which is the most common type for automation, is designed so that the sludge can be pumped between plates that are covered with a filter cloth; the carriage water is forced through the cloth into collection channels (Figure 4-9). When the spaces between the plates are full, the plates are separated and the cake removed (Figure 4-10). Operation usually is at 100 or 225 psig (6.9 or 18 × 10^5 N/m^2), far above the differential attainable in a vacuum filter system. The time to complete a dewatering cycle can be estimated by the relationship developed by Jones (Ref. 4-9).

$$t = \frac{3.15\, R\mu d^2\, (C_f - C_k)^2}{P_v C_f\, (100 - C_f)} \qquad \text{(Eq. 4-3)}$$

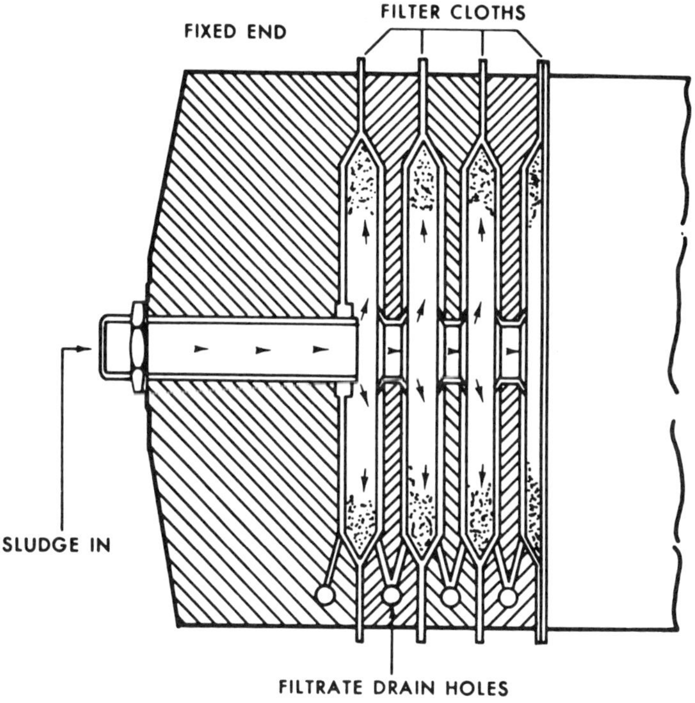

Fig. 4-9. Cutaway view of a filter press. *Source: Process Design Manual for Sludge Treatment and Disposal,* EPA 625/1-74-006, U.S. Environmental Protection Agency, Technology Transfer, Cincinnati, Ohio, 1974.

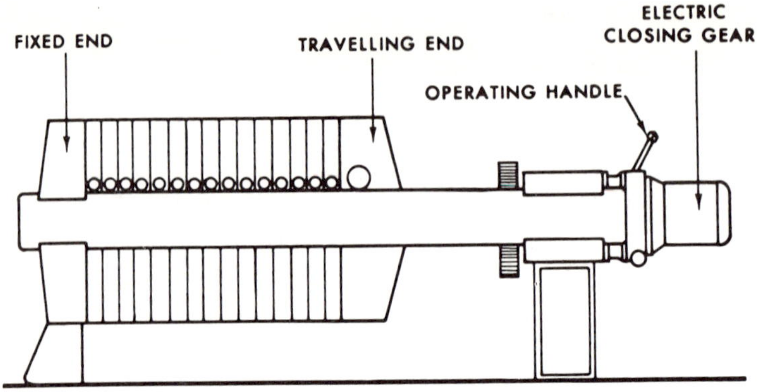

Fig. 4-10. Side view of a filter press. ***Source: Process Design Manual for Sludge Treatment and Disposal,*** **EPA 625/1-74-006, U.S. Environmental Protection Agency, Technology Transfer, Cincinatti, Ohio, 1974.**

where

- t = cycle time, hr
- d = distance between plates, in. (cm X 0.39)
- C_f = feed sludge solids, %
- C_k = cake solids, %
- R = specific resistance in (m/kg) times 10^{-10}
- P_v = filtration pressure, psig (N/m^2 X 0.00014)
- μ = viscosity of filtrate, poises

Specific resistance is obtained through a filtration test using a Buchner funnel under vacuum connected to a graduated cylinder (Figure 4-11); the slope of a plot of time (sec)/volume (ml) versus volume (ml) provides the b term in the equation.

$$r = \frac{2 P_v A^2 b}{\mu w} \qquad \text{(Eq. 4-4)}$$

where

- P_v = pressure differential, N/m^2 (psig X 6.9 X 10^3)
- μ = filtrate viscosity, Ns/m^2 (poise X 0.1)
- w = weight of dry cake solids per unit volume of filtrate kg/m^3 (mg/liter X 10^{-3} (approximately equal to feed solids concentration at low feed solids levels and high capture)
- A = filtration area, m^2 (cm^2 X 10^{-4})
- b = slope of sec/cm^3 versus cm^3, sec/cm^6. Convert to sec/m^6 by multiplying by 10^{12}

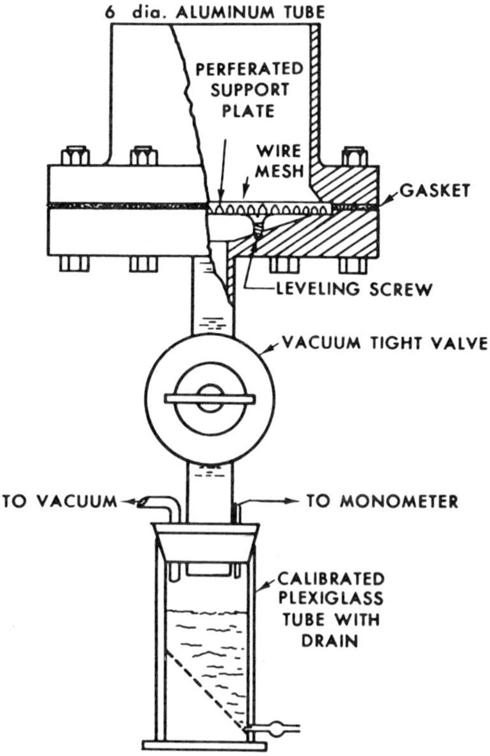

Fig. 4-11. Laboratory vacuum filter apparatus. ***Source: Process Design Manual for Sludge Treatment and Disposal,*** **EPA 625/1-74-006, U.S. Environmental Protection Agency, Technology Transfer, Cincinnati, Ohio, 1974.**

The specific resistance value, as usually obtained, is more applicable in vacuum filtration; however, it provides an indication of the expected pressure filter performance. A pressurized apparatus can be used to obtain more meaningful specific resistance results for application to pressure filters. A value below 10×10^8 sec^2/gm (10×10^{12} m/kg) at 100 psig is desired. Also, pressure filter manufacturers have laboratory and pilot-scale test systems which provide results more amenable to scale-up, as filter media and cake compressibility are taken into account. See Appendix 4-1 for terms and notation.

In field tests with pressure filters, chemical conditioning agents first are selected with the aid of either capillary suction time tests (Figure 4-6) or specific resistance tests at operating pressure. A shear stability test (Ref. 4-10) can be used as a third criterion; no significant change in capillary suction time after 40 sec of shear is desired for pressure filtration. A 10-sec shear time is used for vacuum filtration.

A desired cake thickness then is selected. Filter press tests then are made to determine the precoat needs and cycle time and to define further the conditioning package. Feed and filtrate volume with time, filtrate solids, and cake solids are measured. When of interest, the calorific value as affects incinerability and the amenability to being mixed with soil to form a suitable landfill material are determined, using the filter cake. In one series of pilot tests with an activated sludge, a cake solids level of 20% was attained with a filter press, as compared with 14% with a belt-type pressure filter and 10% with an imperforate bowl, vertical basket centrifuge. Lime dosages can reach 30% or higher on a dry solids basis with pressure filters.

4-3.3 Rotary Vacuum Filtration

Vacuum filters of many types have been used in the dewatering of industrial slurries and waste biological sludges. The most commonly used equipment, the revolving drum vacuum filter, consists of a porous drum covered with a filtering material such as fabric or stainless steel coils (Figure 4-12). The drum is partially immersed in a trough filled with the slurry which is to be dewatered. The drum is internally divided into sections, which allow portions of the drum surface to be subjected to vacuum or pressure as the drum revolves. As each section of the drum is immersed in the liquid, a vacuum is applied which draws the solids to the drum surface, forming a layer or cake. The filtrate passes through the filter cake and into the pipes in the center of the drum. As the drum rotates, the cake moves out of the slurry. After all of the unbound water has been removed, air is drawn through the cake to provide additional drying action.

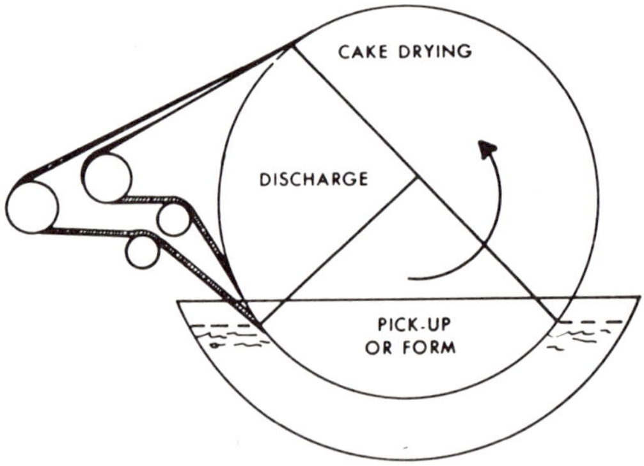

Fig. 4-12. Operating zones of a vacuum filter. ***Source: Process Design Manual for Sludge Treatment and Disposal,*** **EPA 625/1-74-006, U.S. Environmental Protection Agency, Technology Transfer, Cincinnati, Ohio, 1974.**

Pressure can be applied to aid cake release. An alternative to designs in which the filter medium covers the entire drum design is illustrated in Figure 4-12. A belt-type medium is utilized which leaves the drum and passes over a series of rollers where the cake is discharged and the medium is washed.

A rotary vacuum filtration system requires considerable operating attention, sensitive chemical conditioning to prevent filter blinding, and can be odorous due to its open nature. Waste activated sludge by itself is very difficult to dewater using rotary vacuum filters; although, Vesilind reports success with sludges from pure-oxygen systems (Ref. 4-1).

The principal operating variables to be considered in test work with an actual or simulated rotary vacuum filtration system are:

- Feed solids concentration (beneficial to prethicken to as high a level as is pumpable).
- Chemical conditioning (needed to enhance the dewatering rate, to prevent blinding of filter medium pores, and to facilitate cake release; sometimes cake pickup is adversely affected by too large a floc).
- Storage time (usually best to process immediately after withdrawn from stabilization and/or conditioning step, can reaerate to restore filterability) (Ref. 4-4).
- Vacuum level (yield theoretically is proportional to pressure; but too high a pressure can compress the cake and make it less pervious).
- Drum submergence (optimize considering cake thickness and dryness).
- Drum speed (optimize considering filter yield and cake dryness.
- Vat agitation (optimize considering prevention of settling and floc breakup).
- Filter medium (optimize considering blinding and filtrate clarity).

For oily sludges from primary clarification systems a filter precoat such as diatomaceous earth can be used.

Filter leafs, as shown in Figure 4-13, can be used for testing sludge pickup, for accurate determination of the *w* factor in Eq. 4-5, and for preliminary design purposes. A safer method would be to use a standard vacuum flask and to pour the filtrate into a graduate for measurement. The capillary suction time apparatus (Figure 4-6) or jar tests can be used to select chemical conditioner doses used in the more tedious Buchner funnel or filter-leaf tests. Various sized pilot systems are available from equipment manufacturers for use in final design studies.

Initial filtration tests usually are made using a Buchner funnel inserted in a graduated cylinder and connected to a vacuum source. Coarse filter paper is used to simulate a cloth medium. A refinement to the standard glass or porcelein Buchner funnel has been developed to provide a better filter medium support (Figure 4-11); A simpler, plastic funnel equipped with wire mesh screens also is

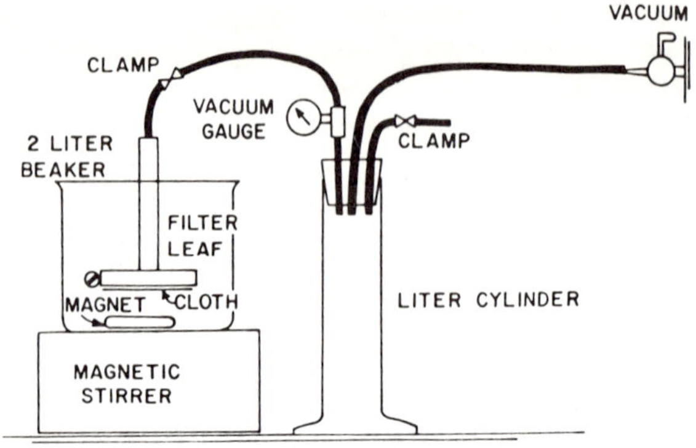

Fig. 4-13. Filter leaf apparatus. ***Source:*** **Vesilind, P. A.,** ***Treatment and Disposal of Wastewater Sludges,*** **Ann Arbor Science Publishers, Inc. Ann Arbor, Michigan, 1974.** ***Note:*** **Vacuum flask is safer.**

available. The conditioned sludge is poured into the funnel, and the vacuum is turned on. Filtrate volume is periodically recorded. Specific resistance values are calculated as previously described. Preliminary information concerning required filter cycle time can be obtained by noting the time to the first vacuum break. A ring then can be used to seal the edge of the cake to prevent air leakage, and additional time provided to simulate a drying period to a second vacuum break through the cake.

Although specific resistance is of prime use in comparing the filtration characteristics of various sludges and in determining optimum chemical conditioning requirements, preliminary sizing of the vacuum filters also can be accomplished using this parameter, as defined in Eq. 4-4. These relationships apply:

$$\text{Filter yield} = \left[\frac{2\,P_v wk}{\mu r t_c}\right]^{1/2} \qquad \text{(Eq. 4-5)}$$

where

Filter yield is in kg/m^2-sec (X 740 = lb/ft^2-hr)
k = form time as fraction of rotation cycle (Figure 4-12)
t_c = filter rotation cycle time, sec
Other terms are as in Eq. 4-4.

Cake compressibility is related to pressure according to

$$r = r' P_v^s \qquad \text{(Eq. 4-6)}$$

where

r' is a constant equal to r at $P_v = 1$
s = compressibility coefficient

The slope of a log-log plot of r versus P_v is s, which is a measure of the degree of cake compressibility. For incompressible cakes of discrete particles, $s = 0$, while for wastewater sludges values can range from 0.5 to slightly over 1.0. If significant, a value of P_v^{1-s} is substituted for P_v in Eq. 4-5. The characteristics of a particular sludge can be taken into account by assigning exponential coefficients other than 0.5 to the solids-deposition level and the form time; techniques for evaluating these have been described by Adams and Eckenfelder (Ref. 2-2).

Filter-leaf tests provide a measure of sludge yield which can be more reliably scaled up. Using the apparatus illustrated in Figure 4-13, the filter cycle (Figure 4-12) is simulated by immersing the filter leaf for the desired cake-form time and then maintaining it vertical for the drying time. The collected solids are removed, weighed, dried, and reweighed.

$$\text{Filter yield} = \frac{(\text{wgt. dry solids, g})(1\ \text{lb}/454\ \text{g})}{(t_c)(\text{test filter area, ft}^2)(1\ \text{hr}/3600\ \text{sec})} \qquad \text{(Eq. 4-7)}$$

Cycle time includes form (immersion) time, drying time, and any time a prototype filter is not under vacuum. Filter yield is calculated as lb/ft^2-hr ($\times$ 4.88 = kg/m^2-hr). The test should be repeated several times; intervening washes of the filter medium should be done if the filter is to be washed in actual plant operation.

Typical performance values for vacuum filtration are difficult to cite due to a high dependence on the nature of the sludge; conservative values are 1 and 3 lb/ft^2-hr and 12 and 30% cake solids at 80% recovery respectively for biological and chemical sludges.

4-3.4 Other Dewatering Approaches

Many dewatering systems are available other than the previously discussed continuous solid-bowl centrifuge, basket centrifuge, disk centrifuge, recessed-plate pressure filter, and rotary vacuum filter.

Where space is available, sludge drying beds sometimes can be used to advantage. Sludge is applied to a bed of sand or other granular filter media which overlies a drain system. The liquid filters through the bed into the drainage system and is removed by gravity. Evaporation results in further dewatering. Dried sludge is removed by hand, clam shells, monorail-mounted rakes, or special motorized bed cleaners. Filter beds may be open to the atmosphere or enclosed in structures similar to a greenhouse. Depending upon the climate, sludge may have to remain on the drying beds for up to three months before it becomes sufficiently dry. This approach generally is limited to small sludge loads. Techniques are available for determining drainage rates using a sand support; evaporative losses are estimated at 75% of the loss from a clean water surface, and for open beds rainfall is added assuming 57% absorption by the sludge (Ref. 3-22).

A solids concentrator, termed a Dual-Cell Gravity filter, consisting of an endless cloth passing over two filter cells has been developed for use with well-conditioned sludges. A roller moves the cloth in a horizontal direction, and also pushes the bottom of the cloth upward, thereby forming two pockets. Sludge injected in one pocket is rotated to remove the free water, and the solids move to the second pocket for further dewatering by self-generated pressure.

A promising type of pressure filter other than the recessed-plate arrangement is the continuous, horizontal-belt filter in which conditioned sludge is squeezed between two horizontally moving belts. A similar continuous, vertical press filter also is available. Performance of these devices is projected to be between that of a rotary drum filter and a batch filter press (Ref. 2-15).

A list of alternative filters available for processing slurries generated in manufacturing operations is as follows (Ref. 4-11, 4-12).

Multicompartment rotary drum vacuum filter
Single-compartment rotary drum vacuum filter
Top-feed drum vacuum filter
Scroll-discharge rotary horizontal vacuum filter
Tilting-pan discharge rotary horizontal vacuum filter
Belt horizontal vacuum filter
Rotary disk vacuum filter
Plate-and-frame filter press
Pressure-leaf filter
Horizontal plate batch filters

Centrifugation alternatives are covered in Figure 4-7.

A novel filtration approach involves the use of a pulverized coal bed as a dewatering medium. The liquid sludge is placed on the bed and allowed to dewater. The top layer of the bed, consisting of a mixture of dewatered sludge and coal, then is removed and incinerated without supplemental fuel. In pilot studies, an industrial waste activated sludge of 1.8% solids reportedly was applied to the bed at 470 liters/m^2-hr with a cationic coagulant dosage on a dry solids weight basis of 0.08%. The solids content of the dewatered sludge was 36%; and 1.0 kg of sludge was treated per kg of coal used (Ref. 4-13).

Another unique approach is an extractive treatment system in which the sludge first is caustic pretreated, stabilized at 50°F, and mixed with a basic solvent. After chilling back to 50°F to remove the heat of solution, the now miscible mixture is fed to a solid bowl centrifuge or to a vacuum filter to separate the solvent-water mixture from the now easily separated solids. The cake is sent to a dryer and the liquid fraction to a decanter. The solvent emitted from the dryer is condensed and returned to the system. The solvent in the liquid fraction, after being heated, becomes immiscible, is separated, and is recycled. The water layer is sent to a distillation column for final cleanup. A

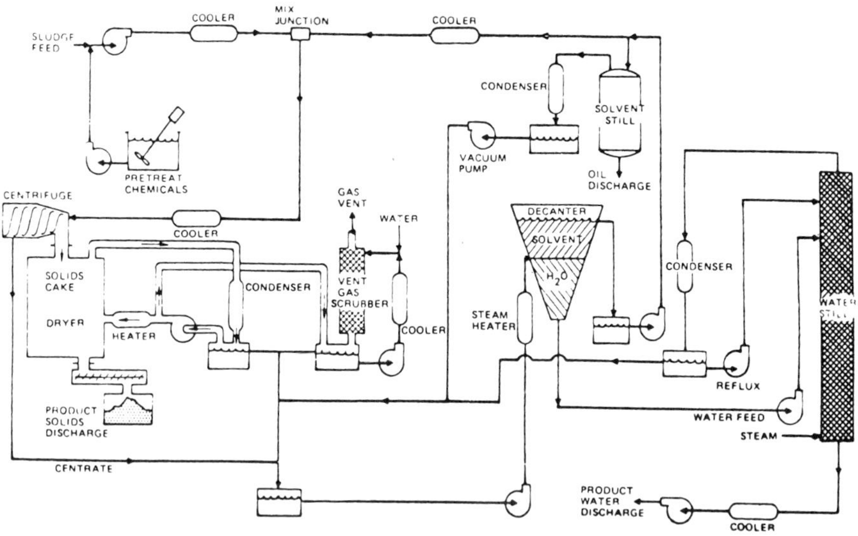

Fig. 4-14. B.E.S.T. solvent extraction schematic, *Source:* Ames, R. K., Van Valkenburg, C. E., and Olson, R. S., "Sludge Dewatering Results with Basic Extractive Sludge Treatment," 30th Purdue University Industrial Waste Conference, Lafayette, Indiana, 1975. *Proc. in Press* (*Courtesy:* Purdue University).

dry solids content of over 95% has been obtained along with clear product water with relatively low residual levels (Figure 4-14) (Ref. 4-14).

REFERENCES

4-1. Vesilind, P. A., *Treatment and Disposal of Wastewater Sludges,* Ann Arbor Science Publishers, Inc., Ann Arbor Michigan, 1974.

4-2. *Process Design Manual for Sludge Treatment and Disposal,* EPA Technology Transfer Publication 625/1-74-006, U.S. Environmental Protection Agency, Washington, D.C., 1974.

4-3. Kumke, G. W., Hall, J. F., and Oeben, R. W., "Conversion to Activated Sludge at Union Carbide's Institute Plant," *J. Water Pollut. Control Fed.,* **40,** 8, pp. 1408-1422, 1968.

4-4. Parker, D. G., Randall, C. W., and King, P. H., "Biological Conditioning for Improved Sludge Filterability," *J. Water Pollut. Control Fed.,* **44,** 11, pp. 2066-2077, 1972.

4-5. Smith, J. E., Jr., Hathaway, S. W., Farrell, J. B., and Dean, R. B., "Sludge

Conditioning with Incinerator Ash," 27th Purdue University Industrial Waste Conference, Lafayette, Indiana, 1972.

4-6. Vesilind, P. A., "Estimation of Sludge Centrifuge Performance," *J. Environ. Eng. Div., Proc. Am. Soc. Civil Eng.,* **96**, SA3, pp. 805-818, 1970.

4-7. Romagnoli, R. J., "Dewatering Characteristics of Sludge from Industrial Oxygen Activated Sludge Systems," 29th Annual Purdue Industrial Waste Conference, Lafayette, Indiana, 1974.

4-8. Stumpf, M. R. and Harper, W. H., "Continuous Feed Centrifuge Replaces Flotation for Removal of Excess Activated Sludge from a Pharmaceutical Wastewater Treatment Plant," Abbott Laboratories, North Chicago, Illinois.

4-9. Jones, B. R. S., "Vacuum Sludge Filtration: II Prediction of Filter Performance," *J. Water Pollut. Control Fed.,* **28**, 9, pp. 1103-1115, 1956.

4-10. "Triton WRC Standard Shear Test Stirrer/Timer, Type 133/131," Triton Electronics Ltd., Bigods Hall, Dunmow, Essex, England.

4-11. "Special Report on Solid-Liquid Separation," *Chem. Eng.* pp. 141-203, June 20, 1966.

4-12. "Solids Separations," *Chem. Eng.,* **48**, 4, February 15, 1971.

4-13. Hirota, M., Okadr, H., Misaka, Y., and Kato, K., "Dewatering of Organic Sludge by Pulverized Coal," *J. Water Pollut. Control Fed.,* **47**, 12, pp. 2774-2782, 1975.

4-14. Ames, R. K., Van Valkenburg, C. E., and Olson, R. S., "Sludge Dewatering Results with Basic Extractive Sludge Treatment," 30th Purdue University Industrial Waste Conference, Lafayette, Indiana, 1975.

4-15. Conway, R. A., Nelson, R. F. and Young, B. A., "High Solubility Gas Flotation", 52nd Water Pollution Control Fed. Conference, Houston, Texas, October, 1979.

4-16. Eikum, A. S. and Paulsrud, B., "Methods for Measuring the Degree of Stability of Aerobic Stabilized Sludges", *Water Research,* **11**, 9, 773 (1977).

APPENDIX 4-1

DEWATERING TERMS AND NOTATION

A	= filtration area, m^2 ($cm^2 \times 10^{-4}$)
b	= slope of plot of time/filtrate volume versus filtrate volume, sec/m^6 sec/m^6
CA	= centrifugal acceleration, expressed as number of times greater than the acceleration of gravity
C_f	= sludge feed solids concentration, %
C_k	= sludge cake solids concentration, %
d	= distance between pressure filter plates, in. (cm $\times$ 0.394)
g	= acceleration of gravity, 384 in./sec^2 (975 cm/sec^2)
μ (mu)	= viscosity of filtrate, poise or Ns/m^2 as specified
ω (omega)	= radial velocity, radians/sec
P_v	= filtrate pressure, psig ($N/m^2 \times 0.00014$)
R	= $r \times 10^{-10}$
r	= specific resistance, m/kg
r'	= constant equal to r at $P_v = 1$
r_1	= radial distance from the centrifuge centerline to the top of the sludge, in. (cm $\times$ 0.394)
r_2	= radial distance from the centrifuge centerline to the bottom of the sludge, in. (cm $\times$ 0.394)
s	= filter cake compressibility coefficient
t_c	= filter rotation cycle time, sec
w	= weight of dry cake solids per unit of filtrate, kg/m^3

5

Activated carbon adsorption

5-1 GENERAL CONSIDERATIONS AND APPLICATIONS

In the adsorptive processes, the contaminants in the wastewater accumulate at the surface of an adsorbent due either to physical or chemical forces. Molecules in the interior of the adsorbent particle are subject to equal forces from all directions, while unbalanced forces acting on molecules at the surface can provide a resultant physical force for attraction. These surface molecules also can have chemically reactive groups. Another major influence is the tendency of hydrophobic materials to congregate at interfaces to lower the interfacial tension. An adsorption equilibrium is established when the concentration of contaminants in bulk solution is in dynamic balance with that at the interface, i.e., an equal number of molecules desorb as adsorb in a given time. Weber in his text on physical-chemical treatment has described the basic nature of the adsorption process in detail (Ref. 5-1).

Activated carbon serves as an excellent adsorbent because of its large interfacial area, due to pores or channels throughout the particles. The total interfacial area, pore size, and surface functional groups depend upon the source of carbon and its mode of activation or reactivation after use. Sources of carbon include coal, bone, wood, and petroleum. Activation involves heating at 800 to 1000°C with steam to remove tars, to enlarge pores, and to partially oxidize the surface molecules.

Some typical applications for activated carbon adsorption in liquid wastewater treatment include:

1. Removal of specific contaminants which are resistant to biodegradation, are toxic, are readily adsorbable, and can be removed from the carbon during regeneration, e.g. phenol and color bodies.
2. Advanced treatment of wastewaters for improved organic removal over that biologically attainable.
3. Recovery of process streams for reuse, e.g., brines.

In addition to the above capabilities, desirable features of the adsorptive processes are compactness and less inherent variability as compared to biological systems. Problems include high costs and the inability to adsorb small polar molecules at reasonable capacities. In any event, as effluent and stream quality criteria have become increasingly stringent, removal of residual organic material by activated carbon is the method beyond biological treatment that has received the most attention.

Carbon treatment can be applied in a typical scheme for treating industrial wastes at two positions. A tertiary treatment step for removing refractory organics following conventional secondary (biological) treatment usually is visualized. However, due to the high impurity concentrations encountered in various individual waste streams contributing to the overall discharge, some of the streams might be treated more effectively at their sources rather than at the terminal facility. This latter approach could be beneficial particularly in handling process unit wastes which produce shock loads to a treatment plant, produce materials inhibitory to biological activity, or produce economically recoverable materials.

The phenomenon of activated carbon adsorption in dilute aqueous systems has been studied rather diligently only in the past few years. Most of the extensive work has been with municipal wastewater systems where the concentration of the impurity is often one to two orders of magnitude less than that to be considered with commonly encountered petrochemical and other industrial wastes. The size and functionality of the organic substances involved are also markedly different between municipal and industrial wastes.

5-2 ADSORPTION OF SPECIFIC CONTAMINANTS

5-2.1 General Studies

The available literature concerning the adsorptive capacities of various carbons for specific compounds is limited. Of importance are the effects of contaminant functionality, chain length, molecular weight, branching, solubility, polarity, as well as the solution pH and carbon surface chemistry. Pertinent literature was compiled by Giusti, Conway, and Lawson, and studies were conducted to fill some of the data gaps as related to synthetic organic contaminants (Ref. 5-2).

Morris and Weber (Ref. 5-3, 5-4) have reported considerable data on the adsorption of phenol, sodium salts of sulfonated organics, and pesticides. Increasing molecular weight was shown to have a favorable effect on carbon capacity while branching had an adverse effect. Also total carbon removal for a mixture was enhanced as compared to that expected from single-solute data. Ward and Getzen (Ref. 5-5) showed that decreasing the pH increased the adsorption of aromatic acids, not only because of an increase in the molecular-ionic ratio but also because of an enhanced specific ion adsorption resulting from an alteration of the carbon surface properties. In their Freundlich isotherm studies of the relationship between adsorption of phenolics and surface chemistry, Snoeyink, Weber, and Mark (Ref. 5-6) found that phenol was adsorbed more extensively by a coal-base carbon than by a coconut-shell carbon.

Data verifying Traube's rule (adsorption increases as a homologous series is ascended) for a polar solvent and fatty acids were reported by Adamson (Ref. 5-7). Kipling (Ref. 5-8) reported pH effects on adsorption of acetic acid and *n*-butylamine and concluded only the undissociated species could be adsorbed on activated carbon. Removals were never greater than 20% for acetic acid and 25% for *n*-butylamine. Conversely, Ward and Getzen (Ref. 5-9) examined pH effects on the adsorption of substituted benzoic acids and found that adsorption was greater than that expected from the ionic-molecular ratio. Weber and Gould (Ref. 5-10) and DiGiano and Weber (Ref. 5-11) investigated organic pesticides and related nitrophenols. The most comprehensive study reported on any group of compounds is the work of Cheldelin and Williams (Ref. 5-12) who obtained Freundlich isotherms for thirty-three amino acids, vitamins, and related substances; important factors noted were the presence or absence of aromatic nuclei and the position of polar groups.

More data have been provided on the adsorption of fatty acids by Landt and Knop (Ref. 5-13) who investigated ash-free sugar charcoal with relatively low concentrations of acid. Coughlin and Ezra (Ref. 5-14) and Coughlin, *et al.* (Ref. 5-15) reported that with increased carbon surface oxides, the adsorption of phenol and nitrobenzene was reduced. Mattson, *et al.* (Ref. 5-16) reported similar data for the uptake of phenol, nitrobenzene, and nitrophenols. Surface chemistry studies are reported by Snoeyink and Weber (Ref. 5-17).

Work with actual petroleum and petrochemical wastes has been reported in several sources. Fox, *et al.* (Ref. 5-18) report activated carbon studies on a waste brine stream containing phenol and acetic acid. Carbon capacities in a continuous system were found to be 0.17 gram phenol per gram carbon and 0.04 to 0.06 gram acetic acid per gram carbon; the carbon was regenerated with sodium hydroxide. Huang and Hardie (Ref. 5-19) employed activated carbon in a physical-chemical treatment process on a refinery waste stream; removals of the chemical oxygen demand (COD) and the total organic carbon (TOC) by activated carbon from a waste which had undergone chemical coagulation were approximately 80 and 90%, respectively. Paulson (Ref. 5-20) reported COD, TOC and biochemical oxygen demand (BOD) removals of from 65 to 100% from primary and secondary effluents of 10 petroleum refineries, while Van Stone (Ref. 5-21, 5-22) presented data for activated carbon treatment of various industrial wastewaters. The average removal of COD from one waste stream containing phenol, benzene, and organic amino compounds was approximately 60%. Normally encountered carbon loadings of 0.5, 0.6, 0.3, and 0.15 gram per gram of carbon were presented for BOD_5, COD, TOC, and phenol, respectively. Kipling (Ref. 5-23) and Mattson and Mark (Ref. 5-24) have compiled a considerable amount of data on the general process of adsorption.

5-2.2 Relative Data

In the study by Giusti, Conway, and Lawson (Ref. 5-2), 93 compounds from 10 functional groups found in chemical plant wastes were screened in single dosage tests to evaluate their relative amenability to activated carbon adsorption. In order to relate removals for compounds within specific functional groups, carbon "loadings" for each compound studied were determined on the basis of grams of sorbate removed per gram of carbon. The carbon used had a wide array of pore sizes, but an exceptionally large number of large pores, so it could effectively adsorb a wide variety of molecular weights with extra efficiency in the high range (Ref. 5-25). Table 5-1 lists the gram per gram loadings as well as percent reductions of sorbate.

A plot of adsorptive capacity versus carbon number for six of the functional groups studied is shown in Figure 5-1. For compounds with less than four carbons in the chain (including functional group carbons) the amenability of adsorption can be summarized as follows: undissociated organic acids > aldehydes > esters > ketones > alcohols > glycols. For compounds with more than four carbons, the alcohols moved ahead of the esters. The shift in degree

TABLE 5-1. Relative Amenability of Typical Organic Compounds to Activated Carbon Adsorption

Compound	Molecular Weight	Aqueous Solubility (%)	Adsorbability* g compound/ g carbon	Adsorbability* Percent Reduction
Alcohols				
Methanol	32.0	∞	0.007	3.6
Ethanol	46.1	∞	0.020	10.0
Propanol	60.1	∞	0.038	18.9
Butanol	74.1	7.7	0.107	53.4
n-Amyl alcohol	88.2	1.7	0.155	71.8
n-Hexanol	102.2	0.58	0.191	95.5
Isopropanol	60.1	∞	0.025	12.6
Allyl alcohol	58.1	∞	0.024	21.9
Isobutanol	74.1	8.5	0.084	41.9
t-Butanol	74.1	∞	0.059	29.5
2-Ethyl butanol	102.2	0.43	0.170	85.5
2-Ethyl hexanol	130.2	0.07	0.138	98.5
Aldehydes				
Formaldehyde	30.0	∞	0.018	9.2
Acetaldehyde	44.1	∞	0.022	11.9
Propionaldehyde	58.1	22	0.057	27.7
Butyraldehyde	72.1	7.1	0.106	52.8
Acrolein	56.1	20.6	0.061	30.6
Crotonaldehyde	70.1	15.5	0.092	45.6
Benzaldehyde	106.1	0.33	0.188	94.0
Paraldehyde	132.2	10.5	0.148	73.9
Amines				
Di-N-Propylamine	101.2	∞	0.174	80.2
Butylamine	73.1	∞	0.103	52.0
Di-N-Butylamine	129.3	∞	0.174	87.0
Allylamine	57.1	∞	0.063	31.4
Ethylenediamine	60.1	∞	0.021	10.7
Diethylenetriamine	103.2	∞	0.062	29.4
Monethanolamine	61.1	∞	0.015	7.2
Diethanolamine	105.1	95.4	0.057	27.5
Triethanolamine	149.1	∞	0.067	33.0
Monoisopropanolamine	75.1	∞	0.040	20.0
Diisopropanolamine	133.2	87	0.091	45.7
Pyridines & Morpholines				
Pyridine	79.1	∞	0.095	47.3
2-Methyl 5-Ethyl pyridine	121.2	sl. sol.	0.179	89.3
N-Methyl morpholine	101.2	∞	0.085	42.5
N-Ethyl morpholine	115.2	∞	0.107	53.3

TABLE 5-1. (Continued)

Compound	Molecular Weight	Aqueous Solubility (%)	Adsorbability* g compound/ g carbon	Adsorbability* Percent Reduction
Aromatics				
Benzene	78.1	0.07	0.080	95.0
Toluene	92.1	0.047	0.050	79.2
Ethyl benzene	106.2	0.02	0.019	84.3
Phenol	94	6.7	0.161	80.6
Hydroquinone	110.1	6.0	0.167	83.3
Aniline	93.1	3.4	0.150	74.9
Styrene	104.2	0.03	0.028	88.8
Nitrobenzene	123.1	0.19	0.196	95.6
Esters				
Methyl acetate	74.1	31.9	0.054	26.2
Ethyl acetate	88.1	8.7	0.100	50.5
Propyl acetate	102.1	2	0.149	75.2
Butyl acetate	116.2	0.68	0.169	84.6
Primary amyl acetate	130.2	0.2	0.175	88.0
Isopropyl acetate	102.1	2.9	0.137	68.1
Isobutyl acetate	116.2	0.63	0.164	82.0
Vinyl acetate	86.1	2.8	0.129	64.3
Ethylene glycol monoethyl ether acetate	132.2	22.9	0.132	65.8
Ethyl acrylate	100.1	2.0	0.157	77.7
Butyl acrylate	128.2	0.2	0.193	95.9
Ethers				
Isopropyl ether	102.2	1.2	0.162	80.0
Butyl ether	130.2	0.03	0.039	100.0
Dichloroisopropyl ether	171.1	0.17	0.200	100.0
Glycols				
Ethlyene glycol	62.1	∞	0.0136	6.8
Diethylene glycol	106.1	∞	0.053	26.2
Triethylene glycol	150.2	∞	0.105	52.3
Tetraethylene glycol	194.2	∞	0.116	58.1
Propylene glycol	76.1	∞	0.024	11.6
Dipropylene glycol	134.2	∞	0.033	16.5
Hexylene glycol	118.2	∞	0.122	61.4

(continued)

TABLE 5-1. (Continued)

Compound	Molecular Weight	Aqueous Solubility (%)	Adsorbability* g compound/ g carbon	Adsorbability* Percent Reduction
Glycol Ethers				
Ethylene glycol monomethyl ether	76.1	∞	0.028	13.5
Ethylene glycol monoethyl ether	90.1	∞	0.063	31.0
Ethylene glycol monobutyl ether	118.2	∞	0.112	55.9
Ethylene glycol monohexyl ether	146.2	0.99	0.170	87.1
Diethylene glycol monoethyl ether	134.2	∞	0.087	43.6
Diethylene glycol monobutyl ether	162.2	∞	0.166	82.7
Ethoxytriglycol	178.2	∞	0.139	69.7
Halogenated				
Ethylene dichloride	99.0	0.81	0.163	81.1
Propylene dichloride	113.0	0.30	0.183	92.9
Ketones				
Acetone	58.1	∞	0.043	21.8
Methyl ethyl ketone	72.1	26.8	0.094	46.8
Methyl propyl ketone	86.1	4.3	0.139	69.5
Methyl butyl ketone	100.2	v. sl. sol.	0.159	80.7
Methyl isobutyl ketone	100.2	1.9	0.169	84.8
Methyl isoamyl ketone	114.2	0.54	0.169	85.2
Diisobutyl ketone	142.2	0.05	0.060	100.0
Cyclohexanone	98.2	2.5	0.134	66.8
Acetophenone	120.1	0.55	0.194	97.2
Isophorone	138.2	1.2	0.193	96.6
Organic Acids				
Formic acid	46.0	∞	0.047	23.5
Acetic acid	60.1	∞	0.048	24.0
Propionic acid	74.1	∞	0.065	32.6
Butyric acid	88.1	∞	0.119	59.5
Valeric acid	102.1	2.4	0.159	79.7
Caproic acid	116.2	1.1	0.194	97.0
Acrylic acid	72.1	∞	0.129	64.5
Benzoic acid	122.1	0.29	0.183	91.1

TABLE 5-1. (Continued)

Compound	Molecular Weight	Aqueous Solubility (%)	Adsorbability* g compound/ g carbon	Adsorbability* Percent Reduction
Oxides				
Propylene oxide	58.1	40.5	0.052	26.1
Styrene oxide	120.2	0.3	0.190	95.3

*Dosage of carbon was 5 g per liter of solution. Initial concentration of compounds was 1000 mg/1 (±3%), except 2-ethyl hexanol (700 mg/1), benzene (416 mg/1), toluene (317 mg/1), ethyl benzene (115 mg/1), styrene (180 mg/1), butyl ether (197 mg/1), and diisobutyl ketone (300 mg/1).

Source: Giusti, D. M., Conway, R. A., and Lawson, C. T., "Activated Carbon Adsorption of Petrochemicals," *J. Water Pollut. Control Fed.*, **46**, 5, pp. 947–65, 1974.

of adsorption above and below four carbons (except for organic acids) is apparently due to a shift in the solubility trends of the compounds. As the solubility decreases, there is a corresponding increase in amenability to adsorption; and as the solubility of one compound in a specific series falls below that of another compound in another series, at a comparable carbon number, the trend is for the first to surpass the second in amenability. Another useful interpretation of the data from Table 5-1 can be gained by relating percent removal and molecular weight as done by Lawson and Hovious in Figure 5-2 (Ref. 5-26).

In interpreting these data it must be remembered that carbon loadings are based on initial concentrations, not equilibrium concentrations, of 1000 mg/liter. Also, results with compounds having solubilities dictating concentrations of less than 1000 mg/liter cannot be related directly to the compounds having solubilities which permitted use of this concentration. A discussion of the results obtained for each functional group follows.

Alcohols. The high polarity and subsequent high solubility, due to the ease with which the hydroxyl radical hydrogen bonds in aqueous solution, best explains the relatively low amenability of alcohols to activated carbon adsorption. However, as molecular weight increases, a corresponding increase in amenability is noted due to the decrease in polarity or increase in hydrophobicity (Figure 5-3). The lower expected capacities for branched-chain alcohols compared to straight chain were borne out in the results. A carbon with a highly polar surface can be used for adsorbing polar molecules; but in the case of aqueous solutions, competition for adsorption sites often limits capacity.

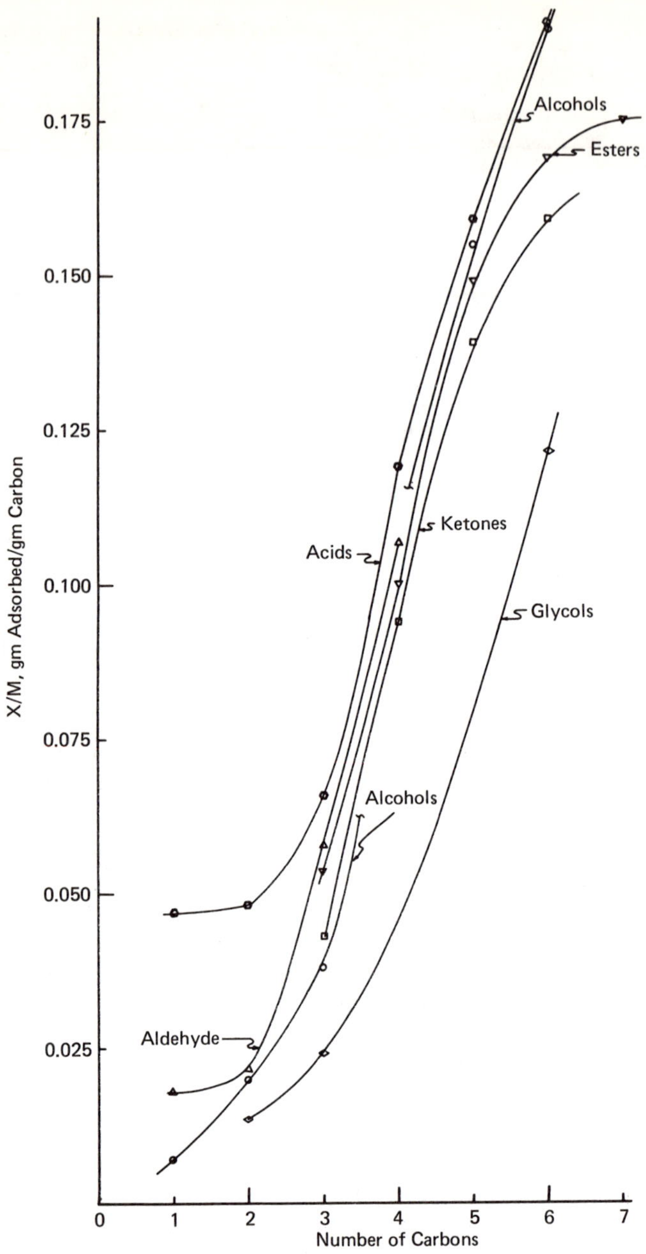

Fig. 5-1. Effect of carbon chain length on relative adsorption of selected organics. *Source:* Guisti, D. M., Conway and Lawson, C. T., "Activated Carbon Adsorption of Petrochemicals," *J. Water Pollut. Control Fed.*, 46, pp. 947-65, 1974.

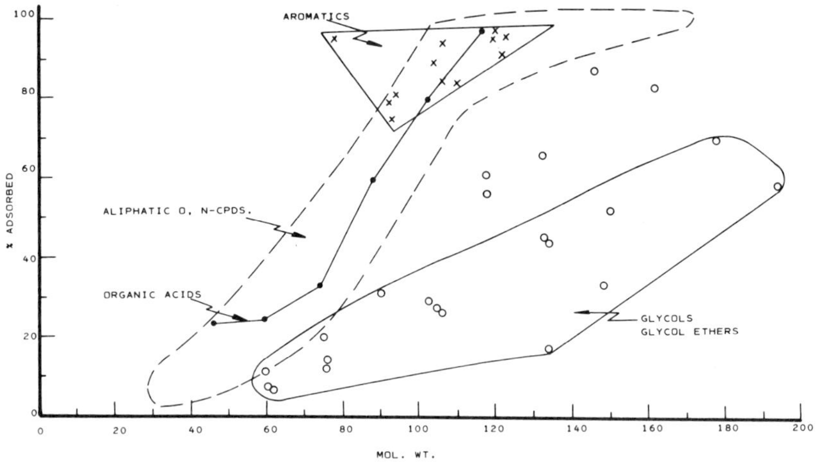

Fig. 5-2. Relationship of compound molecular weight and extent of adsorption. ***Source:*** **Lawson, C. T., and Hovious, J. C., "Caution and Limitations on the Applications of Activated Carbon Adsorption to Organic Chemical Wastewaters," Open Forum on Management of Petroleum Refinery Wastewaters, EPA/API/NPRA/U. of Tulsa, Tulsa, Oklahoma, January 26–29, 1976.**

Aldehydes, Ketones, and Amines. The aldehydes, like the alcohols, are relatively highly polar compounds; however, the polarity of the carbonyl group is less significant at higher molecular weights. The corresponding increase in amenability to adsorption of aldehydes due to increasing molecular weight follows the same trend as the alcohols. The most obvious effect on amenability to adsorption of ketones is the solubility of the compound. The solubility-limiting effect for ketones is as shown for the esters in Figure 5-4. Polarity and solubility are high for the amines, thus, relatively low amounts of these compounds are adsorbed in spite of the fact that the degrees of dissociation due to basicity are relatively low. The fraction undissociated ranged from 0.748 (di-*n*-propylamine) to 0.989 (triethanolamine).

Aromatics, Pyridines, and Morpholines. These three classes are grouped together in this discussion because of their ring structures. The low polarity and subsequent low solubility of most of the aromatics partially explains their ability to be easily removed from aqueous solution by carbon. However, a force aside from the conventional van der Waal forces to which physical adsorption is attributed comes into play. These aromatics are capable of π-bonding with the surface of the activated carbon, which has been found to be aromatic in

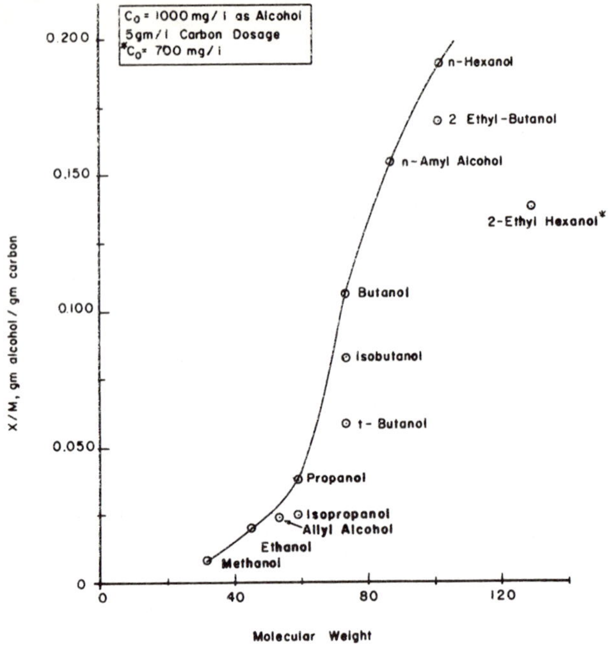

Fig. 5-3. Effect of molecular weight on amenability of adsorption of alcohols. ***Source*****: Giusti, D. M., Conway, R. A., and Lawson, C. T., "Activated Carbon Adsorption of Petrochemicals,"** ***J. Water Pollut. Control Fed.,*** **46, 5, pp. 947-65, 1974.**

nature (Ref. 5-17). The effect is similar for the pyridines and morpholines. The graphical correlation in Figure 5-2 applies best because the initial concentration of most of the aromatics differed due to solubility limitations.

Esters. The solubility-amenability correlation is again obvious for this group of compounds. The rapid change in slope in the curve for the straight chain esters shown in Figure 5-4 indicates that, at primary amyl acetate, solubility limitation is being approached. That is, if a higher molecular weight straight chain acetate were chosen for study, its solubility would limit its concentration to below the 1000 mg/liter used for the acetates indicated on the graph. Only two acrylate esters were tested, so conclusive trends could not be drawn from the data; the apparent high removals were most likely due to the relatively low solubilities of these compounds.

Glycols and Glycol Ethers, Ethers, Halogenated Compounds, and Oxides. Gly-

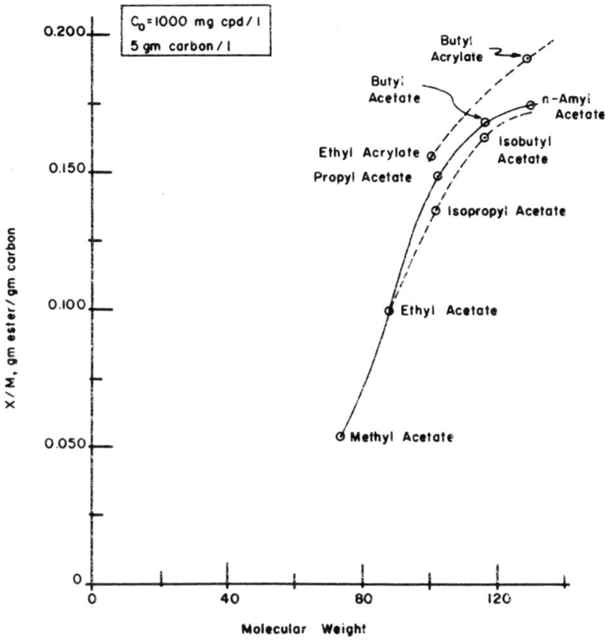

Fig. 5-4. Effect of molecular weight on amenability of adsorption of esters. *Source*: Giusti, D. M., Conway, R. A., and Lawson, C. T., "Activated Carbon Adsorption of Petrochemicals," *J. Water Pollut. Control Fed.*, 46, 5, pp. 947-65, 1974.

cols and glycol ethers showed the lowest affinity for the test carbon of all of the groups studied. The multiple sites on the glycols available for hydrogen bonding give them a hydrophillic nature which carbon cannot overcome except in the case of high molecular weight materials. Due to the limited number of other ethers, halogenated compounds, and oxides studied, no trends can be drawn from the data in Table 5-1 except that the lower the solubility in aqueous solution, the higher the degree of adsorption.

Organic Acids. This class of compounds manifested itself as being the easiest of the straight chain aliphatic compounds to be removed from solution when undissociated. For the straight chain acids, all but formic were above 96% undissociated at the test conditions, indicating that even at a lower test pH level, if adsorption were strictly physical, little increase in efficiency would be attained. It is very conceivable, however, that with the presence of both a carbonyl and a hydroxyl group, adsorption may be chemical as well as physical. Figure 5-5 relates adsorbability to increasing molecular weight.

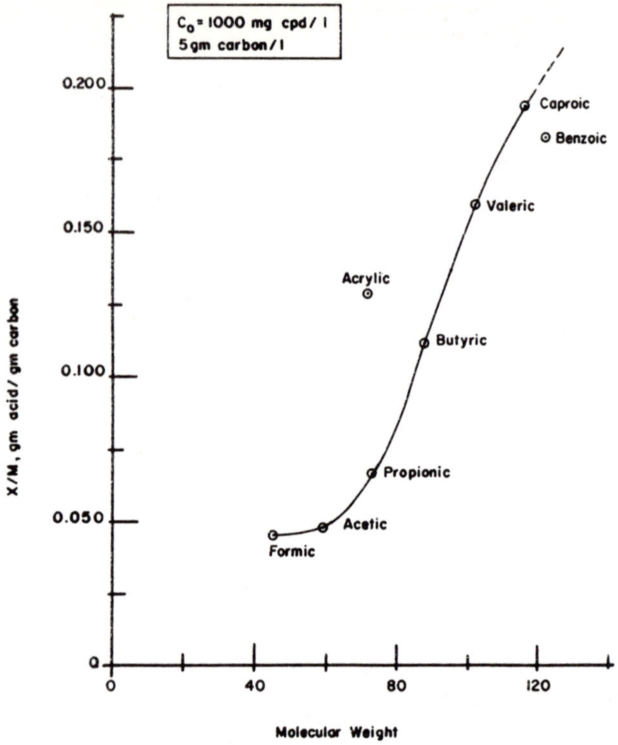

Fig. 5-5. Effect of molecular weight on amenability of adsorption of organic acids. ***Source*****: Giusti, D. M., Conway, R. A., and Lawson, C. T., "Activated Carbon Adsorption of Petrochemicals,"** ***J. Water Pollut. Control Fed.,*** **46, 5, pp. 947-65, 1974.**

5-3 ADSORBENT SELECTION AND ADSORPTION CONDITIONS

Although the selection of a particular commercial activated carbon is tied closely with the nature of the specific problems, the general technical considerations listed in Table 5-2 need to be kept in mind along with costs. Most of the items listed other than surface functionality require little discussion.

The experimental work with various types of carbons using typical petrochemicals under a range of environmental conditions by Giusti, Conway, and Lawson included some elucidation as to mechanisms of activated carbon adsorption and the effects on these mechanisms of easily controlled factors—specifically, pH and surface oxygen (Ref. 5-2).

TABLE 5-2. Activated Carbon Selection for Wastewater Treatment

Specific Area
- Measures area available for adsorption
- 800 to 1200 m^2/g and pore volume of 0.9 cc/g are typical

Pore Size Distribution
- Governs penetration of contaminants of various molecular sizes into the pores, which affects amount of usable area
- One measure is iodine number (pores 10 to 28 Å) versus molassas number (pores >28 Å) *Note:* 1 Angstrom $\simeq 10^{-8}$ cm)
- A broad pore-size range with considerable large-pore area (20 to 50 Å) is typical for most water treatment carbons when a variety of organics are to be adsorbed.

Particle Size Distribution
- Affects rate of adsorption, easier to reach internal surfaces with smaller particles
- Tailor to contactor type, i.e., affects pressure drop in downflow granular carbon columns and degree of expansion (and retention) in upflow columns; powdered carbon needed for stirred reactors where high rates are needed
- 14 × 40 mesh is typical of granular carbons for wastewater treatment (upflow expanded beds)
- Select smallest particles compatible with contact system

Physical Properties
- Hardness or abrasion number (resistance to attrition during handling); apparent density (important in backwashing), moisture content, ash content, wettableness are of interest

Surface Properties
- The nature and level of surface functional groups (oxides) influence the capacity and specificity of adsorption

Ability to be Reactivated
- Attrition losses due to volatilization and grinding often govern process economics (3 to 10% per pass)
- Loss of adsorptive capacity (e.g., iodine number) also is a key factor

Note: Standard methods to determine many of the above properties are presented in Ref. 5-27.

In order to obtain initial, screening-type data on the performance of a given carbon, an equilibrium or isotherm test usually is run. Fornwalt and Hutchins (Ref. 5-28) have outlined the procedure for obtaining a Freundlich isotherm in detail. The equation for this isotherm is:

$$X/M = K(C_e)^{1/n} \qquad \text{(Eq. 5-1)}$$

where X/M is the loading of impurity on a unit weight of carbon; C_e is the equilibrium impurity concentration remaining in solution; K is an empirical constant equal to the intercept of a log-log plot of X/M versus C_e (Figure 5-6) at $C_e = 1$, and $1/n$ is the slope of this log-log plot. By extrapolating this line to the initial impurity concentration (C_o) in the wastewater being tested, the "ultimate capacity" $(X/M)_{C_o}$ of the carbon at this inlet concentration in a

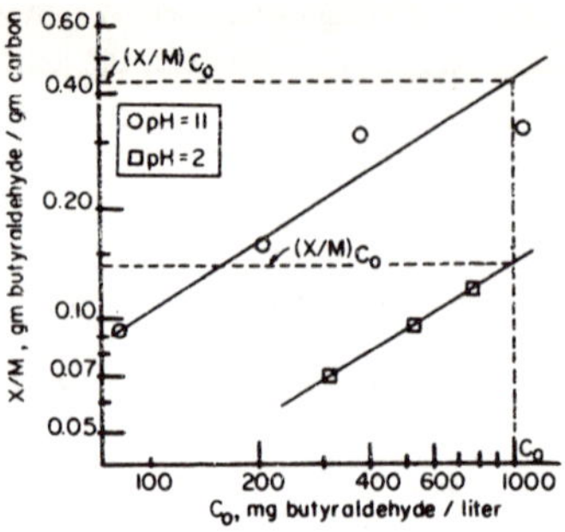

Fig. 5-6. Typical adsorption isotherm. ***Source*****: Giusti, D. M., Conway, R. A., and Lawson, C. T., "Activated Carbon Adsorption of Petrochemicals,"** ***J. Water Pollut. Control Fed.*****, 46, 5, pp. 947-65, 1974.**

continuous system can be found. The lowest level of impurity achievable also is indicated by the isotherm.

Isotherms are obtained using actual or synthesized wastewater samples and the pulverized activated carbon samples of interest. The activated carbon often is soaked in distilled water for 24 hr, dried at 103°C for 24 hr, and ground to -325 mesh in a Waring blender prior to use. The experiment can be carried out by contacting aliquots of the wastewater sample in ground glass stoppered flasks with various dosages of the pulverized activated carbon. Equilibrium is usually achieved by agitating the flasks for 2 hr using a gyratory shaker table or other means. The carbon is removed from solution by vacuum filtration through membrane filter paper, except in tests involving highly volatile sorbates in which case pressure filtration is employed.

Four commercially available water and wastewater treating carbons studied by Giusti, *et al.* (Ref. 5-2) are described in Table 5-3. The performance of these carbons in adsorbing pure components was rated by Freundlich isotherm studies on aqueous solutions of 4-carbon members of several functional groups. Figure 5-6 illustrates an isotherm for the adsorption of a test compound by Carbon A.

Interpretation of the Freundlich isotherm has been discussed by several authors, e.g., Fornwalt and Hutchins (Ref. 5-28). Qualitatively, very steep isotherms, while indicating high ultimate adsorptive capacities, also indicate the possibility that very low effluent concentrations cannot be achieved without massive dosages of carbon. The quantitative interpretation of the isotherm slope as it relates to actual carbon column performance is not fully known, although conceivably it has an effect on the shape of the continuous column breakthrough curve.

The effect of pH is illustrated in Figure 5-7 for several 4-carbon compounds.

TABLE 5-3. Surface Areas and Surface Acidities of Typical Activated Carbons

Carbon[a]	Surface Area[b] (m^2/g)	Surface Groups (me/g)[c] I	II	III	IV
A	1050	0.30	0.00	0.05	0.20
B	1050–1200	0.30	0.00	0.15	0.10
C	1100	0.35	0.00	0.10	0.15
D	1000–1100	0.75	0.15	0.70	0.45

[a]Carbon A–Witco Grade 517 (12 × 30 mesh, petroleum hydrocarbon based); Carbon B–Calgon Filtrasorb 400 (12 × 40 mesh, coal based); Carbon C–Westvaco Nuchar WV-G (12 × 40 mesh, coal based); Carbon D–Nacar G 107 (12 × 30 mesh, coal based).
[b]Data on surface areas were provided in vendor literature.
[c]I–Strongly acidic carboxyl group; II–More weakly acidic carboxyl group; III–Phenolic hydroxyl group; IV–Carbonyl group (as milliequivalents acid per gram of carbon, by volumetric analysis).
Source: Giusti, D. M., Conway, R. A., and Lawson, C. T., "Activated Carbon Adsorption of Petrochemicals," *J. Water Pollut. Control Fed.*, 46, pp. 947–65, 1974.

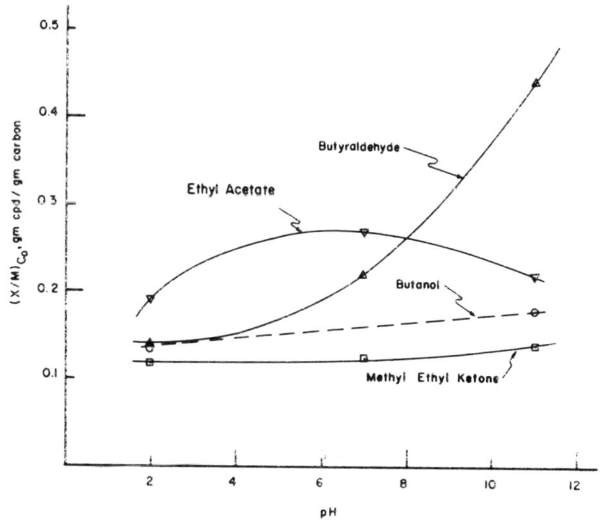

Fig. 5-7. pH Effects on adsorption of typical 4-carbon compounds. *Source*: Giusti, D. M., Conway, R. A., and Lawson, C. T., "Activated Carbon Adsorption of Petrochemicals," *J. Water Pollut. Control Fed.*, 46, 5, pp. 947–65, 1974.

Increasing pH markedly increased the adsorption of butyraldehyde, while the adsorptive capacity for ethyl acetate peaked out at about pH of 7. Butanol and methyl ethyl ketone were more highly adsorbed at high pH than at low pH, but lack of data at neutrality prevented the drawing of a conclusive trend for butanol. Butyric acid isotherms were run only at a pH of 2 where the acid is essentially undissociated.

Although the high surface area of activated carbon long has been recognized as being instrumental in making it a highly effective adsorbent, the surface chemistry of the carbon also affects its adsorptive capabilities. Wolfe (Ref. 5-29) and Heckman (Ref. 5-30) found that in the carbonization process, microcrystallites consisting of fused hexagonal carbon rings or aromatic nuclei were formed similar to the structure of graphite. In the activation process, the oxidizing gases generally attack portions of the structure where there are either vacancies or impurities. According to Snoeyink and Weber, the edges of these microcrystallites are quite often characterized by several types of functional groups as a result of the oxidizing step (Ref. 5-17). These raised regions of the carbon surface are the positions where chemisorption caused by electron-sharing reactions tend to occur (Ref. 5-31).

The oxygen content of various activated carbons ranges from 2 to 25% by weight depending upon the method of activation and degree of chemisorption of oxygen onto the surface of the carbon (Ref. 5-31). Four acidic surface functional groups are formed: a strongly acidic carboxyl group (Group I), a more weakly acidic carboxyl group (Group II), a phenolic hydroxyl group (Group III), and a carbonyl group (Group IV) (Ref. 5-32, 5-33, 5-34).

In an attempt to explain the change in adsorption of the various compounds due to changes in pH, analyses of the acidic surface oxides on the carbons were made. The concentrations of surface oxides measured for the four stock carbons before pretreatment and testing are listed in Table 5-3. Stock carbons were merely crushed to -325 mesh and treated with a base.

The data tended to indicate an inverse relationship between the presence of surface oxides and the ultimate capacity of the carbon (Figure 5-8), perhaps due to a blockage of the micropores. The acidic groups likely are near the edges of the hexagonal layer planes (Ref. 5-31) and conceivably could block the openings of micropores.

Another effect of pH is to control the degree of ionization of the impurities. The adsorption of the unionized fraction is higher. Therefore, for example, lower pH levels would favor adsorption of organic acids. Maximum adsorption was reached at 1 to 1.4 pH units below the pK_a levels, which are 4.7 for acetic acid and 9.9 for phenol (Ref. 5-35). Higher pH levels would favor adsorption of amines. Also, aldehydes and ketones can undergo aldol condensation in the presence of dilute acid or base, which would increase their amenability to adsorption (Ref. 5-2). A summary of factors affecting equilibrium adsorbability is presented in Table 5-4.

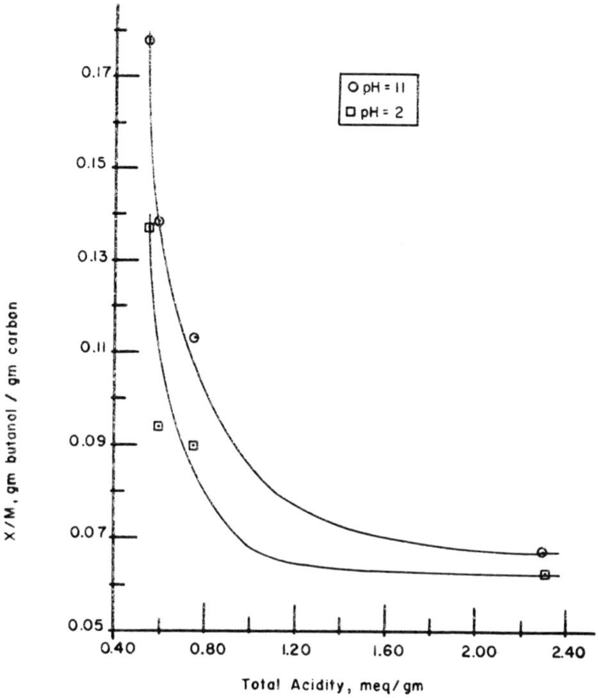

Fig. 5-8. Effect of total surface-oxide acidity on adsorption of butanol. ***Source:*** **Giusti, D. M., Conway, R. A., and Lawson, C. T., "Activated Carbon Adsorption of Petrochemicals,"** ***J. Water Pollut. Control Fed.,*** **46, 5, pp. 947-65, 1974.**

5-4 ADSORPTION SYSTEMS

5-4.1 Design Information

Data from isotherm studies are useful in providing an estimate of the amenability of contaminants to removal by activated carbon under equilibrium conditions, in defining impact of conditions such as pH and component interactions, and in initial selection of the candidate activated carbon(s). Laboratory column studies to provide information on rates and extents of adsorption are the next necessary step in development of a granular-carbon adsorption system. For the design of powdered activated carbon systems the direct use of isotherm data is possible when the capacities at various effluent levels are considered and sufficient time to approach equilibrium is provided.

Since equilibrium is not often achieved in an operating system with granular carbon at the detention times employed, the rate-controlling step from among the following possibilities needs to be considered (Ref. 5-36).

TABLE 5-4. Factors Affecting Equilibrium Adsorbability

Compound Adsorbability Favored by:
- Increasing carbon chain length[a]
- Increasing aromaticity
- Decreasing polarity
- Decreasing branching
- Decreasing solubility
- Decreasing degree of dissociation

Functionality
- Relative adsorbability: acids > aldehydes > esters > ketones > alcohols > glycols when number of carbon atoms is <4

pH Effects
- Undissociated species are more easily adsorbed
  - –low pH favors adsorption of acids (e.g., volatile acids, phenol)[b]
  - –high pH favors adsorption of bases (e.g., amines)
- Other compounds: adsorption can be favored by higher pH
  - Postulated general effect
    - Partial neutralization of surface acidity reduces hydrogen-bonding of surface groups eliminating steric blockage of micropores
  - Aldehydes–condensation products are more strongly adsorbed

Temperature
- Increased temperatures can increase rate of adsorption due to viscosity and diffusivity effects
- Exothermic adsorption reactions are favored by decreasing temperatures, usually a minor effect on equilibrium level

[a]When the rate is controlled by intraparticle transport, decreasing molecular size would result in faster rate all else being equal.
[b]This often is the most significant pH effect, so adsorption generally is increased with decreasing pH.

Transport the sorbate (impurity) through the surface film separating the carbon particle from the turbulent bulk solution
Diffusion of the sorbate within the pores of the sorbent (carbon particle)
Adsorption of the sorbate on the interior surfaces bounding the pore

Maintenance of turbulent conditions aids in speeding the first step. The size and structure of the molecule(s) being adsorbed affects molecular mobility. Smaller pores aid the last step when small molecules are involved. Keeping these mechanisms in mind aids in explaining observed phenomena and in deciding on better alternatives for given situations.

In laboratory column tests, the test unit should be at least 2 ft (61 cm) in height and 1.6 in. (4.1 cm) in diameter for a 8/30 mesh carbon, i.e., carbon that passed through U.S. Standard Mesh Size No. 8 screen and retained on a No. 30 screen. For a 12/40 carbon a 1.0-in. (2.5 cm) column is needed to provide the needed ratio of 25:1 column diameter to particle size to prevent wall effects from significantly affecting results. The length and diameter of column selected should provide a retention time of at least 15 min (25 to 50 min is more typical in full-scale municipal systems) and a minimum application rate of 1 gal./ft^2-min (59 m^3/m^2-day) to simulate large-scale conditions. An application rate of 2-8 gal./ft^2-min would be more typical, but a compromise between desirable application rate and needed detention time often needs to be made. Pilot-scale tests using larger columns are needed for final design of the system.

As described by Joyce (Ref. 5-37), full-scale granular carbon columns contain beds that are 10 to 35 ft (3 to 11 m) in depth and from 1 to 15 ft (0.3 to 4.6 m) in diameter. For municipal systems, surface application rates of 2 to 8 gal./ft^2-min (120 to 470 m^3/m^2-day) and empty-volume retention times of 25 to 50 min are practiced. The lower end of the surface application rate is governed by a decrease in sorption efficiency and the upper by excessive pressure drop in packed beds. If packed beds were used, pressure vessels could be employed to attain the higher surface application rates. Contact times up to 30 min yield proportionate organic removal increase (not linear) at which point this rate increase falls off and becomes negligible at 60 min (Ref. 5-36). Somewhat longer times for industrial application are normally required consistent with higher organic concentrations (Ref. 5-36).

The direct extrapolation of data obtained at a bed depth of 2 ft to predict performance at a 10-ft bed depth cannot be done without considerable risk. Theoretically, it is possible to scale down the depth of the columns and reduce the flow rate to obtain the contact times that would be obtained in deeper beds. There are practical limitations, however, on the minimum volumes of liquid that can be accurately handled and on the balance between surface application rates and contact times. Further, spurious results may be obtained in short columns because of the difficulty in obtaining uniform and reproducible bed packing that would simulate flow patterns in large-scale systems. For these reasons, it is normally advisable to employ bed depths of at least 10 ft (3 m) and preferably deeper if data are to be used in design. Sectioned columns provide a convenient way of obtaining deep beds in the laboratory and are, in fact, preferable since they permit an investigation of the effects of bed depth.

Columns for laboratory or pilot-plant tests can be fabricated from glass or plastic tubing or from metal, plastic, or Pyrex pipe. Connections from the pumps to the columns and between columns can be made with plastic, metal, or rubber tubing. Sampling ports should be provided between columns. Pressure gauges may be installed to follow the pressure-drop characteristics of the system. In order to maintain constant flow rates a small positive displacement pump is desirable. A stainless steel screen at the bottom of the column makes an effective bed support.

The carbon should be charged to the columns in a manner that will avoid air entrapment. This may be accomplished by boiling the carbon in water and charging the hot slurry to the column at a sufficiently slow rate that a layer of water is maintained above the bed. If the system is operated in an upflow direction the carbon may be charged dry, and the waste pumped up through the beds.

After charging is complete, the feed is pumped to the columns at a predetermined flow rate. For the removal of low concentrations of organics from water, starting flow rates as high as 5 gpm/sq ft of cross-sectional bed area (294 m^3/m^2-day) may be used. If necessary, the flow can be reduced until an acceptable effluent is produced. For effluents containing high concentrations of a recoverable material the initial flow rate of 1 gpm/sq ft (59 m^3/m^2-day) or less should be used and subsequently adjusted if necessary.

Once the column system is in operation, effluent samples should be collected from the various bed depths for analysis. Feed samples also should be obtained regularly. In cases where the contaminants in the feed fluctuate in concentration or composition, it is essential to composite the samples by an appropriate method. Information which can be obtained from pilot tests include:

- Pretreatment requirements to reduce suspended solids ($<$ 35 to 50 mg/liter) and oils and grease ($<$10 mg/liter), to adjust pH to optimum level, and to equalize flow and organic concentration
- Performance of various types of carbon under dynamic flow conditions
- Contact time or surface application rate for a given depth (need to develop capacity versus contact time relationship)
- Bed depth
- Carbon dosage requirements
- Adsorption wavefront characteristics (need sample taps at close increments)
- Breakthrough characteristics
- Biological growth effects
- Headloss or bed-expansion characteristics
- Backwash needs
- Regeneration effectiveness (restoration of iodine number, molasses number, COD capacity) and weight loss (2 to 10% even for hard, coal-based carbon; larger regeneration systems are lower in the range due to continuous operation)

Biological growth effects may be either positive in extending bed life or negative in causing odors and plugging problems. Beneficial effects are more likely in up-flow expanded bed operation (Ref. 5-36). Growths may be lessened by pretreatment to lower the BOD and suspended solids content of the influent, by increasing velocity and/or decreasing contact time, or by chlorination. Nitrate addition or influent aeration may be employed to control sulfide generation. Oxygen activated sludge off-gas can be used to advantage to oxygenate carbon columns when enhanced growth is desired, as patented by Conway, Lawson, and Stankewich (Ref. 5-46).

If the concentration of the impurity in the effluent is plotted against throughput volume for any single bed depth employed in a column test, a useful curve will be obtained which will vary in form depending upon the nature of the system under investigation. The point at which the concentration of impurity in the effluent reaches the limit of effluent quality is termed the breakpoint. The part of the curve between the breakpoint and the exhaustion point, where the effluent concentration is the same as the influent, is called the breakthrough curve (Figure 5-9).

During an idealized adsorption cycle in a columnar process, the upper section of the bed will be saturated with impurity, the lower section will be free of impurity, and between these two extremes will be a zone in which the adsorption is actually taking place. The concept of an adsorption mass transfer zone is illustrated in Figure 5-9. At (a) the adsorption zone has just developed. As the column becomes saturated, the adsorption zone moves downward through the

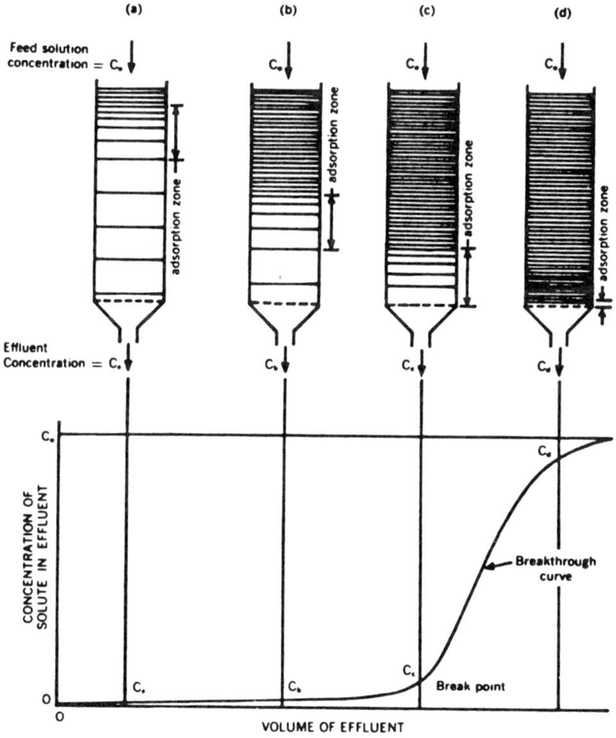

Fig. 5-9. Passage of adsorption zone through a fixed bed. *Source:* Joyce, R. S., "Activated Carbon Adsorption," *Industrial Waste Disposal,* R. D. Ross, ed., Van Nostrand Reinhold, New York, pp. 144-164, 1968.

bed and can be regarded as an adsorption wave front moving through the column. At point (b) approximately half of the bed is saturated, but the effluent concentration is still substantially zero. At (c) in the figure, the adsorption zone has just reached the bottom of the bed, and the breakpoint is attained. The concentration of the impurity in the effluent now rises rapidly as the adsorption zone passes through the bottom of the bed, and at (d) the influent concentration (C_o) almost has been reached.

The shape of the breakthrough curve provides an indication of the relative depth of the adsorption zone. If the breakthrough curve is sharply defined and steep, the zone is short relative to the overall bed depth. If, on the other hand, immediate breakthrough occurs, and considerable time elapses before the effluent concentration approaches the influent value, a relatively deep adsorption zone is indicated. Shallow adsorption zones may be encountered in the carbon treatment of waste effluents containing low concentration of organics, particularly when complex mixtures are not present (Ref. 5-37).

At the exhaustion point, the capacity of the granular carbon in a columnar application mode can be compared to that of the powdered carbon at equilibrium with the influent level, in terms of weight of contaminant adsorbed per unit weight of carbon. These values provide an estimate of the utility of activated carbon adsorption in a given application; treatment of a phenolic waste provides values in the order of 0.3 gram phenol per gram of carbon. Multicomponent columnar studies with butyraldehyde and methyl ethyl ketone indicate adsorptive capacities of 0.13 and 0.19 g/g carbon respectively (Ref. 5-2). These are about 85% of the isotherm equilibrium values. Values in the order of 60 to 80% reportedly are used by consultants in relating isotherm to continuous systems.

The capacity of the granular carbon in the columnar study can be determined by graphically integrating the area above the breakthrough curve and a horizontal line representing the feed level (Figure 5-9). Graphical integration can be accomplished readily by best fitting a rectangle to the initial portion and a triangle to the final portion and summing the area of each in terms of milligrams/liter times liters. The resulting number of milligrams adsorbed, divided by the milligrams of carbon in the column is a measure of the weight capacity of the carbon in terms of milligrams contaminant/milligram carbon. The volumetric capacity of the carbon can be obtained by dividing the gallons (or liters) treated at exhaustion by the weight (pound or gram) of carbon in the test column.

When the breakthrough curve from a test column of suitable length has the classical, ideal shape (e.g., Figure 5-9), it can be used for a preliminary design of a columnar system by constructing a family of breakthrough curves displaced by the horizontal (abscissa) distance to the breakthrough point. Inserting a vertical line down from the exhaustion point shows the required number of columns in series in order to exhaust the first column at the time the last column reaches the breakpoint. Although construction of the graph illustrates the action of each column, the number of columns required can be estimated by directly dividing the volume tested at exhaustion by the volume treated at the breakpoint. At least one additional column is needed to compensate for the column undergoing regeneration.

A first estimate of carbon column sizing can be obtained as per the following example for a hypothetical flow (Q) of 100 gpm, a carbon capacity of 450 liters/kg at exhaustion as determined from a columnar breakthrough test, volume of 400 liters at breakthrough point in test and 1200 liters at exhaustion, and a carbon density of 25 lb/ft^3:

Retention time, $t = 0.5$ hr (based on practice in similar situations)

$$\text{Bed volume rate, } V_r = \frac{1}{t} = \frac{1}{0.5 \text{ hr}} = 2/\text{hr}$$

$$\text{Total bed volume, } V_v = \frac{Q \text{ gal./hr}}{7.48 \text{ gal/cu ft} \times V_r}$$

$$= \frac{100 \text{ gal./min} \times 60 \text{ min/hr}}{7.48 \text{ gal./cu ft} \times 2/\text{hr}} = 401 \text{ cu ft}$$

$$\text{Number series beds} = \frac{\text{volume treated at exhaustion}}{\text{volume treated at breakthrough}} + 1$$

$$= \frac{1200 \text{ liters}}{400 \text{ liters}} + 1 = 4$$

Total carbon weight, $W = 401$ cu ft $\times$ 25 lb/cu ft = 10,025 lb

Carbon volumetric capacity, $C_v = 450$ liters/kg $\times$ 0.454 kg/lb $\times$ 1 gal./3.785 liter = 54 gal./lb

Carbon breakthrough rate (W_r)

$$W_r = \frac{Q \times 60 \text{ min/hr}}{C_v} = \frac{100 \text{ gal./min} \times 60 \text{ min/hr}}{54 \text{ gal./lb}} = 111 \text{ lb/hr}$$

$$\text{Total bed life } (T) = \frac{W}{W_r} = \frac{10{,}025}{111} = 90 \text{ hr}$$

$$\text{Individual bed life} = \frac{90 \text{ hr}}{3 \text{ beds (on line)}} = 30 \text{ hr/bed}$$

Results of laboratory and field tests with small columns only roughly indicate the performance and economics of the adsorptive treatment approach.

On-line large pilot studies are most desirable in obtaining final design information. Data collected at various surface-application rates to columns in series reportedly can provide a firm basis of design (Ref. 5-37). The interrelated influences of the many variables indicates the desirability of using a computer model of the system, such as developed by Weber (Ref. 5-36), to achieve an optimized design. Adams and Eckenfelder have described the design application of a useful model of the system based on the Bohart-Adams relationships (Ref. 2-2).

$$t = \frac{N_o}{C_o v} D - \left[\frac{1}{K_r C_o} \ln \left(\frac{C_o}{C_b} - 1 \right) \right] \qquad \text{(Eq. 5.2)}$$

t = service time, hr
v = hydraulic loading or linear fluid velocity, ft/hr
D = depth of carbon bed, ft
K_r = adsorption rate constant, liters/mg-hr
N_o = adsorptive efficiency, mg/liter
C_o = influent solute (impurity) concentration, mg/liter
C_b = effluent solute concentration, mg/liter

Values of the factors are determined from test runs of columns in series (provides different depths) at various flow rates (Ref. 2-2).

In an alternative approach by Weber, the design of a pulsed bed system is described using a combination of a few basic measurements, a mathematical model, and some typical design parameters proven in practice (Ref. 3-23).

Lawson in more recent studies has found that with petrochemical wastes a direct scale-up from pilot tests based on contact time, with service life corrected by an empirical factor (~0.7), seems to be the best approach for extrapolating typical single-stage performance data to two-stage countercurrent operation, the most likely operating mode for a full-scale tertiary adsorption plant. Three columns in series need to be tested to determine the effects of contact time on service life and organic removal. Scale-up is based on optimum contact time from the three-column test system. Other more complex methods were either conservative or optimistic. (Ref. 5-38).

5-4.2 System Alternatives

Granular carbon is more frequently used than the powdered form, as it is more easily dewatered and regenerated. Alternative columnar contacting systems include (Ref. 5-24):

Series or parallel operation
Downflow or upflow
Pressure or gravity operation (downflow)
Packed or expanded bed operation (upflow)
Pulsed bed operation

Downflow operation results in a contactor serving as a filter as well as an adsorber; however, suspended solids levels above 50 mg/liter are difficult to handle. Backwash capabilities are needed. For upflow packed beds, the influent must be very low in suspended solids. For upflow beds a mode approaching continuous countercurrent operation for most efficient carbon utilization can be achieved by continuously adding new carbon to the top while withdrawing spent carbon from the bottom (Figure 5-10). A number of fixed-beds in series can approximate a countercurrent contacting mode as the lead contactor is removed from service when exhausted, and after being filled with fresh carbon, is placed at the end of the series. Regeneration normally is achieved using a multiple hearth furnace (Figure 5-10).

As reported by Cohen and Westrick, equilibrium with powdered carbon, that passing 300 mesh (85 μm), is achieved in less than 10 min (Ref. 5-39).

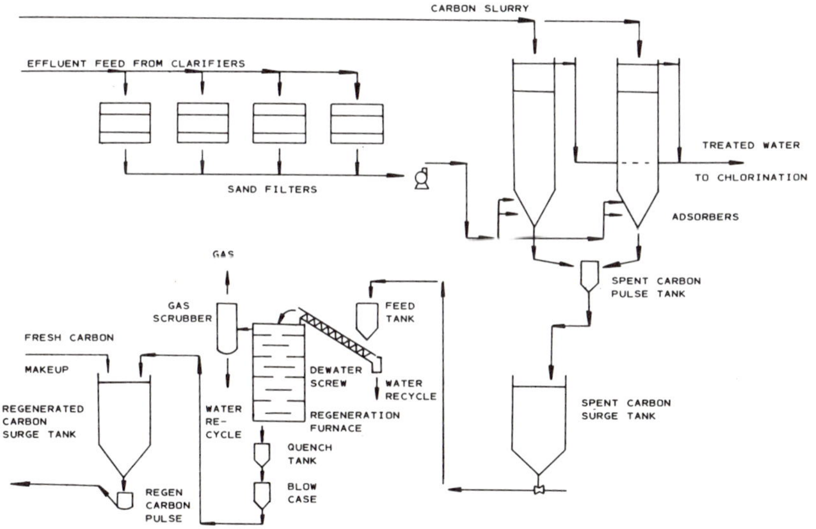

Fig. 5-10. Pulsed, expanded-bed carbon contactor with prefiltration and multiple hearth regeneration. *Source:* Marek, A. C., Jr. and Askins, W., "Advanced Wastewater Treatment for an Organic Chemical Manufacturing Complex," US/USSR Symposium on Physical Chemical Treatment, Cincinnati, Ohio, November 12-14, 1975.

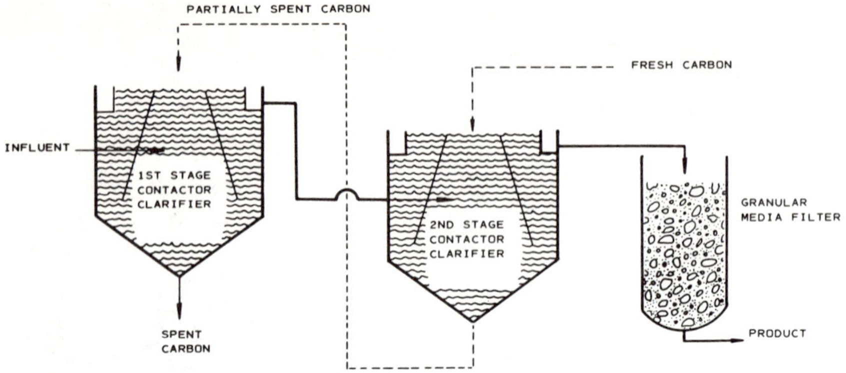

Fig. 5-11. Powdered activated carbon contacting system. ***Source:*** **Peoples, R. F., Krishman, P. and Simonsen, R. N., "Nonbiological Treatment of Refinery Wastewater,"** ***J. Water Pollut. Control Fed.*** **44, 11, pp. 2120-2128, 1972.**

The most common type system consists of staged solids-contact clarifiers, which can be used to mix, flocculate, and settle the powdered carbon, followed by a granular media filter to remove residual carbon (Figure 5-11). Carbon sludge removed from the clarifier is generally gravity thickened, vacuum dewatered to about 25% solids, and regenerated in a fluidized bed reactor, a multiple hearth furnace, a wet-air oxidation system, an upflow refractory-lined reactor, or a downflow spray-tower pyrolysis (oxygen-free) system.

5-4.3 Applications

Activated carbon adsorption in terminal treatment systems applies most appropriately when the levels of contaminants to be treated are low, when equilibrium considerations are such that the level of each contaminant adsorbed is high in relation to that in solution, and the real need for a high level of treatment exists. Some successful applications with petroleum wastes have been reported where soluble constituents are low in concentration and largely aromatic or nonpolar in character (Ref. 5-40). In one case, a 36-min contact time at 18 ft (5.5 m) total depth of four pilot columns, a 3.6 gal./ft^2-min (210 m^3/m^2-day) application rate, resulted in 0.86 lb carbon/1000 gal. (0.1 kg/cu m) at exhaustion and an average TOC reduction of 65% (37 mg/liter reduced to 13 mg/liter).

Capacities for municipal systems are about 0.5 g COD/g carbon (Ref. 5-36) when used in a tertiary mode. The high effectiveness of activated carbon adsorption may be due to the preponderance of fulvic acid-like material in the 100 to 100,000 molecular weight range, which is readily adsorbed (Ref. 5-41).

With multicomponent petrochemical wastes, the relatively high level of mixed oxygenated aliphatic materials remaining even after biological treatment limits the application of adsorptive processes. In studies by Conway, *et al.* usually only 40 to 80% reduction on a COD basis was achieved in column systems and breakthrough was rapid due to capacities in the order of 0.25 g COD/g carbon (Ref. 5-2, 5-42). For these multicomponent wastewaters the lead bed seemed to remove the most adsorbable constituents and subsequent beds were much less effective (e.g., 0.15 g COD/g carbon) perhaps because the less readily adsorbed species block the pores so that the more readily adsorbed materials cannot penetrate (Ref. 5-26). At breakthrough only about 20 gallons of wastewater per pound of carbon (170 liters/kg) were treated, as compared to about 600 gal./lb (5000 liters/kg) for municipal wastewater. Even in equilibrium isotherm studies COD removals with biologically treated petrochemical plant wastewaters range from 45 to 90% at carbon dosages of 20 g/liter (Ref. 5-26). As discussed previously, the equilibrium capacity for a four-component system was about 60% of that which would be predicted from single component data (Ref. 5-2). In this case study, continuous column tests showed only 90% of the capacity of equilibrium batch tests; thus, use of a factor of about 0.5 where converting batch data on individual compounds to continuous systems treating multicomponents seems appropriate.

In such cases, the utility of activated carbon adsorption would seem to be in application to streams at specific manufacturing units for which carbon either has an economically useful adsorptive capacity or can remove materials which are nondegradable in or inhibitory to subsequent biological treatment systems. At the source units the organic levels are higher, which results in higher carbon loadings.

As reported by Himmelstein, Fox, and Winter, recovery of contaminants when treating a waste at its source sometimes is feasible using solvents, acids, or bases to regenerate the carbon (Ref. 5-43). In a follow up study by Fox, *et al.*, phenol and acetic acid were adsorbed from a brine under acid conditions and desorbed with caustic. Organic solvents also have been evaluated as regenerants by these investigators, as well as by Skriba and Purcupile (Ref. 5-44). With chemical regeneration, capacities can drop considerably after several cycles.

An adsorption process developed to upgrade a biologically treated effluent from a chemical plant is shown in Figure 5-10. This design was selected from 17 alternatives based on technical performance, capital costs, and operating costs (Ref. 5-45). Design conditions are:

20,000,000 gal./day flow

Downflow, trimedia filters at 5 gal./ft^2-min (204 liters/ m^2-min), 64 to 84%

suspended solids removal at 16 to 37 mg/liter feed level, up to 24-hr runs between backwash, 15 gal./ft^2-min (611 liters/m^2-min) backwash rate with air scour provisions

Upflow, expanded, pulsed-bed carbon adsorbers at 8 gal./ft^2-min, 30-min retention, 30-ft deep, parallel beds, 5% carbon pulse/day

Regeneration in multiple-hearth furnace, 100,000 lb/day, 7% loss per cycle

TOC reduction from 40-100 mg/liter feed to 13 mg/liter effluent were achieved; associated color, odor, foam, and toxicity standards presumably also were met.

REFERENCES

5-1. Weber, W. J., Jr., *Physicochemical Processes for Water Quality Control,* Wiley-Interscience, New York, 1972.

5-2. Giusti, D. M., Conway, R. A., and Lawson, C. T., "Activated Carbon Adsorption of Petrochemicals," *J. Water Pollut. Control Fed.* **46**, 5, pp. 947–65, 1974.

5-3. Morris, J. C. and Weber, W. J., Jr., *Adsorption of Biochemically Resistant Materials from Solution,* Part I, AWTR-9 Report No. 999-WP-11, U.S. Public Health Service, May 1964.

5-4. Morris, J. C. and Weber, W. J., Jr., *Adsorption of Biochemically Resistant Materials from Solution,* Part II, AWTR-16, Report No. 999-WP-33, U.S. Public Health Service, March 1966.

5-5. Ward, Thomas M., Getzen, and Forrest W., "Influence of pH on the Adsorption of Aromatic Acids on Activated Carbon," *Environ. Sci. Technol.,* **4**(1), p. 64, 1970.

5-6. Snoeyink, Vernon L., Weber, Walter J., Jr., and Mark, Harry B., Jr., "Sorption of Phenol and Nitrophenol by Active Carbon," *Environ. Sci. Technol.,* **3**(10), p. 918, 1969.

5-7. Adamson, A. W., *Physical Chemistry of Surfaces,* 2nd Ed. Academic Press, New York, 1967.

5-8. Kipling, J. J., "The Adsorption of Acetic Acid and *n*-Butylamine from Aqueous Solutions by Activated Charcoals," *J. Chem. Soc.,* p. 1483, 1948.

5-9. Ward, T. M. and Getzen, F. W., "Model for Adsorption of Weak Electrolytes on Solids as a Function of pH," *J. Colloid Interface Sci.,* **31**, p. 441, 1969.

5-10. Weber, W. J., Jr. and Gould, J. P., "Organic Pesticides in the Environment," *Ad. Chem. Ser.,* **60**, R. F. Gould, Ed., ACS Publishing, Washington, D. C., p. 280, 1966.

5-11. DiGiano, F. J. and Weber, W. J., Jr., *Technical Publication T-69-1,* Dept. of Civil Engineering, University of Michigan, Ann Arbor, Michigan, 1969.

5-12. Cheldelin, V. H. and Williams, R. J., "Adsorption of Organic Compounds I. Adsorption of Ampholytes on an Activated Charcoal," *J. Am. Chem. Soc.,* **64**, p. 1513, 1942.

5-13. Landt, E. and Knop, W., *Z. Phys. Chem.,* A162, p. 331, 1932.

5-14. Coughlin, R. W. and Ezra, F. S., "Role of Surface Acidity in the Adsorption of Organic Pollutants on the Surface of Carbon," *Environ. Sci. Technol.,* 2(4), p. 291, 1968.

5-15. Coughlin, R. W., Ezra, F. S., and Tan, R. H., "Influence of Chemisorbed Oxygen in Adsorption onto Carbon from Aqueous Sollution," *J. Colloid Interface Sci.,* **28**(3/4), p. 386, 1968.

5-16. Mattson, J. S., Mark, H. B., Jr., Malbin, M. D., Weber, W. J., Jr., and Crittenden, J. C., "Surface Chemistry of Active Carbon: Specific Adsorption of Phenols," *J. Colloid Interface Sci.,* **31**(1), p. 116, 1969.

5-17. Snoeyink, V. L. and Weber, W. J., Jr., "The Surface Chemistry of Active Carbon–A Discussion of Structure and Surface Functional Groups," *Environ. Sci. Technol.,* **1**, 3, p. 228, 1967.

5-18. Fox, R. D., Keller, R. T., Pinamont, C. J., and Severson, J. L., "Purification of a Waste Brine by Carbon Adsorption with Emphasis on Wastewater Reuse," Environmental Research Laboratory, The Dow Chemical Company, Midland, Michigan.

5-19. Huang, J. and Hardie, M. G., "Reclamation of Refinery Waste by Physicochemical Treatment Processes," presented at the Fourth Annual Mid-Atlantic Industrial Waste Conference, University of Delaware, November 17-19, 1970.

5-20. Paulson, E. G., "Adsorption as a Treatment of Refinery Effluent," presented at 25th Midyear Meeting of the American Petroleum Institute's Division of Refining, Houston, Texas, May 14, 1970.

5-21. Van Stone, Gary, "Activated Carbon in Pollution Control," Water Management Division of Calgon Corporation, Pittsburgh, Pennsylvania.

5-22. Van Stone, Gary, "A Solution to Industrial Wastewater Treatment–Granular Activated Carbon," Water Management Division of Calgon Corporation, Pittsburgh, Pennsylvania.

5-23. Kipling, J. J., *Adsorption from Solutions of Non-Electrolytes,* Academic Press, New York, 1965.

5-24. Mattson, J. S., and Mark, H. B., Jr., *Activated Carbon–Surface Chemistry and Adsorption from Solution,* Marcel Dekker, Inc., New York, 1971.

5-25. Product Data Bulletin, Nuchar Granular Active Carbon, Grade: WV–G 12 X 40, Westvaco Chemical Division, New York.

5-26. Lawson, C. T. and Hovious, J. C., "Cautions and Limitations on the Application of Activated Carbon Adsorption to Organic Chemical Wastewaters," Open Forum on Management of Petroleum Refinery Wastewaters, EPA/API/NPRA/U. of Tulsa, Tulsa, Oklahoma, January 26–29, 1976.

5-27. *Process Design Manual for Carbon Adsorption,* Technology Transfer Publication, U.S. Environmental Protection Agency, Washington, D.C., 1973.

5-28. Fornwalt, H. J. and Hutchins, R. A., "Purifying Liquids with Activated Carbon," Part I, *Chem. Eng.,* **73,** p. 179, 1966.

5-29. Wolfe, W. F., "A Model of Active Carbon," *J. Phys. Chem.,* **63,** p. 653, 1959.

5-30. Heckman, F. A., *Rubber Chem. Technol.,* **37,** p. 1245, 1964.

5-31. Zettlemoyer, A. C., "Modern Techniques for Investigating Interactions with Surfaces," *Chem. Rev.,* **59,** p. 937, 1959.

5-32. Barrer, R. M., "Specificity in Physical Sorption," *J. Colloidal Interface Sci.,* **21**(4), p. 415, 1966.

5-33. Garten, V. A., Weiss, D. E., and Willis, J. B., *Australian J. Chem.,* **10,** p. 295, 1957.

5-34. Studebaker, M. L., Huffman, K. W. D., Wolfe, A. C., and Nabors, L. G., "Oxygen Containing Groups on the Surface of Carbon Black," *Ind. Eng. Chem.,* **48,** p. 162, 1956.

5-35. Wang, L. K–P. *et al.,* "Effect of pH Adjustment upon Activated Carbon Adsorption of Dissolved Organics from Industrial Effluents," 27th Purdue University Industrial Waste Conference, Lafayette, Indiana, 1972.

5-36. Weber, W. J., Jr., "The Role of Activated Carbon in Physicochemical Treatment," US/USSR Symposium on Physical Chemical Treatment, Cincinnati, Ohio, November 12–14, 1975.

5-37. Joyce, R. S., "Activated Carbon Adsorption," *Industrial Waste Disposal,* R. D. Ross, (ed.), Van Nostrand Reinhold, New York, pp. 144–164, 1968.

5-38. Lawson, C. T., Union Carbide Corporation, South Charleston, West Virginia, personal communication.

5-39. Cohen, J. M. and Westrick, J. J., "Overview of Physical–Chemical Treatment," US/USSR Symposium on Physical Chemical Treatment, Cincinnati, Ohio, November 12–14, 1975.

5-40. Peoples, R. F., Krishman, P., and Simonsen, R. N., "Nonbiological Treatment of Refinery Wastewater," *J. Water Pollut. Control Fed.*, **44**, 11, pp. 2120–2128, 1972.

5-41. DeWalle, F. B. and Chian, E. S. K., "Removal of Organic Matter by Activated Carbon Columns," *J. Envrion. Eng. Div., Proc. Am. Soc. Civil Eng.*, **100**, EE5, pp. 1089–1102, 1974.

5-42. Conway, R. A., Hovious, J. C., *et al.*, "Treatability of Wastewater from Organic Chemicals and Plastics Manufacturing–Experience and Concepts," presented to EPA Guidelines personnel 2/9/73 and to the Effluent Standards and Water Quality Advisory Committee to the EPA 4/19/73 (revised September 1974).

5-43. Himmelstein, K. J., Fox, R. D., and Winter, T. H., "In-Place Regeneration of Activated Carbon," *Chem. Eng. Prog.*, **69**, 11, pp. 65–69, 1973.

5-44. Skriba, M. C. and Purcupile, J. C., "Powdered Carbon Regeneration by Immiscible Fluid Displacement," 28th Purdue University Industrial Waste Conference, Lafayette, Indiana, 1973.

5-45. Marek, A. C., Jr. and Askins, W., "Advanced Wastewater Treatment for an Organic Chemical Manufacturing Complex," US/USSR Symposium on Physical Chemical Treatment, Cincinnati, Ohio, November 12–14, 1975.

5-46. Conway, R. A., Lawson, C. T. and Stankewich, M. J., Jr., "Activated Carbon Treatment of Oxygenated Wastewater," U.S. Patent 4,080,287, March 21, 1978.

APPENDIX 5-1

ACTIVATED CARBON ADSORPTION TERMS AND NOTATION

C_b = breakthrough or allowable effluent impurity concentration, mg/liter
C_e = equilibrium impurity concentration remaining in solution, mg/liter
C_o = initial impurity concentration, mg/liter
C_v = carbon volumetric capacity, gal./lb (X 8.34 = liters/kg)
D = bed depth, ft (X 0.305 = m)
K = an empirical constant in the Freundlich isotherm
K_r = adsorption rate constant, liters/mg–hr
n = exponential coefficient in Freundlich isotherm
N_o = adsorptive efficiency, mg/liter
Q = flow rate, gal./min (X 0.0631 = liters/sec)
T = carbon bed life, hr
t = retention time (empty bed basis), hr
v = linear fluid velocity, ft/hr (X 0.305 = m/hr)
V_r = bed-volume rate, bed volume/hr
V_v = bed volume, cu ft (X 0.028 = m^3)
W = weight of carbon in column, lb (X 0.454 = kg)
W_r = carbon breakthrough rate, lb/hr (X 0.454 = kg/hr)
X/M = equilibrium loading of impurity on a unit weight of carbon, mg/mg
X/M_{C_o} = extrapolated equilibrium loading at initial impurity concentration, mg/mg

6

Special separations and reactions

Prior chapters have covered the unit operations and processes which commonly are considered for treatment of various multicomponent industrial wastewaters before discharge. This section presents an overview of more specialized treatment processes which usually are considered for reducing specific problem components in wastes often from individual production units. Alternatives to be discussed are:

Membrane separation
Ion exchange
Physicochemical oxidation and reduction
Liquid-liquid extraction
Stripping
Evaporation
Other phase-change operations
Precipitation
Magnetic separation
Foam separation

Resin adsorption
Electrochemical
Heat removal

Figure 6-1 presents the useful separation ranges for many of the processes discussed. An overall reference for these types of operations is Perry's *Chemical Engineers' Handbook* (Ref. 6-33).

6-1 MEMBRANE SEPARATION

The membrane separation processes of reverse osmosis, ultrafiltration, electrodialysis, and dialysis are attractive for treating wastewaters as the energy costs associated with a change in phase (e.g., liquid to vapor) are not involved. However, membrane fouling, which can reduce flux rates below economic levels, is a major concern.

6-1.1 Reverse Osmosis

Osmosis is the spontaneous flow of solvent (e.g., water) from a dilute solution through a semi-permeable membrane (impurity or solute permeates at a much slower rate) to a more concentrated solution. *Reverse osmosis* is the application of sufficient pressure to the concentrated solution to overcome the osmotic pressure and force the net flow of water through the membrane toward the dilute phase. This allows the concentration of solute (impurity) to be built up in a circulating system on one side of the membrane while relatively pure water is transported through the membrane. Ions and small molecules in true solution can be separated from water by this technique.

Transport of materials is described by the solution-diffusion theory. The molecules diffuse through the liquid-like solid membrane. Water flux or solvent flow per unit area is given by (Ref. 5-1, 6-1)

$$F_w = M_w (\Delta P - \Delta \pi) \qquad \text{(Eq. 6-1)}$$

where

F_w = water flux through the membrane
M_w = membrane constant dependent upon membrane thickness, temperature, diffusion coefficient of water in the membrane, and other factors
ΔP = pressure difference across the membrane
$\Delta \pi$ = osmotic pressure difference across the membrane

Solute (impurity) flux is given by

$$F_i = M_i \Delta C_i \qquad \text{(Eq. 6-2)}$$

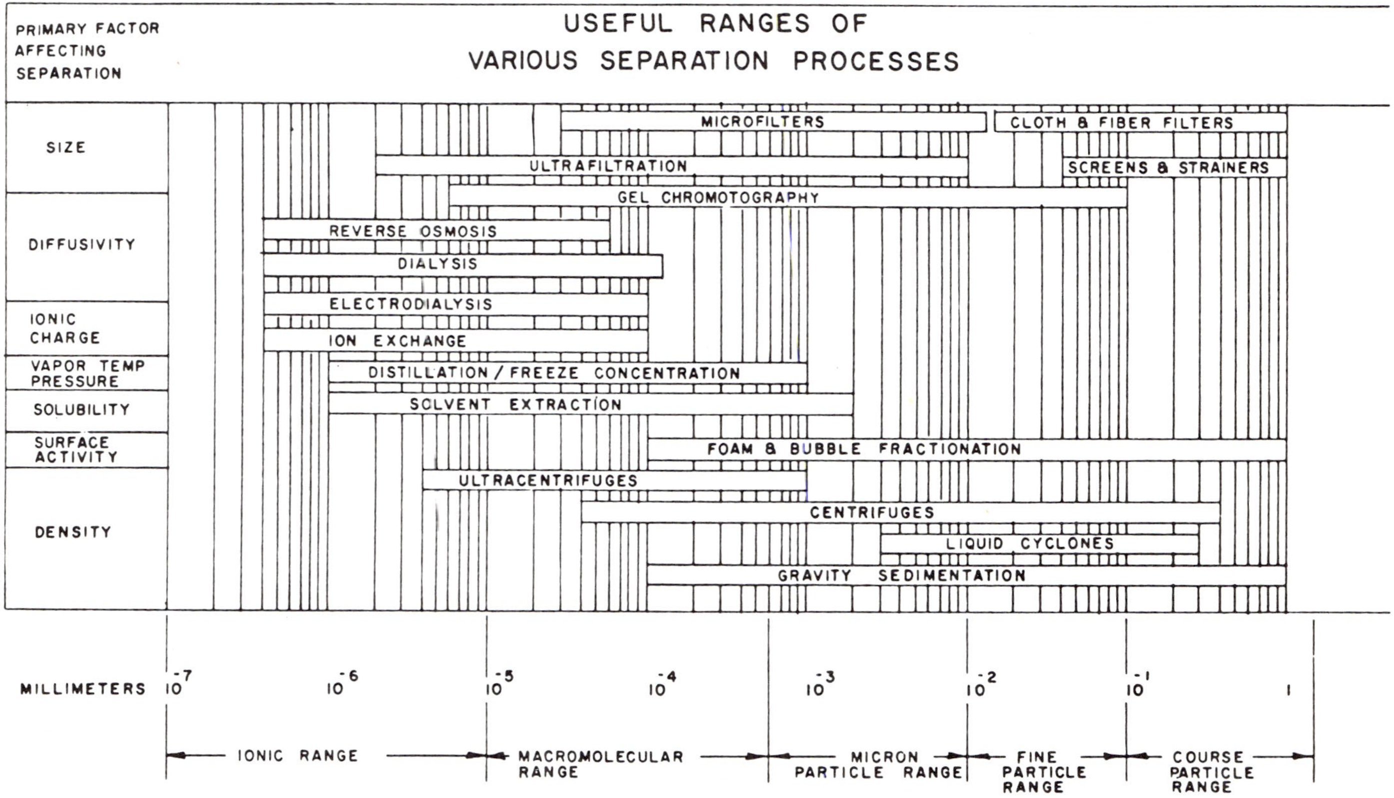

Fig. 6-1. Useful ranges of various separation processes. (Copyright 1969. Dorr-Oliver, Inc.).

where

F_i = impurity flux through membrane
M_i = membrane constant dependent upon membrane thickness, diffusion coefficient for impurity (usually salt) in the membrane, and other factors
ΔC_i = impurity concentration difference between solutions across the membrane

Pressures on the order of 600 to 1000 psig (4140 to 6900 kN/m^2) are required to overcome osmotic pressure and attain adequate solvent flux. The relationships in Eqs. 6-1 and 6-2 indicate that as pressure is increased both the amount and quality of purified water increases, as the solute flow does not change. At a given pressure, the quality of product water decreases as the feed-water impurity concentration increases.

Cruver has summarized typical solute rejection for high-selectivity cellulose acetate membranes as follows (Ref. 5-1):

Multivalent ions are rejected more effectively than are univalent ions.
Coions affect the rejection of a particular ion, e.g., sodium is rejected to a greater extent as sodium sulfate than as sodium chloride.
Undissociated or partially dissociated substances are poorly rejected.
In general, low molecular weight, water soluble organic molecules are poorly rejected.

Other membranes, based on polyamides and various cellulose acetate copolymers are available.

In addition to allowing the achievement of the required degree of separation at an economic flux level under ideal conditions, the membrane must be incorporated in an operating system which satisfies these practical requirements:

Minimum concentration polarization, i.e., ratio of impurity concentration at the membrane surface to that in the bulk stream.
High packing density, i.e., membrane surface area per unit volume of the pressure module.
Ability to handle any particulate impurities (by prefiltration if necessary).
Adequate support for the membrane and other physical features such as effectiveness of seals, ease of membrane replacement, and ease of cleaning.

Some typical reverse osmosis systems are shown in Figure 6-2; selected performance is cited in Table 6-1. Another available reverse osmosis system aimed at longer membrane life utilizes a porous matrix of resin-bonded sand as a structure upon which the membrane is integrally cast in a tubular configuration. Operation is at about 350 psig (2400 kN/m^2); turbulent flow is maintained to allow the handling of solids (Ref. 6-2).

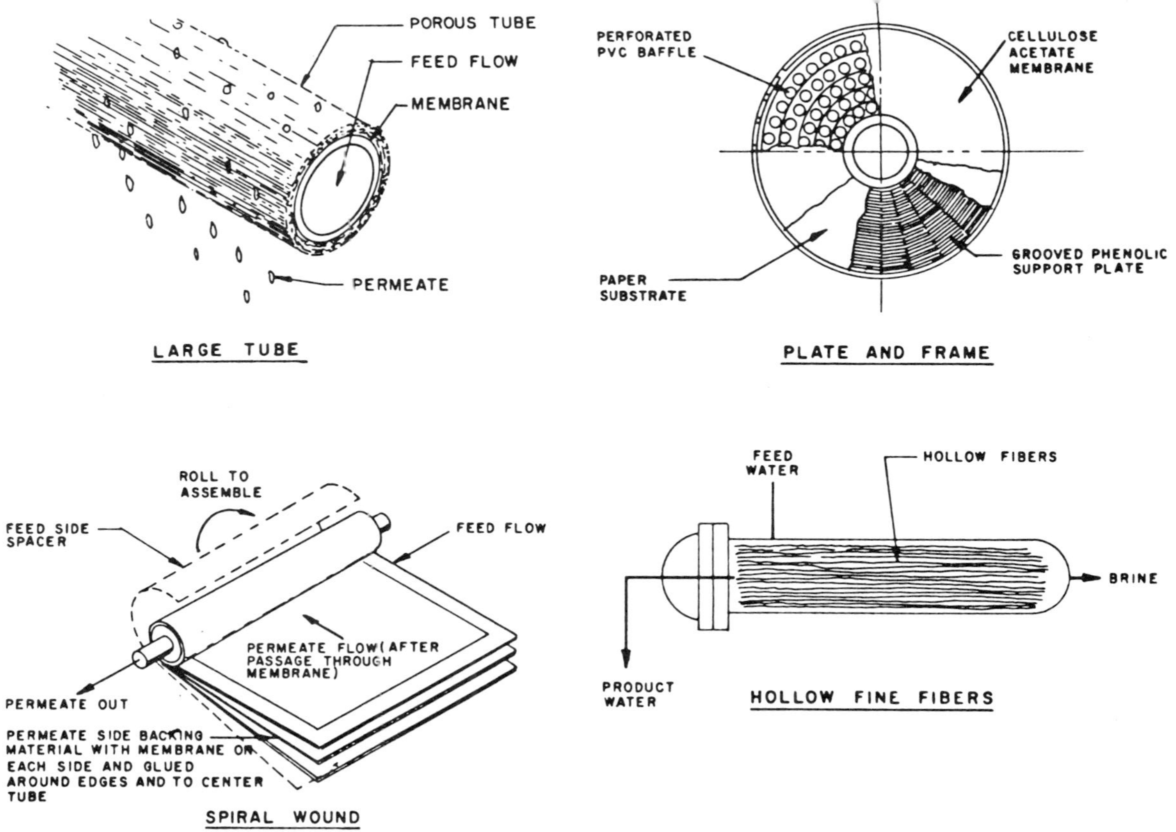

Fig. 6-2. Reverse osmosis module designs. *Source:* Weber, W. J., Jr., *Physicochemical Processes for Water Quality Control,* Wiley-Interscience, New York, 1972.

TABLE 6-1. Features of Reverse Osmosis Modules

	Packing Density (ft^2/ft^3)	Water Flux at 600 psi (gal./ft^2-day)	Water Output per Unit Volume (gal./ft^2-day)	Sodium Chloride Rejection	Useful pH Range	Ease of Cleaning
Plate-and-frame	150	10	1500	Very good	2–8	Fair
Large tubes	50	10	500	Very good	2–8	Very good
Spiral wound	250	10	2500	Very good	2–8	Good to very good
Polyamide hollow fine fibers	5000	1 (400 psi)	5000	Fair	0–12	Fair
Cellulose acetate hollow fine fibers	2500	3 (250 psi)	7500	Good	3–7	Fair

Conversions: $ft^2/ft^3 \times 3.28 = m^2/m^3$
$gal./ft^2\text{-day} \times 0.04 = m^3/m^2\text{-day}$
$gal./ft^3\text{-day} \times 0.13 = m^3/m^3\text{-day}$

Source: Weber, W. J., Jr., *Physicochemical Processes for Water Quality Control,* Wiley-Interscience, New York, 1972.

In regard to industrial wastewater treatment some potential uses of reverse osmosis are: (1) as a pretreatment step immediately prior to ion-exchange systems in wastewater reuse applications; and (2) recovery of valuable or reusable impurities from whey, pulp and paper wastes, and metal finishing wastes. As the need for water reuse and by-product recovery increases, reverse osmosis systems will find more application.

6-1.2 Ultrafiltration

The ultrafiltration technique involves pore flow through the membrane and can separate organics of a sub-colloidal (not in true molecular dispersion) nature, as well as colloidal and particulate matter. It involves viscous flow through tortuous pores in a microporous solid membrane. The molecular weight cutoff in pore flow is 500 to 1000. The rejection distribution experienced is dependent upon the organic molecules imvolved, their state of dispersion, the nature of the membrane, and the nature of the impurities overlaying the membrane. A useful membrane must have a relatively narrow molecular weight cutoff in order to prevent pore plugging. A high solvent flux at low pressure differential is another basic requirement. The ultrafiltration process discriminates on the basis of molecular size, shape, and flexibility.

The flux in ultrafiltration systems is related to the pressure differential by

$$F_u = \frac{\Delta P}{R_m} \qquad \text{(Eq. 6-3)}$$

where

F_u = solvent flux through membrane
ΔP = pressure differential across membrane
R_m = membrane resistance involving pore radius, pore tortuosity, solvent viscosity, porosity, and other factors

The rejection mechanism is selective sieving. Differential pressures in the order of only 20 to 100 psig (140 to 690 kN/m^2) are effective in this process, as salts pass through and an osmotic pressure difference is not generated. Fluxes generally range from 20 to 50 gal./ft^2-day (0.8 to 2.0 m^3/m^2-day) with cast membranes and sometimes higher with particulate membranes.

The same considerations cited for reverse-osmosis modular systems apply for ultrafiltration; however, boundary layer effects are more pronounced due to the slower diffusion of the macromolecules away from the membrane. Mass transfer through the gel layer formed adjacent to the membrane often controls the flux rate above moderate pressure or concentration levels in the circulating feed stream (Ref. 5-1).

$$F_u = \frac{\Delta P}{R_m + R_g} \qquad \text{(Eq. 6-4)}$$

where R_g = gel layer resistance depending on rate at which the gel material can back-diffuse into the mainstream of flow.

When the gel layer controls flux, the flux usually decreases logarithmically with increasing retained impurity concentration, as described by Michaels (Ref. 6-1).

Typical ultrafiltration module arrangements are as in reverse osmosis (Figure 6-2). Ultrafiltration hollow fibers, however, generally are 10 to 30 times larger in diameter and can be cleaned by reversing the direction of flow (Ref. 6-3). In addition to the cast membrane systems, a particulate membrane also has been developed. The ultrafiltration module in the Ucarsep system is made up of specially designed and assembled inert, porous carbon tubes, the inner surfaces of which have been covered with one of a number of specially developed inorganic compositions (Figure 6-3). This surface is the filtration layer, while the carbon tubes act as supports. This combination provides a greater degree of

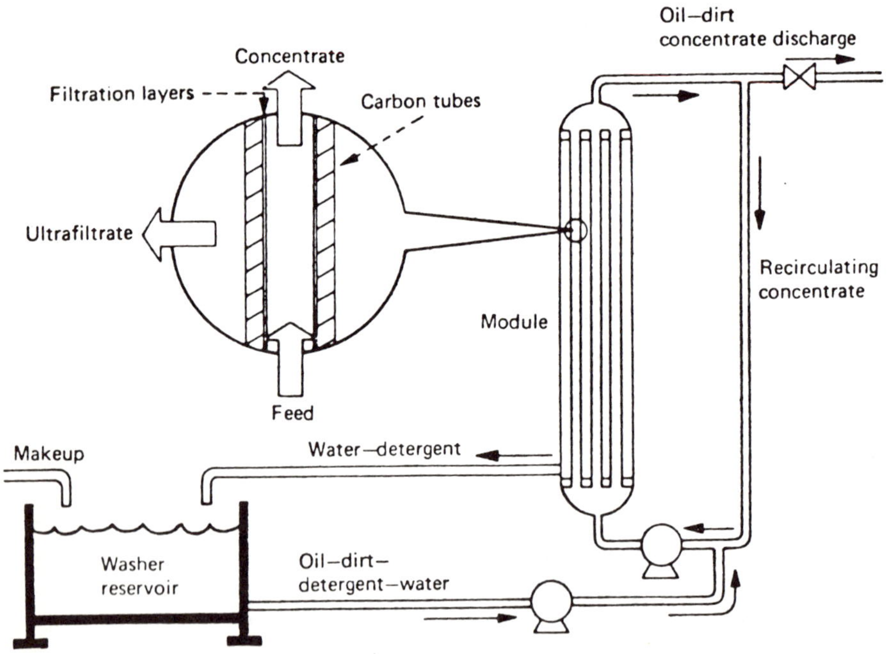

Fig. 6-3. Ucarsep ultrafiltration system in a metal-washer cleanup application (*Courtesy:* Union Carbide Corp., New York, N.Y.).

chemical resistance and thermal stability than is possible with conventional cellulose ester membranes (Ref. 6-4, 6-5).

A laboratory-scale approach for preliminary evaluation of an ultrafiltration system has been described by Nelson (Ref. 6-4). First, various membranes are tested in a single-tube module while returning both the permeate and the concentrate to the feed reservoir (Figure 6-3). From membrane flux and rejection data, a membrane choice is made. Then a batch-wise concentration test is conducted to determine the effect of feed concentration level on membrane flux and rejection. Finally, differential tests are run at various operating pressures, temperatures (affects viscosity), and circulating rates (affects gel-layer thickness) while holding the feed concentration at a suitable level as previously determined.

In applying results of such screening tests to guide the pilot-scale Ucarsep ultrafiltration of a mixed polyglycol waste, Nelson found that operation at 35 to 40°C and 90% water recovery, 13 ft/sec (4 m/sec) cross-flow velocity, and 100 psig (690 kN/m^2) resulted in a stable flux of 50 gal./ft^2-day (2 m^3/m^2-day) (Ref. 6-4). In tests of this particulate membrane system with other wastes, fluxes above 100 gal./ft^2-day (r m^3/m^2-day) reportedly have been achieved (Ref. 6-5).

Applications of ultrafiltration have been in the areas of:

Recovery of protein fraction from cottage cheese whey
Recovery of macromolecules, such as polyglycols
Separation and concentration of paint pigment from electro-painting bath liquids
Recovery of polyvinyl alcohol size from textile wastes
Separation of stabilized oil/water emulsions
Pretreatment of reverse-osmosis or ion-exchange feed streams to remove particulate and colloidal matter
Clear water recovery from papermill black liquors or from rolling mill coolants

6-1.3 Electrodialysis and Dialysis

In electrodialysis, the membrane functions for selective ion transport with the principal driving force being an electrical potential gradient. In dialysis, the membrane functions for selective solute transport; a concentration difference is the driving force.

Desalination of brackish water (< 1% salt) is the major use for electrodialysis. Wastewater demineralization by this technique has been studied; however, considerable pretreatment is required. The practice has been to use diatomaceous earth filters and carbon adsorption to remove suspended matter and soluble organics. Recovery of carboxylic acids, pulping chemicals, and chromates from plating wastes are other applications.

Dialysis has been applied to the recovery of rayon and electroplating wastes (Ref. 6-6) and to the cleanup of mercerizing wastes in the cotton industry as described by McCarty and Sawyer. Sodium and hydroxide ions pass through the dialysis membrane; the water then is evaporated to recover the sodium hydroxide. The contaminating organic material is retained by the dialysis membrane for separate disposal (Ref. 6-7).

6-2 ION EXCHANGE

Ion exchange can be defined as the reversible interchange of ions between a liquid and a solid, with no change in the physical structure of the solid. Ion exchange materials include natural zeolites, synthetic zeolites, and various organic resins. A typical cation exchange reaction can be represented as follows:

$$RB + AY \rightleftharpoons RA + BY \qquad \text{(Eq. 6-5)}$$

where R represents the base ion exchange material, and B its exchangeable ion; A is the cation in solution and Y its associated anion.

When all of the "useable" exchange sites on the ion exchange material have been exhausted in forming the more stable RA compound, the material can be regenerated by washing the material with a concentrated solution of B, thereby forcing, through mass action, a reversal of the above equilibrium:

$$RA + BY_1 \rightleftharpoons RB + AY_1 + BY_1 \qquad \text{(Eq. 6-6)}$$

In effect, A has only been removed from one stream and concentrated in another. In addition to producing a stream free of contamination by A, A has, hopefully, been concentrated to such an extent that ultimate disposal will be more readily accomplished. In some cases A will be in a reuseable form.

Similar equilibrium expressions for exhaustion and regeneration can be written for anion exchange. There are quite a broad range of ion exchange materials available. Table 6-2 summarizes the characteristics of representative materials.

The theoretical ion exchange capacity of the resin is the total number of ionic sites per unit weight or unit volume. Since ion exchange resins will swell when wet and are employed wet, the wet volume capacity is generally the best figure for comparing materials with respect to capacity. Operating capacity is the number of sites utilized in a given volume of wet resin in a given cycle, i.e., given flow and/or stirring rate, resin particle size, solution concentration, rate of regeneration, and weight of regenerant employed per unit volume of resin.

When an ion exchange resin is contacted with a solution containing ex-

TABLE 6-2. Ion Exchange Materials

Cation Exchangers

Natural zeolites

Glauconite or "green sand," a silicate containing iron and potassium.

Low capacity, resistant to physical abrasion, moderate to high resistance to chemical attack.

Synthetic zeolites

Composition $Na_2O:Al_2O_3:SiO_2$, ranges 1:1:4 to 1:1:15

Resistance to chemical attack increases with increased silica content, exchange capacity decreases with increased silica content.

Synthetic organic resins

Strongly acidic

Mostly styrene-divinylbenzene (DVB) copolymer beads with the active group being sulfonate ($R\text{-}SO_3H$ or $R\text{-}SO_3Na$).

Capacity varies with degree of cross-linking, i.e., % DVB. Higher % DVB yields higher dry weight and lower wet volume capacities.

Resistance to chemical attack also determined by degree of crosslinking, useable through entire pH range.

Hydrogen form regenerated with HCl or H_2SO_4; sodium form with NaCl.

Weakly acidic

Mostly styrene-divinylbenzene copolymer beads with the active sites being carboxylic (R-COOH or R-COONa) and phenolic groups.

Subject to same effects from crosslinking, generally useable only above pH 7.0.

Hydrogen form regenerated with HCl or H_2SO_4; sodium form with NaCl.

Anion Exchangers

Organic

Strongly basic

Primarily styrene-divinylbenzene copolymer beads with active sites being quaternary ammonium structures ($R\text{-}R'_3NOH$ or $R\text{-}R'_3NCl$).

Irreversibly adsorbs organic materials, leading to fouling.

Some difference in selectivity and chemical stability based on type of substituent groups or amine.

Useable through entire pH range, generally less rugged and subject to greater fouling than cation resins.

Hydroxide form regenerated with NaOH; chloride form with NaCl or HCl.

Weakly basic

Polyamine resins containing 1°, 2°, and 3° amine groups ($R\text{-}NH_3OH$ or $R\text{-}NH_3Cl$).

Capable of adsorbtion of free acids from solution through a neutralization type reaction:

$$RNH_2 + HCl \longrightarrow RNH_3Cl$$

or

$$RNH_3OH + HCl \longrightarrow RNH_3Cl + H_2O$$

Ion exchange activity below pH 7.0.

Liquid ion exchange materials also fall in this category and are employed as either the free amine or in the hydrated form. These materials are employed in a solvent and are contacted with the liquid to be treated in equipment similar to that employed in solvent extraction.

(continued)

TABLE 6-2. (Continued)

Free base or hydroxide form regenerated with NaOH, NH_4OH, or Na_2CO_3; chloride form with HCl.

Weak-base resins will reversibly adsorb organics. Hence, they are useful as organic scavengers ahead of other resin beds.

Typical Ion-Exchange Reactions

$$2R\text{-}SO_3H + Ca^{2+} \rightleftharpoons (R\text{-}SO_3)_2Ca + 2H^+$$

$$2R\text{-}NH_3Cl + SO_4^{2-} \rightleftharpoons (R\text{-}NH_3)_2SO_4 + 2Cl^-$$

Note: Matrixes and functional groups of resins commonly used for water purification are shown in Figure 6–17.

Source: Price, K. S., Clark, Dietz, and Associates-Engineers, Inc., Urbana, Illinois (with modifications).

changeable ions, the resin will show a selectivity for various ions depending upon the type of resin used. According to Schmitt, at low ionic concentrations the extent of exchange increases with increasing atomic number of the exchanging ion. A strong-acid cation resin, for example, shows an affinity or selectivity series for univalent cations in dilute solutions as follows:

$$Ag > Cs > Rb > K > NH_4 > Na > H$$

For divalent ions the order of selectivity is as follows:

$$Ba > Pb > Ca > Ni > Cd > Cu > Zn > Mg$$

For anion resins, univalent anions are exchanged as follows:

$$I > Br > Cl > F$$

At low concentrations the affinity of exchange increases with increasing valency of the exchanging ion. Thus, the order of affinity for a conventional cation exchanger is $Th^{+4} > Al^{+3} > Ca^{+2} > Na^{+1}$. Many special resins with highly specific selectivities have been developed (Ref. 6-8).

Several alternative modes of operation have been developed; the following usually have the most effectiveness:

Downflow with countercurrent regeneration (Figure 6–4)
Upflow fluidized bed
Upflow with countercurrent regeneration
Continuous countercurrent (Figure 6–5)

Several commercial systems are available for high-volume, continuous ion exchange treatment. In these systems, the spent resin is pulsed out of the bed and subjected to backwashing, regeneration, and rinsing in a sequenced operation

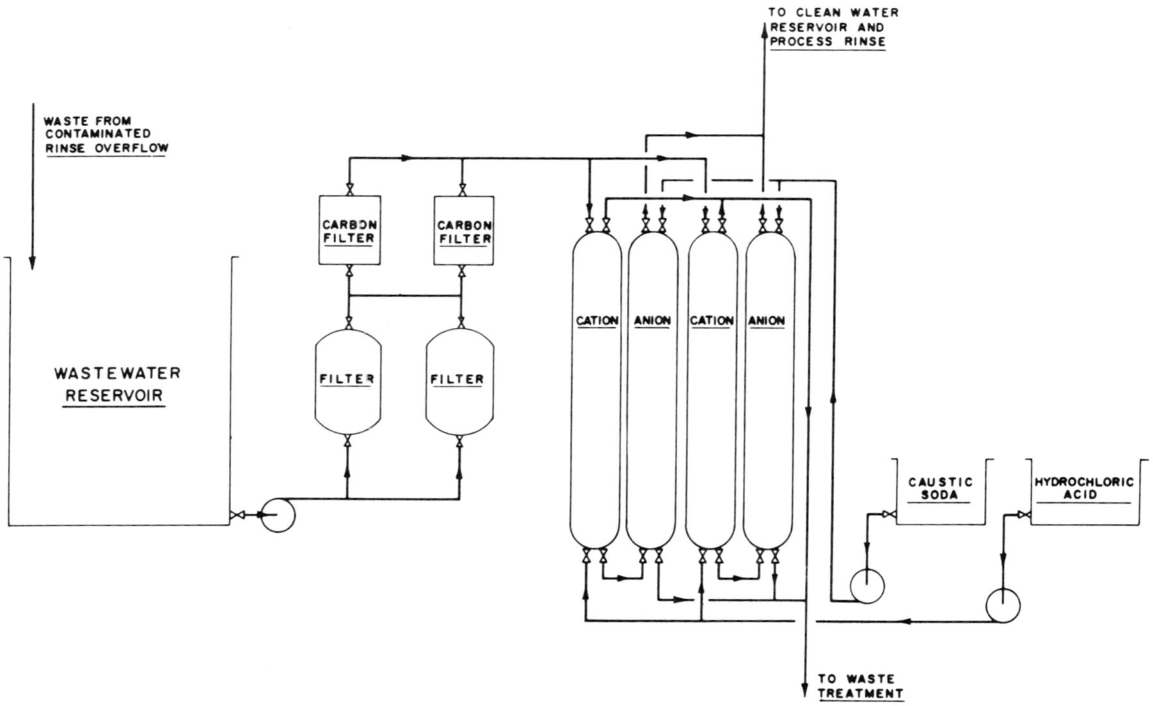

Fig. 6-4. Ion-exchange system for metal finishing wastes. *Source: Waste treatment, Upgrading Metal-Finishing Facilities to Reduce Pollution. Part 2,* EPA 625/3-74-002 Environmental Protection Agency Technology Transfer Seminar Publication, U.S. Environmental Protection Agency, Cincinnati, Ohio, 1974.

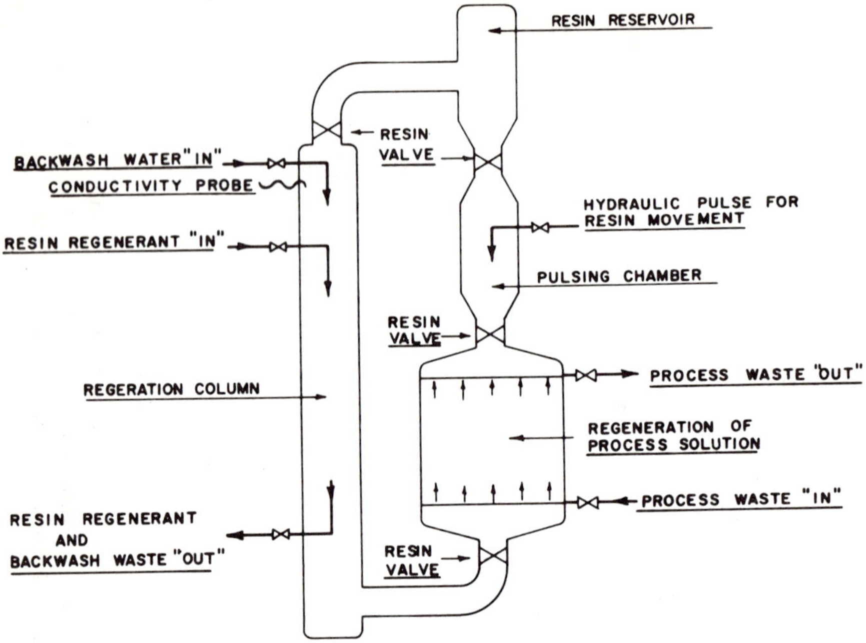

Fig. 6-5. Moving-bed ion exchanger. *Source: Waste Treatment, Upgrading Metal-Finishing Facilities to Reduce Pollution. Part 2,* EPA 625/3-74-002, Environmental Protection Agency Technology Transfer Seminar Publication, U.S. Environmental Protection Agency, Cincinnati, Ohio, 1974.

(Ref. 6-9). Complete removal of cations and anions ("demineralization") can be accomplished by using the hydrogen form of a cation exchange resin and the hydroxide form of an anion exchange resin in either series beds or in a single mixed bed.

Experimental evaluation of the ion-exchange process proceeds much as described in Chapter 5 for carbon adsorption. Detailed sizing and costing techniques have been described by Sanks (Ref. 3-23) and by Weber (Ref. 5-1).

Potential applications for ion-exchange in wastewater treatment include:

Metal finishing (Au^{3+}, Ag^{+}, CN^{-}, CrO_4^{2-}, Cr^{3+}, Ni^{+} recovery)
Cooling tower blowdown (chromate removal)
Photographic (silver recovery)
Petrochemical (amino acids, organic acids, alkaloids)
Nuclear (radionuclides)
Coking (phenol, ammonia)

Pulp and paper (color)
Mining (copper, uranium)
Municipal (ammonia)

In the ammonia recovery and removal process (ARRP) biologically treated municipal wastewater is prepared for ion exchange through the use of dual-media filtration and activated carbon adsorption. The wastewater is passed over clinoptilolite, a natural zeolite with an affinity for ammonium ions. Upon regeneration, the spent NaCl regenerant is treated with NaOH, air stripped in towers, and the evolved ammonia absorbed in sulfuric acid to produce 40% ammonium sulfate, a fertilizer (Ref. 6-10).

In addition, ion exchange has been used as a polishing step in recovery of a range of process waters for reuse and as a step in removing heavy metals from various process streams before discharge to a central treatment system.

6-3 PHYSICOCHEMICAL OXIDATION AND REDUCTION

While biological treatment provides an economical tool for enzymatically conducting a multitude of desired conversions at ambient temperatures and pressures, many instances arise that suggest the use of physiocochemical approaches. These include:

- Conversion of a toxic material to a less objectionable form.
- Removal of last traces of organic matter or ammonia.
- Oxidation of highly concentrated organic wastes.

A summary of physicochemical oxidations is presented in Table 6-3; an example of a reduction process is the use of sulfur dioxide to convert hexavalent chromium ions to the trivalent form which is less toxic and can be removed by precipitation.

Chemical oxidation is a process in which the oxidation state of the substance of interest is increased. In reduction, the oxidation state is decreased. In the case of inorganic compounds, oxidation is equivalent to a loss of electrons. For organic compounds, oxidation can mean the addition of oxygen, the withdrawal of hydrogen, or the withdrawal of electrons with or without the withdrawal of protons (Ref. 5-1).

The handling of many oxidizing agents is potentially hazardous, and suppliers' instructions should be carefully followed. A comparison of the economics of applying various oxidizing agents for treating cyanide wastes was reported by Goldstein (Ref. 6-11).

TABLE 6-3. Summary of Physicochemical Oxidations

Oxidant	Typical Application	Comments
Chlorine	Cyanide oxidation Ammonia removal Disinfection	Residual chlorinated organics can present problems. Oxidations can be catalyzed with UV light.
Ozone	Decolorization Disinfection Trace organic removal Cyanide removal Phenol oxidation	Normally "no problem by-products" are produced. Oxidations can be catalyzed with UV light or activated carbon. Generated on site.
Hydrogen peroxide	Oxidation of sulfides and mercaptans Cyanide (K, Zn, Cd) oxidation	Preferential sulfide oxidation. Metal catalysts aid phenol and cyanide oxidation.
Chlorine dioxide	Phenol Cyanide	Generated on-site by sodium chlorite and chlorine solutions
Oxygen (liquid-phase)	High pressure oxidation of concentrated wastes Radiation-enhanced oxidations Catalytic oxidation of sulfides and cyanide	For reaction to proceed at economical rate, catalysts, radiation, or high temperatures and pressures are necessary.
Oxygen (vapor-phase)	Catalytically oxidize organics in a vaporized stream	Not a frequent practice; vaporization costs are high when large volumes of water are to be treated.
Potassium permanganate	Specialized partial oxidations	Amenable are aldehydes, amines, aromatic alcohols, keto acids, alkenes, and phenol

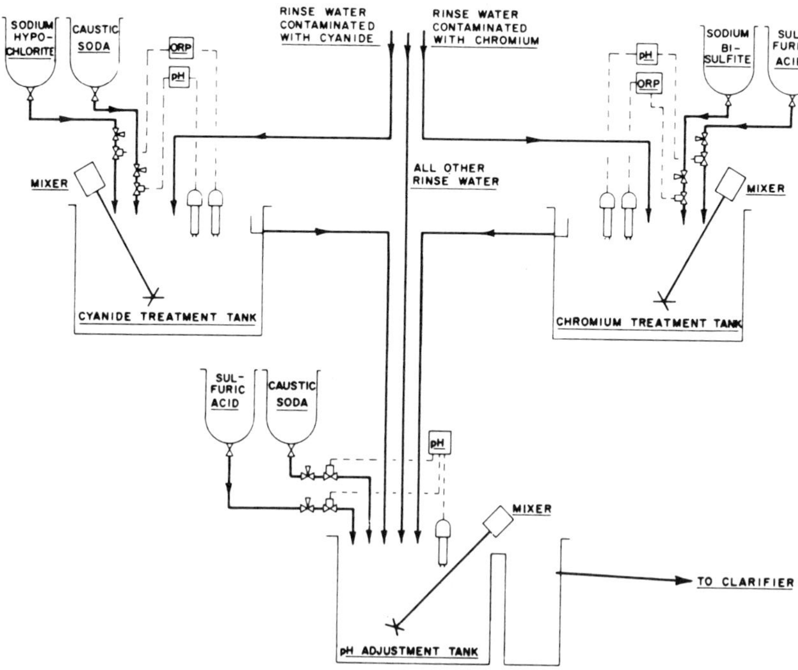

Fig. 6-6. Continuous treatment of cyanide and chromium. *Source: Waste Treatment, Upgrading Metal-Finishing Facilities to Reduce Pollution. Part 2,* **EPA 625/3-74-002, Environmental Protection Agency Technology Transfer Seminar Publication, U.S. Environmental Protection Agency, Cincinnati, Ohio, 1974.**

6-3.1 Chlorine

A major application for chlorine in industrial wastewater treatment is the alkaline (pH $\geqslant$ 10) chlorination of cyanide (CN^-) to cyanate (CNO^-) and then sometimes to nitrogen and carbon dioxide at a lower pH level (Figure 6-6) (Ref. 5-1, 6-12):

$$Cl_2 + 2OH^- \rightleftharpoons OCl^- + Cl^- + HOH \qquad \text{(Eq. 6-7)}$$

$$CN^- + OCl^- \rightleftharpoons CNO^- + Cl^- \qquad \text{(Eq. 6-8)}$$

$$CNO^- + 1.5\ OCl^- + 0.5\ HOH \rightarrow 0.5\ N_2 + HCO_3^- + 1.5\ Cl^- \qquad \text{(Eq. 6-9)}$$

Chlorine also has been used for the oxidation of amonia in municipal wastewater following organic removal by biooxidation; a subsequent dechlorination step using carbon columns or, more commonly, reduction with sulfur dioxide is utilized (Ref. 6-13):

$$NH_4^+ + HOCl \rightarrow NH_2Cl + H_2O + H^+ \quad \text{(Eq. 6-10)}$$

$$2\,NH_2Cl + HOCl \rightarrow N_2 + 3\,HCl + H_2O \quad \text{(Eq. 6-11)}$$

For acetic acid in a brine solution, chlorine oxidation catalyzed by ultraviolet radiation was used for conversion to methylene chloride, a potentially valuable product (Ref. 6-14).

$$CH_3COOH + Cl_2 \xrightarrow{\text{(UV rad.)}} CH_3Cl + CO_2 + HCl \quad \text{(Eq. 6-12)}$$

Treatment of textile dyes with chlorine in the presence of gamma radiation caused more decolorization than did either of the two agents individually applied (Ref. 6-15).

Chlorine can add to many organic materials to form bio-resistant end products which can be odorous and toxic to aquatic life. The possibility of this undesirable side reaction needs to be considered when using chlorine for disinfection or oxidation of industrial wastewaters (Ref. 6-16).

Chlorine is shipped and stored as a liquid under pressure in cylinders and is mixed into the wastewater preferably by a vacuum water injector. Since chlorine gas is highly toxic, care must be taken in handling. A safer form, sodium hypochlorite solution, can be purchased at concentrations up to 15% available chlorine. Hypochlorite solutions also can be prepared using calcium hypochlorite, which is sold in powdered or granular form with about 70% available chlorine.

6-3.2 Ozone

Ozone is a very strong oxidant for most organics in water, and, normally, aqueous by-products present no problem. Production is on-site, usually from oxygen using electrical energy. Ozone in aqueous solution is subject to decomposition (half-life of about 20 min) which dictates specially designed reactors (Fig. 6-7) (Ref. 6-17). Economic considerations usually limit applications to the removal of very low concentrations of contaminants or to the treatment of problem materials not amenable to other methods. In these cases, however, it is a uniquely effective tool.

Ozonation has been found to render dye wastes more biodegradable and nontoxic (Ref. 6-18). When applied prior to activated carbon adsorption, ozone was found to produce a synergistic effect with the carbon in treating dye wastes (Ref. 6-19). Ozone also is effective in cyanide oxidation to either cyanate or

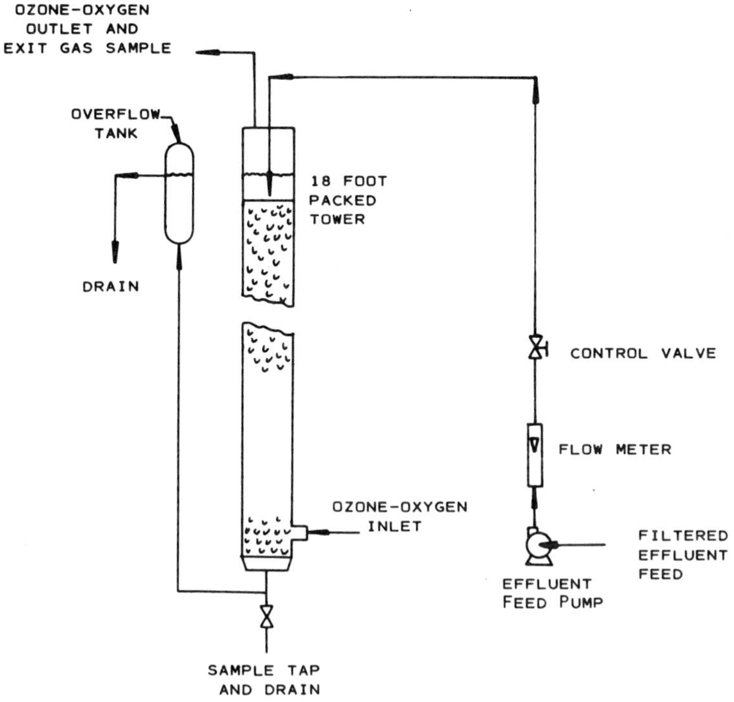

Fig. 6-7. Continuous countercurrent ozonation flow schematic. ***Source: Ozone Treatment of Secondary Effluents from Wastewater Treatment Plants;*** **TWRC-4, U.S. Environmental Protection Agency, Cincinnati, Ohio, 1969.**

nitrogen and carbon dioxide. Complete oxidation of phenol by ozone is possible:

$$C_6H_5OH + 14\ O_3 \rightarrow 6\ CO_2 + 3\ H_2O + 14\ O_2 \qquad \text{(Eq. 6-13)}$$

although partial oxidation is much more economical (Ref. 6-20). The utilization of ultraviolet light has been shown to enhance the effectiveness of some ozonation reactions (Ref. 6-21). Gould and Weber showed virtually complete removal of phenol and its aromatic degradation products when about 5 moles of ozone were consumed (in the presence of UV light) per mole of phenol originally present; glyoxal and glyoxylic acid were the end products. At this point, TOC reduction was about 67% and COD reduction about 75% (Ref. 6-22).

Disinfection without formation of undesirable by-products is a major advantage of ozone. A high oxygen residual after ozone treatment is a further attraction.

6-3.3 Hydrogen Peroxide

Hydrogen peroxide is manufactured commercially by either an electrolytic process or an organic-air oxidation process. It is shipped to the site of use. Hydrogen peroxide was found to be specific to the oxidation of sulfides in the presence of high organic concentrations (Ref. 6-23).

$$H_2S + H_2O_2 \rightarrow S + 2\,H_2O \qquad \text{(Eq. 6-14)}$$

Under alkaline conditions sulfates are usually produced.

$$NaSH + 4\,H_2O_2 \rightarrow NaHSO_4 + 4\,H_2O \qquad \text{(Eq. 6-15)}$$

Sulfites and sulfur dioxide also can be readily oxidized with hydrogen peroxide. Phenols are oxidized with a ferrous ion catalyst. Cyanides can be oxidized; a pH of 8.5 to 10.0 is specified. The rate can be increased by traces of catalytic metals such as copper. The Kastone process (du Pont) utilizes hydrogen peroxide and formaldehyde to oxidize cyanides in zinc and cadmium plating wastes.

6-3.4 Chlorine Dioxide

Like ozone, chlorine dioxide is an unstable gas. It is prepared at the point of application by the reaction of a solution of sodium chlorite with a solution of chlorine:

$$2\,NaClO_2 + Cl_2 \rightleftharpoons 2\,ClO_2 + 2\,NaCl \qquad \text{(Eq. 6-16)}$$

Phenols and cyanides have been treated with this oxidant. In the gaseous form, chlorine dioxide is subject to explosion; however, aqueous solutions reportedly are sufficiently stable for safe handling (Ref. 5-1).

6-3.5 Oxygen

Molecular oxygen in aqueous solution is a weak oxidant in comparison with ozone and hydrogen peroxide; however, it can oxidize aqueous sulfides at reasonable rates in the presence of trace metal catalysts (Ref. 6-24). An induction period for sulfide oxidation and a pH dependence with maxima at pH 8 and 11 also were reported by Chen and Morris (Ref. 6-25). Cyanides have been oxidized by dissolved oxygen in an activated carbon column; copper is added to the incoming stream to first form a copper cyanide complex (Ref. 6-26).

Concentrated phenol and other coking by-products were oxidized in the vapor phase by oxygen over a copper catalyst at 300 to 400°C at atmospheric pressure; at 1000°C phenols were oxidized completely without a catalyst (Ref. 6-27). An approach in which a stream containing 0.5 to 1.0% organics is vaporized and then catalytically oxidized at 575°C was reported by Hyde (Ref. 6-28). High pressure wet oxidation (the Zimpro process) can be used to oxidize wastes high in organics. Economics are attractive only when the organic content is very high (>> 1%), the constituents are toxic or marginally biodegradable, or the real estate required for biotreatment is prohibitively expensive (Ref. 6-29).

Studies on radiation-enhanced oxidation of pulp mill effluents indicate exposure to 10^7 Roentgens in the presence of oxygen at 500 psi (3500 kN/m^2) can achieve 90% decolorization (Ref. 6-30). Data on high pressure radiolysis of textile wastes was summarized by Garrison, *et al.* (Ref. 6-31). Condren and Etzel found COD reductions of about 75% with an aerated, settled synthetic municipal wastewater using gamma irradiation; a free radical mechanism was described (Ref. 6-32).

6-3.6 Reductants

Hexavalent chromium from metal-finishing operations and in cooling-tower blowdown usually is treated by first reducing it to the trivalent form with sulfur dioxide (SO_2) or sodium bisulfite ($NaHSO_3$) (Figure 6-6).

$$H_2Cr_2O_7 + 3\ SO_2 \rightarrow Cr_2\ (SO_4)_3 + H_2O \qquad \text{(Eq. 3-17)}$$

Trivalent chromium then can be precipitated as the hydroxide. Ferrous sulfate as well as sulfur dioxide has been used as a reductant of chromic acid and to decolorize some dyes contained in textile wastes. Catalyzed hydrogenation also is practiced.

6-4 LIQUID-LIQUID EXTRACTION

In the liquid-liquid extraction process, one or more impurities are removed from the wastewater by intimate contact with a second liquid having low aqueous solubility and for which the impurities have a high affinity. The separation can be based either on physical differences which affect differential solubility between the solvents or on a definite chemical reaction such as complexation of metals with an amine-type solvent. In the solvent extraction contactor, the solvent with the preferential affinity for the impurity flows countercurrent to the aqueous feed, such as in the rotating disk contactor shown in Figure 6-8. Other contactor designs, as well as a discussion of distribution coefficients (ratio of solubilities of impurities in the two solvents), separation factors, and wastewater treatment applications have been summarized by Mayhue (Ref. 6-34). The extraction of phenolic materials from wastewater before discharge to a biological treatment facility is a major application of this separation process.

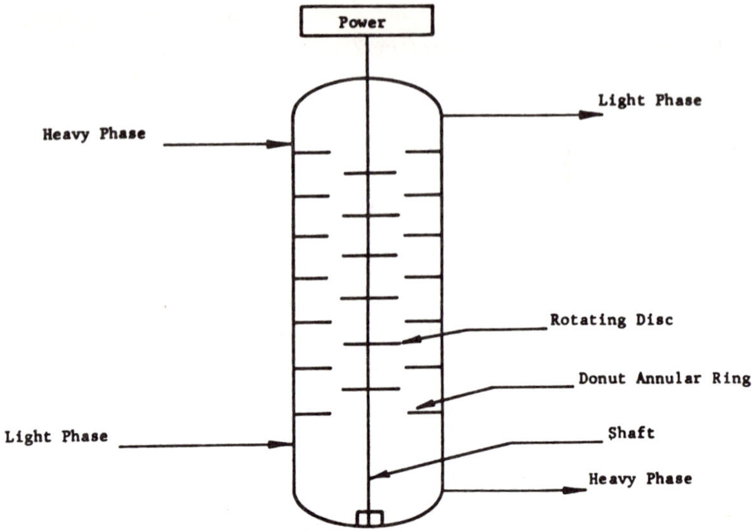

Fig. 6-8. Rotating disk contactor used in solvent extraction. *Source*: Mayhue, L. F., *Solvent Extraction Status Report,* EPA-R2-72-073, U. S. Environmental Protection Agency, Washington, D.C., 1972.

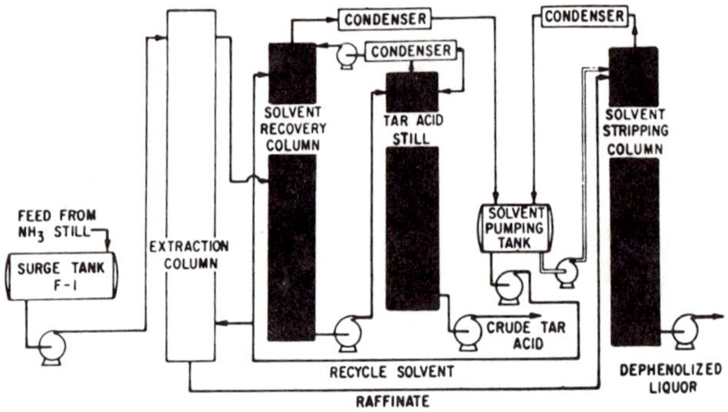

Fig. 6-9. Flow diagram of a dephenolization plant. *Source:* Reid, G. W. and Streebin, L. E., *Evaluation of Wastewaters from Petroleum and Coal Processing,* EPA-R2-72-001, U.S. Environmental Protection Agency, Washington, D.C., 1972.

A basis for selecting solvents for phenol extraction was developed by Kiezyk and MacKay (Ref. 6-35). A flow diagram of a dephenolization system is shown in Figure 6-9 (Ref. 6-36). Initial phenol levels must be high before the extraction process is more attractive than biological treatment alone; carbon adsorption with chemical regeneration is an emerging alternative (Ref. 5-18). At very high phenol levels, incineration becomes feasible.

6-5 STRIPPING

Dissolved gases and volatile materials can be removed from aqueous streams by employing steam, air, or an inert gas as a stripping medium. An example of this operation is the removal of about 20 mg/liter ammonia from biologically treated municipal wastewater using stripping towers that are structually similar to cooling towers, usually with a cross current air flow. The pH first is raised to 9 to 11 to shift the equilibrium toward gaseous ammonia.

$$NH_4^+ + OH^- \rightleftharpoons NH_3 + H_2O \qquad \text{(Eq. 3-18)}$$

Design details have been described by Smith (Ref. 3-23), by Adams and Eckenfelder (Ref. 2-2), and by Tchobanoglous (Metcalf and Eddy, Ref. 3-5). Cold temperatures, which increase ammonia solubility and can cause freezing and calcium carbonate scaling of the tower packing present problems, as may ammonia release to the atmosphere in the future. An alternative is first to pass the wastewater over ion-exchange beds and then, after regeneration, to strip the ammonia from the spent regenerant. The ammonia in this relatively low volume of air can be absorbed in an acid scrubber (Ref. 6-10).

The stripping of ammonia and hydrogen sulfide from "sour condensates" in petroleum refining is a widespread application of the stripping process. The overall equilibrium can be depicted as follows:

$$NH_4^+ + SH^- \rightleftharpoons NH_3 + H_2S \qquad \text{(Eq. 6-19)}$$

The equilibrium given in Eq. 6-16 applies, as does

$$SH^- + H^+ \rightleftharpoons H_2S \qquad \text{(Eq. 6-20)}$$

Thorough analyses of this operation and other treatment methods involving refinery wastes have been developed by Beychok (Ref. 6-37). Most of the stripping methods involve the downward flow of sour water through a trayed or packed tower, while the ascending flow of stripping steam, flue gas, or fuel gas removes the H_2S and, depending upon the pH, the NH_3 (Figure 6-10).

Some of the key variables influencing stripping operations are:

Stripping medium (air, steam, inert gas)
Waste characteristics (pH, other solutes)
Volume ratio of stripping gas to liquid
Area of gas-liquid interface
Temperature
Turbulence
Concentration gradient
Equilibrium relationships (Henry's law)

Air stripping of volatile components from petrochemical wastes under various temperatures, unit air flow rates, and tank geometries was studied by Gaudy,

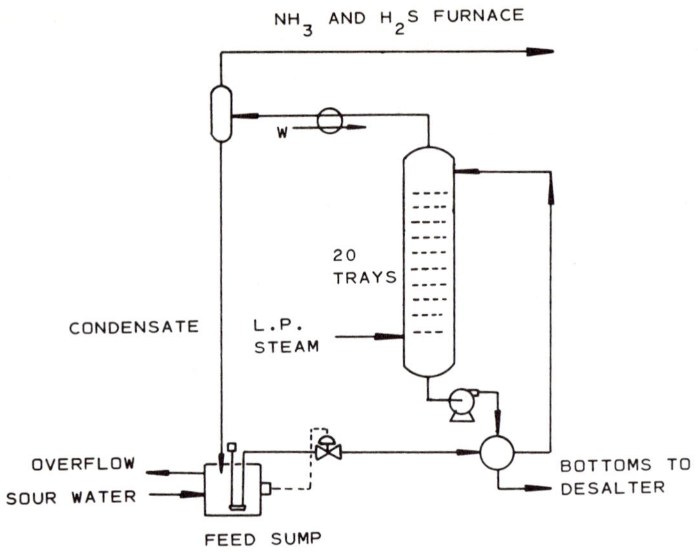

Fig. 6–10. Typical sour water stripper used in petroleum refining. (Ref. 6–37).

Englebrecht, and Turner. A linear relationship between the transfer rate constant and the gas flow rate was found to best describe the system, although some components fit a first-order relationship (Ref. 6-38). Stripping applications in the pulp and paper industry have been summarized by Willard (Ref. 6-39). The low portion of a typical fractional distillation unit has been used successfully by Butyrun and Ayers to stream strip odorous materials and methanol kraft mill condensate (Ref. 6-40). The odorous materials and methanol are incinerated; a stream ejector is needed as a flame arrestor in the transport line to allow a sufficiently high temperature to be used in the distillation column to strip out the methanol.

6-6 EVAPORATION

The process of evaporation is used both to recover useful by-products and to concentrate wastes prior to incineration. Typical applications are:

Concentration of pulping liquor prior to incineration,
Concentration of chromium-bearing wastes for recovery (Figure 6–11),
Concentration and recovery of spent solids from distilleries and breweries,
Concentration of radioactive wastes prior to storage,
Solar evaporation as an ultimate disposal means of process streams.

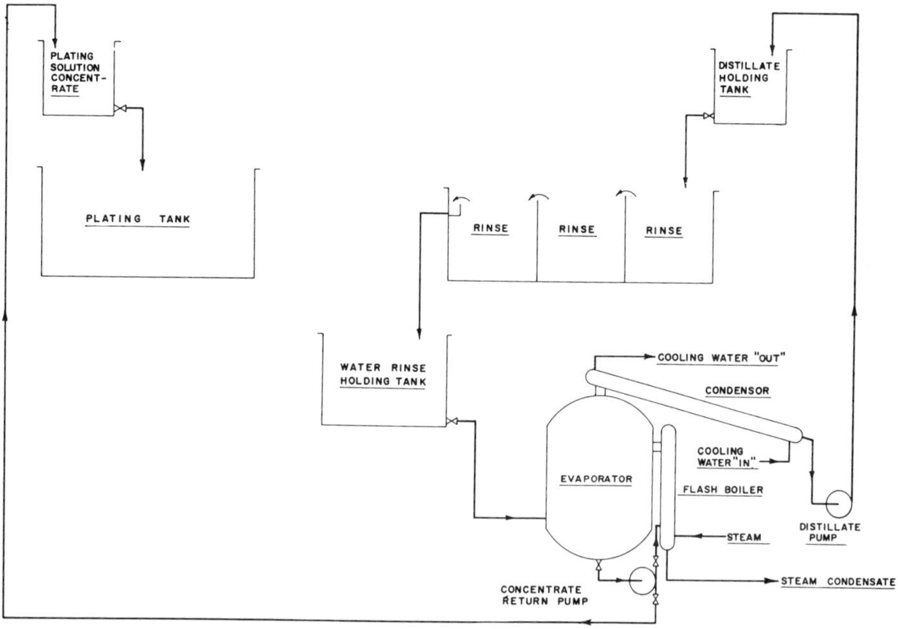

Fig. 6-11. Evaporative recovery in closed-loop metal plating. ***Source: Waste Treatment, Upgrading Metal-Finishing Facilities to Reduce Pollution. Part 2,*** **EPA 625/3-74-002, Environmental Protection Agency Technology Transfer Seminar Publication, U.S. Environmental Protection Agency, Cincinnati, Ohio, 1974.**

Commercial evaporator types as described by Rich include jacketed, horizontal tube, vertical tube, long tube, and forced circulation (Ref. 6-41). Key factors in addition to costs in evaporator selection are: fluid viscosity, concentration of foam formers, corrosiveness, and scaling tendency. Systems generally are operated under a vacuum to decrease the waste boiling temperature. In multiple effect evaporation, modules of decreasing pressures are operated in series, utilizing vapor from the preceding stage as steam in any given stage, to increase steam economy. A submerged combustion evaporator was found by Stein to be economically attractive when fuel costs were relatively low (Ref. 6-42). A vertical-tube, falling film evaporator with vapor compression recycle and seed-slurry scale prevention has been developed for treating cooling-water blowdown and chemical process water (Resources Conservation Co.) (Ref. 6-43).

The quantity of heat required to evaporate water from solution is the sum of:

Sensible heat to raise temperature to the boiling point

Latent heat of vaporization
Heat of concentration
Heat of crystallization of any products formed
Convective and other heat losses

The heat transfer rate is proportional to the heating surface area, the overall transfer coefficient (dependent on properties of solution, evaporator, degree of fouling, etc.), and the mean temperature difference between the heating medium and the boiling liquid.

The quantity of water to be evaporated to obtain a desired concentration effect can be calculated through mass balance. Knowing the required evaporation rate, the required heat transfer rate can be calculated. Then the evaporator and operating conditons can be selected to achieve the required heat transfer rate.

6-7 OTHER PHASE-CHANGE OPERATIONS

Flash distillation, an operation in which an appreciable proportion of a liquid is quickly converted to vapor in such a way that the final vapor is in equilibrium with the final liquid, has been applied in a multistage mode to a fine chemical plant waste to allow recycling of 88% of the process water and 95% of the contained catalysts and salt (Ref. 6-44). The blowdown from the distillation unit is sent to a crystallizer which recovers the rest of the water in the waste.

Crystallization also can be used to treat pickle liquors and brines. Various types of distillation (e.g., extractive distillation) can be used to recover low-boiling organics.

Freezing was studied as a means of recovering contaminant-free ice crystals; economic limitations were found (Ref. 3-5). Another proposed system involving phase change utilizes membranes selectively permeable to gases; the wastewater is passed through the tubes and gases (e.g., ammonia) pass through the membrane and remain attached to the tube exterior, from which they are periodically purged (Ref. 3-5).

6-8 PRECIPITATION

The process of precipitation in waste treatment has been studied mostly with regard to the removal of phosphates and heavy metals. Also, the formation of aluminum and iron hydroxide precipitates is important in coagulation and flocculation processes. The equilibrium that exists between crystals of a compound in the solid state and its ions in solution is amenable to consideration under the mass action law and can be treated mathematically as if the equilibrium were homogenous in nature (Ref. 6-7), as per the following example

$$Cu(OH)_2(s) \rightleftharpoons Cu^{2+} + 2\ OH^- \qquad \text{(Eq. 6-21)}$$

$$[Cu^{2+}]\ [OH^-]^2 = K_{sp} \qquad \text{(Eq. 6-22)}$$

where K_{sp} is the solubility product, which has a value in this case of 2.2×10^{-20} according to Bard (Ref. 6-45). The brackets indicate use of the molar concentrations of the ions.

The precipitation equilibrium can be shifted by several means. The common ion effect is when the concentration of one soluble constituent is raised by addition from an outside source; the other constituent(s) must decrease and the precipitate increase for the K_{sp} to remain the same. High pH levels with associated high hydroxyl ion concentrations will decrease the solubility of copper according to Eq. 6-22. A diverse ion effect is when unrelated ions through charge interactions decrease the activity (effective concentration) of the slightly soluble ions and shift the equilibrium to put more in solution. Chelation and the formation of complex ions also can serve to keep more of the metal ions in solution.

For phosphate precipitation, Leckie and Stumm identified the involvement of 36 solubility and complex-formation equilibria and prepared the solubility diagram in Figure 6-12 (Ref. 6-46). These are equilibrium relationships. Kinetic

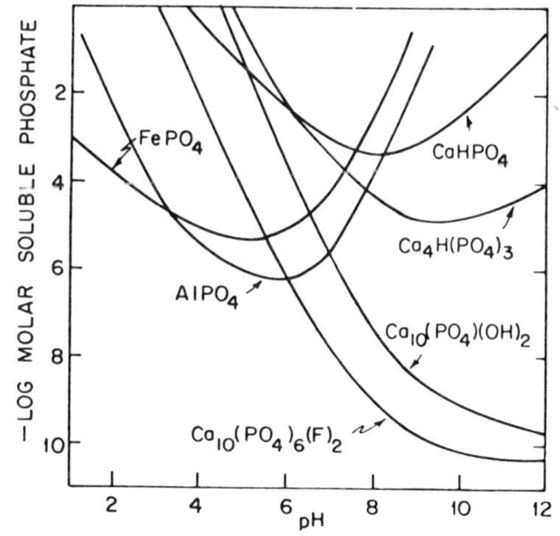

Fig. 6-12. Solubility diagrams for solid phosphate phases. ***Source*****: Leckie, J. and Stumm, W., "Phosphate Precipitation,"** ***Water Quality Improvement by Physical Chemical Means,*** **E. F. Gloyna and W. W. Eckenfelder, Jr., eds., University of Texas Press, Austin, Texas, 1970. (Copyright © 1970 by the University of Texas Press, All Rights Reserved.)**

considerations include the provision of adequate time under the reaction conditions both for the precipitation to occur and, more importantly, for the agglomeration of the colloidal precipitate to a settleable size. In some cases, seeding with other particles can provide more active sites for crystal growth (Ref. 6-46).

Lancy and Rice have compiled useful data on heavy metal precipitation (Figure 6-13, Table 6-4) along with clarifier design criteria (Ref. 6-12). Clarification systems also are discussed in Chapter 2-6.

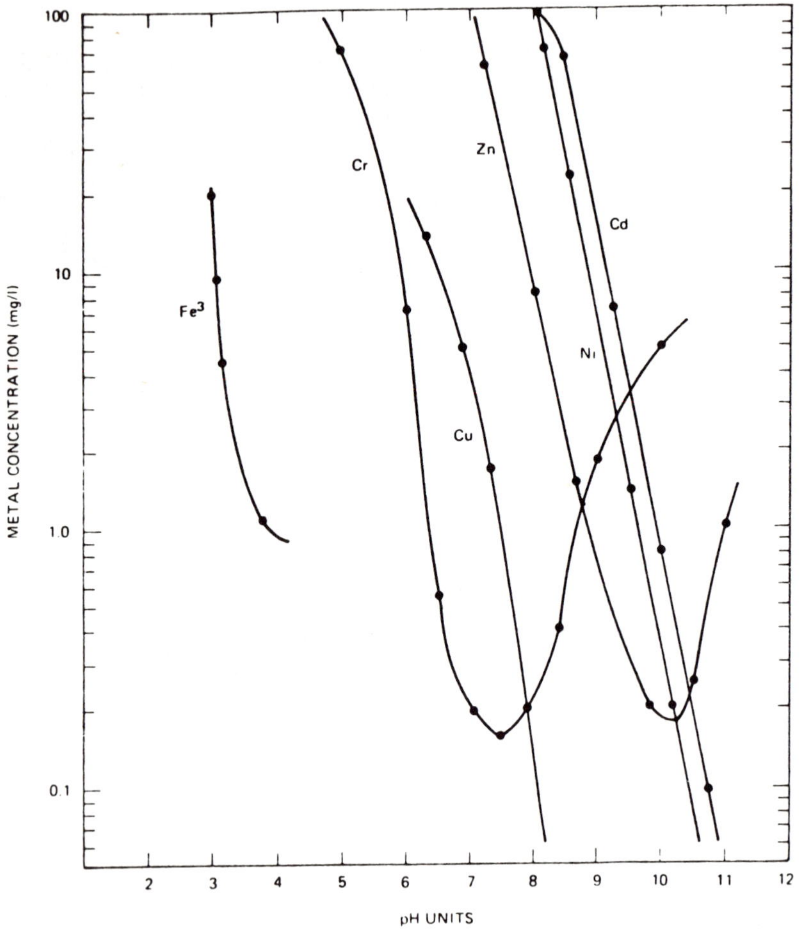

Fig. 6-13. Precipitation of metal salts as a function of pH. ***Source: Waste Treatment, Upgrading Metal-Finishing Facilities to Reduce Pollution. Part 2,*** **EPA 625/3-74-002, Environmental Protection Agency Technology Transfer Seminar Publication, U.S. Environmental Protection Agency, Cincinnati, Ohio, 1974.**

TABLE 6-4. Soluble Metal Content after Precipitation in Distilled and Hard Water

pH	Precipitation											
	Fe^3		Ni		Cr^3		Zn		Cd		Cu^2	
	Distilled, mg/l	Hard, mg/l	Distilled, mg/l	Hard, mg/l	Distilled, mg/l	Hard, mg/l	Distilled, mg/l	Hard, mg/l	Distilled, mg/l	Hard, mg/l	Distilled, mg/l	Hard, mg/l
6.5	0	0.8	18	19.2	18	17.8	19	18.5	19	19.2	8	11
7.0	0	.4	16.5	18.9	15	13.7	18	17.8	18.5	18.4	1.3	5.8
8.0	0	0	12.5	10.8	5.3	7.1	1.3	9.1	15	15.2	.5	2.4
8.5	0	0	3.5	2.3	1.5	5	0	1.6	3.5	4.8	.2	1.7
9.0	0	0	2	.6	.4	3.4	1.5	1.5	.7	.9	0	1.2
10.0	0	0	0	0	0	.3	19.5	8.4	.3	0	0	.4

Source: Waste Treatment, Upgrading Metal-Finishing Facilities to Reduce Pollution. Part 2, EPA 625/3-74-002, Environmental Protection Agency Technology Transfer Seminar Publication, U.S. Environmental Protection Agency, Cincinnati, Ohio, 1974.

In a process developed by the U.S. Department of Agriculture for heavy metal removal, starch xanthate is used for precipitation. After dewatering the resulting sludge to at least 50% solids, concentrated nitric acid is added to yield metal nitrates, which are suited for metal recovery by conventional means. As compared to hydroxide-precipitation techniques, lower metal levels have been achieved and metal recovery was facilitated.

Some organic species (e.g., color bodies) can be removed by high lime precipitation. This process has been applied mainly to pulp and paper and textile wastewaters.

6-9 MAGNETIC SEPARATION

For insoluble materials difficult to separate by gravity or conventional filtration techniques, high gradient magnetic separation (HGMS) may be an alternative The process has the advantages of very high throughput rates and capacities. In this process the wastewater generally is seeded with fine ferromagnetic powder (magnetite, Fe_3O_4). This seed creates nucleation sites on which emulsified oil agglomerates and to which globular oil attaches. A chemical coagulant can be added to bind the seed and any particulates into a floc. The mixture is then passed at application rates up to 50 gal./ft^2-min (2 m^3/m^2-min) through a stainless-steel-wool filter which is magnetized by a surrounding electrified coil (Figure 6-14). The insoluble material collects on the magnetized matrix, from which it can be periodically flushed by turning off the magnet coil (Ref. 6-47, 6-48). Other types of HGMS have been described by Oberteuffer (Ref. 6-49).

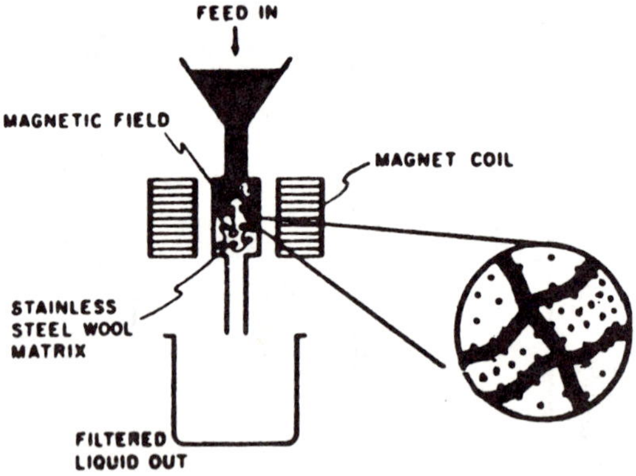

Fig. 6-14. High gradient magnetic separator schematic. *Source*: Oberteuffer, J. A., "High Gradient Magnetic Separation," *IEEE Transactions on Magnetics, MAG-9,* 3, pp. 303-306, 1973.

High gradient magnetic separation treatment of steel mill wastewaters, the major particulate pollutant of which is itself ferromagnetic, was reported by Marston to achieve nearly complete removal of the "tramp oil" which agglomerates around the metal particles (Ref. 6-47). For other wastes, the weight of seed required is about twice the weight of solids entrained in the floc. Thus, those solids when flushed from the filter need to be processed for seed recovery and recycle in a second unit having a much higher magnetic field strength and flow velocity. Marston, in small-scale tests, has shown HGMS to be effective in treating wastewater from oil refineries and paper and pulp processing mills, as well as steel rolling mills.

DeLatour in laboratory studies of phosphate removal has shown application rates of 150 gal./ft^2-min (6 m^3/m^2-min) through the magnetized matrix, as it has a 0.95 void fraction and provides a low resistance to flow (Ref. 6-50). The addition of bentonite clay, as well as magnetite and aluminum sulfate, was required to achieve effective phosphate removal. A magnetic filter aid developed by Bolts, *et al.* allows longer filter runs than with diatomite and can be recovered and reused (Ref. 6-51).

6-10 FOAM SEPARATION

A classification scheme for the many adsorptive bubble separation methods which have been developed is shown in Figure 6-15. Foam fractionation separates dissolved substances from homogenous solutions, while froth flotation separates insoluble substances from heterogeneous systems. In the nonfoaming techniques, dissolved or suspended matter is carried up by the rising bubbles to the pool surface where it is deposited as the bubbles exit; in the solvent sublation mode, an immiscible liquid is added on top of the main liquid to capture the material carried up (Ref. 6-52). The nonfoaming techniques would be used in wastewater treatment only to recover very low levels of problem materials at their production source.

The underlying principle of foam fractionation is that the hydrophobic (antagonistic to water) group on a surface active molecule migrates to the bubble-liquid interface and forms a stable entity with the dispersed air (Ref. 6-53). The hydrophilic (affinity for water) ends of the molecules remain in the aqueous phase, and the hydrophobic ends protrude into the gas phase. During continuous operation the bubbles float to the surface and form a foam layer, which is removed as shown in Figure 6-16. Solutes (impurities) with insufficient foaming ability are fractionated after addition of a surfactant which can either form electrostatic bonds or chelate with the solute. Inorganic cations can be removed in this manner by using an anionic surfactant (Ref. 6-53).

In froth flotation, the hydrophobic suspended matter attaches to minute

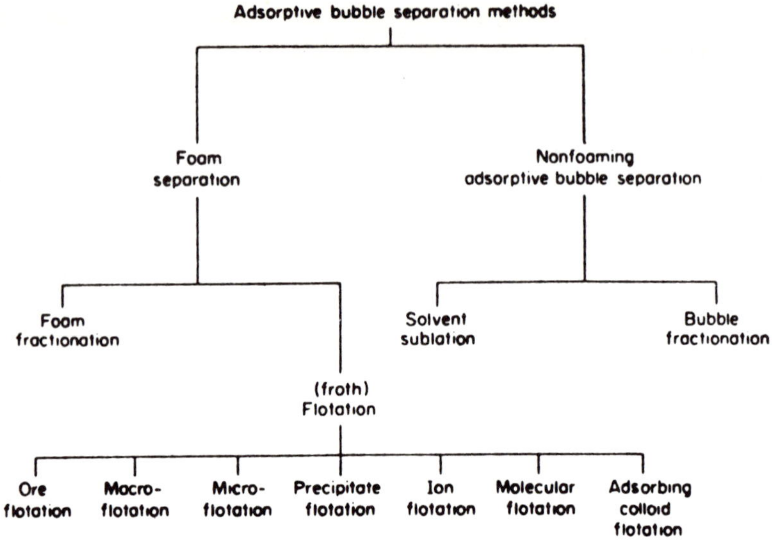

Fig. 6-15. Classification of adsorptive bubble separation techniques. *Source: Adsorptive Bubble Separation Techniques,* R. Lemlich, ed., Academic Press, New York, 1972.

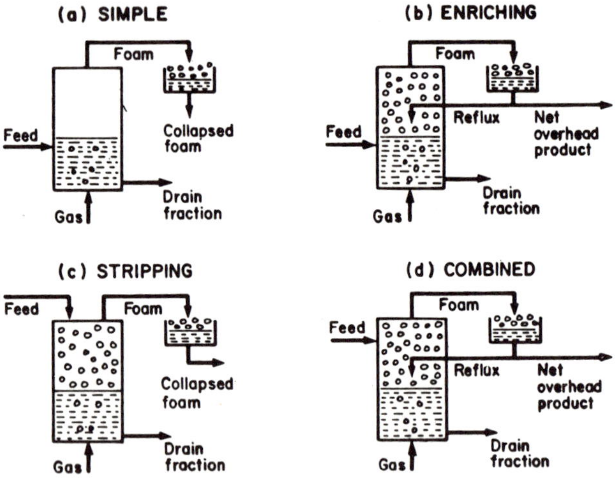

Fig. 6-16. Foam fractionation operating modes. *Source:* Ng, K. S. and Mueller, J. C., "Foam Separation–A Technique for Water Pollution Abatement," *Water Sewage Works,* pp. 48-55, June 1975. (Reprinted with permission of *Water Sewage Works*).

bubbles generated usually by releasing pressurized dissolved air into near atmospheric conditions. The bubble-particle agglomerate rises to the top, and the concentrated suspended matter is removed in a froth layer. Mechanical air dispersion systems also can be used. The foam must be stable and drain well for foam separation processes to be effective.

Applications as described by Ng, *et al.* (Ref. 6-53) for the foam separation procedures include:

- Removal of problem surfactants
- Concentration of proteins, grease, and suspended solids from food processing wastes
- Removal of resin acids and unsaturated fatty acids from bleached kraft mill effluents
- Recovery of oil from refinery wastewater
- Removal of heavy metal ions
- Recovery of fiber in pulp and paper mill effluents
- Removal of precipitated lignin from pulp mill effluents (Ref. 6-54)
- Removal of bacterial floc and algae from biological treatment plant effluent (Ref. 6-55)
- Removal of colloidal metal oxides and clays from mineral processing wastes

When conventional methods fail to meet effluent standards, foam separation can be considered either for removing problem constituents at their sources or as a polishing treatment step.

6-11 RESIN AND OTHER ADSORPTIONS

Primarily due to the ease of regeneration, synthetic resins are considered as an alternative to activated carbon for adsorption in certain applications. The adsorption of organic compounds by synthetic resins has been reviewed by Kim, Snoeyink and Saunders (Ref. 6-56). The various matrixes and functional groups of common resins are presented in Figure 6-17. Nonionic resins, those composed of the matrix only without functional groups, are regenerated with solvents, such as methanol, acetone, or isopropanol. The weak-base anion exchange resins have considerable adsorptive capacity for organic materials; however, development of a recoverable regenerant to use in place of sodium hydroxide was thought by Kim, *et al.* to be an important requisite for their broader usage. In one study of phenol adsorption by Pollio and Kunin, the free base form of a weak-base resin (tertiary amine) could be regenerated either with polar solvents such as methanol or with sodium hydroxide solution (Ref. 6-57).

The materials adsorbed on weak-base anion exchange resins include: phenol, *p*-nitrophenol, anionic surfactants, carboxylic acids, proteins (bovine serum

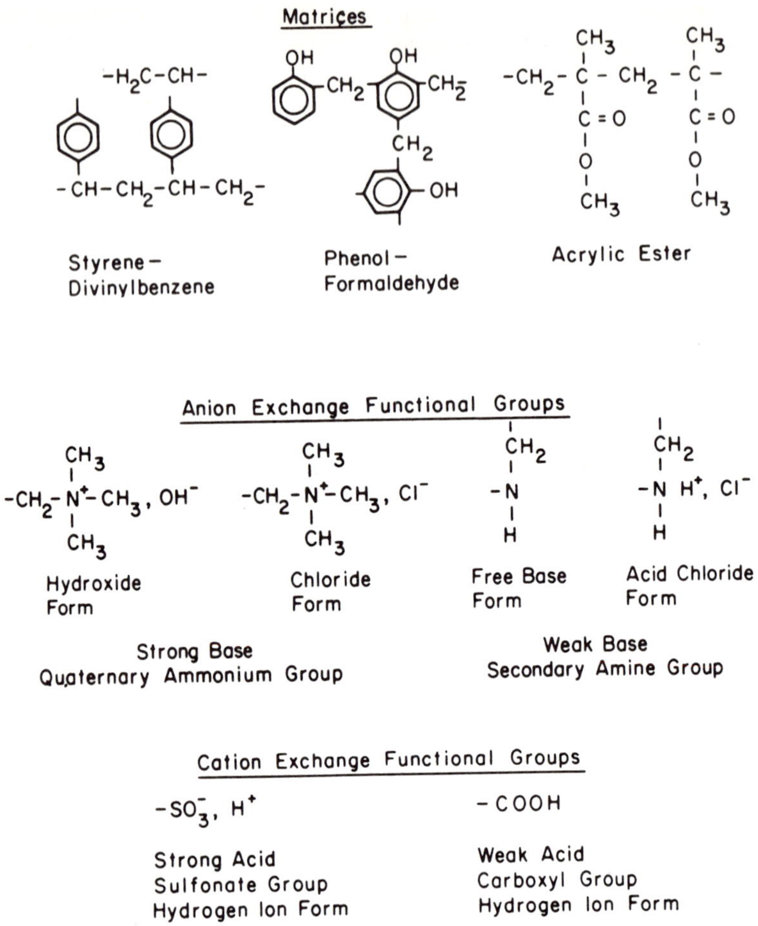

Fig. 6-17. Matrixes and functional groups of resins commonly used for water purification. ***Source:*** **Kim, B. R., Snoeyink, V. L., and Saunders, F. M., "Adsorption of Organic Compounds by Synthetic Resins,"** ***J. Water Pollut. Control Fed.,*** **48, 1, pp. 120-133, 1976.**

albumin), anionic dyes in textile wastes (Ref. 6-58), and color bodies in kraft paper wastes. In the case of *p*-nitrophenol, adsorption capacity was sensitive to pH level with a maximum at about pH 5. The effectiveness of the resins in some applications was reasonably close to that of activated carbon. Weakly acidic cation exchange resin readily removed cationic dyes. Nonionic resins were effective in removing chlorinated pesticides, detergent components, emulsifiers, wetting agents, dispersants, and all types of textile dyes. Materials with

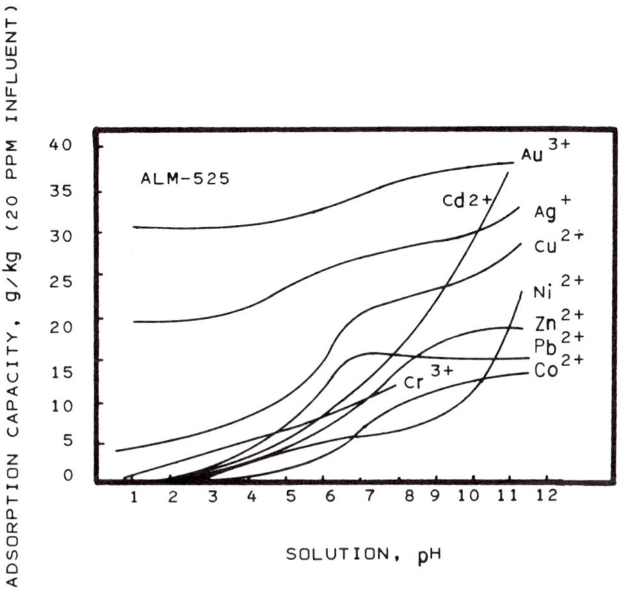

Fig. 6–18. ALM resins used in metal adsorption and sludge fixation. ***Source:*** **"Adsorbents Win Heavy Metals from Process Streams and Wastes,"** ***Chem. Eng.*****, pp. 49–50, September 29, 1975.**

hydrophilic and hydrophobic components are amenable to adsorption by the nonionic resins as the hydrophobic portion of the molecule is attracted to the hydrophobic polystyrene surface (Ref. 6-59).

Another application for the macroporous weak-base resins is in a precolumn to protect ion-exchange demineralizer resins from irreversible organic fouling. Also, Kim, *et al.* suggest the use of a weak-base resin preceding an activated carbon adsorption system, as the resin could be regenerated *in situ.* The frequency of thermal regeneration of the activated carbon would be reduced, as it would serve only as a polishing step.

The ALM series of adsorbents and resins (Nippon Soda Co.) is an example of a specific design to adsorb heavy metals (Figure 6–18). Although adsorption varies with concentration and pH, typically, these high molecular weight polymers with sulfur- and nitrogen-based chelating functional groups can reduce Hg levels to < 0.1 μg/liter, Cd to < 0.5 μg/liter, Pb to < 1 μg/liter, and Cu and Cr to < 10 μg/liter (Ref. 6-60). Nut wastes have been shown to be effective adsorbents for mercury (Ref. 6-61). Pine bark has been used for removal of hexavalent chromium and uranium from aqueous media (6-62).

6-12 ELECTROCHEMICAL

Electrochemical processes for pollution control have been applied for disinfection, odor control, color removal, coagulation, destruction of ammonia, electroflotation, and recovery of heavy metals. Several chemical and physical effects that occur through electrochemical treatment have been described by Rhees (Ref. 6-63).

Chemical oxidation occurring at the anode can result in pollutant removal in several ways. When consumable anodes are used, such as aluminum or iron, the generated ions precipitate phosphate (Ref. 6-64) and coagulate suspended solids, aiding in their physical removal. When nonconsumable anodes are used, such as lead dioxide, pollutants are affected by direct oxidation at the anode surface or by formation of oxidants (*e.g.,* chloride anions migrate to the anode and lose electrons to become chlorine) which subsequently oxidize pollutants or disinfect the wastewater.

Hydrogen is the primary product of cathode reaction although conversion of some pollutants to elemental form (e.g., copper cations gain electrons to become elemental copper) can be important. Secondary chemical effects occur which are related to high pH near the cathode and low pH near the anode.

According to Rhees, physical changes promoting separation of suspended solids also occur during electrolytic treatment. Solids subjected to the electric field on passing through the cell tend to coagulate due to polarization attraction.

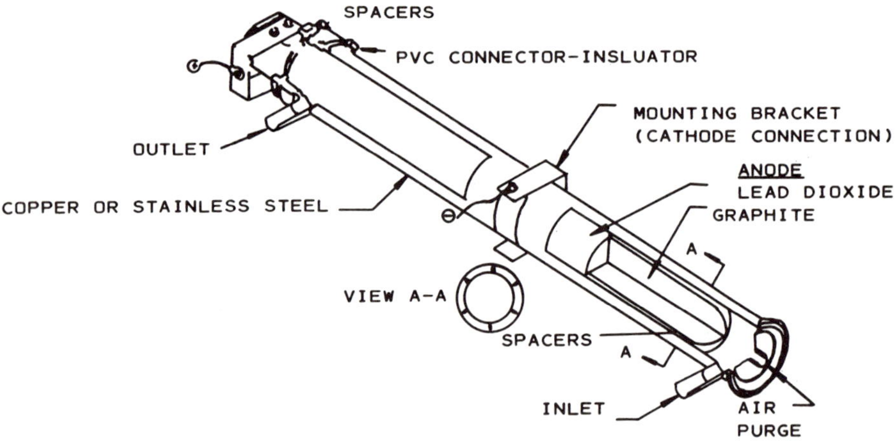

Fig. 6-19. Electrochemical Pepcon cell for *in-situ* chlorine generation. *Source:* Rhees, R. C., "Interface of Electrochemical and Conventional Methods for Pollution Control," Pacific Engineering and Production Co. of Nevada, Henderson, Nevada 89015, 1975.

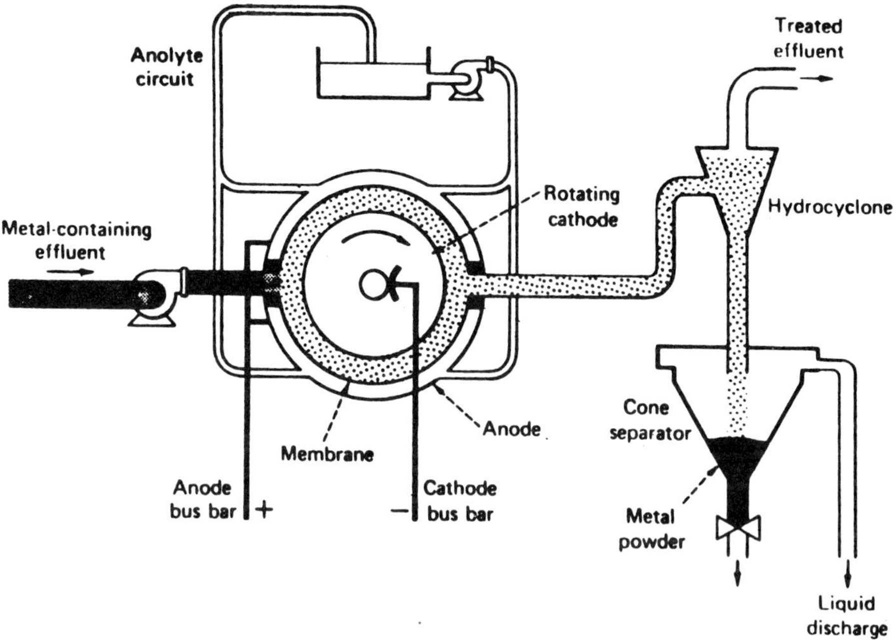

Fig. 6-20. Electrochemical Eco-cell for metal recovery. *Source:* **"Heavy-Metals Recovery Promises to Pare Water-Cleanup Bills,"** *Chem. Eng.*, **pp. 29-32, December 22, 1975.**

Partial oxidation of organic substances contributes to increased polarity which further enhances coagulation. Cell gases promote flotation of suspended impurities.

One type of unit consists of Pepcon cells (Pacific Engineering) grouped in modules (Figure 6-19). The metal tube, which serves as the cathode and container, is mounted on and supported by the cathode bus. The anode is a graphite rod electroplated with a dense, hard, impervious lead dioxide coating. The anode is inserted into the metal tubing and sealed in place with nonconducting supports and spacers. The graphite rod projects above the anode-cathode assembly and is electrically connected to the anode bus. The liquid waste or brine stream is pumped through the cell. Direct current is supplied from a suitable source.

Various streams within a conventional treatment system (e.g., supernatant from an anaerobic digester) can be passed through a module of electrolytic cells to achieve the effects previously cited. Brine must be added to achieve a suitable concentration of electrolyte. Another potential application is the oxidation of ammonia in brines used to regenerate ion-exchange systems. Also, solutions containing high levels of cyanide and heavy metals were electrolytically treated to form, primarily, ammonium, and cyanate ions, urea, and sodium carbonate (Ref. 6-65). Use of a semiconductive carbonaceous bed of charged particles (Resource Control) providing a relatively large effective surface area resulted in more complete cyanide oxidation (Ref. 6-66).

Electrolysis also can be used in metal recovery (Ref. 6-67). The continuous Eco-Cell (Ecological Engineering) reportedly overcomes the principal weaknesses of conventional electrolytic metal recovery devices by providing much higher mass transfer rates and using an ion-exchange membrane to shield the anode from corrosive impurities. As illustrated in Figure 6-20, the turbulent flow sweeps away the dendritic metal crystals that grow on the rotating cathode. Another approach, called extended Surface Electrolysis (du Pont), is based upon use of compact porous, sandwich-like electrodes to provide greater surface area than do conventional cells. The dilute wastewater flows through the porous electrodes, and metals plate out on the cathodes. Intermittent leaching removes the metals, which then can be recovered by flat-plate electrolysis.

6-13 HEAT REMOVAL

The wet process industries have been reported to be responsible for about 25% of the heat discharged to the environment; the steam-electric generating industries are associated with much of the remainder. Effects of a heated discharge on the receiving water biota include:

Depressed dissolved oxygen level due to increased microbial activity,
Rare, acute lethal effects in immediate area of discharge,
Rapid temperature fluctuation effects,
Increased sensitivity to toxic materials,
Shift in dominant species,
Increased yields of food fish.

Proper mixing and dispersion should be employed to utilize as fully as possible the capacity of the receiving water for acceptance and dissipation of excess heat. As an alternative, various approaches can be used to dissipate heat before discharge:

Cooling pond–large area required, heat transfer reportedly is about 35 Btu/ ft^2-hr (400 kJ/hr-m^2);
Spray pond–water sprayed 6 to 8 ft (1.8 to 2.4 m) into air;
Atmospheric towers–packed or spray, no mechanical inputs, low maintenance, long life, high initial cost, high pumping head, high wind losses, long and narrow;
Hyperbolic towers–high initial cost, low power, low maintenance, long life,

some concern that may not be efficient in hot summer months when most needed;

Mechanical draft—close control, small land area, low pumping head, lower capital cost than natural draft, high operating costs, high maintenance.

Krenkel and Parker have covered in detail the biological and engineering aspects of thermal pollution (Ref. 6-68, 6-69).

REFERENCES

6-1. Michaels, A. A., "New Separation Technique for the CPI," *Chem. Eng. Prog.,* **64**, 12, pp. 31-43, 1968.

6-2. Swope, G., "Westinghouse Enters the Reverse Osmosis Field," *Water Sewage Works,* pp. 252-253, August 1971.

6-3. Breslau, B. R., *et al.,* "Hollow Fiber Ultrafiltration," *Chem. Eng. Prog.,* **71**, 12, pp. 74-80, 1975.

6-4. Nelson, R. F., "Ultrafiltration for Polyglycol Removal," 2nd AIChE/EPA National Conference on Complete Water Reuse, Chicago, Illinois, May 4-8, 1975.

6-5. "Ultrafiltration Sets Sail for New Process Services," *Chem. Eng.,* pp. 88-9 February 18, 1974.

6-6. Rickles, R. N., "Waste Recovery and Pollution Abatement," *Chem. Eng.,* **72**, p. 133, September 27, 1965.

6-7. Sawyer, C. N. and McCarty, P. L., *Chemistry for Sanitary Engineers,* McGraw-Hill Book Co., New York, 1967.

6-8. Schmitt, C. R., "Don't Overlook Ion Exchange," *Water Wastes Eng.,* **43**, pp. 32-33, December 1971.

6-9. *Water Eng.,* **7**, 2, pp. 18-38, 1970.

6-10. "Converting Sewage into Savings," *Chem. Week,* p. 47, January 14, 1976.

6-11. Goldstein, M., "Economics of Treating Cyanide Wastes," *Pollut. Eng.,* pp. 36-38, March 1976.

6-12. *Waste Treatment, Upgrading Metal-Finishing Facilities to Reduce Pollution. Part 2,* EPA Technology Transfer Seminar Publication, U.S. Environmental Protection Agency, Washington, D. C., 1974.

6-13. Atkins, P. F. and Barnes, R. A., "Ammonia Removal in a Physical Chemical Treatment Plant," 27th Purdue University Industrial Waste Conference, Lafayette, Indiana, 1972.

6-14. Moyer, J. R. and Parmele, C. S., "Acetic Acid Removal from Brine by U. V. Catalyzed Chlorination," American Chemical Society's National Meeting, Boston, April 1972.

6-15. *Dyestuff Color Removal by Ionizing Radiation and Chemical Oxidation,* EPA-R2-73-048, U.S. Environmental Protection Agency, Washington, D. C., 1973.

6-16. *The Effect of Chlorination on Selected Organic Chemicals,* EPA 12020 EXG 03/72, U.S. Environmental Protection Agency, Washington, D. C., 1972.

6-17. Masscheiein, W., Fransolet, G., and Genot, J., "Techniques for Dispensing and Dissolving Ozone in Water," *Water Sewage Works,* pp. 57-60, December 1975.

6-18. Netzer, A. and Mijamoto, H. K., "The Biotreatability of Industrial Dye Wastes Before and After Ozonation and Hypochlorination-Dechlorination," Canada Centre for Inland Wastes, Burlington, Ontario, Canada, 1975.

6-19. "Ozone-Carbon Dye-Waste Treatment," *Text. Ind. (Japan)*, pp. 143-145, October 1974.

6-20. Eisenhauer, H. R., "Dephenolization by Ozonolysis," *Water Res.,* **5**, pp. 467-472, 1971.

6-21. Mauk, C. E., and Prengle, H. W., "Ozone with Ultraviolet Light Provides Improved Chemical Oxidation of Refractory Organics," *Pollut. Eng.,* **8**, 1, pp. 42-43, 1976.

6-22. Gould, J. P. and Weber, W. J., Jr., "Oxidation of Phenols by Ozone," *J. Water Pollut. Control Fed.,* **48**, 1, pp. 47-60, 1976.

6-23. Kibbel, W. H., Jr., Raleigh, C. W., and Shepherd, J. A., "Hydrogen Peroxide for Industrial Pollution Control," 27th Purdue University Industrial Waste Conference, Lafayette, Indiana, May, 1972.

6-24. Chen, K. Y. and Morris, J. C., "Oxidation of Aqueous Sulfides by O_2: General Characteristics and Catalytic Influences," *Proc. 8th Int. Water Pollut. Res. Conf.,* Pergamon Press, New York, 1971.

6-25. Chen, K. Y. and Morris, J. C., "Kinetics of Oxidation of Aqueous Sulfides by Oxygen," *Env. Sci. Technol.,* **6**, 6, p. 529, 1972.

6-26. Van Stone, G. R., "Treatment of Coke Plant Waste Effluent," *Ind. Wastes,* p. 23, July/August 1972.

6-27. Borkowski, B., "The Catalytic Oxidation of Phenols and Other Impurities in Evaporated Effluents," *Water Res.,* **1**, p. 367, 1967.

6-28. Hyde, A. C., "Chemical Plant Waste Treatment by Ten Methods," *J. Water Pollut. Control Fed.*, **37**, 11, pp. 1486–1494, 1965.

6-29. Ely, R. P., "Wet Air Oxidation," *Pollut. Eng.*, pp. 37–38, May 1973.

6-30. Gellman, I., "Physical–Chemical Effluent Treatment in the Paper Industry," U.S.-U.S.S.R. Symposium on Physical Chemical Treatment, Cincinnati, Ohio, November 12–14, 1975.

6-31. Garrison, A. W., Case, F. N., Smiley, D. E., and Kau, D. L., "The Effect of High Pressure Radiolysis on Textile Wastes, Including Dyes and Deldrin," *Proc. 5th Int. Water Pollut. Res. Conf.*, Pergamon Press, New York, 1971.

6-32. Condren, A. J. and Etzel, J. E., "Destruction of Dilute Organic Pollutants by Gamma Irradiation," *Effluent Water Treat. J.*, **7**, pp. 528–532, 1976.

6-33. *Chemical Engineers' Handbook*, 5th Ed., R. H. Perry, C. H. Chilton, (eds.), McGraw-Hill Book Co., New York, 1973.

6-34. Mayhue, L. F., *Solvent Extraction Status Report*, EPA-R2-72-073, U.S. Environmental Protection Agency, Washington, D. C., 1972.

6-35. Kiezyk, P. R. and MacKay, D., "The Screening and Selection of Solvents for the Extraction of Phenol from Water," *Can. J. Chem. Eng.*, **51**, pp. 741–745, December 1973.

6-36. Reid, G. W. and Streebin, L. E., *Evaluation of Wastewaters from Petroleum and Coal Processing*, EPA-R2-72-001, U.S. Environmental Protection Agency, Washington, D. C., 1972.

6-37. Beychok, M. R., *Aqueous Wastes from Petroleum and Petrochemical Plants*, 2nd ed, John Wiley and Sons, New York, 1978.

6-38. Gaudy, A. F., Jr., Engelbrecht, R. S., and Turner, B. G., "Stripping Kinetics of Volatile Components of Petrochemical Wastes," *J. Water Pollut. Control Fed.*, **33**, 4, pp. 382–392, 1961.

6-39. Willard, H. K., "Pulp Mill Wastewater Stripping by Air and Steam: Current Efforts for BOD and Carbon Removal," 27th Purdue University Industrial Waste Conference, Lafayette, Indiana, 1972.

6-40. Butyrun, G. L. and Ayers, K. C., "Mead Experiences in Steam Stripping Kraft Mill Condensates," *TAPPI*, **58**, 10, pp. 78–80, 1975.

6-41. Rich, L. G., *Unit Operations of Sanitary Engineering*, John Wiley and Sons, New York, 1961.

6-42. Stein, J. L., *Submerged Combustion Evaporator for Concentration of Brewery Spent Grain Liquor*, EPA-660/2-74-059, U.S. Environmental Protection Agency, Washington, D. C. 20460, 1974.

6-43. Stickney, W. W. and Fosberg, T. M., "Chemical Waste Treatment by

Evaporation," 80th National Meeting of American Institute of Chemical Engineers, Boston, Massachusetts, September 7–10, 1975.

6–44. "Distillation Goes to Work on Waste," *Chem. Week,* p. 51, August 5, 1970.

6–45. Bard, A. J., *Chemical Equilibrium,* Harper and Row, Publishers, New York, 1966.

6–46. Leckie, J. and Stumm, W., "Phosphate Precipitation," *Water Quality Improvement by Physical Chemical Means,* E. F. Gloyna and W. W. Eckenfelder, Jr., (eds.) University of Texas Press, Austin, Texas, 1970.

6–47. Marston, P. G., "The Application of High Gradient Magnetic Separation to Municipal and Industrial Wastewater Treatment," 47th Annual Water Pollution Federation Conference, Denver, Colorado, October 1974.

6–48. Oberteuffer, J. A., "High Gradient Magnetic Separation," *IEEE Transactions on Magnetics, MAG–9,* 3, pp. 303–306, 1973.

6–49. Oberteuffer, J. A., "Magnetic Separation: A Review of Principles, Devices, and Applications," *IEEE Transactions on Magnetics, MAG–10,* June 1974.

6–50. DeLatour, C., "Magnetic Separation in Water Pollution Control," *IEEE Transactions on Magnetics, MAG–9,* 3, pp. 314–316, 1973.

6–51. Bolts, B. A., *et al.,* "Magnetic Filter Aids," *Chem. Eng. Prog.,* **71,** 12, pp. 47–50, 1975.

6–52. *Adsorptive Bubble Separation Techniques,* R. Lemlich, (ed.), Academic Press, New York, 1972.

6–53. Ng, K. S. and Mueller, J. C., "Foam Separation—A Technique for Water Pollution Abatement," *Water Sewage Works,* pp. 48–55, June 1975.

6–54. Wang, M-H., *et al.,* "Removal of Lignin from Water by Precipitate Flotation," *J. Environ. Eng. Div., Proc. Am. Soc. Civil Eng.,* **100,** EE 3, pp. 629–639, 1974.

6–55. Becker, E. R., *BOD, Solids and Nutrient Removal by Foam Flotation,* EPA PB–232184, U.S. Environmental Protection Agency, Washington, D. C., 1974.

6–56. Kim, B. R., Snoeyink, V. L., and Saunders, F. M., "Adsorption of Organic Compounds by Synthetic Resins," *J. Water Pollut. Control Fed.,* **48,** 1, pp. 120–133, 1976.

6–57. Pollio, F. Y. and Kunin, R., "Sorption of Phenol by Ion Exchange Resins," *Environ. Sci. Technol.,* **1,** p. 160, 1967.

6–58. Snyder, A. J. and Alspaugh, T. A., *Catalyzed Bio-Oxidation and Tertiary*

Treatment of Integrated Textile Wastewaters, EPA-660/2-74-039, U.S. Environmental Protection Agency, Washington, D. C. 20460, 1974.

6-59. "New Resins Adsorb Water Soluble Organics," *Chem. Eng. News,* p. 54, January 17, 1966.

6-60. "Adsorbents Win Heavy Metals from Process Streams and Wastes," *Chem. Eng.,* pp. 49-50, September 29, 1975.

6-61. Waiss, A. C., Jr., Kuhule, J. A., and Wiley, M. E., "Removal of Mercury from Water Using Nut Wastes," PAT-APPL-154369, U. S. Dept. of Agriculture, Washington, D. C., March 6, 1973.

6-62. Cook, G. T., "Use of Pine Bark for the Removal of Metal Ions from Aqueous Media," Tech. Report KY-L-790, Union Carbide Corp., Paducah, Kentucky 42001, 1976.

6-63. Rhees, R. C., "Interface of Electrochemical and Conventional Methods for Pollution Control," Pacific Engineering and Production Co. of Nevada, Henderson, Nevada 89015, 1975.

6-64. Sudek, S. E., *An Electrochemical Method for Removal of Phosphates from Wastewater,* EPA 17010-02/70, U.S. Environmental Protection Agency, Washington, D. C., 1970.

6-65. Dart, M. C., Gentiles, J. D., and Trenton, D. C., "Electrolytic Treatment of Strong Cyanide Waste," *J. Appl. Chem.,* **13**, p. 55, 1963.

6-66. "New Device for Electroplaters," *Chem. Week,* pp. 22-23, February 4, 1970.

6-67. "Heavy-Metals Recovery Promises to Pare Water-Cleanup Bills," *Chem. Eng.,* pp. 29-32, December 22, 1975.

6-68. *Biological Aspects of Thermal Pollution,* P. A. Krenkel and F. L. Parker, (eds.), Vanderbilt University Press, Nashville, Tennessee, 1969.

6-69. *Engineering Aspects of Thermal Pollution,* P. A. Krenkel and F. L. Parker, (eds.), Vanderbilt University Press, Nashville, Tennessee, 1969.

APPENDIX 6-1

MEMBRANE FILTRATION TERMS AND NOTATION

ΔC_i = impurity concentration difference across *RO* membrane
F_i = impurity (solute) flux through *RO* membrane
F_u = water flux through *UF* membrane
F_w = water flux through *RO* membrane
M_i = *RO* membrane constant incorporating impurity (usually salt) diffusion coefficient and other factors
M_w = *RO* membrane constant incorporating water diffusion coefficient and other factors
ΔP = pressure difference across membrane
$\Delta \pi$ = osmotic pressure difference across membrane
R_g = boundary-layer gel resistance in *UF* system
R_m = *UF* membrane resistance involving pore radius, pore tortuosity and other factors
RO = reverse osmosis
UF = ultrafiltration

7

Ultimate disposal methods for wastewater, sludges, and residues

An overall analysis of the origin of various types of wastes, treatment alternatives, and disposal approaches is presented in Section 1-1 and summarized in Table 1-1 and Figure 1-1. Section 1-2 covers means of characterizing the wastes, while the establishment of treatment requirements is discussed in Section 1-3. A detailed list of treatment operations and processes is presented in Table 1-2; the ones of major utility are discussed in Chapters 2 to 6. In this section the ultimate disposal of the treated wastewater and by-products (e.g., sludges, brines) is covered. Table 7-1 presents an overview of the ultimate disposal alternatives. More details concerning operations and processes also applicable to solid wastes are presented in subsequent chapters.

TABLE 7-1. Major Ultimate Disposal Alternatives for Treated Industrial Wastewaters, Organics and Solids

Surface Waters
Publicly Owned Treatment Systems
Land
Liquid wastes
solar evaporation ponds
irrigation
Solids
soil cultivation and amendment (land farming)
landfill
Incineration/Evaporation/Oxidation
Incineration
multiple-hearth
fluidized bed
rotary kiln
multiple chamber
vortex burners and furnace
thermal pit
special design/multipurpose
off-shore incineration
Evaporation/drying
Oxidation (organic stabilization)
chemical
biological
physicochemical
electrochemical
Deep-well injection
Deep ocean disposal
Outfall diffusion
Barge transport
Reuse and recovery
Water reuse
Product recovery
Conversion to useful material
Energy recovery
Pyrolysis

7-1 SURFACE WATERS AND PUBLICLY OWNED TREATMENT SYSTEMS

7-1.1 Surface Waters

Most dilute wastewaters eventually are discharged into surface waters after appropriate treatment. Proper dispersion of this effluent in the receiving river, lake, estuary, or ocean is frequently neglected in the design of disposal facilities.

Both the location of the discharge point and the type of dispersion mechanism used are of considerable importance in protecting other water users and in avoiding unsightly conditions. Rapid dispersion does not increase dilution, but speeds up the process and thereby avoids localized undesirable conditions.

Surface discharge via a pipe or ditch leading to the shore-line is the least expensive approach, but problems can be expected, e.g., the formation of foam due to air entrainment if the wastewater spills down the bank during low-water conditions and the incomplete dilution for several miles downstream due to low horizontal mixing rates and density differences. A boom arrangement has been developed to divert surface pollutants from ditches leading to final retention ponds or receiving waters (Figure 7-1).

A properly designed subsurface dispersion system can allow the full assimilative capacity of the receiving body to be utilized. Submerged discharge devices include: open-end pipes, nozzle-end pipes, diffuser systems consisting of a closed-end pipe with slots or holes along it, and split discharges through a branched-pipe system (Ref. 7-1). The discharge can be directed at the best angle to effect rapid dispersion. Dispersion devices have been installed specifically to protect against shoreline contamination, to protect the plant intake water, to prevent oil slicks, and to prevent fog formation.

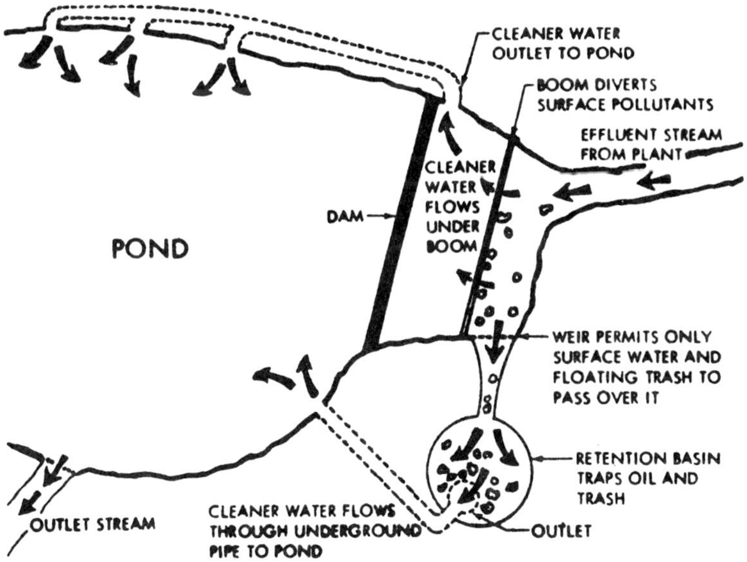

Fig. 7-1. Boom for diverting surface pollutants at Union Carbide's Oak Ridge, Tennessee, Plant.

The design of the dispersion system is dependent upon the uses of the receiving water body and the location of any nearby water intakes, the flow and turbulent nature of the receiving water, and the relative physical and chemical characteristics of the effluent and the receiving stream. The vertical and horizontal mixing action in the receiving rivers should be defined, as should the tidal action in estuaries and the wave action and thermal currents in lakes into which the wastes are to be discharged. Current patterns in relation to water intakes should be established. Density differences, due to the temperature levels and solids contents of the wastewater and receiving water, tend to resist mixing. A comprehensive review of design considerations for dispersion systems has been published by the American Petroleum Institute (Ref. 7-1).

The amount of self-purification that could occur in the receiving water depends on its flow rate, oxygen content, and reoxygenation rate. The portion of this natural purification capacity that can be utilized in assimilating municipal and industrial wastes primarily is a decision to be made by regulatory agencies and political bodies under constraints of existing laws with input from all interested parties. Standards can be set either on the basis of maintaining a given set of conditions in the receiving water ("receiving water standards") or on the quality of the effluent allowed to be discharged ("effluent quality standards"). The process of standard setting, including means of predicting the impact of a given waste load on a lake, river, or estuary has been described by Tchobanoglous (Metcalf and Eddy, Ref. 3-5). Biodegradable material, as well as conservative pollutants and heat, need to be considered. Low flow augmentation through the use of impoundments with controlled release is a means of increasing the assimilative capacity of a river. Artificial reaeration of streams has been practiced in Germany (Ref. 3-17).

7-1.2 Publicly Owned Facilities

With regard to the discharge of industrial waste to publicly owned treatment facilities, both the projected advantages of the joint treatment and the pretreatment requirements need to be considered. Potential technical advantages include:

Increased dilution of constituents
Utilization of nutrients and seed source in domestic wastewater
Improved location as to final discharge of treated effluent
Improved operation at a larger facility

Economic factors, political entanglements, and regulatory agency relationships also need to be considered. Standards for pretreatment by various industries to remove pollutants incompatible with a publicly owned treatment system have been specified by the U.S. Environmental Protection Agency (Ref. 7-2). Typical pretreatments are equalization, neutralization, cyanide removal, chromium reduction, precipitation of heavy metals, solids separation, oil separation, and spill protection.

7-1.3 Monitoring

Whether discharge is to surface water or to a publicly owned treatment facility, effluent monitoring is essential. Information of concern is listed in Table 1-3. Laboratory test methods have been tabulated by Cox along with sampling information (Ref. 7-3). Information concerning on-stream sampling, monitoring, and analytical instrumentation has been compiled by Chermisinoff (Ref. 7-4). Benefits of a comprehensive sampling and monitoring program, before and after the treatment system, include:

Provision of data reflecting efficiency of manufacturing units,
Aid in improving operation of treatment system,
Maintenance of record of final discharge,
Aid in ensuring that environmental standards set by regulatory agencies are being met and to allow variances to be detected early and controlled.

Continuous on-stream analysis should be augmented by laboratory analysis of samples composited proportional to flow, as well as by grab samples collected whenever one of the continuously monitored parameters exceeds the norm.

Accurate flow measurement in pressure and gravity sewers can be made continuously through the use of proportional or rectangular weirs, orifices and venturi tubes, prop meters, flumes, and impact meters. Spot checks or periodic measurements are accomplished with varying degrees of accuracy by using portable weirs and flumes, velocity meters, timing of floats over known distances, falling head measurement, and dye dispersion. Caution should be exercised when using the dye-dispersion technique because certain organics present in process sewers tend to react with the dyes. Where continuous flow measurement is incorporated, indicating recorders are easily installed to provide permanent records of instantaneous and total flow, to actuate alarms, and to trigger sampling devices.

Where gravity sewers are used, samples may be obtained continuously with peristaltic pumps, steam or air ejectors, motorized bucket or dip samplers, and vacuum pumps. Samples composited in proportion to flow are obtained by interconnecting the actuating mechanism of the sampler to the flow totalizer of an indicating flow recorder. Each time a given quantity of waste flows past the weir, a constant-volume sample is obtained. Common sampling problems include stripping of volatiles, dilution by steam, disproportionate collection of solids and immiscible liquids, and degradation after collection.

Pressure sewers may be continuously sampled by drip sampling through a cracked valve or by using electrically operated valves with an actuator such as a timer or flow totalizer.

Continuous analyses of pH, total carbon, conductivity, and specific constituents by gas chromatography are used for controlling processes and protecting treatment facilities. Total carbon, pH, and conductivity warn against spills or abnormal fluctuations in waste characteristics, and let treatment plant operators bypass flow to storage basins until normal conditions are resumed. Stored waste is later bled back into the system for treatment, thus, eliminating shock loads and reduced plant efficiencies. Gas chromatograms point directly to the source of the spilled material. Specific ion electrodes can be used for on-stream monitoring of problem inorganic materials. Knowing normal concentrations of specific chemicals in the waste stream and knowing the origin of each chemical, the operator can quickly analyze the problem and notify the correct production unit of the increased discharges. Product savings can pay for the installation of such monitoring systems.

Grab and composite samples from individual units can be analyzed in the laboratory for many of the components shown in Table 1-3. Records are kept, and a trend in waste discharge is obtained. Upward variations from this trend indicate a potential problem, and a closer than normal watch should be maintained. In this way, many small problems are corrected before they become major ones. Continuous carbon analyzers can be installed on cooling water outfalls with benefits that are readily recognizable. Small heat-exchanger leaks that previously went undetected can be discovered in early stages, thus reducing costly chemical losses, minimizing maintenance expense, and protecting the receiving stream from discharges of untreated organics. For complex plants computerized monitoring stations can be used. Using computers, flows are totalized, multiple samplers are actuated, carbon content is continuously totalized, alarm systems are operated, and summation of data is available upon demand.

Any worker wastes can be treated most economically along with the process streams, but state regulations frequently require separate treatment facilities, unless adequate proof is acquired to show that the pH or constituents in the mixed waste reliably serve for disinfection.

As a final note, the matter of data handling and reporting needs consideration. At the outset of the monitoring program, a system for efficiently reducing the measurement data to a useable form needs to be conceived. Such data should be transmitted promptly to the interested parties. The industrial personnel involved in in-plant control work and treatment of the wastewater need to be provided with a continual measure of their performance in meeting the established goals. These data also provide a basis for distributing charges to the source units. When discharge is to a publicly owned facility, those operating personnel need to be kept informed both of abnormal discharges in order to make process adjustments and of trends in order to plan long-term system modifications. Data should be available for the prompt answering of inquiries concerning involvement in any observed adverse effects on the receiving system. Regulatory agencies need to be informed promptly of reportable spills. Adequate communications constitute the cornerstone of successful control programs.

7-2 LAND

Whenever practical, some dilute industrial wastewaters are best disposed of on land, as are many residues, such as process filter cakes, oily sludges, and wasted biomass from biological treatment systems. Land disposal techniques for dilute wastewaters include solar evaporation ponds and various types of irrigation, while slurries can be worked into the ground (soil cultivation) or landfilled after being dewatered.

7-2.1 Solar Evaporation Ponds

In situations where the evaporation rate exceeds rainfall, the use of solar evaporation ponds for handling small or problem wastestreams should be considered. Factors to be included in such an evaluation include:

Standard evaporation rate, as determined in a U.S. Weather Bureau Class A Pan filled to a depth of 8 inches (20.3 cm);
Supporting meteorological information regarding precipitation, air temperature, wind velocity, relative humidity;
Proposed reservoir design in terms of surface area and depth;
Concentration of surfactants and other evaporation suppressants in the wastewater;
Infiltration characteristics and need for ground-water protection;
Oxidation of organic matter by biological, chemical, and photochemical means; volatilization (Ref. 7-5);
Water salinity and temperature.

This method can provide an inexpensive means of achieving zero discharge to surface waters. Evaporation ponds are used extensively in the southwesternly part of the United States.

7-2.2 Irrigation

For wastestreams of relatively low flow rate containing readily biodegradable constituents, irrigation offers an alternative disposal means. In the spray irrigation process, the waste is intermittently applied through raised sprinklers to acreage covered with various types of vegetation, and the wastes percolate through the soil. Another land disposal means is overland flow which is used for soils high in clay; slopes are 2 to 6%. An infiltration method is used where ground water recharge is a prime concern. These procedures are depicted in Figure 7-2 and have been described by Godfrey (Ref. 7-6).

The wastewater is applied to the land at such a rate as to nurture the cover vegetation, since the root systems provide substantial increases in percolation rates and the leaf systems consume additional water by evapotranspiration. Inert suspended matter can clog the soil, while extreme pH levels and sodium ion content may harm the cover crop or change the flocculent nature of the soil. Stabilization of the organic matter contained in the wastewater is accomplished by bacteria attached to the soil particles through which the wastewater percolates.

Some important considerations as described by Snyder (Ref. 7-7) for spray irrigation of paperboard wastes include:

Selection of an appropriate site which is large enough, close to the manufacturing facility, and as a safety factor, has expansion possibilities. Normal loading rates would be in the order of 0.7 to 1.0 in./day (1.8 to 2.5 cm/day) and 50 to 200 lb BOD/acre-day (56 to 230 kg/ha-day); although some situations fall with a broader range of 0.25 to 2 in./day (0.6 to 5.1 cm/day) and 25 to 550 lb/acre-day (28 to 620 kg/ha-day).

High soil permeability is preferred but is not essential; an underdrain system may be installed with provisions for effluent disposal depending upon local water quality standards.

Grasses that can survive under high moisture and the contained impurities need to be selected for cover.

Sufficient storage needs to be provided for periods of cold weather when spraying is impractical.

Sprinklers can be laid out on a triangular pattern, 100 ft (30 m) apart; each covers a wetted area with a radius of 62 ft (19 m) at a spraying pressure of 49 psig (340 kN/m^2). The quality of groundwater, runoff, and percolate collected in underdrains needs to be monitored.

7-2.3 Soil Cultivation and Amendment (Landfarming)

Many sludges from municipal facilities are spread or sprayed from tank wagons on croplands. An alternative approach used for land disposal of oily industrial solids is direct cultivation into the soil (Ref. 7-8, 7-9). Some concerns over solids disposal on land include:

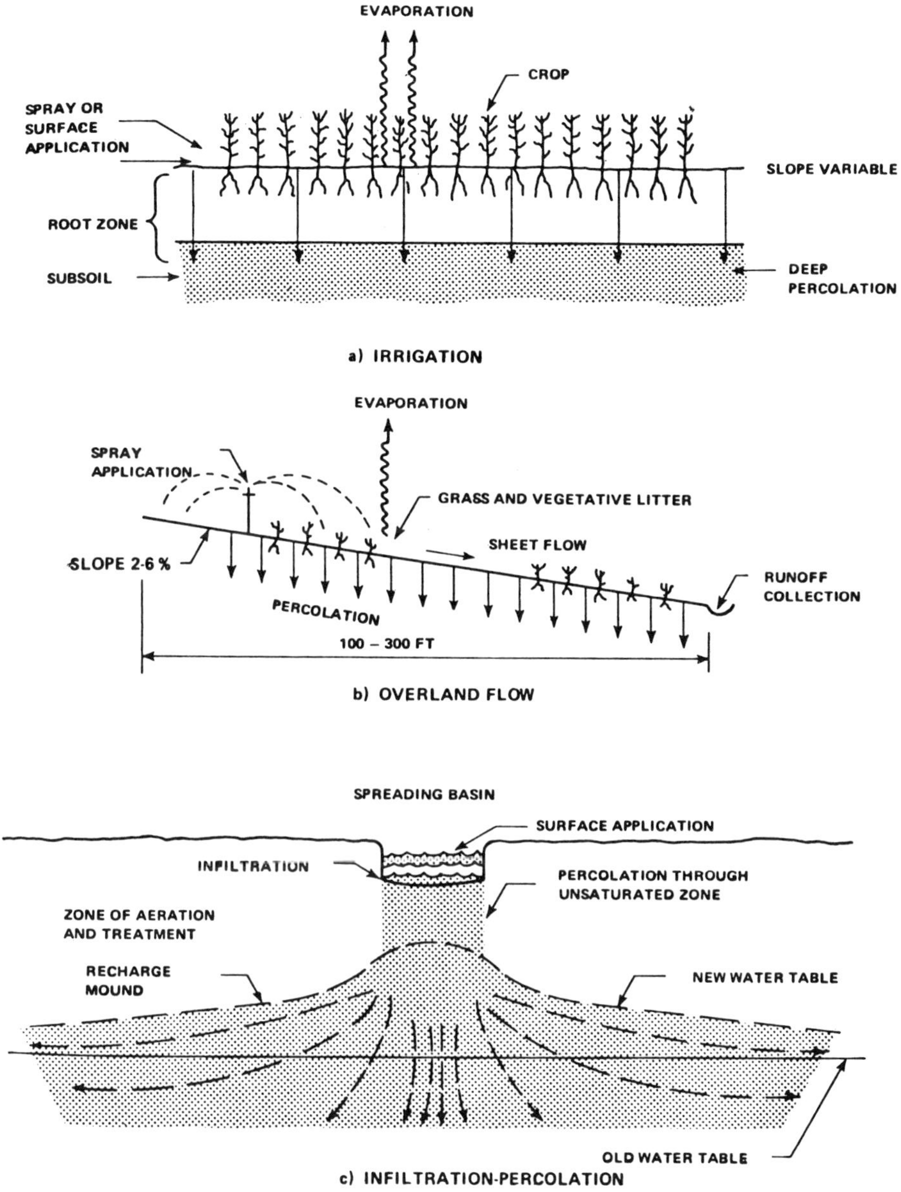

Fig. 7-2. Three irrigation approaches to land treatment *Source:* Alternative Waste Management Techniques for Best Practicable Waste Treatment, EPA-430/9-75-013, U.S. Environmental Protection Agency, Washington, D.C., 20460, 1975.

Metals in sludge being released to ground or surface waters and/or accumulating in cover crops,
Other conservative toxic constituents of the sludge,
Ammonia conversion to nitrate ions which can cause problems in groundwater,
Pathogens,
Hydraulic overloading of soil,
Effects on vegetation growth rate,
Runoff and erosion during rainfall,
Aesthetic objections,
Ultimate use of the land.

Allowable levels and control means have been described by the U.S. Environmental Protection Agency in their basic reference on sludge disposal (Ref. 4-2), as well as in more current guidelines (Ref. 7-10, 1-20). A review of landspreading of liquid municipal sewage sludge was made by Carroll, *et al.* (Ref. 7-11).

Application methods include ridge and furrow irrigation, spray irrigation, and mechanical incorporation of the sludge into the soil. Application rates depend on sludge composition, soil characteristics, climate, vegetation, and cropping practices. A rough guide for materials similar to a stabilized (equivalent to anaerobic digestion) municipal sludge is 10 tons dry solids/acre-year (22,000 kg/ha-year) when applied to cropland. Applications to low-quality land can be higher but may lead to water contamination if drainage and runoff controls are not installed. Leachates that are unsuitable for groundwater recharge may be intercepted by tile drains for subsequent treatment. Surface runoff also can be impounded and treated (Ref. 4-2).

A total, all-time limit dependent upon the soil's cation exchange capacity and the metal content of the sludge may be imposed (Ref. 7-12, 7-13):

$$\text{dry tons/acre} = \frac{32{,}700 \times CEC}{\text{mg/kg Zn} + 2\,(\text{mg/kg Cu}) + 4\,(\text{mg/kg Ni}) - 200}$$

where CEC = cation exchange capacity of the unsludged soil in meq/100 g

$$\text{dry tons/acre} \times 2245 = \text{kg/ha}$$

This tentative criterion may be altered according to research findings. Maximum metal contents for cropland application are listed in Table 7-2. These are to be updated in RCRA regulations, along with other criteria on waste compatibility, height above water table, soil porosity and pH, and monitoring (Ref. 1-20).

The soil cultivation process has been applied to oily wastes by Kincannon at Deer Park, Texas (Ref. 7-8). The rate of microbial decomposition of the oil was about 0.5 lb of oil per cubic foot of soil per month (8 kg/m^3-month) without fertilizer and twice that with fertilizer added. Infiltration problems were not

TABLE 7-2. Tentative Maximum Metal Content of Sludges for Cropland Application

Metal	Maximum Content of Dry Sludge (mg/kg)
Cadmium (Cd)	10
Mercury (Hg)	10
Copper (Cu)	100
Nickel (Ni)	200
Lead (Pb)	1000
Zinc (Zn)	2000

Source: Acceptable Methods for the Utilization of Sludge, Technical Bulletin, U.S. Environmental Protection Agency, Washington, D.C., November 1974.

encountered; rainfall runoff contained up to 100 mg/liter naphthenic acids and 150 mg/liter ammonia-nitrogen. Francke and Clark report a rate of 1.5 lb oil/ft^3-month (24 kg/m^3-month) (Ref. 7-14.)

As reported for another refinery by Dotson, *et al.* (Ref. 7-9), oily wastes and emulsions were applied to a depth of 6 in. (15 cm), and a bulldozer with a scraper blade mixed the sludge with 6 in. (15 cm) of soil. The mixing was repeated 2 to 4 times per month until the hydrocarbons were consumed by bacterial action. Decomposition required 3 to 9 months, depending upon ground temperature, soil moisture content, and type of hydrocarbon. Oily decomposition rates of 5 to 60 lb/ft^3-month (80 to 960 kg/m^3-month) were reported to be possible in an initial small pilot study.

7-2.4 Landfill

Landfilling of dewatered, stabilized excess biomass can be accomplished either alone or with refuse by conventional techniques. The solid wastes must be systematically deposited and covered with earth to control environmental impacts within defined limits. Leachate and runoff should be minimized and, when necessary, collected and suitably treated to prevent pollution of ground and surface waters. Wastes should be spread in layers not over 2 ft (0.6 m) thick, followed by compaction. A minimum of 6 in. (15 cm) of suitable compacted earth must be added as cover at the end of each working day. When each portion or cell of the landfill is completed, a minimum of 2 ft (0.6 m) of compacted earth should be placed over it and suitable grass planted to prevent erosion (Ref. 4-2). For hazardous solid wastes, additional requirements are imposed under the Resource Recovery and Conservation Act of 1976, as discussed in a subsequent chapter (Ref. 1-20).

A chemical landfill has been described by Slover for handling pitch, tars,

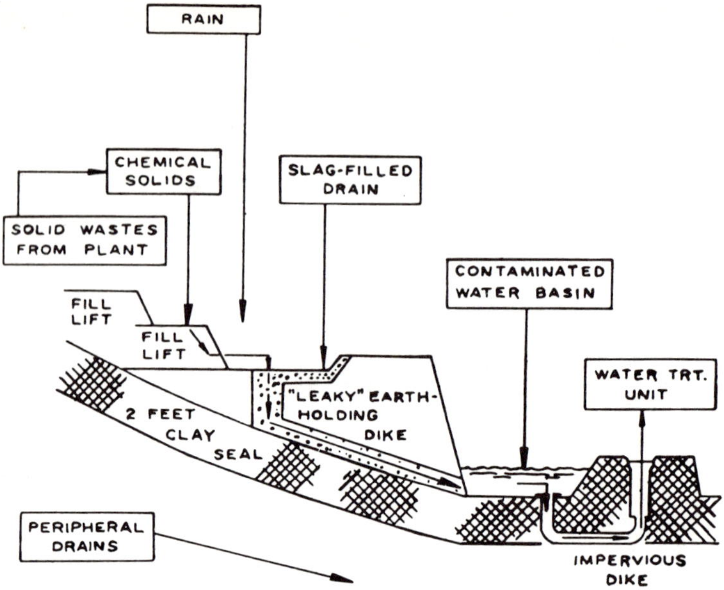

Fig. 7-3. Chemical Landfill at Union Carbide's Institute Plant. ***Source:*** **Slover, E. E., "Solid Waste Disposal in a Multi-Product Chemical Plant,"** ***The Textile Industry and the Environment,*** **American Association of Textile Chemists and Colorists, Research Triangle Park, North Carolina, 1973.**

polymers, inorganic solids, and viscous, insoluble organic products trapped in filter papers and filter aids (Ref. 7-15). Some key features of this operation, depicted in Figure 7-3, are:

- Clay-sealed bottom
- Peripheral interception of rainfall
- Runoff and leachate collection for treatment
- Pervious earth and rock basin to stabilize fill mass
- Wastes neutralized and stabilized before landfilling
- Chemical wastes blended with equal volume of relatively porous earth before compaction
- pH maintained at near neutral by limestone addition

The intent is to promote rapid microbial conversion of the chemical wastes to inert end products. The cell-type construction used in conventional landfills did not promote rapid degradation, prevented venting of gases, and caused miring of the equipment.

Detailed information of this and other landfill techniques for solid waste disposal can be found in Chapter 11 of this book.

7-3 INCINERATION

Incineration is a controlled process for oxidizing solid, liquid, or gaseous combustible wastes primarily to carbon dioxide, water, and ash. Any contained sulfur, nitrogen, metals, halogens, and impurities of other elements are converted to various end products, the identity of which depends on the degree to which equilibrium (end products contain minimum amount of free energy) conditions are approached in the particular incineration device. Liquid wastes and treatment plant residues considered for disposal by this process include:

Wasted biomass from biological treatment systems
Slurry streams and oils from clarification systems
Wastewater high in organic content ($>$ 5 to 10%)
Hazardous or toxic materials
Waste solvents and tars

Preprocessing steps include liquid-liquid separation as described in Chapter 4 and flash drying of solids. Means for calculating oxygen and heat requirements have been described by Tchobanoglous (Metcalf and Eddy, Ref. 3-5) and by Monroe (Ross, Ref. 7-23).

Descriptions of the incineration processes are presented in the solid-wastes chapters of this book; mainly comments concerning the applicability of these alternatives to liquid-waste disposal are offered in this section based largely on descriptions by Witt (Ref. 7-16). Incineration processes also have been reviewed by Dunn (Ref. 7-17).

Multiple Hearth Furnace. This incinerator consists of vertically stacked hearths raked by rabble arms rotated from a common shaft. Dewatered biomass and other sludges are fed to the top hearth; progressive drying and then combustion occurs as the solids drop to successively lower hearths. Air emissions need to be controlled, e.g., with an afterburner for odors and a high-pressure-drop venturi scrubber for particulates.

Fluidized Bed. Sand fluidized by the combustion air is commonly used as the heat-transfer agent in a fluidized-bed incinerator. Temperature in the bed is maintained at 1200 to 1400°F (650 to 760°C). The gases above the bed are hotter, which serves to control odors. When a combustible particle is introduced, the transfer of energy is rapid, and the particle quickly reaches its ignition temperature. This system is the major alternative to the multiple hearth furnace for incineration of dewatered sludges. Ferric chloride conditioning agent reportedly can present slagging problems. Capital costs would be expected to be 20 to 50% higher with a fluidized bed than with a multiple hearth incinerator, but operating costs would be 30 to 50% lower. After suitable modification, its successful use for disposal of spent caustic and oily sludge from a petroleum refinery was demonstrated by American Oil Co. (Ref. 7-18).

Rotary Kiln. This type incinerator consists of a refractory-lined, inclined cylinder which is slowly rotated. It is versatile in being able to handle a range of low-ash liquid residues and solid chemical wastes, such as plastics, powders, and filter cakes.

Vortex Furnace and Burner. The vertical, cylindrical vortex furnace is fired tangentially with a modified oil burner; as the gases rise, preheated high-velocity secondary air is introduced from tangential tuyeres maintaining the vortex. The heat release of this unit is unusually high, about 100,000 Btu/ft^3-hr (37×10^6 kJ/m^3-hr), which lowers capital costs but can cause refractory problems (Ref. 7-16). The Vortex burner (Trane Thermal Company) is a tangentially fired device with a short, open combustion chamber. This burner is designed to achieve complete or equilibrium combustion in a short retention period. Liquid wastes either high in organic content or containing problem materials are handled in this type facility. Where air quality criteria can be met, clean burning materials can be handled in a pedestal-mounted, oil burned firing horizontally into the atmosphere; a capacity of 300 gal./hr (1140 liters/hr) of organic liquid containing up to 25% water was processed (Ref. 7-16). If wastes contain more than 8000 Btu/lb (440 cal/g), their combustion will be self-sustaining (Ref. 7-19). An integrated system involving a submerged exhaust quench has been developed for recovery of waste heat and chemicals (Ref. 7-20).

Thermal Pit. This type of incinerator consists of a refractory lined pit with an open top; along the top and to one side is an air manifold with a number of nozzles angled toward the bottom of the pit. Largely clean-burning solids are handled in this facility with only incidental addition of residues derived from liquid waste treatment operations. Air quality standards may preclude its use.

Multiple Chamber. The two fundamental types of multiple-chamber incinerators are the retort-type with a U-shaped path for combustion gas for smaller installations and the in-line type for capacities above 1000 lb/hr (450 kg/hr). Successive chambers are provided for ignition, mixing with secondary air, and combustion. These are not specifically designed to handle liquids or dewatered sludges, but could be used to process incidental quantities when built for other purposes.

Special Design and Multipurpose Incinerators. A survey of successful industrial waste incinerators was conducted by the ASME Research Committee on Industrial Wastes (Ref. 7-21). Among the operations described were:

Tars of various compositions blended and burned in a long furnace followed by a spray chamber, high energy venturi scrubber, and demister (Dow Chemical, Midland, Michigan).

Facility for incineration of a chlorinated liquid waste, followed by cooling and absorption of HCl for subsequent recovery (du Pont Co., Louisville, Kentucky).

Special burner for handling a mixture of oils and solid particles up to 0.6 in. (1.6 cm) in diameter (Rohm and Haas Co., Deer Park, Texas).

A conventional Babcock and Wilcox recovery boiler for the reuse of chemicals contained in black liquor of a Kraft paper mill, but preceded by sextuple effect evaporators to bring the liquor to the correct boiling temperature (American Can Co., Halsey, Oregon).

Offshore Incineration. Incineration at sea on a contract basis using an incinerator ship has been shown to be a viable alternative for problem materials (Ref. 7-22). Organochlorine wastes were successfully incinerated at a maximum rate of 25 metric tons per hour with a 1200°C minimum and a 1350°C average flame temperature. Tests indicate above 99.9% oxidation of the wastes. Hydrogen chloride, carbon dioxide, and water are discharged directly to the atmosphere with no measurable adverse environmental impacts.

7-4 DEEP-WELL INJECTION AND DEEP OCEAN DISPOSAL

Deep-Well Injection. Increased efforts to reduce contamination of surface and groundwater resources and the accelerated production of diverse types of relatively untreatable wastes result in a continued interest in deep-well injection, a process for permanently storing waste in underground strata. Deep wells have been used successfully for many years in oil producing operations to return large volumes of saline water, removed during oil production, to the subsurface. Deep wells have, however, received only limited use for industrial waste disposal primarily due to legal and regulatory agency restraints and uncertainties, as well as technical considerations. Warner has summarized the technology of deep-well disposal in considerable detail (Ross, Ref. 7-23).

The requirements for a suitable injection stratum include:

Isolation by impervious strata (aquicludes) from usable underground water supplies and mineral resources.

Sufficient capacity: depth, porosity (percent interconnected voids), areal extent, saturation (percent voids occupied by connate water).

Permeability sufficient to accept waste at a desired rate and pressure.

Formation and its natural fluid compatible with waste to be injected, or waste can be economically treated to achieve compatibility.

Injection strata generally are sandstone or limestone. Aquicludes may be clays, shale, marl, crystalline limestone, siltstones, impervious sandstones, and gypsum.

A geologist familiar with the geographical area of interest can review nearby well logs and estimate the probability of locating a suitable stratum. A test well (or wells) generally is necessary to provide verification of aquiclude locations, ground-water levels, and depths of potential injection strata. Samples of injection stratum material and connate water can be obtained, and pump-in tests to determine flow at various injection pressures can be performed. Compatibility tests of wastes with the stratum samples should be performed at the bottom-hole temperature and pressure; formation of precipitates is the main concern.

A typical injection well is shown in Figure 7-4. More than one casing often is necessary in deep wells to protect fresh waters aquifers from contamination, to close off zones of soft rock and salt water aquifers, and to protect gas- and oil-bearing strata. State and federal laws may dictate the methods of sealing, cementing, and pressure testing of such strata (Ref. 7-23).

Deep-well disposal costs are near those of biological treatment when the well can handle 10,000 to 500,000 gal./day (38 to 1900 m^3/day) and the site is within 8 miles (13 km) from the plant (Ref. 7-24).

The approval of new injection wells for industrial waste disposal in the United States is dependent upon these guidelines (Ref. 7-24): (1) examination of alternatives to make certain that all are "less satisfactory environmentally," (2) drilling of preinjection test wells, (3) designing wells for maximum environmental protection, (4) developing provisions for monitoring the injection wells, (5) contingency plans for failures and shutdowns of wells, (6) provisions for supervision of plugging of wells when they are shut down. For certain industries and in certain locations the disposal of wastes and the storage of fluids in the subsurface by use of well injection may be the most environmentally acceptable practice available. Acid wastes from titanium dioxide production at a new plant is a cited example of a stream suited for deep-well disposal. In 1973, 270 industrial waste disposal wells were in operation in the United States, 57 of which were used by chemical, petrochemical, and pharmaceutical plants.

Underground injection programs come under the provisons of the Safe Drinking Water Act of 1974 (Public Law 93-523). Regulations have been proposed by U.S. Environmental Protection Agency for guiding State programs in this area (Ref. 7-25, 7-26). Regulated "injections" can range from deep well at high pressure to gravity flow injection through dug wells near the surface. The scope of the proposed regulations include:

1. Establishment of a permit system administered by the states.
2. Protection of all underground drinking water sources of 3000 mg/liter

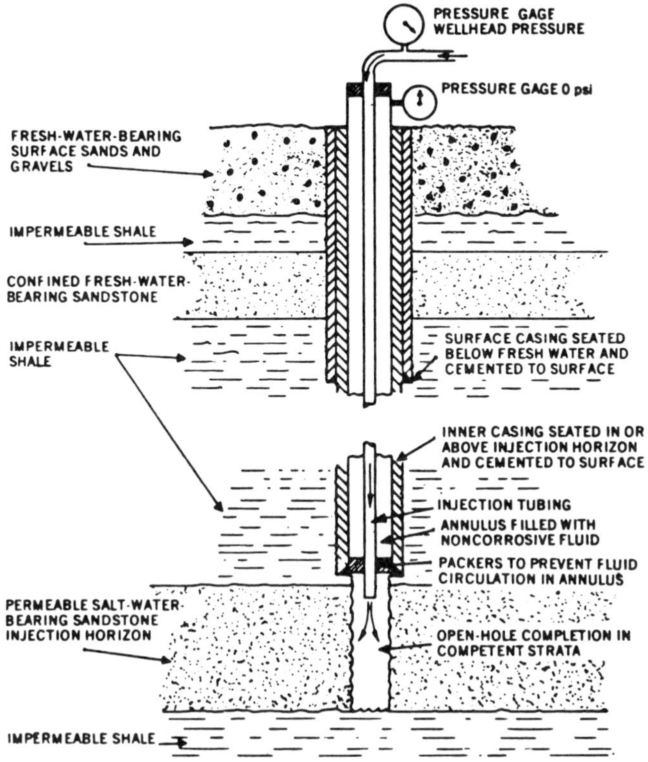

Fig. 7-4. Schematic diagram of a waste injection well completed in competent sandstone. ***Source:*** **Warner, D. L., "Deep Well Disposal,"** ***Industrial Waste Disposal,*** **R. D. Ross, ed., Van Nostrand Reinhold, New York, pp. 245-260, 1968.**

total dissolved solids (TDS) or less by a casing cemented to the surface. Where underground injection endangers a present or potential source of drinking water up to 10,000 TDS, the EPA expects the states to protect such a source by cemented casings unless an acceptable alternative is developed or the state can demonstrate that the aquifer is not a potential source for a public water system.

3. Prevention of upward migration of fluid by cementing the long string (well casing) above the injection zone.
4. Maintenance of injection through tubing with a suitable packing set immediately above the injection zone to isolate that zone from potential drinking water sources.
5. Prohibition of "annular injection," e.g., between casings or between

tubing and casing, except for currently operated wells when compelling evidence is presented that underground water sources are not endangered.

Development of extensive data needs to be accomplished on any proposed new underground injection operations, including:

1. Detailed engineering data on casing and cementing programs.
2. Map showing all wells within 2 miles (3.2 km); completion and plugging reports are required.
3. Operating plan for injection zone, pressure levels, and fluids involved (types, source, characteristics).
4. Geologic data on injection zones and confining beds.
5. Underground sources of drinking water which may be affected.
6. Electric log on all new wells and on existing wells when available.

These requirements are not intended to interfere with underground water or brine injection in oil or gas production, or for secondary or tertiary oil recovery, so long as fresh-water aquifers are unaffected. Endangerment of an underground water source includes both increasing the required treatment of that water and adversely affecting the health of those who might use it.

The need for careful study of waste compatibility with the water in the injection zone is demonstrated by experiences of a well operation abandoned after 3.5 years of operation (Ref. 7-27). The organic constituents were unstable and reactive in the aquifer containing very saline water. The formation of iron hydroxide, iron carbonate, and terephthalic acid precipitates, and the production of carbon dioxide were factors in plugging of the two injection wells. Certain organic constituents apparently were degraded by microorganisms under anaerobic conditions to form methane gas and cause iron and sulfur reduction.

On the other hand, another chemical plant waste has been successfully injected into isolated salt-water sands 0.5 mile (0.8 km) below potable aquifers for over 5 years (Ref. 7-28). Pretreatment is practiced to adjust the pH to between 5.0 and 8.0 and to reduce suspended solids below 10 mg/liter retained on a 5-micron filter paper. The lower pH level was set to prevent corrosion of the carbon steel used throughout the disposal system; the upper pH level limits swelling of clays in the injection zone. The suspended solids limit prevents physical plugging of the pores. The solids removal system consists of sedimentation, skimming, filtration through deep-bed sand filters, and polishing filtration through 3-micron cartridges.

Well construction details include:

1. 12.4-in. (31-cm) surface casing to 200 ft (61-m) below the fresh water strata, cemented back to the surface.
2. 9.6-in. (24-cm) well casing (long string) from surface to the top of injection strata, cemented full length.

3. 7-in. (18-cm) injection tubing sealed inside well casing top and bottom. The annular space is liquid full, and the annular pressure is continuously recorded.
4. Completion is with a 6.5-in. (17-cm) diameter screen and liner gravel packed.

Total investment in the collection, pretreatment, and disposal system facilities exceeds 2 million dollars (1972); operating costs exceed $250,000/yr. Over a 4.5-yr period 1.1 billion gal. were injected in the 3-well system, 2 at 3500 ft (1100 m) and one at 5900 ft (1800 m).

7-4.2 Deep-Ocean Disposal

Either outfalls or barge transport can be used within environmental constraints for ocean disposal of industrial wastes. The design of an outfall is based on receiving water standards. As described by Tchobanoglous (Metcalf and Eddy, Ref. 3-5), the important factors are the initial dilution of the waste, the dispersion into surrounding waters, and the decay rate of the waste constituents. The initial dilution rate is a function of the depth of the discharge port (Y_o in Figure 7-5), the diameter of the discharge orifice (D), and the Froude number for the liquid-liquid system (takes into account specific gravity differences). Dispersion due to eddy diffusion mainly is a function of current velocity and diffuser system length. Decay rate of degradable constituents usually follows a first-order relationship. The disposal of industrial pollutants in the sea has been reviewed in a *Chem. Eng.* Staff Report (Table 7-3, Ref. 7-29).

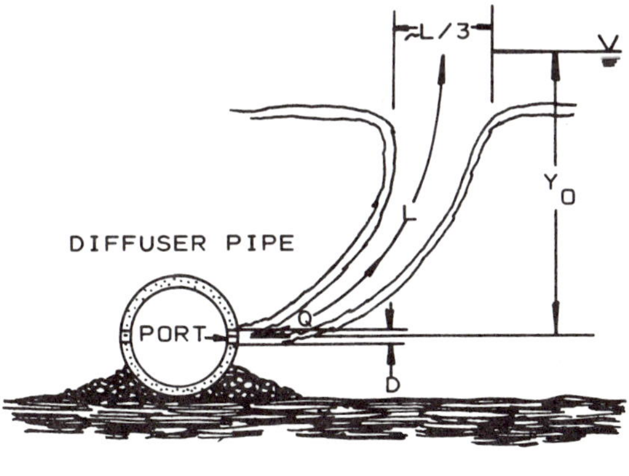

Fig. 7-5. Sketch for turbulent jet mixing analysis with a diffuser. ***Source*****: Rawn, A. M., Bowerman, F. R. and Brooks, N. H., "Diffusers for Disposal of Sewage in Seawater",** ***J. Env. Eng. Div., Proc. Am. Soc. Civil Eng.*****, 86, SA2, pp. 65–105, 1960.**

TABLE 7-3. Factors to be Considered in the Design of Marine Waste-Disposal Systems (after E. A. Pearson)

I. Beneficial Uses

1. Water contact sports
2. Marine recreation, boating
3. Marine working environment
4. Fishery, propagation, migration, etc.
5. Economic fishery, propagation, harvesting
6. Industrial, commercial use, cooling
7. Waste disposal

II. Water Quality Criteria To Protect Beneficial Uses

1. Public health
   a. Coliform
   b. Other
2. Fishery
   a. Toxic substances
   b. Antagonistic substances
   c. Oxygen depressants
   d. Stimulants, fertilizers
   e. Transparency, turbidity
   f. Suspended and settleable debris
3. Nuisance
   a. Grease and oil films
   b. Floating debris
   c. Settleable debris
   d. Odors
4. Aesthetic
   a. Sleek areas
   b. Colors
   c. Turbidity or transparency
   d. Floating debris
   e. Plankton bloom
   f. Other
5. Economic and other

III. Oceanographic Characteristics of Outfall Sites

1. General water circulation system
2. Current
   a. Surface and subsurface
   b. Strength and direction as a function of time
   c. Effect of wind, wave, tide, littoral drift
3. Eddy diffusivity or dispersion characteristics
4. Density structure, salinity, temperature-depth relationship
5. Wave and swell effects
6. Submarine topography
7. Submarine geology

IV. Waste Dispersion Considerations

1. Initial mixing–diffuser
   a. Jet mixing
   b. Buoyancy–gravitational mixing
   c. Density gradients–thermoclines
   d. Diffuser orientation
   e. Waste dilution–flow continuity
   f. Port selection, area–spacing

2. Waste transport–dispersion
   a. Current regimen
   b. Eddy diffusion
   c. Mixing depth, effective
   d. Rational dispersion equations
      (1) Concentration dilution only–conservation waste
      (2) Concentration including decay–nonconservative waste, (bacteria, radioisotopes, BOD, etc)

V. Economic Analyses

1. Various types of treatment and effluent characteristics.
2. Length, depth, and cost of outfall systems for each type of effluent to meet water quality criteria requirements.
3. Selection of optimum and least-cost combination of treatment and outfall system to protect beneficial uses.

Source: "Ocean Pollution and Marine Waste Disposal," *Chem. Eng.*, pp. 60–67, February 8, 1971.

Barging industrial wastes to sea was at a rate of 5 million tons (4.5 million metric tons) in 1968 and 10 million tons in 1971. Of the 1968 total, 58% were waste acids primarily from steel mills, 12% were refinery wastes, and 7% were pesticide residues. Included among the other materials were caustic, chlorinated hydrocarbons, black liquor, ammonium sulfate, titanium pigment wastes, phenols, cyanides, and sodium metal.

Disposal into the sea at more-remote locations is primarily by discharge in an allowed area after transportation in towed or self-propelled barges. Bulk wastes are normally discharged while the barges are underway. Containerized wastes are weighted and sunk, or ruptured by axes or rifle fire. Barges transporting liquid wastes must be certified for ocean waters by the Coast Guard, and rather stringent regulations on the manner of shipment must be followed. Barges are double-skinned and range from 1000 to 5000 tons (Ref. 7-29). Discharge rates vary between 4 and 20 tons/min through hoses trailing the barge at a depth of 1 to 2.5 fathoms while underway at a speed of 3 to 6 knots (Ref. 7-16).

Ocean dumping of industrial waste is closely regulated. In the United States, clearance for this operation has been centralized with the Federal Environmental Protection Agency. Before permits are issued several studies need to be conducted: bioassays of the waste, *in situ* bioaccumulation studies, and *in situ* dispersion studies. Dispersion study results have been reviewed by Ball, Reynolds, and Beckett (Ref. 7-30). The rate at which the waste was diluted with sea water appeared to be mainly a function of the difference in specific gravity between the waste and sea water. The dilution concentration at an arbitrary time of one minute after dumping was mainly a function of the rate of release and secondarily of the speed of the barge.

7-5 REUSE AND RECOVERY

7-5.1 Water Reuse

The concept of recycling treated wastewater stirs the imagination because of its apparent potential for minimizing the discharge of contaminated water to the environment. However, many technical and economic limitations need to be overcome before this concept achieves wide application (Ref. 7-31).

The recycle of aqueous process streams within individual production units is practiced extensively in the petrochemical and other industries. Frequently, when water is used as a process fluid for scrubbing organic vapors, recycle of the waste is practiced with the objective of recovering the contained organic materials. Other common recovery operations involve decantation to remove floatable oils, distillation to remove dissolved organics, and filtration to remove suspended solids. Good design practice and economic considerations are dictating the need for recovering usable chemicals and water at the source units in an ever increasing number of circumstances.

Even for a relatively straightforward application of water reuse, technical limitations must be considered. Within a multi-unit complex, contaminants may be transferred from one process to another. This cross-contamination has the potential for causing quality problems, and has resulted in serious safety hazards. Uncontrollable chain reactions of certain reactive chemicals may be initiated by a very small amount of certain contaminants.

The practice of recycling cooling water is used where water must be conserved. Closed cooling water systems are, in general, economical only when fresh water is in short supply. A considerable investment is required to install cooling towers or ponds for temperature control if the necessary volume of water is large. When cooling water recycle is practiced, a 3 to 5% purge of the recycle stream characteristically is required to control the build-up of inorganic salts, concentrated by evaporative losses. Losses from the closed cycle from leaks and miscellaneous water requirements may approach 30%. To prevent biological growth in the recycled water, inhibitors must be added on a continuous basis and are necessarily present in the purge stream. Serious obstacles to the conversion of existing facilities to recycle cooling-water systems include space limitations and costs associated with repiping and upgrading heat exchange equipment originally designed for once-through cooling.

Although conceptually attractive, effluent from secondary treatment of most industrial wastewaters is not near a suitable quality for direct reuse. By-products of chemical reactions in the form of soluble inorganic salts (usually chlorides and sulfates) are contained in the process wastewater; no cost effective means is now available to remove these salts along with refractory organics and suspended solids, and thereby qualify these waters for reuse. Considerable reuse research now in progress, however, may lead to improved process steps and sequences (Ref. 7-37).

7-5.2 Recovery and Conversion

The operations of recovery of original product or conversion to other useful products need to be tailored to individual situations. Some approaches include:

Hydrogenation of organic wastes to form oil and gas for fuel (Ref. 7-16).
Acetylation of waste cellulose to form cellulose acetate (Ref. 7-16).
Crosslinking of cellulosic wastes to form construction board (Ref. 7-16).
Thermal reforming of scrap plastic film into lower molecular weight polymers in the form of waxes, greases, and oils (Ref. 7-16).
Multieffect evaporation of pharmaceutical and food processing wastes by the

Carver-Greenfield process, which involves first adding a small quantity of oil to prevent fouling of heat-transfer surfaces (Ref. 7-32).

Recovery of calcium hydroxide and aluminum sulfate from lime-softening and alum sludge (Figure 7-6) (Ref. 7-33).

Fox has described several approaches for product recovery involving activated carbon adsorption with chemical regeneration of the carbon (Figure 7-7), liquid-liquid extraction of process tars, ion-exchange for recovery of ammonia and nitrates, and hydrolysis of aromatic by-products (Figure 7-8) (Ref. 7-34).

7-5.3 Pyrolysis and Energy Recovery

Pyrolysis is the process of destructive distillation carried out in an oxygen-free atmosphere. Through pyrolysis, organic matter can be readily converted to gases, liquids, and inert char. This process has several advantages over incineration as a means for disposal of concentrated carbonaceous wastes (Ref. 7-16):

Reaction takes place in a closed system which minimizes air-pollution problems.

Capital cost is lower at larger throughputs.

Operating costs are lower.

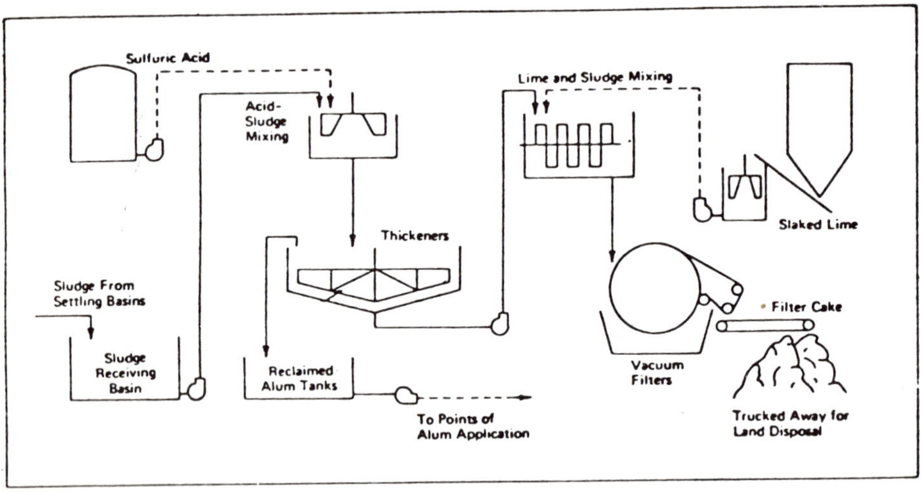

Fig. 7-6. Process diagram for typical alum-recovery installation. ***Source*****: "Disposal of Water Treatment-Plant Wastes,"** ***J. Am. Water Works Assoc.*****, pp. 814-820, December 1972. Reprinted from** ***J. Am. Water Works Assoc.*****, 64, by permission of the Association. Copyrighted 1972 by the American Water Works Association, Inc.**

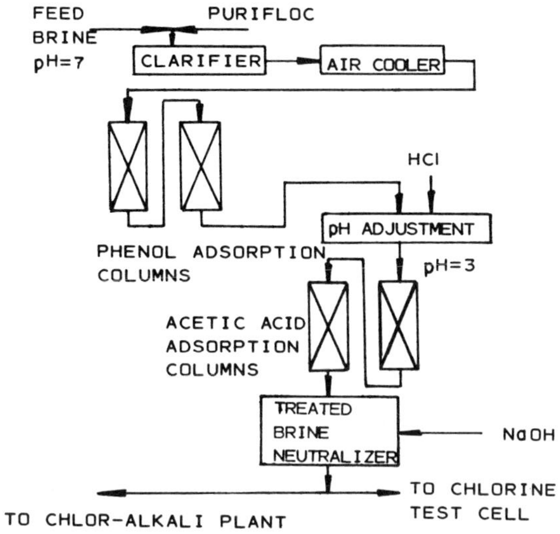

Fig. 7-7. Activated carbon adsorption process for brine and product recovery. *Source:* Fox, R. D., "Pollution Control at the Source," *Chem. Eng.* pp. 72-82. August 6, 1973.

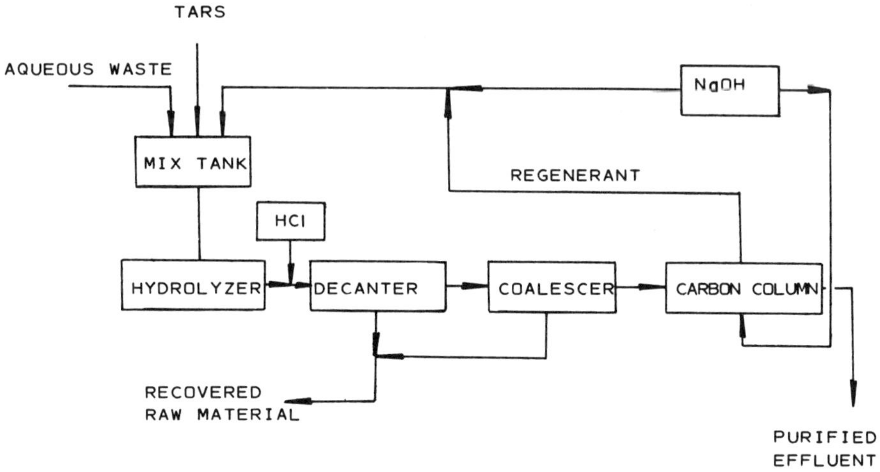

Fig. 7-8. Hydrolysis process for recovery of useful by-products. *Source*: Fox, R. D., "Pollution Control at the Source," *Chem. Eng.*, pp. 72-82, August 6, 1973.

System provides its own fuel.
Potential exists for recovering chemicals, charcoal, fuel gas, or synthesis gas.

One system, the Lantz converter, is a sealed, airtight retort cylinder with an insulated shell. The gas-fired retort, about 20 ft (6 m) long, revolves slowly on a slight incline like a rotary kiln. Waste is injected through a star valve. The retort operates at about 1200°F (650°C), and the wastes are pyrolyzed, in the absence of oxygen, to steam, volatile material, carbon dioxide, and charcoal (Ref. 7-16). A pyrolysis process developed by the Jet Propulsion Laboratory (Pasadena, California) converts a slurry from a municipal treatment operation to activated carbon for use back in the treatment scheme (Ref. 7-35). The Purox system (Union Carbide Corp.) is another type of available pyrolysis approach. In this system, oxygen is introduced into the combustion zone at the bottom of a vertical shaft furnace where it reacts with carbon char from the overlying pyrolysis zone. A fuel gas is produced. The potential for energy recovery also can be realized by burning liquid residues and sludges in coal-fired boilers (Ref. 7-36).

REFERENCES

7-1. "Dispersion of Refinery Effluents in Receiving Waters," *Man. Disposal Refinery Wastes,* **1**, American Petroleum Institute, New York, 1963.

7-2. "Pretreatment Standards," *Federal Register,* **44**, 123, 27736–27773, June 26, 1978 (to be modified in settlement of Chemical Manufacturer's Association *vs.* U.S. Environmental Protection Agency suit).

7-3. Cox, G. V., "Industrial Waste Effluent Monitoring," *Am. Lab.,* pp. 36–40, July 1974.

7-4. Chermisinoff, P. N., "Water Pollution Sampling, Monitoring, and Analysis Instrumentation," *Pollut. Eng.,* pp. 36–48, April 1974.

7-5. MacKay, D. and Wolkoff, A. W., "Rate of Evaporation of Low-Solubility Contaminants from Water Bodies to Atmosphere," *Environ. Sci. Technol.,* 7, 7, pp. 611–615, 1973.

7-6. Godfrey, K. A., "Land Treatment of Municipal Sewage," *Civ. Eng. ASCE,* pp. 103–109, September 1973.

7-7. Snyder, C. D., "Disposal of Liquid Effluent by Land Irrigation at Lagro Celotex Plant," 27th Purdue University Industrial Waste Conference, Lafayette, Indiana, 1972.

7-8. Kincannon, C. B., *Oily Waste Disposal Cultivation Process,* EPA–R2–72–110, U.S. Environmental Protection Agency, Washington, D.C. 20460, 1972.

7-9. Dotson, G. K., Dean, R. B., Cooke, W. B., and Kennar, B. A., "Land Spreading, A Conserving and Non-Polluting Method of Disposing of Oily Wastes," *Proc. 5th Int. Water Pollut. Res. Conf.*, Pergamon Press, New York, 1971.

7-10. *Municipal Sludge Management: Environmental Factors, Technical Bulletin,* U.S. Environmental Protection Agency, *Fed. Reg.*, pp. 57420–57427, November 2, 1977.

7-11. Carroll, T. E., Maase, D. L., Genco, J. M., and Ifeadi, C. N., *Review of Land Spreading of Municipal Sewage Sludge,* U.S. Environmental Protection Agency, Cincinnati, Ohio 45268.

7-12. *Acceptable Methods for the Utilization and Disposal of Sludges,* Draft Document EPA 430/9-75-XXX, U.S. Environmental Protection Agency, Washington, D.C., 1975.

7-13. Walker, J. M., "Sewage Sludges—Management Aspects for Land Application," *Compost Sci.*, pp. 12–21, March-April, 1975.

7-14. Francke, H. C. and Clark, F. E., *Disposal of Oil Wastes by Microbial Assimilation,* Oak Ridge Y-12 Plant, Oak Ridge, Tennessee 37830, 1974.

7-15. Slover, E. E., "Solid Waste Disposal in a Multi-Product Chemical Plant," *Text. Ind. Environ.*, American Association of Textile Chemists and Colorists, New York, 1973.

7-16. Witt, P. A., "Disposal of Solid Wastes," *Chem. Eng.*, pp. 62–69, October 4, 1971.

7-17. Dunn, K. S., "Incineration's Role in Ultimate Disposal of Process Wastes," *Chem. Eng.*, pp. 141–150, October 6, 1975.

7-18. *Fluid Bed Incineration of Petroleum Refinery Wastes,* U.S. Environmental Protection Agency, Washington D.C., 1971.

7-19. "Incinerators—Pro and Con," *Mod. Manuf.*, p. 112, February 1971.

7-20. Kiang, Y-H. K., "Liquid Waste Disposal System," *Chem. Eng. Prog.*, pp. 71–77, January 1975.

7-21. *ASME Survey of the Present State of the Art in Disposal of Industrial Wastes by Combustion,* American Society of Mechanical Engineers, New York, 1971 (Vol. 1), 1972 (Vol. 2).

7-22. Wastler, T. A., Offutt, C. K., Fitzsimmons, C. K., and DesRosiers, P. E., *Disposal of Organochlorine Wastes by Incineration at Sea,* U.S. Environmental Protection Agency, Washington, D.C., 1975.

7-23. Warner, D. L., "Deep Well Disposal, " *Industrial Waste Disposal,* R. D. Ross, (ed.), Van Nostrand Reinhold Co., New York, pp. 245–260, 1968.

7-24. "Deep Wells are Down but Not Out," *Chem. Week,* p. 40, January 9, 1974.

7-25. U.S. Environmental Protection Agency Proposed Regulations on State Underground Injection Control Programs; *Fed. Reg.,* **41**, p. 36730, August 31, 1976.

7-26. Hernandez, J. W., "Underground Injection Program Set Under Safe Drinking Water Act," *Water Sewage Works,* pp. 67–69, October 1977.

7-27. Leenkeer, J. A., Malcolm, R. L., and White, W. R., "Investigation of the Reactivity and Fate of Certain Organic Components of an Industrial Waste after Deep-Well Injection," *Environ. Sci. Technol.,* **10**, 5, pp. 445–51, 1976.

7-28. Dugas, R. S. and Reed, P. E., "Successful Pretreatment and Deep-Well Injection of Chemical Plant Wastewater," Uniroyal Chemical Co., Geismar, Louisiana.

7-29. "Ocean Pollution and Marine Waste Disposal," *Chem. Eng.,* pp. 60–67, February 8, 1971.

7-30. Ball, J., Reynolds, T. D., and Beckett, D. E., "Deep Sea Ocean Disposal of Petrochemical Wastes," Texas A and M University, College Station, Texas.

7-31. Conway, R. A., "Status of Technology for Water Pollution Abatement at Petrochemical Plants," Statement before the Panel on Environmental Science and Technology, Senate Subcommittee on Air and Water Pollution, Washington, D.C., May 14, 1971.

7-32. "Making Hard-to-Treat Chemical Wastes Evaporate," *Chem. Week,* pp. 31–32, May 9, 1973.

7-33. "Disposal of Water Treatment-Plant Wastes," *J. Am. Water Works Assoc.,* pp. 814–820, December 1972.

7-34. Fox. R. D., "Pollution Control at the Source," *Chem. Eng.,* pp. 72–82, August 6, 1973.

7-35. "Sludge Pyrolysis Schemes Now Head for Tryouts," *Chem. Eng.,* p. 90, December 8, 1975.

7-36. Qwane, D. E. and Stumpf, M. R., "Coal-Fired Boilers Burn Waste Sludges and Odors in an Integrated Pollution Control System," 46th Annual Conference of Water Pollution Control Federation, Cleveland, Ohio, 1973.

7-37. Sherm, M., *et al.* (Union Carbide Corp.), *Treatment of Organic Manufacturing Wastewater for Reuse,* Report on EPA Grant No. 5–801398, U.S. Environmental Protection Agency, Ada, Oklahoma 74820.

8

Preparation, recovery, and reuse of solid waste

8.1 NATURE OF SOLID WASTES

Of the various types of waste material that industry can generate, gaseous waste is perhaps the easiest to handle and solid waste is undoubtedly the most difficult. Perhaps this is because of the various possibilities that exist in a solid waste inventory. Solid waste may be combustible or noncombustible. It may be comprised of large, bulky items or very finely divided, almost airborne types of material. It may be free-flowing or immobile. It may be wet with moisture or other materials, or absolutely dry. It may have heating values from 0 to 20,000 Btu per pound or higher. And, the limitations for its disposal are few.

This Chapter will discuss the problems associated with the preparation of solid waste materials to make them suitable for disposal, and the options for recovery and reuse.

There are great possibilities for resource recovery from industrial solid wastes, and this area should first be examined before disposal methods are considered. There are very few ways available to dispose of solid waste. If it will burn, it can be incinerated. If it can yield valuable gases or liquids as a by-product, then pyrolysis might be suitable. If it will biodegrade, then composting is a possibility. If it will not lend itself to these processes, it must be landfilled or dumped at sea with the hope that it will remain there forever. Therefore, in order to properly define a treatment system or a treatment mode for the disposal of solid waste, several things must be known: (1) the quantity of the solid; (2) its physical and chemical characteristics, noting its possible hazards; and (3) the identification of recoverable materials within the waste. Once these are defined, then a means of handling and disposing of the waste can be developed.

The quantity of the solid waste is reasonably easy to determine. Densities for most types of solids can be determined from the appropriate literature; the material to be disposed of should be stated in both volume as well as total weight. This information aids both in the handling as well as the disposal of the waste material.

The second determination is equally as important as the first. The physical form of the waste should be discerned, whether it is solid or a sticky, tarry substance, or a mixture of liquid and solids, etc., and if it is contained in drums, fiber packs, or in bulk. It is also important to know as much as possible about the waste, such as heating value, moisture, amount of volatile material, even its chemical formula, since these may be important in selecting a solid waste disposal system. For example, if the solid waste material contains sulfur, it is extremely important to know how much sulfur, especially if it is to be incinerated because there are limitations in the amount of sulfur dioxide which can be emitted from an incinerator into the atmosphere. This does not prelude incineration, but it suggests that if incineration is going to be used, some type of subsequent scrubbing or absorption of the sulfur dioxide will be required.

The third consideration is resource recovery. Metals of all kinds, glass, chemical, and heat values are present in solid wastes. By recovering and recycling these entities, two important objectives are achieved: (1) a valuable resource has been made available for reuse and (2) a valuable space in a landfill or ocean resource has been left for other wastes which offer no recovery possibilities.

For the large quantities of industrial solid wastes, growing larger every year, it is incumbent upon industry to select not only the most economical but the best method environmentally for the disposal of its waste material.

8.2 TREATMENT METHODS

Solid waste materials are difficult to handle because they are not uniform. Gaseous wastes and liquid wastes conform to the shape of the container. Solid wastes, however, may be anything from a very small particle to a large beam or even a refrigerator or automobile. In the industrial field the variation may not be as dramatic but, nevertheless, there will be a large variation in size and in material.

Solid wastes may be treated to make them more acceptable for their final method of disposal. Since the treatment methods are somewhat limited, the treatment methods will generally conform to the type of final disposal method used. For example, if landfill is to be the final disposal method, it is generally desirable to compact the raw wastes in some manner so that they take up as little room as is possible in the landfill. In the landfill itself, raw waste material is often compacted by the bulldozer running over the top of the fill, but if the material can be compacted prior to sending it to the landfill, then the transportation is less expensive because it requires fewer vehicles and the amount of handling at the landfill is usually reduced.

On the other hand, if refuse materials which are going to an incinerator are compacted, it may not be to the advantage of the incineration process. An incinerator works best when air comes in contact with the material to be burned. Thus, it more difficult for air to get to the combustible material of compacted waste; consequently, the burning rate is slower. Another example of where compaction would not be desirable is when the final disposal method is composting. In a compost operation, it is desirable to have the organic material finely divided so that there is uniformity in the composting system. It would therefore appear that shredding waste material would be more desirable for composting and incineration, and compaction would be the most desirable preparation or treatment method for landfill,

Another type of pretreatment which is often used in the solid waste handling industry is separation by specific gravity. This might involve the separation of glass, metals or heavy plastics from paper and other cellulose trash for either reclamation of these materials or because they would not properly react to the prescribed treatment method. Generally, in landfill operations, no separation is made. Only large items which would not be satisfactory for compaction are removed before the compaction step is instituted, but for the feed to an incinerator or composting system, it is desirable to remove glass and metal pieces of much smaller size because they can damage the grates in an incinerator or the conveyors in a composting operation.

Metals, if they are ferrous in nature, can be removed by magnetic means. There are many different types of industrial magnets which can be located in conveyors or other types of feeding apparatus which will remove both large and small iron and steel scraps from a solid waste feed to an incinerator or composting operation. However, there are many waste metals which are not ferrous and therefore not magnetic. Materials such as glass which have melting temperatures that may be exceeded in an incinerator will cause heavy slagging and subsequent refractory damage. It is desirable to remove glass before it reaches the high temperature zone. Certain plastic materials, because of their high flammability or because they will not biodegrade easily in a composting system, must be removed before the processing step. In general, all of these materials are heavier than other combustible or biodegradable materials which are normally being handled in either an incinerator or composting operation. They can be separated from the waste by means of an air separation device which will lift the light materials and leave the heavy materials at the bottom of the chamber from where they can be removed for separate handling.

Industrial solid waste disposal problems are unique and decidedly different from the solid waste disposal problems which face the municipal community. Municipalities must handle a wide variety of materials from the trash and garbage generated by the individual citizen to the disposal of abandoned automobiles, used applicances, broken furniture, building material, tools, and almost anything conceivable, including such difficult materials as automobile tires. The industrial community, on the other hand, is usually concerned with the following classes of material from the perspective of this chapter.

1. Office paper and general office trash, garbage from company cafeterias, and individual lunch pails.
2. Cartons and boxes constructed of cardboard, both with and without plastic liners, in which raw materials are received.
3. Wood, in many cases in large quantities, consisting of material from crates and pallets on which incoming material is received.
4. Multiwall bags often with plastic liners.
5. Fiber drums and containers often with plastic linings.
6. Off-specification products which are solid in nature.

The first five categories are relatively easy to understand and appreciate. Category #6, however, could offer a multitude of possibilities. Plants producing such plastic materials as polyethylene, polypropylene, polyvinyl chloride, and a variety of other resins and plastics, will often have materials which must be disposed of because they do not meet product specifications. Companies which manufacture containers made of these materials will often have off-specification products which must be destroyed. Sometimes these materials can be recycled, but, in general, they are usually disposed of. The disposal method, therefore, is one that must be carefully considered, and the treatment method must be similarly considered.

8.3 COMPACTION

There are many places in an industrial plant where compactors may be used. If the plant has its own final disposal operation, then the compactor may be situated so that the material is dumped into it prior to going to a landfill or other disposal method. If the plant is using an incinerator, then often the compactor is located at the incinerator, and the raw trash is carried to the compactor, and the compactor not only compacts the material but then charges the compacted material into the incinerator. Compaction can create incineration problems.

If, on the other hand, the plant utilizes contract disposal of its waste, often it will be compacted when the disposal contractor comes to pick it up at the plant. In many cases, the disposal contractor will supply a stationary compactor and holder for the compacted waste to simplify his operation, especially if he is taking the waste to a landfill beyond the plant's boundary.

Compactors may also be either stationary or mobile; however, mobile compactors or compactor trucks which receive waste material and compact it instantaneously and then move on to another spot are not generally used in industrial applications. This does not, however, preclude their use.

A compactor generally will reduce waste volume by a factor of from 3 to 1 to 10 to 1, depending upon the material to be compacted and the force required to compact it. Since there are over 150 different models of stationary compactors manufactured by almost 40 different companies, it is difficult to select the proper compactor for the proper job. The compactor should never be used unless there is an advantage to the ultimate compaction of the waste in its final disposal or in its transportation.

Compactors generally consist of a container unit and a compaction unit. The container unit accepts the raw trash and feeds it into the compactor. The compactor consists of a hydraulically or pneumatically operated ram which squeezes the waste into a more dense form by utilizing high pressure.

Industrial wastes that can be adequately compacted are materials such as trash, loose refuse, bags, cartons, etc. Large pieces of wood, metal, glass, and heavy plastics should never be compacted since they will cause problems with the compaction equipment by bending or breaking the ram. Compactors should not accept tar, sludges, or liquid material since, of course, they will not compact and will squeeze out of the compaction unit and cause many operational problems. Garbage and cafeteria waste materials can be compacted wherever there is sufficient dry trash to absorb the water or other liquids in the waste, but if putrescible materials are to be fed into a compactor, then greater maintenance and cleaning will have to be employed.

The majority of compactors used in industrial applications are stationary compactors. This is a compaction unit having a hydraulically-operated ram which moves in a horizontal direction (Figure 8-1). Wastes are fed into a receiving hopper and the ram, which can be actuated either manually or by optical or sonic devices, then compresses the wastes into a steel container. This container is generally square or rectangular and has an opening in the lower half of one end of the box. This opening allows the compactor ram to operate inside the container during the compaction cycle. The receiving box or container can

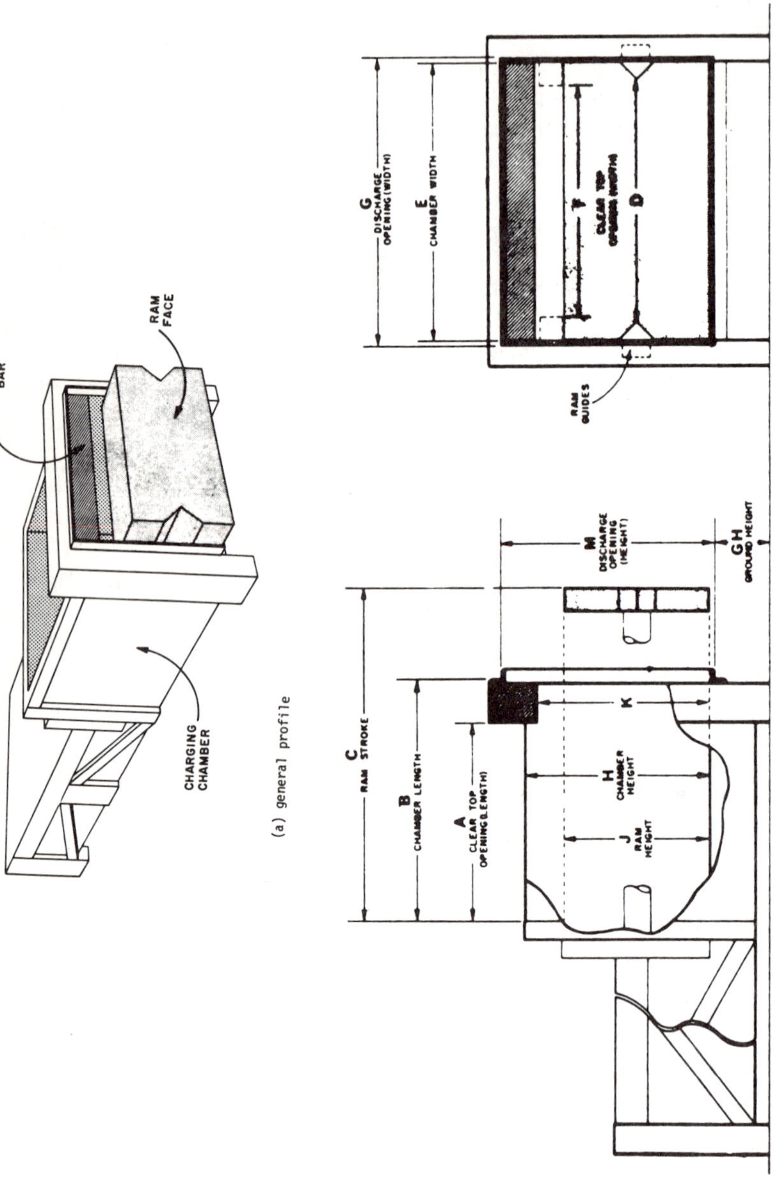

Fig. 8-1. Horizontal ram compactor. *Source*: National Solid Wastes Management Association, Copyright 1977.

be detached from the ram once it is full and hauled to a disposal site for dumping. The loading end of the box or container is hinged to swing fully open to permit dumping of the refuse at the point of disposal. When a container is completely full, the compactor ram is entirely withdrawn and retaining cables, and usually a tarpaulin, are put into place. The container is lifted onto a tilt-frame hoist type truck to be hauled away, and a new empty container is then strapped or locked to the compactor making the unit once again ready for operation. The sides and the top of the container are tapered slightly to facilitate the refuse removal. Pivoting ram compactors where the ram is pivoted at one end of the container and actuated by a hydraulic cylinder are also used for smaller, lighter materials (Figure 8-2).

When purchasing or renting stationary compactors, several things should be considered. First, the size of the charging area. If it is not large enough to accommodate, without any difficulty, the largest piece of refuse generated by the user, then it is of little value. If the compactor bin is to be fed from trash carts, then the compactor should be selected to at least handle the load from one full trash cart or container load. Otherwise, it would be inconvenient. Charging areas in compactors usually range in size from 1 to 12 cu yd (Table 8-1).

The compactor should be selected to fit the place where it is going to be used. This may appear obvious, but the purchaser should certainly consider that he has to fit not only the compactor but the removable container into a given area. He must also provide for the removal of the container in such a way that the pick-up vehicle can easily get into a position to attach itself and lift the container. One way that a building may gain dock space as well as increased plant security is by installing the compactor machine through the wall so that the charging area as well as the working mechanism remains inside the building while the container is attached to the machine through a chute which passes through the wall. Also, chutes from upper floors may run directly to the charging area, simplifying indoor transportation and reducing the need for dock space.

A second important consideration in the selection of a compactor is the cycle time. This refers to the time in seconds required for the ram face, which is fully retracted, to pack the refuse from the charging box into the container and return to its originally fully retracted position ready for another load of refuse. Cycle times vary widely from as low as 20 sec to over 60 sec. Rapid cycle times are important if the application under consideration requires the capability of accepting refuse very quickly. Usually short cycle time compactors are lighter in weight and are designed to compact smaller quantities of material per cycle; consequently, they may cost less but are less rugged than compactors with a longer cycle time. Short cycle times, however, do not necessarily mean that high compaction ratios will be obtained.

Another consideration in the selection of the compactor is the pressure on the ram face; this should be ascertained by finding the force rating of a specific

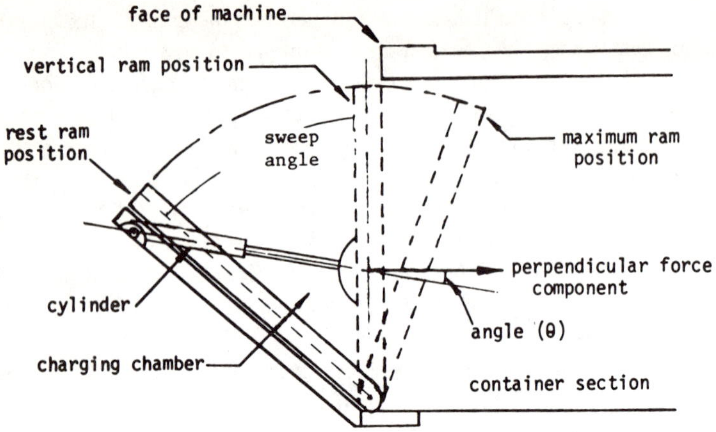

Insert (a). Side View Depicting Defined Ram Positions

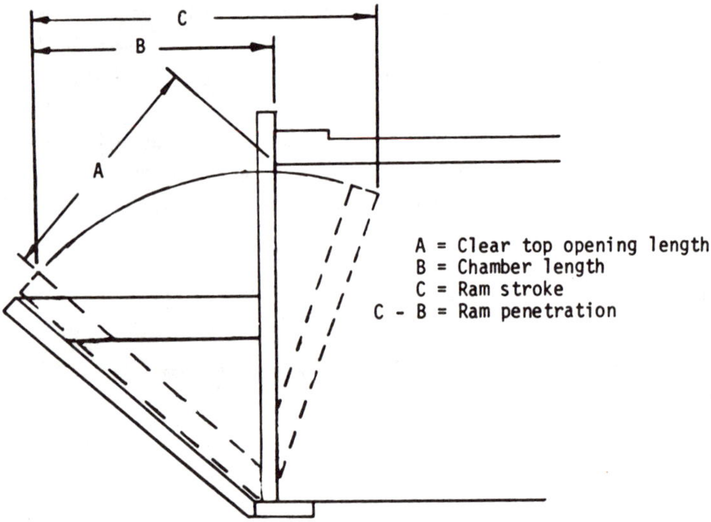

Insert (b). Side View Showing Rating Dimensions

Fig. 8-2. Pivoting Ram Compactor. ***Source:*** **National Solid Wastes Management Association, Copyright 1977.**

compactor. The NSWMA (National Solid Wastes Management Association) has rated compactors and gives each a force rating defined as the pounds per square inch on the ram face. This includes the full width and the full height of the ram and will measure how much pressure the compactor generates, expressed in psi (pounds per square inch) of the ram face. Generally, compactors of the stationary type range from 15 to 50 psi.

Also important in the selection of a compactor for industrial applications is the penetration of the ram face into the container, or the length of the stroke. As the container begins to fill up with compacted waste, there is a tendency for the portion of the material closest to the ram to fall back or to bulge out and upward into the charging area which may leave a substantial amount of the refuse in the charging space. This can be a problem which will interfere with the loading operation. Therefore, the further that a ram penetrates into the compaction container, the easier it is to load the container in a clean and efficient manner. Virtually all compactors are built with penetrations ranging from 4 to 26 in.

The machine volume displacement is another important consideration in the selection of a compactor. This is determined by multiplying the cubic volume that the ram can displace each time it pushes a load of refuse into the container by the number of cycles the machine makes in 1 hr. This is merely a measure of the rate at which waste can be compacted into a container and relates to the rate at which the waste is being generated.

Finally, the compactor and the container should be from the same manufacturer so that the pressure developed by the compactor, its cycle time, its cubic displacement, and all of the other major factors are compatible. For example, a light gauge container which may not be able to withstand high compaction pressures would tend to buckle and bulge under a high pressure compaction operation. In the selection of a compactor, ruggedness is the prime consideration. The machine and the containers should be designed and fabricated for heavy duty. The ram face should have a heavy shear or breaker bar which cuts any material that extends above the charging hopper. The hydraulic system should be rugged and made of reliable components, and the motor should have the necessary overload portection in the event that the operator overloads the system.

The National Solid Wastes Management Association rates compactors by various criteria, shown in Table 8-1. This is typical for one of many manufacturers.

The cost of compactors will vary as to their capacity. A rough cost range is 0.2 to 0.3 cu yd capacity per $1000 cost.

In addition to stationary compactors, there is a wide variety of compactors known as bag compactors. These machines pack single bags which must be removed when full and replaced by empty ones. They would be applicable to certain uniform types of waste in an industrial plant but, in general, are not suitable for a wide variety of types and sizes of waste. Compaction ratios are usually in the range of 3 to 1 to 7 to 1. Packing densities range from 18 to 60 lb/cu ft depending upon the composition of the feed. Based on the density of 6 lb/cu ft and a compaction ratio of 3 or 4 to 1, a density of 18 to 24 lb/cu ft may be expected.

The advantage of the bag compactor is that the compacted material can be

TABLE 8-1. Typical NSWMA Compactor Ratings

Manufacturer: E-Z Pack, Peabody Solid Wastes Management

Model Number	NSWMA Base Size (cu.yds.)	Clear Top Opening Length (L) Width (W) (inches)	Chamber Length (inches)	Ram Stroke (inches)	Ram Penetration (inches)	Force Rating Normal (N) Maximum (M) in pounds)
AP-27	0.18	L=19.0 W=21.5	24.5	31.5	7.0	N=17.1/5900 M=21.3/7400
.5 FPDP	0.31	L=14.5 W=28.5	25.5	30.0	9.5	N=22.3/12700 M=24.8/14100
1 FPLC	0.64	L=25.0 W=46.0	32.5	46.0	13.5	N=24.9/22600 M=27.7/25100
2.3 FPLC	0.99	L=36.0 W=58.5	40.5	65.0	24.5	N=44.7/50900 M=49.6/56600
2.5 FPLC	1.61	L=39.0 W=58.5	41.5	53.0	11.5	N=19.5/35300 M=21.7/39200
3 FPLI	1.71	L=42.0 W=58.5	44.5	65.0	20.5	N=28.2/50900 M=31.4/56600
3.8 FPLI	2.41	L=60.0 W=58.5	62.5	83.0	20.5	N=28.2/50900 M=31.4/56600
4 FPLC	2.66	L=66.5 W=58.5	69.0	84.0	15.0	N=28.2/50900 M=31.4/56600
5 FPHI	3.24	L=80.0 W=58.5	84.0	108.0	24.0	N=42.6/76900 M=50.0/90000
6.7 FPHI	4.32	L=77.0 W=78.5	83.5	108.0	24.5	N=31.8/76900 M=37.5/90000
11 FPTP	8.74	L=114.5 W=78.0	121.0	153.0	32.0	N=30.4/10200 M=38.0/127500
15 FPTP	11.18	L=126.5 W=79.5	133.5	165.0	31.5	N=26.1/10200 M=32.6/127500

* Roll-off container only

Note: The National Solid Wastes Management Association presents this technical information on commercial/industrial stationary compaction equipment to assist users in the selection of equipment. The National Solid Wastes Management Association does not present this data as a warranty for equipment performance. The ratings represent data supplied by manufacturers that have been computed according to the NSWMA criteria and certified for accuracy by a registered professional engineer, selected by the manufacturer.
Source: National Solid Wastes Management Association, Copyright 1977.

Date: August 1977

System Pressures Normal (N) Maximum (M) (psi)	Ram Face (inches)	Volume Displacement Rate (cu.yds./hr.)	Rated Motor Size power)	Cycle Time (seconds)	Discharge Opening Width (W) Height (H) (inches)	Ground Height (inches)	Base Unit Weight (pounds)
N=1200 M=1500	W=23.0 H=15.0	22	1.5	30	W=24.5 H=19.0	8.0	1600
N=1800 M=2000	W=30.0 H=19.0	53	5	21	W=30.5 H=26.5	15.5	2100
N=1800 M=2000	W=48.0 H=19.0	74	5	31	W=60.5* H=42.0	16.0	3900
N=1800 M=2000	W=60.0 H=19.0	85	10	42	W=62.0 H=42.0	16.5	6150
N=1800 M=2000	W=60.0 H=30.0	156	7.5	37	W=61.5 H=42.0	16.0	5500
N=1800 M=2000	W=60.0 H=30.0	146	10	42	W=62.0 H=44.0	16.5	7600
N=1800 M=2000	W=60.0 H=30.0	155	10	56	W=62.0 H=44.0	16.5	8200
N=1800 M=2000	W=60.0 H=30.0	196	10	49	W=62.0 H=42.0	16.0	7200
N=2000 M=2340	W=60.0 H=30.0	307	30	38	W=62.5 H=44.0	16.5	14900
N=2000 M=2340	W=80.5 H=30.0	408	30	38	W=82.0 H=43.0	16.0	16900
N=1300 M=1625	W=80.0 H=42.0	700	60	45	W=83.5 H=60.5	11.0	31300
N=1300 M=1625	W=81.5 H=48.0	1059	60	38	W=83.5 H=60.0	13.5	34900

handled and removed by one person. All of the compacted material is uniform in size and can be easily disposed of in a landfill.

Equipment known as a console compactor can be used in a similar manner to a bag compactor. These units compress waste into a corrugated box container or plastic or paper bag. Some operate horizontally and some vertically. The containerized packages are generally 3½ cu ft in volume, although some manufacturers make larger packages. Densities range between 12 and 30 lb/cu ft.

Rotary type compactors consist of a ram mechanism which packs loose waste into plastic or paper bags, which are fixed in an open position on a rotating platform. When the bag under the packing ram is filled, the platform indexes one position, moving the full bag and replacing it with an empty one. The plastic or paper bag is held in place in a metal compartment which confines and prevents the bag from rupturing during packaging. Rotary type compactors may have from 8 to 20 individual compartments.

While compaction is not usually recommended prior to final disposal by incineration, there are many incinerator systems, especially of the small, two-chamber, controlled air variety, which use a compactor feeder mechanism. In this case, both the compactor and the container are entirely stationary. The container is between the ram and the incinerator. Material fed into the charging hopper is compacted to about 3 to 1 in volume ratio, and then a slide door at the far end of the compactor opens, and the ram, actuated a second time, pushes the charge into the incinerator. The ram then retracts, and the door closes and the system is ready to compact once again. Because compacted material does not burn as fast or as freely as uncompacted material, this is not the best arrangement.

8.4 SEPARATION OF GLASS AND METALS

Today, more than ever before, it is desirable to separate materials which can be reclaimed or recycled from both municipal and industrial trash. The municipal industry, of course, is much larger and much more completely formed than the industrial waste industry. However, the industrial segment has been reclaiming metal for many years, mainly by the segregation of the wastes before they are considered to be a waste material. Most manufacturing operations keep scrap bins for their metal cuttings and these materials are sold to a scrap dealer who reprocesses them and sends them back to the basic metal industry. Therefore, a discussion of separation techniques from an industrial point of view perhaps is not entirely pertinent. On the other hand, certainly it must be taken into consideration that many industries do have separation problems, more often to protect equipment downstream than to reclaim the material.

As indicated in the previous section, large or, in many cases, even small pieces of metal in a compactor can cause serious problems by damaging the compactor in one way or another. Therefore, it is desirable that metal be removed. Glass, by the same reasoning, is better uncompacted with other trash whether its final destination is landfill or incineration.

The separation of metals and glass from other wastes is accomplished by so many different types of systems that it would be impossible to describe each of them here. In general, the systems are not modular in concept but are devised specifically for the particular situation in which they have to operate. For example, at the present time, there are between 10 and 15 companies manufacturing magnetic separators. These magnetic separators may be used on the feed chutes to incinerators, hammer mills, or other types of shredding devices, or they may be used on conveyors, storage bins, dryers, at a point in the preprocessing of solid wastes. The thing to remember about magnetic separators is that they can remove tramp iron and steel or any materials which are magnetic in nature by merely attracting them to their surface. These magnetic separators are usually electromagnets which can be deenergized so that the material can be easily removed from them. Their purpose in most industrial applications is not recycling or reclamation but protection of downstream equipment.

In many applications where shredders are used to prepare wastes for subsequent landfill or incineration, magnetic separation takes place after rather than before the shredder because it then is dealing with smaller particles. On the other hand, removing the metallic materials ahead of the shredder would have saved considerable horsepower if much metal were present in the original feed.

Aside from magnetic separation, there are a number of different types of density or gravity separation devices. These can be used to remove metal, glass, and any other heavy materials which are not generally suitable for incineration. The classifier which is used for this purpose may be considered to separate the waste into an organic and an inert fraction and then eventually separate the glass from the metal.

Density or specific gravity separation may be accomplished in a number of different ways. One is to use air separation techniques. This usually involves feeding material onto a screen (or a perforated plate) at a relatively low rate and blowing air at low pressure up through the screen at sufficient velocities to carry the light particles, usually paper, cardboard and similar material, up and off of the screen and into a second chamber. The heavy material, such as the glass and metals, is discharged from the end of the screen into a hopper for further separation.

Another approach which is used more often to separate glass and metals is the use of a vibrating screen which by the frequency of the screen oscillations separates materials of different specific gravity. This separation occurs on the surface of a sloped screen causing one material to proceed from the inlet to the discharge end of the screen while the other material remains on the screen.

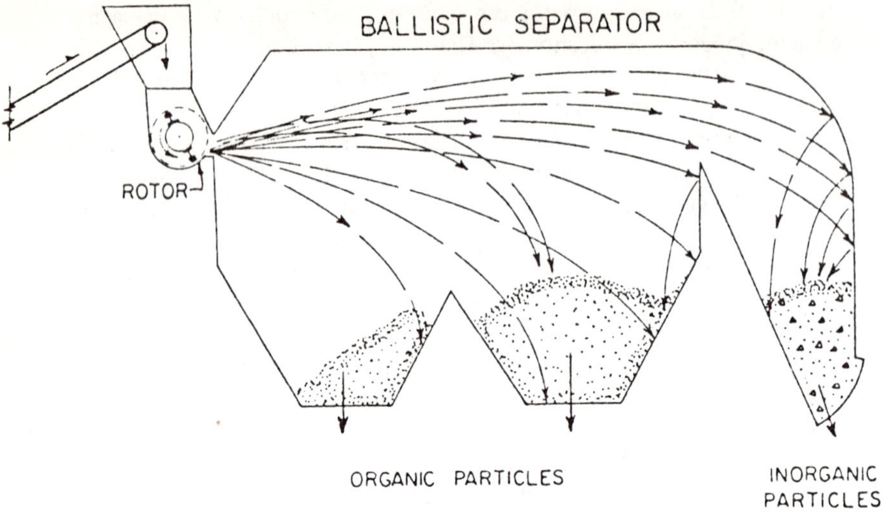

Fig. 8-3. Ballistic separator. *Source: Compost Sci.*, Spring 1963.

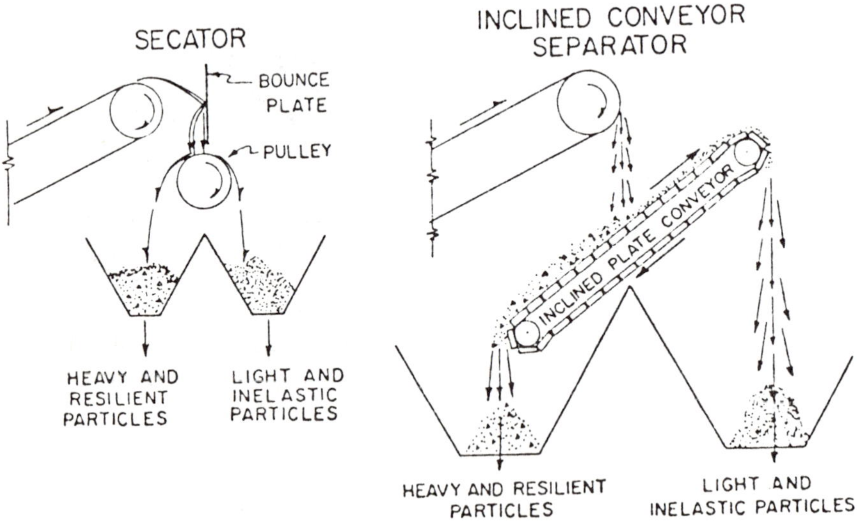

Fig. 8-4. Secator. *Source: Compost Sci.*, Spring 1963.

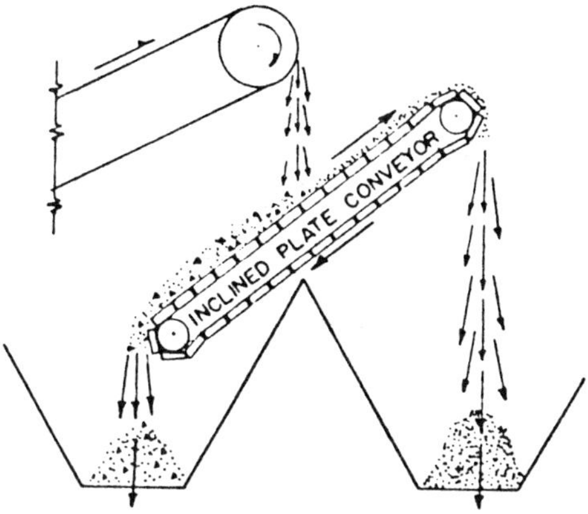

Fig. 8-5. Inclined conveyor separator. *Source: Compost Sci.,* **Spring 1963.**

Intertial separation also utilizes density difference as shown in Figures 8-3, 8-4 and 8-5.

In the same manner, centrifugal separations can be used or employed to remove a heavier material from a lighter material. A circular screen or track will remove the heavier materials at the outside and the light materials at the inside of the rotary or centrifugal action. Flotation techniques have occasionally been used to separate bottles and metal using a different medium for the separation. In every case, the separator is usually designed specifically for the application, and its purpose is to provide a more uniform waste feed to the downstream processing operation. Metals separated in this manner are usually clean and free of other refuse and available for immediate resale to metal scavengers. Glass material may either be landfilled or recycled by selling to a potential user.

It is important to emphasize that in the industrial operation this separation is most generally desirable for the protection of downstream equipment. However, it may also represent a worthwhile investment in terms of resource recovery.

8.5 SHREDDING

Another consideration in the preparation of solid waste is to use some method of size reduction to provide not only a reduction in the size of the particles entering an incinerator, landfill, composting system, or other device, but also to provide uniformity in the size of the particles. This has considerable advantages in all three areas and is almost essential for a good composting operation.

In a landfill in which the material has been properly shredded, compaction densities can be much more uniform or, if desirable, the shredded feed can be placed in a compactor and compacted to a uniform density prior to placing in the landfill. This is not normally done because it is an expensive step in terms of power usage, but it may have some desirable aspects in terms of the life of the landfill.

The use of a shredded feed to an incinerator is very desirable. Incineration systems operated on a consistent feed can work at their optimum conditions. A shredded feed provides a tremendous increase in surface area; combustion is a surface reaction. Air can reach all the particles in the feed and, therefore, they can be burned to completion much more easily than random or compacted trash. In many cases, a finely shredded material will burn so rapidly that it is difficult to supply sufficient combustion air for clean combustion. Therefore, the combustion system or incinerator must be designed for the type of feed supplied to it. It is for this reason that boilers are fired on atomized liquid fuel or pulverized coal wherever possible. The fine atomization or pulverization creates a greater surface area for the combustion reaction to occur. The same is true of shredded solid waste being fed into an incinerator.

A traveling grate type incinerator is specially designed to accept a shredded or prepared, presized, and screened waste material, just as traveling grate or vibrating grate stokers are used for boilers, and boilers require a carefully sized coal.

There are a number of different types of size reduction machines which can handle industrial solid wastes. They are rolling ring crushers, jaw crushers, hammer mills, shredders, and hogs. The type of waste to be processed and the desired end product greatly determine the selection of the machine for the particular application. Each type of machine achieves a size reduction by different types of crushing action. The rolling ring crusher and roll crushers use impact shearing and compression actions. Jaw crushers use impact and compression, and hammer mills use impact, attrition, and shearing. Perhaps it would be prudent to discuss these various types of actions before discussing the machines themselves.

An impact action can take two forms: gravity and dynamic impact. Gravity impact is the dropping of an object onto a hard surface, such as breaking a bottle by dropping it on concrete. Dynamic impact occurs when the feed material strikes a rapidly revolving surface which is harder than it is, such as a hammer mill. The feed material in this case is unsupported, and the force of the impact accelerates the shredded particle toward the breaker plate and toward other hammers and the outlet of the machine.

Attrition is the action which occurs when two hard surfaces rub together with a softer material between them. Hammer mills often operate with very close clearances between the hammers and the discharge grates to produce an attrition action to give further size reduction than the impact action which they also impart. Attrition requires greater power consumption and causes wear on the surfaces which are used to carry out the attrition. It is often applied to materials which have a low abrasion characteristic. Attrition also takes place between the particles themselves as they rub and grind each other in the size reduction process.

The compression action involves the squeezing of material between two hard surfaces with either one stationary surface and one moving surface or with both surfaces moving. This action works best when the feed material is hard, brittle, and abrasive.

Shearing action represents cleaving or splitting and works best on materials which are friable with a low silica content.

Often to avoid excessive wear on a machine, the size reduction of industrial solid waste will take place in more than one stage. Machines are designed for use in a primary, a secondary, or a tertiary application where primary milling might reduce the size of the material to 3 in., a secondary milling would take it down to 1 in., and a tertiary milling would reduce it to approximately 1/8 in.

When selecting size reduction equipment, several things must be considered. Obviously, the feed size is the first thing of importance because if the feed hopper is not as big as the material which it must handle, the size reduction process will not work. Also, the required throughput capacity in tons per hour is important to establish, and the final product size and shape will ultimately affect the size and the type of milling action required.

Some mills will produce a sliver-shaped product while others will produce cube-shaped pieces, so the mill and the product must be closely related. If the waste is wet, it is going to gum up the mill. If it is exceptionally dry, it will have little or no lubrication. If it is hard or soft, cold or hot, it will have different reactions in different types of mills. It is important to know whether the material is friable, and its degree of abrasiveness, because friable materials are easy to mill, but highly abrasive materials will require expensive machines made out of hardened surfaces.

Hammer Mills. The most common type of industrial size reduction equipment is the hammer mill, Figure 8-6. The hammer mill consists of a large motor-driven rotor to which are hinged a number of heavy hammers that pivot on this hinge and turn with the rotor like so many blades of a saw. The mill is enclosed with a hardened cover, and at one side of the mill is a hardened plate known as a breaker plate. Material enters the feed hopper, is struck by the mill and thrown against the breaker plate, and the material is ground to smaller size by the impact of the hammer, the attrition between the particle and the breaker plate and the particle and other particles, and a shearing action caused by the hammer.

Each machine consists of several major components: the feed hopper, the main frame, the bearings, the rotor assembly, the breaker plates, and the dis-

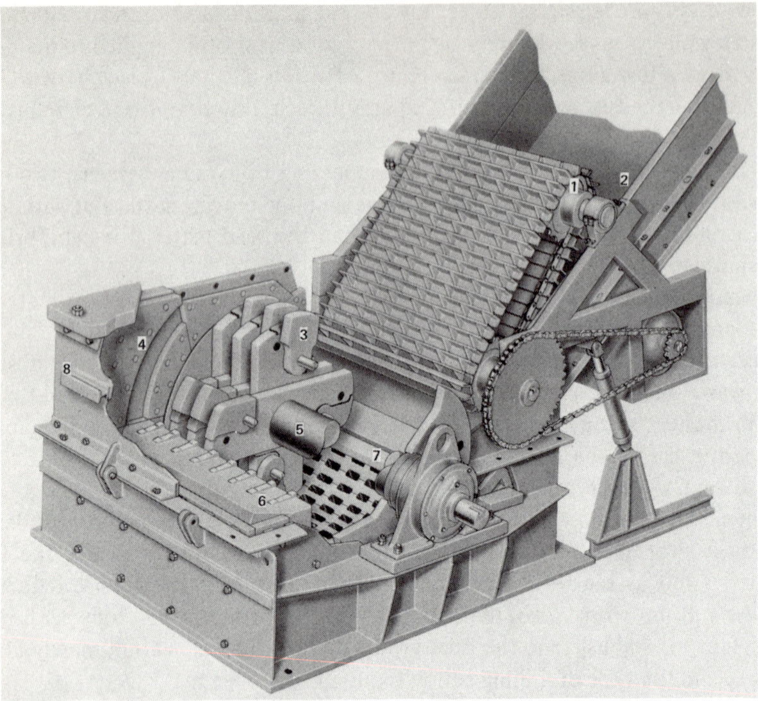

Fig. 8-6. Hammermill. ***Source:*** **Hammermills, Inc., Cedar Rapids, Iowa 52405.**

1. **Compression feed device**
2. **Side feed**
3. **Reversible hammers**
4. **ARS liners**
5. **Rotor**
6. **Tangentially opening grate**
7. **Notched, reversible cutting bar**
8. **Hydraulic opening back**

charge grates. The machine should also have protective devices and adjustment features.

The feed hopper should be large enough to permit the feeding of large particles without blockages due to bridging. The size of the feed material is used to determine the hopper size.

The rotor is usually mounted on a heat treated alloy steel shaft with self-aligning roller bearings enclosed in dust tight housing. The rotor shaft should have a larger diameter in the center where the strain is the greatest, and the design of the rotor should be suited for the type of solid waste it is going to handle.

Hammer mills can have rigid hammers or flexible hammers. Flexible hammers are pinned to the rotor by a hammer pin. These clearances must be large enough to provide ample wearing area on the pin in the pin hole and yet still make it possible to change or move hammers. As hammers wear at one end of the mill, they can be relocated for even wear throughout. Mills which are not in balance usually are noisy, cause vibration and use greater horsepower than new mills which are properly balanced.

In a hammer type rotor, the hammers do the actual size reduction. Heavy hammers operate at slow speeds, and light hammers operate at high speeds. Heavy hammers are best for coarse crushing, and light hammers, operating at high speeds, are better for fine products. The hammer type rotor can sometimes use reversible hammers. When double and reversible hammers are used, four working tips can be made available by reversing and inverting the hammer.

The breaker plate is used to absorb the force of the crushing action since the material is usually hurled against it by the hammers. It must absorb large shocks and, therefore, it is normally cushioned by heavy springs. Breaker plates can be replaced and can be made of abrasion resistant steel or of regular steel with special abrasion resistant liners. Some designs provide for the use of a cutting bar in place of the breaker plate, providing a shear action rather than an impact cutting action.

The discharge cage plays an important part in the product size because the cage holds the milled material in the chamber until it can be squeezed through the cage opening. This means that the material must be ground fine enough to go through the discharge cage before it can be released. In general, two types of discharge cages are used: perforated and bar type grates. The perforated grate is used for fine products, and the bar type grate is used for heavier materials. The bar grate is made up of a number of individual bars; the desired product size determines the discharge grate arrangement. Fewer or a greater number of bars can be installed depending upon the size required.

The discharge section of the machine is as important as the feed hopper because if it is not sized properly and does not allow for complete release of the milled product, the waste can back up into the hammer mill increasing wear and horsepower requirements, and eventually the product will clog the mill and shut down the operation. Therefore, protective devices should be built into the mill to minimize possible damage from unshreddable feed items. Most units are equipped with a metal trap which will catch heavy metal materials which have been struck by the hammer. Magnetic methods are, of course, also used as described before.

Another protective device is the use of a hinged breaker plate. This plate is adjustable and held in place on the lower end by shear pins so that excessive impact against the breaker plate will cause the shear pins to fail and allow the hinged breaker plate to swing open preventing damage. Breaker plates can also be adjusted to compensate for wear and control the product size if the adjustment feature is present on the mill.

The ultimate size of the product is dependent upon the size of the discharge grate opening as well as the speed of the mill. However, increased speed of the mill will not always guarantee a finer product. A high speed mill requires more power than a low speed mill, and it is more difficult to get the feed material between the hammers in a high speed operation.

For maintenance, the mill should be accessible. Most hammer mills have a cover which can be removed exposing the hammers and the rotors. The hammers can then be removed from the rotor (unless it is a fixed hammer design) by removing the pins, and can be replaced when they are worn.

Crushers. Crushing action is usually used on metal wastes for several reasons. Crushing reduces machine turnings into a uniform size. The smaller, more uniform size permits fast, economical, and easy handling by either conveyor or shovel, and requires far less storage space. Crushing of turnings or chips requires only 1/3 to 1/5 the storage space needed by the raw turnings; therefore, the greatly reduced volume cuts the shipping costs and storage space. It also enables the reclamation of valuable cutting oils and other compounds because they are more easily extracted and reclaimed after crushing. Once they are crushed, they can be put into a centrifuge, and the oil can be removed from them by centrifugal force.

Both ring and hammer type crushers are used for metal turnings; though, the ring type of hammer is better suited for metal turnings because it has a large number of shear faces. Hammer mills, however, are used for handling metal chips, especially when smaller sizes are required.

The metal handling shredder should be made of heavy, rolled plate steel welded to form a sturdy frame. The frame should have renewable manganese steel liner plates, and it should be mounted on a strong, structural steel base. Large capacity crushers are 20 to 30 tons in weight and occupy a plant area of 75 or 80 sq ft. Individual hammers may weigh as much as 100 lb. They are made of manganese steel coated with wear-resistant metal and should have multiple edges for maximum shredding efficiency. The centrifugal force in the mill keeps the hammers in their proper positions for crushing, and they are free to swing back when they make contact with an occasional difficult-to-shred piece of metal which can accidently be fed into the crusher.

After crushing occurs, metal turnings are further reduced in size by passing them over and through the dishcarge grates in the bottom of the crusher. These grates usually have a rectangular cross-section and are made of high carbon steel. The sections are usually riveted together.

The metal turnings crusher should have several adjustments to control the product size and to compensate for the different types of turnings to be processed. For example, some alloys are easily crushed while others need longer processing. Ductile materials are much more difficult to crush than brittle ones.

Crushers are available in capacities from 1 to 50 tons per hour. The corresponding horsepower would range from 25 to 500, and motor speeds would be around 800 rpm. For large items such as automobiles, the horesepower per ton is much greater, and mills are available up to 4000 hp.

Paper, Wood, and Glass. For paper type wastes, an entirely different type of system should be used. Generally, waste paper is shredded either prior to incineration or as a pretreatment for baling. For example, preshredding of paper waste makes it possible to increase the weight of a bale by approximately 35%, which reduces storage and freight rates. Paper mills often shred material prior to making corrugated or pasteboard cartons because in this form they can be most easily processed. In paper shredding, the hammers are usually rigid and are fixed to the rotor by means of a key. Nevertheless, they are removable so that they can be turned around to provide for wear on their cutting edges.

Waste paper can also be pulped, which is shredding in a wet condition. This process is used when the waste paper is to be reintroduced into the paper-making process.

Shredding of wood waste is often called hogging. This is done so that the waste can be manufactured into secondary products such as pressed board or used as fuel in the boiler. Here again, size reduction facilitates transportation and storage, and it permits the material to be easily pressed or compacted into particle board, fireplace logs, or whatever other products may be desirable. In a wood hogging operation, bark, waste, cuttings, wood chips, and planer shavings are all introduced into the shredder and reduced to a material similar to sawdust.

Hogging machines are a swing-hammer type of hammer mill but with knife-like hammers. These hammers work against a perforated plate or series of spaced bars through which the shredded wood is forced. In wood hogging, it is often necessary to shut down the unit to sharpen the knives since they can become dull after some period of operation. Wood hogs can accept logs up to as much as 30 in. in diameter and 2 ft long and shred them to very small sizes in a single pass. The hog's capacity will vary with the type of wood, whether it is hardwood or soft wood, whether bark is present or not, the amount of moisture content, and the uniformity of the rate of feed. In most wood wastes, the average moisture content is as high as 50% but a large capacity hog can produce over 150 tons per hour of below 3-in. particles from bark and wood refuse.

Shredders are often used in the making of glass cullet, which is glass which has been recycled to make new bottles and cheap glass. Cullet can be made from bottles or other types of glass waste, but it must be uniform in size to be introduced into the glass furnace. Here, a jaw or roll crusher is most often used because a hammer mill would wear excessively because of the abrasion characteristics of the glass. Hammer mills are occasionally used when a fine glass product is required but generally are not suitable. A jaw crusher that uses compression action with no attrition is the most suitable for the reduction of glass into cullet.

Cost. Shredders and size reduction equipment are expensive items. Their initial cost is high, and their maintenance is high. Shredders can be purchased for a wide variety of prices, depending upon their size. Installation normally runs about 20% of the initial cost of the shredder. Maintenance cost for most mills is expressed in dollars per ton, and it greatly depends upon the nature of the feed material and the final size. Maintenance cost for wood processing can be high or low depending upon whether the wood contains much sand or gravel. For cardboard, it would be considerably less. The maintenance on a metal turnings crusher would be higher than either, and a glass waste crusher generally has a high maintenance cost.

The operating costs, which are essentially power costs, are best expressed as horsepower per ton. These vary with the type of feed material and the required product size. Some typical horsepowers are shown in Table 8-2 for specific applications.

Bulky Wastes. Occasionally, industry runs into applications which require bulky waste crushing or disposal. When this operation cannot be handled by a hammer mill or similar device, it usually is accomplished by a machine which is similar to a jaw crusher that operates and looks like a series of large shears, that will cut up virtually anything from large demolition lumber to refrigerators or

TABLE 8-2. Shredder Horsepower

Capacity (tons/hr)	Approximate Shredder Horsepower
10	500
20	625
30	750
40	850
50	1050
60	1200
70	1400

Source: National Solid Wastes Management Association, Copyright 1977.

Fig. 8-7. Bulky waste crusher. *Source:* Von Roll Ltd., Montreal, Canada.

similar size industrial waste. These machines consist of two multiple steel frames which are movably connected and when open are arranged in a V fashion. The fixed frame comprises seven parallel steel section members, 1-ft wide, which are spaced 1 ft apart and to which the edge cutters are bolted. The movable frame comprises two rake sections of three seams each, which are rigidly connected by a cross-member at their upper end and pivot on a shaft below. The width of the beams and their spacing correspond to those of the fixed frame. Their edges are also armed with cutters.

The pivoted rake sections are moved back and forth by two double-acting hydraulic cylinders. When closing, the beams of the movable system swing into the spaces of the fixed frame and, in the process, the bulky waste material is caught between the two frames and cut up by the cutting edges. The force exerted by the hydraulic cylinders can be established at almost any level desired. Usually, it is on the order of 350,000 lb. The horsepower for this type of operation is relatively low because the mechanical advantage is high.

Each stroke of the crushing machine reduces the size of the material and has it drop further down into the V section. ultimately, the width of the pieces discharged from the crusher will be approximately 1 ft because this is the space between the cutters. The length of the pieces will depend upon the original size of the objects being crushed and the positions they assume in the crusher during the operation.

In this type of system, the actual cutting is done by wear resistant, replacable cutters each having four cutting edges of which only one is engaged. Thus, the cutters can be turned three times before regrinding becomes necessary. The cutters are made of tool steel and generally are subject only to slight wear.

This type of system has been developed by Von Roll, Ltd., of Montreal, and is probably more useful in municipal waste than in industrial waste but is mentioned here because occasionally industrial waste disposal requires the handling of bulky materials, Figure 8-7.

8.6 RESOURCE CONSERVATION AND RECOVERY

On September 30, 1976, the Congress passed the Resource Conservation and Recovery Act, a law which changed the entire solid waste management field. This law had been two years in the making, and, in its final form, it combined a number of important areas: first, the area of resource recovery and the conservation of natural resources, and secondly, the treatment of hazardous waste. It made the private waste management area a much more complex business since it provided for administrative reporting, licensing, permits, and planning.

Perhaps one of the most important things that this law accomplished was to institute a ban on open dumping, which meant that the area of sanitary landfill became a highly specialized and highly regulated activity.

Actually, the Resource Recovery Act of 1970 gave the U.S. Environmental Protection Agency the general mandate to promote the demonstration, construction, and application of solid waste management and resource recovery systems which would preserve and enhance the quality of air, water, and land resources. The act also specifically directed the EPA to conduct and assist programs related to methods to eliminate adverse health and welfare effects of the release of material present in solid waste into the environment. Also, this act directed the EPA to recommend guidelines for solid waste recovery and disposal systems, including those for private use, consistent with public health and welfare.

8-6.1 Incentives for Resource Recovery

During the past three decades, the United States has become the most affluent and also the most wasteful society in the world. We have thrown away billions of dollars in valuable resources which could have been recycled because it appeared to be more economical to bury waste in a landfill or burn it in an incinerator than to reclaim it. While this may have been essentially correct at one point in our history, several vital facts are now obvious which are beginning to change that situation. The first of these began in the Winter of 1973–74 when the first fuel shortages occurred, and prices of liquid fuels jumped to almost double their previous levels virtually overnight. Concerned about alternative types of fuels, the nation suddenly realized that much of the material which was being placed in landfills, or burned in municipal incinerators, or even in plant incinerators, had calorific value. While industry hasn't dramatically shifted over to burning refuse to generate heat because of a number of technical problems to be solved, it has, nevertheless, begun to look in this direction, and within a few years, fuel-rich waste materials will be used for power generation on a large scale.

The second incentive for resource recovery is the fact that the cost of land is rising rapidly and landfills are becoming expensive to purchase and to maintain. Landfills are also under much more stringent regulations than they were in the past. Leachate recovery systems, provision for methane generation, etc., all have made landfill operation a different business than it was in the past. Add to this the sophistication of the operation in terms of cover requirements and the equipment required to do it; this type of land use has a much greater economic impact than it did a few years ago.

Third, many materials, such as steel scrap, which have been recycled for years are even more attractive with the increased cost of energy because greater power and fuel usage is required to produce virgin steel from ore than to recycle scrap materials. This is also true of other waste materials such as glass, which heretofore had not been recycled to any great extent but now recycling appears increasingly attractive.

We shall probably continue to throw away many valuable resources for a number of years to come, but our attitudes are changing. We are beginning to move in the direction of recovery on a grand scale.

8-6.2 Recoverable Materials

The industrial complex of our nation has a number of things to consider concerning recovery. This is true of our municipal operations. Specifically, there are three major areas:

1. Recovery for reuse (conservation). Such recovery would involve the direct reuse of the resource by some treatment process that would convert it back into its initial form so that more of the same type of products could be made from it rather than using virgin material. An example is the scrap automobile business where the scrap automobile is almost completely recycled to the basic metals from which it was made. Another example is the water treatment industry where water which has been used for process and sanitary purposes is treated and recycled back to the initial source.

2. Recovery of heat. This might involve the recovery of heat from fossil fuel

generating systems, such as boilers, furnaces, ovens, etc., or it might involve the use of plastic or cellulosic waste materials as a fuel in an existing boiler, or the burning of methane from the anaerobic degradation of cellulosic materials in a landfill with the subsequent use of the methane as a fuel.

3. Recovery for other purposes. This might mean that the waste material is entirely changed in character but, nevertheless, is useful because of some reprocessing. Such would be the case in the destructive distillation of automobile ties to manufacture oils and other hydrocarbon materials. Another application is the manufacture of a road aggregate from various types of waste and trash, especially the slag from some pyrolysis-type furnaces. Composting could also be considered a form of reuse if the compost is used as a soil conditioner or fertilizer.

8-.3 Industrial Recycling Operations

Industry itself will internally recycle those items which are readily recyclable within an individual plant or plant complex. Industry usually will sell, give away, or pay to have hauled away by-products or wastes that are easily recycled by other people. The recycling industry, while it is very large, cannot be considered to be close-knit or well-organized except in certain categories. The oldest, because of economic incentive, is the metal recycling industry; metal scrap always has a value even though this value varies with the price of the initial commodity and also the availability of the scrap. Shops which manufacture items made of steel always have a scrap dealer to whom they can sell their cuttings, turnings, etc. The value of such scrap changes from day to day, and its price is published in financial periodicals.

While the iron and steel recycling industry is by far the oldest and probably the largest in the metals field, there are many other metals such as lead, copper, aluminum, and tin which are widely used and which have high recycle value. Aluminum recycling has grown into a sizable industry in the last ten years.

It would be impossible for the average plant, unless in the basic metals industry, to recycle its own waste products within the plant's boundaries. By using organizations which specialize in the recycling of such materials, the value of the material is recovered and put back into circulation, thus conserving the supply of the virgin material. There are tremendous incentives today in terms of power and fuel consumption which can be seen from Table 8-3. For example, it requires about 3% of the original energy commitment to make a ton of aluminum from recycled metal rather than mined ore. At the present time, a little more than 1 million tons of aluminum are recycled, and this represents a tremendous energy saving to the nation. What is more important is the fact that well over 2 million tons of aluminum are not recycled at the present time; and recyclable aluminum represents less than 30% of our domestic use of this metal.

TABLE 8-3. Energy Requirements—Virgin *vs.* Recycle

	Virgin Material Requirement (kwh/ton)	Recycled Materials Requirement (kwh/ton)	kwh *Savings* For Each Ton of Recycled Material	Barrels of Oil *Saved* for Each Ton of Recycled Material
Aluminum	51,379	2000	49,379	29.1
Copper	13,532	1727	11,805	7.0
Iron	4270	1666	2604	1.5
Magnesium	90,821	1875	88,946	52.2
Paper	6730	2520	4210	2.5
Titanium	126,115	52,416	73,699	43.3

Based on Data from U.S. Environmental Protection Agency and Oak Ridge National Laboratory (Atomic Energy Commission).

Therefore, the doubling of our current aluminum recycling rate from 1 million to 2 million tons per year would represent an approximate saving of 49 billion kwh or 29 million barrels of oil per year.

Another example is the recycling of steel. It requires two to three times the energy to manufacture steel products with virgin ore rather than with the recycled metal. Every time a ton of steel is produced with virgin ore rather than recycled scrap, it costs the nation 8.8 million Btu of energy. Yet each year, only about one-third of the recycled steel available to us in the United States is recovered and reused. Each million tons of scrap that is lost as a raw material costs this nation over 8 trillion Btu of energy and over 1,500,000 barrels of oil. Similar astounding figures are also shown for copper, magnesium, and titanium.

Studies made by the EPA comparing the environmental impact utilizing recycled waste paper rather than virgin pulp show energy savings in the range of 60 to 70% in favor of the recycling method. In the manufacture of products utilizing low grade waste, a study shows that over three times the energy was required to make the same product with virgin pulp. In another paper product area, it required two and a half times the energy to manufacture with virgin fiber as opposed to deinked and recycled waste paper.

Currently, only about 13 million tons out of an annual paper production of over 60 million tons are derived from recycled sources. Yet, it has been confirmed that over 35 million tons of additional waste paper are recoverable for raw material use. The doubling of our current use of recycled paper would represent an energy saving of 55 billion kwh or 32.5 million barrels of oil.

Paper comprises almost half of the nation's collected solid waste. It represents another important energy source after recyclable materials have been extracted for new raw material uses. The EPA indicates that about 80% of this nonrecyclable solid waste is combustible and could be recovered in the form of energy.

Glass accounts for approximately 9% by weight of total municipal solid waste. In recent years, over 13 million tons per year of Glass products have been discarded and less than 3%, or 350,000 tons, were recovered and recycled. Because glass is made of an inexpensive raw material, the recycled cost in dollars per ton is just about equivalent to the cost of virgin materials.

8-7 METAL RECOVERY

There are very few industries which can afford a complete classification system for the separation and reuse of metals; therefore, the material generally is sold to other processors. However, it is possible to improve the separation of metallic wastes in a plant and also possible to reprocess within the plant if the company wants to do so.

8-7.1 Ferrous Metals

The first consideration is the separation of ferrous metals because they are easily separated magnetically, Figure 8-8. Ferrous metal recovery systems usually involve shredding of the metal, even if it is of the size of an automobile body, in a hammer mill. This waste is then separated into light and heavy fractions. The heavy fraction is separated on a magnetic belt and passed through a nuggetizer which compresses the waste into briquettes and then over a metal drum to separate the ferrous metals from nonferrous metals. In an industrial application, it may be possible to separate ferrous and nonferrous metals at the point of the generation of the waste by visually or magnetically segregating them into separate bins. This will make the waste more valuable because it will require less processing downstream.

Where an extermely small amount of ferrous metal finds its way into the general refuse discharged from the plant and where incineration is used to destroy the refuse, magnetic separation might be accomplished after incineration.

The functions of a magnetic separator are to recover ferrous metals for their value, to protect the equipment from damage, to provide an iron-free, non-magnetic fraction and to reduce the volume of refuse which might get to an incinerator or landfill.

The recycling of ferrous materials from a manufacturing facility is relatively straight-forward since a company manufacturing equipment made of steel will have steel scrap to send to the scrap dealer which is probably noncontaminated by other metals and, therefore, not even magnetic separation is required. Domestically, the two largest sources of scrap steel are tin cans and automobile bodies, and each of these areas represents a very large portion of the total recycled steel in the United States which is nonindustrial. Over 10 million cars a year are recycled at an estimated overall gross revenue to the scrap industry of over 6 billion dollars.

Often, the problem of scrap metal handling is not easy to overcome. There is a wide mixture of both ferrous and nonferrous metals present in industrial wastes, in slags and smelter by-products, or in sludges, or in incinerator ash which can be reclaimed in most instances with a profit. Very substantial stock piles of these wastes are often available for processing in industrial areas. Some type of concentration is the most applicable method for the recovery and treatment of these wastes. The flow sheet in Figure 8-9 shows the composite of treatment methods which have been successfully applied in the recovery of metal alloys and such valuable materials as brass, bronze, copper, lead, zinc, ferrochrome, silver, gold, tin, silicon carbide, and abrasives.

The material is put through a crusher where size reduction tends to make the particles more uniform. Large pieces of metal, if malleable, are normally sorted out on the feeder ahead of the crusher or put through an impact type crusher. This material is then sent to a grinder to give final size reduction and to clean

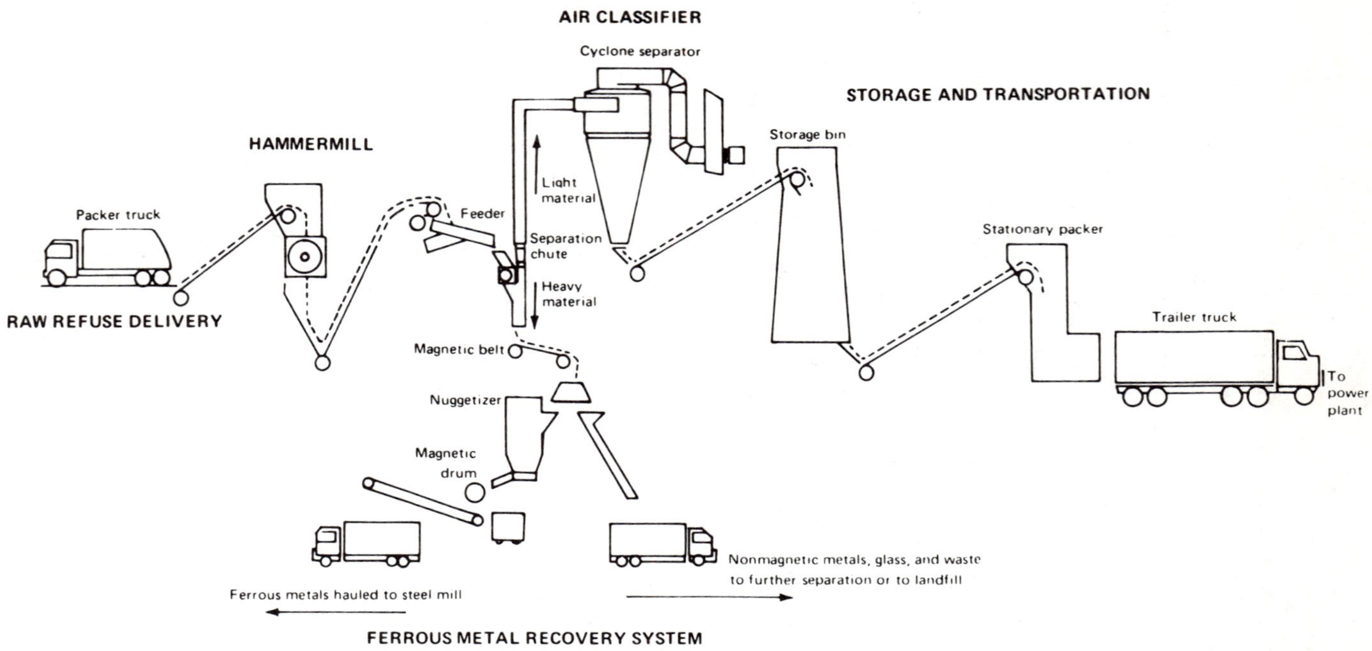

Fig. 8-8. Ferrous waste recovery system. *Source:* Third Report to Congress, Resource Recovery and Waste Division, SW 161, U.S. EPA, 1975.

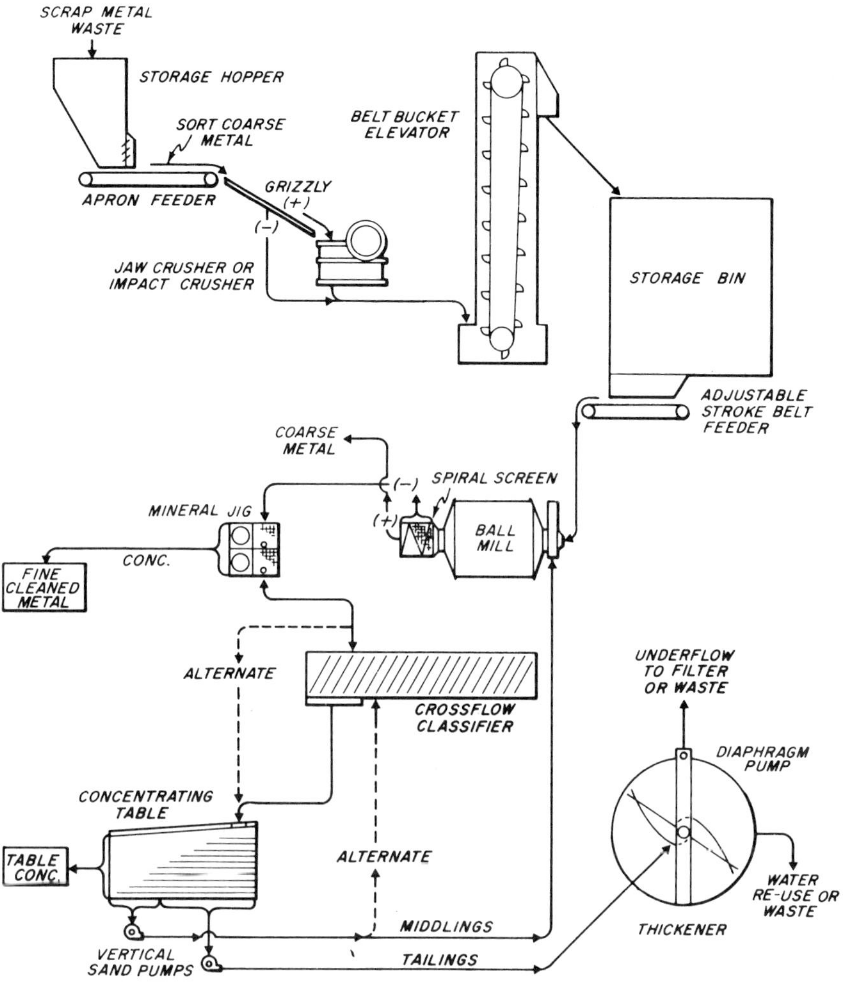

Fig. 8-9. Composite flowsheet–metal separation. *Source: Pollut. Eng.*

and free the metal from other nonmetallic wastes. Grinding is normally accomplished in a ball mill. The product from the ball mill is then screened and classified; it is passed over a spiral screen which is attached directly to and revolves with the mill; and the product size is usually in the 3/16 to 1/4 in. range. This is then a finished high grade metallic product requiring no further treatment. However, the coarse product from the spiral screen, if it is not clean, is up-graded in a jig. Here, a clean metallic product is removed through the jig side draw-off.

The waste discharges over the end of the jig and can be sent back to the grinding circuit if it is oversized. A jigging operation is a wet classifying system. The purpose of the jig is to recover fines which are not recovered in the screening and classification operations.

Fine materials from the jig are treated over one or more classifying tables. The amount of metal recoverable is usually small; the table serves as a pilot for the operation giving a visual indication of the plant efficiency. If larger sizes are present, they can be returned by a pump to the classifier and, in some cases, to the hill feed. It may be necessary to reclaim the water and the tailings before final disposal; this can be handled in a thickener.

8-7.2 Aluminum

Aluminum is slightly more difficult to reclaim from mixed waste than ferrous metals becuase the material is essentially nonmagnetic in the usual sense. However, an aluminum "magnet" has been developed which will process such materials. Table 8-4 shows the technique for the recovery of aluminum from solid waste. The basic approach is to use gravity separation, electric or magnetic separation using an electrostatic system or an aluminum magnet, and chemical or thermal separation of the material. Aluminum scrap today is worth approximately $300/ton so it is certainly worth the reclaiming effort. Gravity separation methods are shown in Figure 8-10 where a zig-zag separator using high pressure air will separate aluminum from heavier materials. Another method is the heavy media or sink-float separation. In this system, a heavy liquid is used as the medium. When a mixture of aluminum and other nonferrous metals is placed in the heavy liquid, the particles with a greater specific gravity than the liquid will sink; the lighter particles will rise to the top and will be skimmed, Figure 8-11. Tabling and jigging as described for ferrous metals are also applicable as shown in Figure 8-12.

It is possible to separate aluminum from other materials magnetically since the development of the Eddy current or aluminum magnet. When an electromagnet generates a magnetic field that passes through a nonferrous conductive metal surface, Eddy currents are produced. These Eddy currents are similar to ripples produced when a rock is dropped into water. The interaction of these currents on the magnetic field exerts a repellant force on the metal and thus separates it from other materials. The aluminum magnet can be used for separating aluminum from copper and zinc, but it is a sensitive process because it is complicated by other factors such as the shape of the metal particles and their distance from the field, which could alter the amount of force exerted on a given piece of metal. Normally, the separation operation takes place after the ferrous materials are removed by a ferrous magnet.

Another possibility is the use of electrostatic separation, Figure 8-13. When

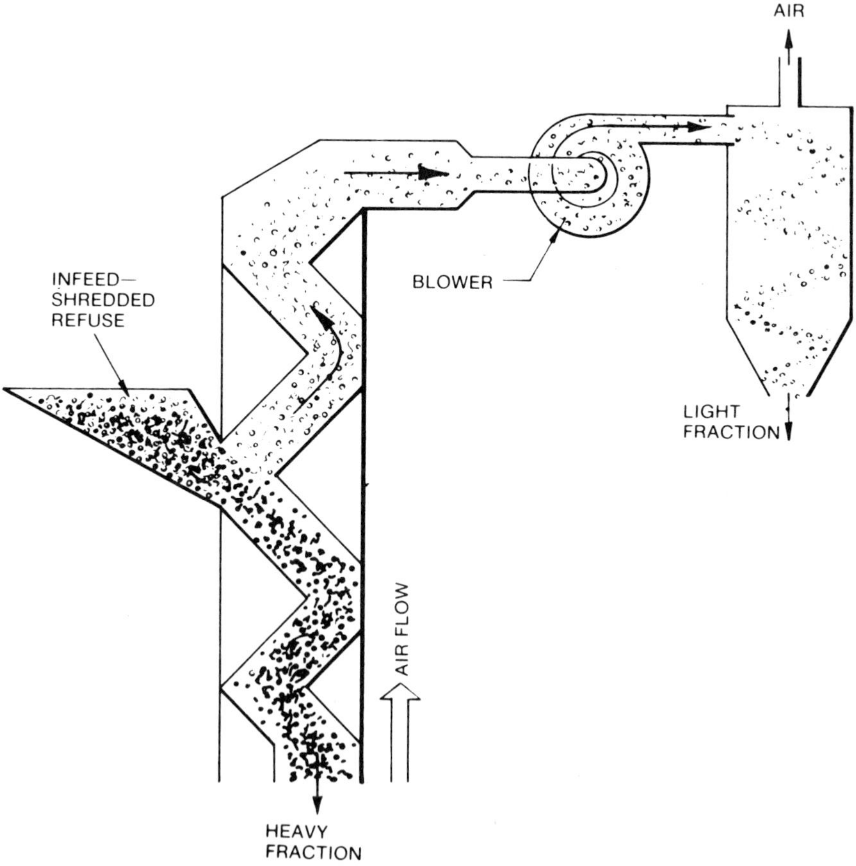

Fig. 8-10. Zig-zag separator.

feed material is exposed to an electric charge, the conductor quickly loses the charge while the nonconductor retains it. A revolving drum with an opposite electrical charge will attract and retain the nonconductor, separating it from the conductor which drops off immediately. High intensity electrostatic separators are now being used in the food and mineral processing industries, although not to any great extent in the metals processing industry.

There are also chemical methods for the separation of aluminum from other materials. Sweating is one of the precesses which uses the different melting points of metals to separate them. Sweating is widely employed, for example, to separate lead and zinc from aluminum. This process typically takes place in a continuous furnace with several zones in which temperatures are varied to melt specific metals. The metals then flow out of individual openings in the furnace

TABLE 8-4. Techniques for Aluminum Separation from Solid Wastes

Alternative	Method	Technique	Separation Procedure	Removal of Heavy Metals
Gravity separation	Sink/float separation	Particles with greater sp.gr. than the liquid sink and lighter particles rise	Skimming from surface of liquid	Removed as sludge from bottom of container
	Tabling	Flowing a pulp across a riffled table on a slant (2–5°) in a direction perpendicular to the riffles (150–175 vibrations per minute)	Light materials buoyed by water wash over the riffles and removed at low end of table	Migrate along the riffles toward the end of the table
	Jigging	Pulsation of a liquid through a bed of shredded refuse	Stratification of material in layers	Heavy layers removed by baffles or spigots
Electric or magnetic separation	Aluminum magnet	Eddy currents produced by magnetic field	Particles repelled to the side from other materials	
	Electrostatic separation	Electric Charge —Aluminum is a conductor and glass is not	A revolving drum oppositely charged	Conductor (Aluminum) drops off, but nonconductor remains

	Sweating	Uses different melting temperatures of metals	Metal drawn from various temperature zones in furnace	
Chemical/ thermal separation	Froth flotation	2 or more solids are separated by floating one of them in a foam on the surface of a liquid	Separate materials by surface characteristics and not relative weight	Removed from bottom of vessel
	Cryogenic separation	Mixed materials are placed in a cold media; such as liquid nitrogen	Aluminum remains malleable, but other metals become brittle and shatter	Followed by screening and flotation–heavy metals pass screen or settle on tank bottom

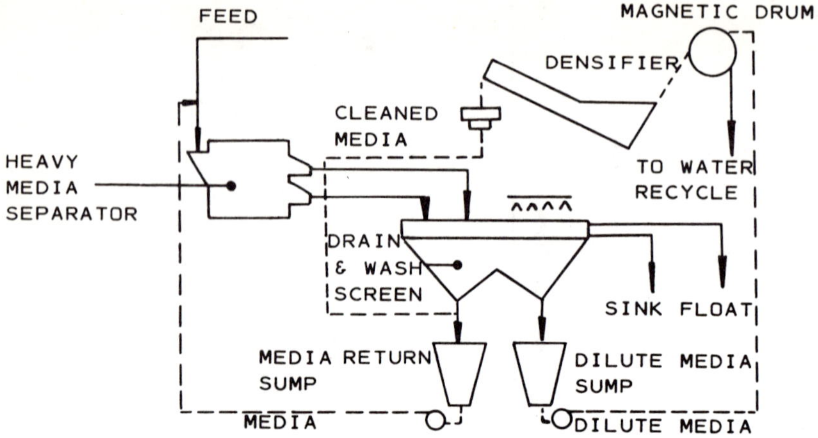

Fig. 8-11. Heavy media spearator.

shell; the other method is a batch process using a furnace setting that allows metals to melt and collect at the furnace bottom.

The U.S. Bureau of Mines has used froth flotation as a means of recovering metals from incinerator residues by a process also involving successive grinding and size segregation by screening. In this process, two or more solids are separated by floating one of them in a foam on the surface of a liquid. This process will separate certain meterials regardless of their densities because it depends on the surface characteristics, not just on the relative weight. A flotation tank is filled with a slurry of mixed materials and water and when required, a special flotation agent. Air is constantly circulated through the slurry. The material to be floated away has a greater surface affinity for air than for water. It is assisted by the special flotation agent. It gathers air bubbles around it causing it to float to the surface in a light froth that flows over the walls of the tank; the heavier material, or tailings, become wetted and sink to the bottom of the vessel.

The metals industry has also used cryogenic separation. Some materials when exposed to a very cold environment remain malleable, while others, such as steel, rubber, zinc, and some plastics become very brittle. In cryogenic separation, mixed materials are placed in an extremely cold substance such as liquid nitrogen and then treated in a hammer mill or crusher. The aluminum and copper remain intact while the heavier materials shatter and then can be separated by air or flotation classification. This is particularly applicable to the treatment of automobile waste metals.

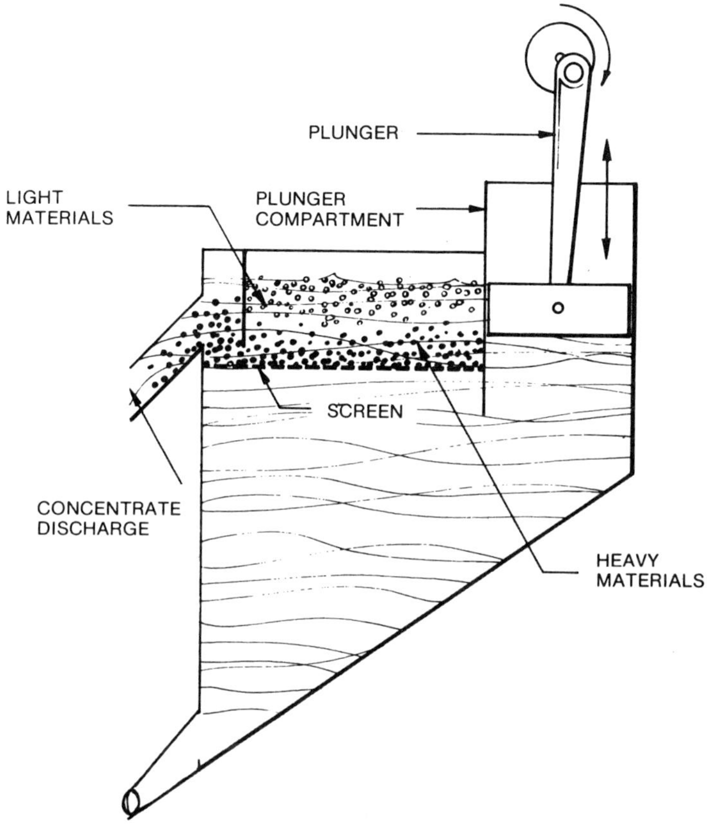

Fig. 8-12. Jigging system. *Source: Resource Recovery.*

8-7.3 Copper

Copper, a very valuable resource and one which has a high reclamation value, usually finds its way into industrial wastes in the form of wire or electrical components. Often, it is coated with plastic or cellulosic insulating material, or, in some cases, rubber. In order to reclaim the copper, it is necessary to remove the insulation from the surface of the wire; this may be done either mechanically or thermally. Mechanical methods chop up the wire into fine pieces as well as the insulation and then separate the fine copper particles from the lighter insulation by some type of classification. Thermal methods utilize a furnace to burn

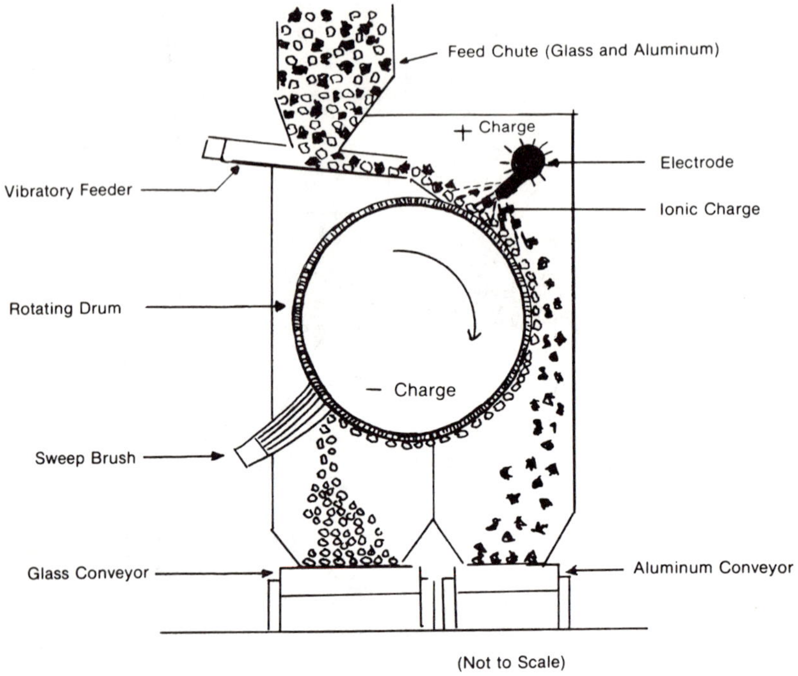

Fig. 8-13. **Electrostatic separator.** *Source: Resource Recovery.*

off the insulation without melting the wire. Copper in forms other than wire is generally separated by cutting, sawing, or melting of soldered joints.

8-7.4 Lead

The lead recycling industry is a large one and one which has serious problems with air pollution. The primary source of lead in the lead recycling industry is the automobile storage battery which has an antimony-lead plate. Used batteries returned by industrial concerns and private consumers are crushed to remove the hard rubber or plastic casing, and they are then fed to a high temperature lead smelter which burns off the organic materials and evaporates the residual sulfuric acid which is generally present on the battery plates even though the bulk of the acid has been drained off. This lead is then re-refined into new battery plates. The relatively low melting point of lead makes thermal recovery the best arrangement but creates some severe air pollution problems.

8-8 GLASS RECLAMATION

From an industrial viewpoint, there are very few applications where industry can successfully reclaim glass except for the glass industry itself. Used glass products comprise an average of 8% of the residential waste mix; only a small portion of this (less than 5% of the glass consumed) is eventually recycled, although this proportion is growing. Nevertheless, glass is inert and when properly disposed of in a sanitary landfill presents no environmental or health hazards so there is less incentive for the reclamation of glass either economically or environmentally than other materials.

There are three basic categories of glass products: containers, flat glass, and pressed or blown glass. Glass containers account for 75% of the glass produced in the United States. The principal raw materials used in making glass are sand, soda ash, and limestone. Scrap glass, which is known as cullet, also is used in the production of new glass. The key technical consideration favoring the use of cullet is the fact that it liquifies at a slightly lower temperature than the basic raw materials so the net effect is a lower melting temperature for the batch and reduced fuel requirements. This, of course, saves money and also extends the life of the furnace lining. Theoretically, cullet can comprise a large percentage of the input materials used in producing new glass. Up to 100% cullet can be used in the production of some types of glass; however, it is currently estimated that cullet comprises less than 20% of the materials used in making new glass.

The glass industry manufactures cullet of its own which is known as home cullet. However, the glass recycling industry also supplies large quantities of cullet for the glass industry. This purchased cullet is derived from beverage bottles and glass salvaging operations. Glass manufacturers have relied on internally generated cullet versus purchased cullet, but this situation is beginning to change, even though the primary sources of cullet have been found in the pressed or blown glass or flat glass segments of the industry, as compared with the container segment. While glass containers account for over 75% of the glass produced in this country, less than 20% of the purchased cullet used in the production of these containers comes from the postconsumer waste mix.

The cost of recovering glass in the residential waste mix or industrial waste mix has increased to the point where many scrap dealers have found it unprofitable to recover cullet. Glass manufacturers have imposed stringent quality control requirements for cullet which significantly adds to the price of processing. For used glass from consumer waste, the cullet must be: (1) color sorted into clear glass with 98% accuracy from glass which is amber or green; (2) clean and free of contaminants such as metals and rings with 99% accuracy. Chemically acceptable glass must be used, and the cullet must be crushed into small pieces. The sorting process has the most significant impact on the cost of processing used glass.

Only those industries which use large quantities of glass containers really have

TABLE 8-5. Glass Separation Methods

Method	Technique	Separation Procedure	Comments
Froth flotation	Two or more solids separated by floating one to the surface for removal	Separate materials by surface characteristics and not relative weight	• Potential for 95 percent recovery • Glass is not color sorted
Dense media separation	Materials of different densities can be separated by sinking or floating	Material separate by stages	Particles are large enough for color sorting
Color sorting	Use of electronic color-sorter	• Passes through transparency sorting device first to remove stones, ceramics, and residual metal • Color sorter matches particles with background for flint, amber, and green. • Air jets deflect particles into bins	• Improves marketability of recovered glass • Constant calibration is required • Organic residuals may cloud readings • Undersized pieces will not work in unit (under 3/16 in.)

a recycling problem. Such industries are the beverage and food industry and pharmaceuticals. Methods for separation are shown in Table 8-5.

8-9 PAPER WASTE

Of the waste products generated by American industry which are recyclable, over 50% is paper, paper products, or packing waste by weight. Sources of these products are paper from shipping and receiving areas which has been used for packaging, cafeteria paper waste and office paper waste, cartons, bags, and other shipping enclosures. Since most industrial waste is random in size and in consistency, it is a doubtful proposition to consider internal recycling of any type unless the material is fed to an incinerator to generate heat for some process or for comfort heating. Even the domestic paper recycling industry has severe economic difficulties because paper mills are not located close to cities, and cities are where the greatest amount of paper is generated. It is usually cheaper and more efficient to produce new pulp from trees than it is to recycle paper. Coated paper, for example, contains large quantities of ink and clay which cannot be recycled and which end up as a sludge requiring another disposal process. Newsprint, which is easily recyclable, is not normally a product of industrial operations.

Only those plants that manufacture paper or wood or wood products have an opportunity for internal recycle; there, the problem is often a waste disposal and handling problem for which they must find a solution. In such plants, paper scraps are usually collected manually and by air handling systems, segregated by type of paper, and then baled for reprocessing.

In the fiber reclamation process, paper is repulped in water with a mixture of 3 parts of paper to 97 parts of water at about 100°F. This is stirred in a huge tub by a hydropulper or beater. The stirring force is so great that the sheets of paper are rapidly broken down to a slurry which can be drained from the bottom of the tub. As many nonfibrous contaminants as possible are removed. This liquid is then forced through a coarse screen which defibers the paper even further, and the accepted stock then returns to the pulper where it is thickened for refining and pulp storage. Then it goes back through the paper-making process to manufacture a wide variety of finished papers, see Figure 8-14.

Most waste paper is so contaminated that it is impractical for reuse, so separation at the point of collection is possibly the only reliable way to avoid excessive contamination of paper stock. This is why only paper companies with an internal recycle system can actually do a good job on paper waste. Industrial waste paper contains grit, sand and glass, wire, paper clips, tin straps, aluminum foil, and staples, as well as many different types of plastic and hot-melt coatings,

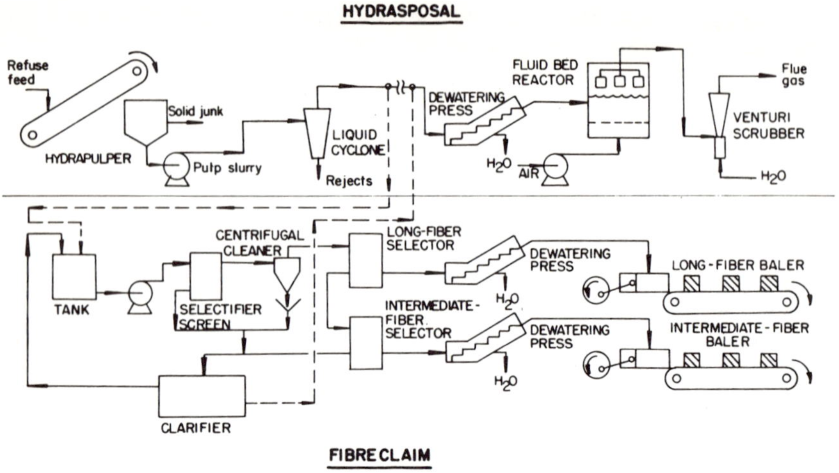

Black-Clawson developed wet disposal system for recovering PAPER FIBER in a Franklin, Ohio, disposal plant.

Fig. 8-14. Fiber recovery process. ***Source: Chem. Eng.,*** **p. 86, December 14, 1970.**

rubber bands, pressure sensitive adhesives, etc. In a paper mill, however, this can easily be segregated and the types of paper can be treated separately.

The recycling process for fiber recovery presents certain problems. Difficulties are encountered in collecting and sorting the stock paper. Also, the removal of contaminants proposes problems of operation such as plastic and wax removal and deinking processes, and related to this, of course, is the matter of economics. Certainly there is a profit in recycled paper, and it is becoming environmentally attractive to advertise the aspect that recycled paper is being used to print this or that particular document. But only as the cost of forest products increases is the true economic incentive going to drive more companies towards using recycled material.

Environmentally, it may be said that each ton of paper takes 17 trees, and considering the over 60 million tons of paper used in the United States each year, of which 13 million tons are recycled, one might deduce that 200 million trees could be saved annually. This, however, is a fallacy of statistics and not a reality. Forestation and reforestation of timber areas is functionally an industry like agriculture where old trees must be removed and new ones grown. The simple admonition to save trees by recycling paper is not sufficient to explain the problem. The recycle of paper is only one tool in our conservation effort; there are many other ways that we may preserve our resources.

8-10 REUSING PLASTICS

Since more and more of our industrial wastes are plastic materials, both in the packaging area and in the waste or reject product area, plastic recycling is an extremely important technology. While most plastic materials, because they are high in organic compounds have high calorific values and presumably could be used as fuels, there are a number of shortcomings to this type of recycle. Plastic materials burn very rapidly and, consequently, good combustion is difficult. They have high burning temperatures which would not be compatible with most boiler refractories, and some plastics like polyvinyl chloride contain materials which would create air pollution problems.

It is also important to consider that the feed stock for most plastic manufacturing operations is either natural gas or light fuel fractions which are decreasing in availability. Therefore, any recycling of plastic materials is decidedly advantageous. With new chemical processes comes a promise to break down flexible and rigid polyurethane foams into their basic building blocks, which then can serve to remake the polymers. Other approaches aim to recover polyvinyl chloride either by itself or compounded with additives. Fortunately, the amount of waste in most plastic operations is small. Thermoplastic scrap can be reground and made into powder or pellets which can be remolded.

One of the largest plastic scrap industries in the United States is the automotive industry which uses flexible polyurethanes for seat cushions at a rate greater than 300 million pounds per year. Furniture makers also consume large quantities of this same material. Since the foam is not thermosetting, it cannot be reground and remolded, so most of this material now goes to land fill. There is an alternative to this disposal which involves hydrolyzing the scrap materials to cause polymer breakdown. The decomposition process takes place at atmospheric pressure and about 500°F in the presence of steam. The resulting polyols and diamines can then be fed to a polyurethane production circuit. Ford Motor Company, which began to work on this in 1972, still does not have a production unit for the recycle of urethane waste, but the process appears to be promising. General Motors currently has a similar process that employs hydrolysis of the flexible foam with steam at atmospheric pressure and about 600°F, Figure 8-15.

The rigid polyurethane foams have been approached in a slightly different manner by the Upjohn Company. These much higher molecular weight foams require a different decomposition approach. The Upjohn patent feeds pulverized scrap into a nitrogen purged reactor containing a liquid medium of 90 to 95% aliphatic diol and 5 to 10% dialkanolamine, where an ester interchange takes place yielding a homogeneous liquid that can serve without any further treatment, as the polyol ingredient in the preparation of a new foam.

An even larger area for reclamation is polyvinyl chloride materials. Several

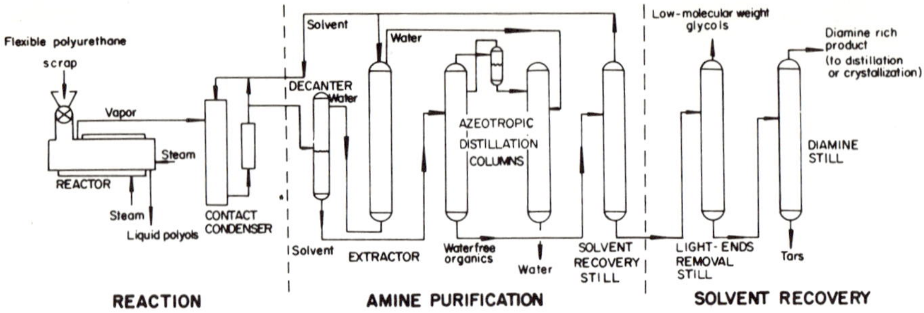

Fig. 8-15. GM's polyurethane recovery system. *Source: Chem. Eng.*, **p. 54, February 16, 1976.**

companies, Gerwain Chemical Corporation, Hafner Industries, and Polymer Recovery Corporation have devised processes. These entail shredding the feed and dissolving the material in a solvent and recovering these materials after filtering off the fibers by evaporation and drying

Similar processes are being studied for the recycling of ployethylene, polypropylene, and polystyrene materials. Total decomposition of polyethylene and polystyrene have been used to a limited extent, and combustion of these plastics with a deficiency of air has been able to achieve the production of carbon black.

8-11 SECONDARY RECOVERY APPLICATIONS

In the event that a material cannot be recycled back into its own industry, it may be possible to recycle the material to produce other compounds or materials that can be useful. An example is the destructive distillation of used tires, which is more completely described in Chapter 10 under the section dealing with pyrolysis. It is a pyrolytic decomposition. This is not the only means of recycling tires. Tires are recycled back into rubber by stripping out the steel cords, mascerating the body of the tire, and then dissolving the rubber in a solvent under high temperature and pressure which in effect devulcanizes the material. The result is a thin sheet of reprocessed, devulcanized rubber that can be recycled back into the tire industry. Unfortunately, the recycle of rubber back into tires is only about 10% of the new rubber production. If, however, the pyrolytic decomposition is as successful economically as it is technically, then a ton of scrap could yield as much as 140 gal. of liquid oils (similar in composition to conventional coal chemicals), and about 1500 cu ft of gas (with a heating value equivalent to natural gas), and a solid residue of high carbon content, see Table 8-6.

Burning of tires does not seem to be desirable because of the sulfur content which produces sulfur dioxide. Other pyrolysis projects which are discussed in Chapter 10 presume to take municipal garbage and trash and make a usable by-product gas from it, although none of these have been eminently successful to date.

There are literally dozens of processes where waste has been combined with aggregate concrete or some other type of silicate cement material to make building blocks or road aggregate. All of these have both positive and negative aspects but may certainly be a better place to put our waste than in a landfill, see Figure 8-16.

A research project at Texas A & M University involved paving a 380-ft section of a Houston, Texas, roadway in 1974 with a garbage-derived aggregate called, "Littercrete," and at last appearance, it seemed to be holding up as well as ordinary pavment at the same site. Littercrete contains glass, bottle caps, and other pieces of metal, including blobs of aluminum melted when garbage is incinerated. The material is treated with lime to foster adhesion with asphalt.

Other projects are much more extensive. Combustion Equipment Associates is constructing a plant in New York City to handle 3000 ton/day of city garbage and convert this garbage into a fine brown powder called "Ecofuel." Metals and other heavy materials are mechanically removed from the garbage before the remaining material, mostly cellulose, is treated with chemicals and then pulverized. This technique permits the fuel to be stored without decomposing. The powder can be burned more efficiently than raw garbage and can be used with oil, coal or natural gas. A similar plant in East Bridgewater, Massachusetts, converts 1200 tons of garbage/day into Ecofuel which is shipped 160 miles to Waterbury, Connecticut, and burned with oil to generate steam in a power plant.

The Ecofuel will sell for about the same price as coal or natural gas and gives the United States an opportunity to supplement its dwindling fuel supply. It has been estimated that if all the garbage in the country were converted to powdered fuel, it would add the equivalent of 2 million barrels/day to the nation's oil supply or about the same amount of oil as will flow through the Alaskan pipeline at peak capacity.

One of the largest producers of wastes in the United States is the animal industry. Animal wastes, manure, total nearly 2 billion tons/year which greatly dwarfs the 250 million tons of household, commercial, and municipal solid wastes generated and creates a disposal problem of great magnitude.

The U.S. Bureau of Mines has developed a process for recycling manure with any cellulosic waste. The waste is placed in a reaction vessel with carbon monoxide at an initial pressure of 1200 psi and heated at 380°C for 20 min. Based on dry manure, the yield of oil is 47% or 3 barrels per ton. The oil is largely paraffinic with virtually no aromatics present and has a nitrogen content of about 2% and an oxygen content of about 9%, making it nearly impossible to

TABLE 8-6. Yield of Rubber from Pyrolytic Decomposition

	Yields of Carbonization Products Yields, Weight Per Cent						Properties of the Solid Residue			
Test	Temp., °C	Residue	Heavy Oil	Light Oil	Gas	Total	Per cent Fixed Carbon	Per cent Ash	Per cent Sulfur	Heating Value, BTU/lb.
PT-1	500	42.0	45.2	4.2	5.0	96.7	86.3	9.6	2.0	13,470
PT-3	900	52.3	52.3	6.5	20.0	97.3	90.5	8.3	1.7	13,500
PT-4	500–564	41.8	41.3	4.3	5.5	97.4	87.7	9.2	1.9	13,490
PT-2	500	44.8	45.6	1.7	3.2	95.6	74.8	10.2	2.3	13,720
PT-5	500	40.3	44.6	3.3	4.8	98.0	83.1	12.7	2.0	13,020
PT-6	500–900	42.1	42.7	5.5	7.2	100.0	85.4	13.6	2.5	12,640
TT-3	500	36.5	48.7	4.3	5.0	95.7	80.1	16.5	3.2	12,250

Source: Pollution Engineering Nov./Dec. 1970, pp. 42–44.

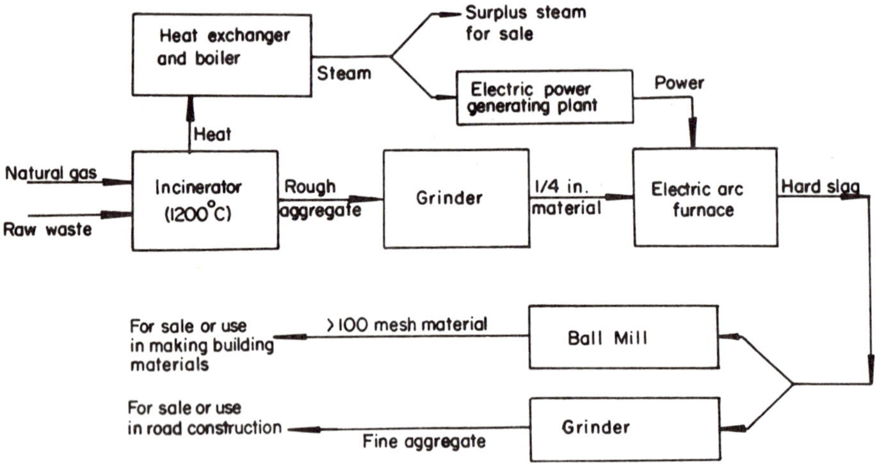

Fig. 8-16. Two-step burning converts solid waste into building materials. *Source: Chem. Eng.*, p. 58.

Neutral Oil, Volume Per Cent				Gas Properties		
Per cent Residue	Olefins	Aromatics	Paraffins and Naphthenes	Specific Gravity	BTU/per cu. ft.	Per cent Hydrogen
39.6	15.5	51.5	33.0	0.628	922	54.97
30.1	13.0	84.8	2.2	.460	765	54.29
35.5	17.5	51.6	30.9	.658	835	47.83
39.8	16.0	54.2	29.8	.599	890	49.80
30.4	19.0	50.6	30.4	.739	921	54.52
35.2	12.0	56.8	31.2	.479	762	58.93
27.6	25.0	51.4	23.6	.781	1,145	49.48

refine it into gasoline. It also has a very low sulfur content of less than 0.35%. At the moment, this process does not seem to be economically attractive, but other uses of manure to produce methane in anerobic systems may be of more interest.

There have been a number of applications where food has been made from waste materials, although the practical and economic application is in question. One which has been successful is the concentration of fish stickwater from the processing of menhadden. This stickwater, although dilute, is high in protein and, if concentrated and later dried, can make a fish meal suitable for human consumption. One problem is the aesthetic aspect of the material, but the protein value is unquestioned.

Heat and the addition of 1/2% sulfur dioxide has been used to make activated sludge from biological waste treatment plants into an organic molasses-like animal feed which has a high protein content. Yeast has been grown in a Porteus-process sewage treatment plant. The yeast growing on the sludge reportedly sterilizes the sludge and dissolves a substantial portion of the solids. A 1% yeast content, which is feasible, would make the process economically attractive. A number of processes to convert cellulosic waste into food have been piloted,

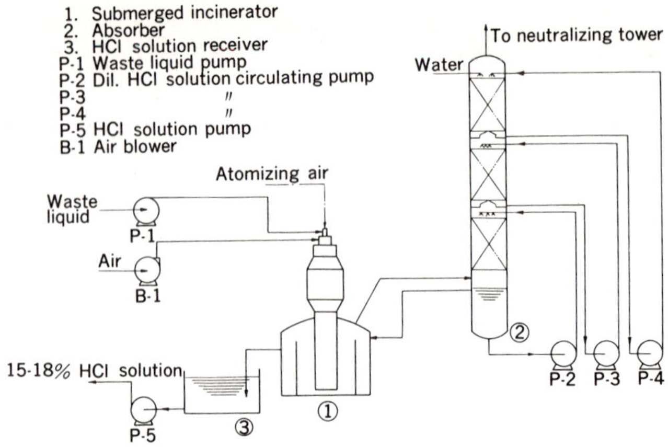

Fig. 8–17. Flow sheet for HCl recovery process. (15–20% HCl recovery process) (Courtesy NCE Corp. USA. Jenkintown, Pa.)

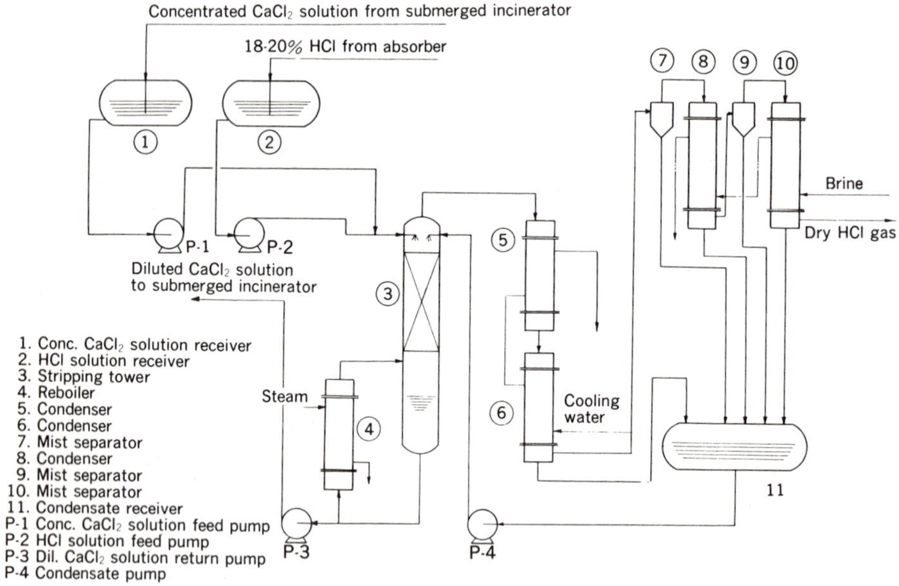

Fig. 8–18. Flow sheet for anhydrous HCl recovery. (Courtesy NCE Corp., USA. Jenkintown, Pa.)

some with more success than others, but it is feasible and practical to convert cellulosic waste into a food by bioprocessing.

It has often been said that the hog farmer saves everything but the squeal of the pig in his processing and reprocessing of various parts of the animal; this is true in many industries. The citrus fruit industry, for example, after extracting the juice for concentration and possible freezing, takes the hulls of the oranges and grapefruits and puts them through a hammer mill to make a pulp, after which the pulp is drained. The liquid drained from the hulls is concentrated into a heavy molasses-like material, and the hulls themselves are dried and ground and both materials are currently used for cattle feed. Many industries have this type of high usage factor for their products.

The chemical industry has a number of recycling systems which reuse waste products or try to capture the hydrocarbon value in wastes so that they can be reprocessed. There are far too many to mention here, but typical examples would be the recovery of hydrochloric acid from chlorinated wastes by a variety or processes, several of which are shown in Figures 8-17 and 8-18, and the recovery of sulfuric acid from waste pickling liquor, also by a variety of processes.

The Claus process, which has been used for about 40 years, captures sulfur from sour gas streams. All of these processes are examples of recycling in the chemical industry where an otherwise waste product would be thrown away but is captured and reused. The motivation, however, in the chemical industry today is not always economics. It may be motivated by antipollution legislation which could require greater expenditures for prevention of pollution than recycling.

8-12 RECOVERY OF HEAT

Today one of the most valuable resources in our national arsenal is the heat which is lost by industrial operations which, in many cases, could be recovered and reused. Unfortunately, there is no accurate measure of the heat lost by industry to the atmosphere in the United States, although it can be reasonably estimated that this is in terms of millions of billions of Btu's per hour. Every process heating application which includes metallurgical furnaces, processing ovens, petrochemical furnaces, and even comfort heating and air conditioning applications discharges heat to the atmosphere in large quantities.

Because of a cheap fuel source, the United States has been well behind the rest of the world in reclaiming heat from processing applications. In Europe, heat reclamation has been a way of life because of higher fuel costs. In the United States, we have only begun to realize the incentive in the last five years because of rising fuel costs and the possible reduction in the availability of fuel such as natural gas and light fuel oil. Considering the fact that fuel has tripled in price in five years and that we may not have certain fuel sources previously available to us, a national drive for energy recovery is underway. The President in his National Energy Plan developed in April of 1977 proposed the development of a number of alternative fuel sources as well as suggesting some tax benefits for energy saving, but the incentive for industry to develop significant heat recycling plans was not there. It is unfortunate that the President's advisers did not see the possibilities that exist in recycling of current energy losses rather than the strong emphasis in the plan being placed on alternative fuel and power sources.

Today in the United States there are fantastic opportunities for reclaiming both low level and high level heat through the use of well-known devices which have never been employed because the cost of the device heretofore has not shown adequate amortization. Perhaps even today, the economic incentive is not entirely justified but, in a few short years, it will be and those industries which have made a concerted effort to inventory their energy losses and convert these losses into savings will be ahead of the game.

A complete book could be written on the techniques of energy recovery, but it is not the purpose of this chapter to delve into individual applications except to say that there are literally hundreds of applications for energy recovery in every process and manufacturing operation. What is given in this chapter is a discussion of the various levels of energy loss and suggested systems for recovery.

The first and most logical type of heat recovery device is the waste heat boiler. This can take the form of either a fire tube or a water tube boiler installation which recovers heat from a high temperature processing application, usually a metallurgical type furnace. Generally, waste heat boilers are designed to operate at flue gas temperatures of 1400°F or greater because at lower temperature the efficiency is quite poor, but steam generated from 1400°F temperatures (and even lower in some cases) can be used for process application or for comfort heat for factories or office buildings. Steam generated in this manner can also be converted into electrical energy, although at this low level it is usually not as practical as with higher temperature superheated steam; however, it should not be ignored. Most plants, being entities in themselves, do not generate waste power for other users. Therefore, the application for reuse of the heat at the moment has to be within the plant itself. It is conceivable in the future that a number of plants in a metropolitan area may generate heat or power for surrounding residential areas, but this of course, has a number of social and economic problems which must be solved first.

Boilers and process furnaces can also utilize economizers to preheat feed streams, water, or feed water to a boiler, recovering heat at a lower temperature level but nevertheless recovering heat. Most boilers have stack temperatures in the range of 300 to 600°F; recovery of this heat is becoming more and more practical.

Refinery furnaces use air preheaters to supply reheated combustion air to the burner. These preheaters merely exchange the heat from high temperature exhaust gases to the combustion air used in the furnace.

At lower energy levels in the range of 400 to 1000°F, it may be more expeditious to heat air or water for comfort heating or for process requirements than to try to generate low pressure steam. Regenerative type rotary systems, heat pipes, intermediate fluid exchangers, and extended surface shell and tube type units may be used for this application, as well as plate type gas-to-gas heat exchangers. All have specific advantages depending upon the materials of construction and the corrosivity or cleanliness of the flue gas from the process heating device. It is obvious that the lower the temperature of the waste gas, the larger the heat exchanger required to recover the waste heat.

At low level temperature conditions up to 300°F, plate-to-plate type gas heat exchangers, rotary regenerative units, and heat pipes may be used. It is conceivable today to take a large quantity of low temperature gas and preheat either a heat transfer fluid or water in a heat exchanger and then employ (a typical plant heat recovery system is shown in Figure 8-19) a Rankine cycle device using Freon or some similar fluid which can then be expended through a turbine to generate power. The point is that heat at almost any level can be recovered. The economic impetus to do this will only be apparent as the cost of virgin fuel increases.

The alternate fuel industry today is growing at a fantastic rate. Spent lubricating oils from automotive applications, waste solvents, etc., are all being used as primary fuels for boilers and other heating applications. The entire area of heat recuperation from the process of incineration is a possibility. Industrial wastes consisting of paper and cardboard and other cellulosic materials have heating values of 7000 to 9000 Btu/lb. Plastics may vary from 13,000 Btu/lb to over 18,000 Btu/lb, and where proper burning devices can be developed, these materials can be incinerated and heat can be recuperated from many process uses. The major problem in industry is that the amount of waste heat generated by a given process plant cannot be adequately reused within the same facility within the time span required. Waste heat must be considered a dependent variable in the sense that it can only be captured when the processing requirements demand that fuel will be burned. The waste heat cannot be stored but must be utilized immediately so that if there is no immediate demand for the waste heat, there is little possibility for its recovery. In this regard, the large municipal heat recovery incinerator installations in St. Louis, Chicago, and Nashville have an advantage over industry because there is always a need for steam and power in a municipal operation.

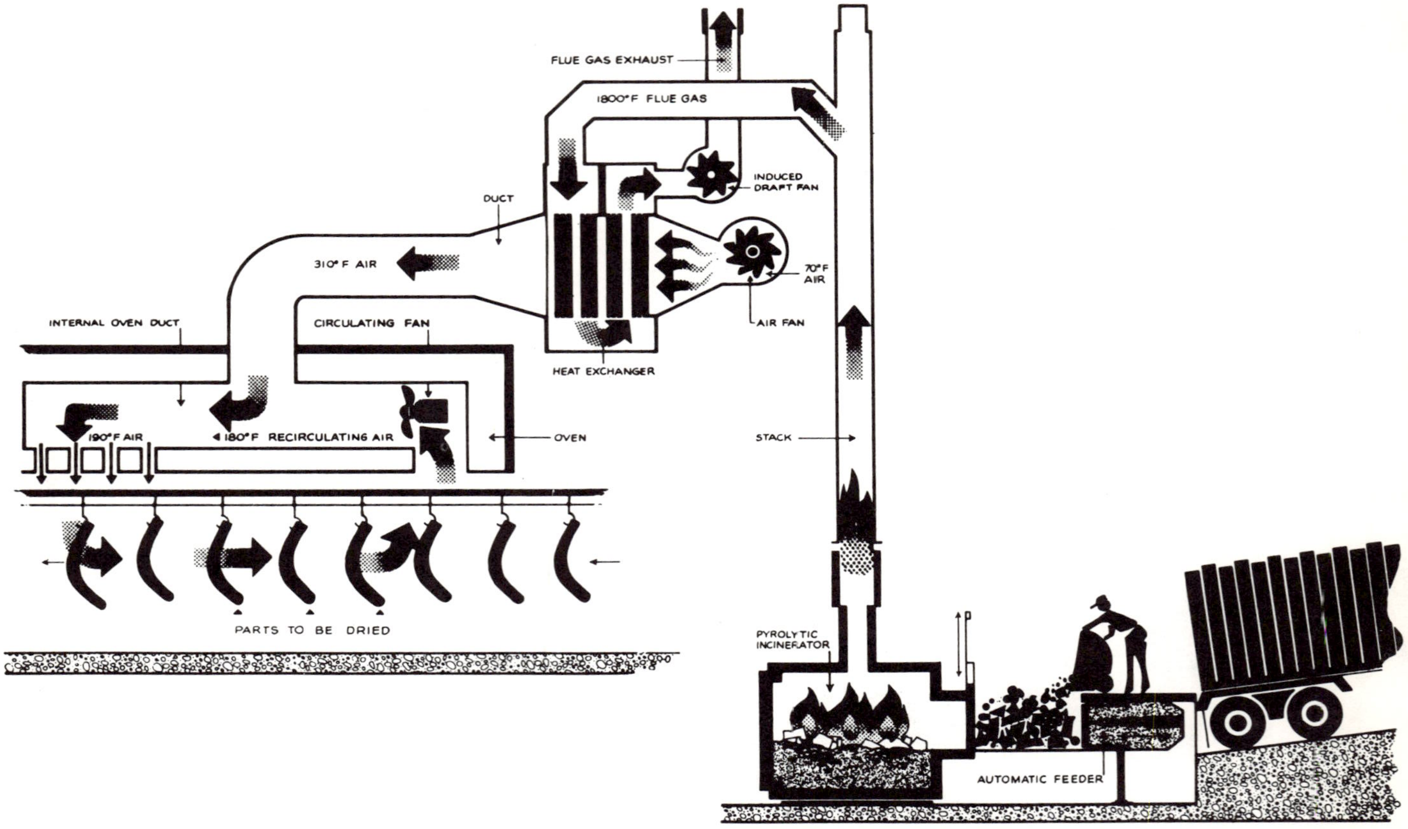

Fig. 8-19. Heat recovery system.

8-13 METHANE RECOVERY

The fact that existing landfills, which have been in operation for a number of years, generate large quantities of methane has stimulated an entirely new industry in the energy recycling area. Many landfills are now being tapped for their methane production, and the quality of the methane is equivalent to that of pipeline gas once it has been properly treated. Functionally, this methane production is nothing more than a bioconversion of cellulose and similar products in a natural state under anaerobic conditions to produce a usable fuel. Other types of fuel are also possible. For example, in Brazil, the production of ethanol from sugar cane is a major project. By 1980, they hope to produce 4.7 billion liters/yr. of ethanol in this manner.

Chicago's Institute of Gas Technology is working on the production of gas by anaerobic digestion of a number of biomass materials, including urban refuse, kelp from the sea, water hyacinths, and a variety of grasses. Cattle manure is another possibility for conversion into methane. In these systems, carbon dioxide has to be separated from the product stream leaving a pure methane gas. Waste Management, Inc., under an ERDA contract has begun operation of a 100 ton/day waste plant at Pompano Beach, Florida, to produce 300,000 SCFD of methane from municipal trash.

In one application, the manure generated by 50,000 head of cattle should produce gas at a rate of 1.8 million cu ft/day, which after cleaning will give approximately 1.2 million cu ft/day of fuel at about 850 Btu/cu ft at costs which

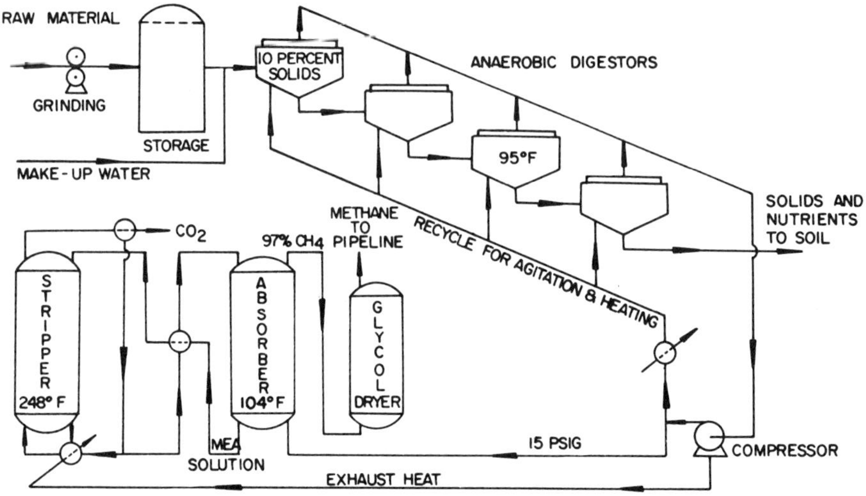

Fig. 8-20. Bioconversion process for producing methane.

run between $1.75 and $2.40/million Btu which, in the very near future, should be competitive with pipeline gas.

Anaerobic digestion studies at the University of Missouri with materials, such as hay, corn stalks, and oak leaves, have demonstrated the feasibility of producing methane from numerous materials. These studies show that up to 6 cu ft of methane can be produced per pound of dry crop matter. The disadvantages of such systems are that they usually require heat and pressure to accelerate the generation of the methane and the economics of such processes have not been fully tested, see flowsheet, Figure 8-20.

TABLE 8-7. Transportation Costs—Scrap *vs.* Virgin Material

Comparison of revenues from transporting virgin and secondary materials, 1970

Mode	Virgin material ($/net ton)	Secondary material ($/net ton)
Rail transport—direct comparison		
iron ore vs. ferrous scrap	2.31	4.83
roundwood vs. waste paper	1.92	8.01
woodchips and residue vs. waste paper	2.58	8.01
Rail transport—chemically weighted[a]		
coal basis vs. ferrous scrap[b]	7.34	4.83
coke basis vs. ferrous scrap[c]	7.79	4.83
woodchips and residue vs. waste paper	9.47	8.81
pulp vs. waste paper	16.81	8.81
Motor transport—direct comparison		
woodchips and residue vs. waste paper	2.44	9.68
pulp vs. waste paper	5.10	9.68
Motor transport—chemically weighted[a]		
woodchips and residue vs. waste paper	8.95	10.65
pulp vs. waste paper	5.10	10.65

[a]Chemically weighted by the trasnportation costs for component amounts needed to produce one ton of raw steel or one ton of paperboard.
[b]The coal basis assumes that coal is moved to the blast furnace for conversion into coke; transport of limestone is included.
[c]The coke basis assumes that coke is moved to the blast furnace from distant coking ovens; transport of limestone is included.
Source: RPI study

8-14 SALES AND TRANSPORTATION

In the recycling or reuse of industrial products, the very obvious and perhaps the simplest approach for the waste producer is to find another user for his by-products. There are literally dozens of potential users for off-speculation products, or waste products, from every industry. Just as the metal recycling industry has built up well-defined guidelines for the recycle of various metals, similar applications are present for chemical wastes. There are brokers in every major city looking to find applications or buyers for another's waste products. This is the first area which should be examined before any reprocessing moves are considered—how can my waste products be sold to another or recycled back into my business for profit?

Finally, one of the major considerations in the waste industry, especially when reprocessing is concerned, is transportation. The logistics of transportation of a waste from a producer to a processor to a user often times make the economics of the entire system infeasible. Instead of getting preferential transportation rates, the scrap producer is often faced with higher rates for the scrap material than for virgin material. There is a crying need for legislation in the United States to provide preferential freight rates for the reuse and recycle of existing waste and for the hauling of material to contract disposal sites to provide the impetus to both recycle and dispose of difficult waste properly. The ICC rate structure was not based on the handling of waste material; but it should have a special category to provide incentive for reprocessors and waste disposers, see Table 8-7.

REFERENCES

8-1. National Solid Wastes Management Association Industrial Compactor Ratings—1977.

8-2. Wiley, J., "Some Specialized Equipment Used in European Compost Systems," *Compost Sci.*, 1963.

8-3. Young, R. A., "Make Stockpiled Scrap Metal Waste a Product," *Pollut. Eng.* April 1975.

8-4. Dale, J. C., "Recovery of Aluminum from Solid Waste," *Resource Recovery*, January/February/March 1974.

8-5. Editorial, "Papermakers Study Recycling," *Chem. Eng.*, December 14, 1970.

8-6. Iammartino, N. R., "New Routes tackle Tough Plastics—Recycling Jobs," *Chem. Eng.*, February 16, 1976.

8-7. Beckman, Dr. J. A. and Laman, J. R., "Destructive Distillation of Tires," *Pollut. Eng.,* November/December 1970.

8-8. Editorial, "New Twist in Waste Use," *Chem. Week,* January 14, 1970.

8-9. Clausen, E. C., Sitton, O. C., and Gaddy, J. L., "Converting Crops into Methane," *Chem. Eng. Prog.,* January 1977.

8-10. Albrecht, O. W., "Shipping Wastes to Useful Places," *Envir. Sci. Technol.* Vol. **10**, No. 5, May 1976.

8-11. Dempster, J. H. and Gomaa, H. M., "Recycling of Waste Acid from Steel Pickling Operations," *Ind. Wastes,* March/April 1973.

8-12. Hokanson, A. E. and Kateen, R., "Chemicals from Wood Waste," *Chem. Eng. Prog., January 1978.*

8-13. Jones, J., "Converting Solid Wastes and Residues into Fuel," *Chem. Eng.,* January 2, 1978.

8-14. Editorial, "Prime Contractors for Resource Recovery Systems," *Resource Recovery Energy Rev.,* March/April 1977.

8-15. Glass Container Manufacturers Institute Inc., personal communication, Washington, D.C.

8-16. NSWMA Technical Bulletin, "Resource Recovery—Glass and Metal Containers."

8-17. Lauber, J. D., Conley, F. W., and Barshied, R. D., "Air Pollution Control of Aluminum and Copper Recycling Processes," *Pollut. Eng.,* December 1973.

8-18. Cross, F. L., Jr., "Compactor Sizing and Selection," *Pollut. Eng.,* July 1973.

8-19. Herb, J. H., "Solid Waste Shredders—Applications and Limitations," *Pollut. Eng.,* January/February 1971.

8-20. Franconeri, P. J., "Size Reduction of Industrial Solid Wastes," *Pollut. Eng.,* March 1976.

8-21. National Compactor—American Baler Machine Co., personal communication, Jacksonville, Florida.

8-22. Pavoni, J. L., Heer, J. E., and Hagerty, D. J., *Solid Waste Management,* Van Nostrand Reinhold Co., New York, 1973.

8-23. Schroering, J. B., "Applications and Limitations of Waste Compactors," *Plant Eng.,* March 6, 1969.

8-24. Pollock, K. M., "The Use, Selection and Economics of Stationary Compactors," *Solid Wastes Management,* February 1972.

8-25. Twichell, E. S., "One Company's Approach to Ferrous Extraction," *Solid Wastes Management,* November 1975.

8-26. Mackay, B. B., "Wastes Exchange Confounds Environmentalists Locked in Tennysonian Biases," *Solid Wastes Management,* September 1977.

8-27. Cutler, H. and Goldman, G. S., "Transportation: Bugaboo of Scrap Iron Recycling," *Environ. Sci. Technol.,* Vol **7**, No. 5, May 1973.

8-28. Editorial, *Chem. Eng.,* October 13. 1975.

8-29. Editorial, "Process Converts Animal Wastes to Oil," *Chem. Eng.,* August 16, 1971.

8-30. Editorial, "Moving to Garbage Power," *Time,* January 9, 1978.

8-31. Young, L. J. and Purcupile, J. C., "Food From Sewage? It Can Be Done," *Water Wastes Eng.,* June 1973.

8-32. Hagerty, D. J., "Refuse into Power," *Solid Wastes Management,* November 1975.

8-33. Coutant, C. C., "How to Put Waste Heat to Work," *Environ. Sci. Technol.,* Vol. **10**, No. 9, September 1976.

8-34. Kohn, P. M., "Biomass: A Growing Energy Source," *Chem. Eng.,* January 30, 1978.

8-35. Editorial, "Closing Plastic's Cycle," *Chem. Eng.,* June 15, 1970.

8-36. Spatz, D. D., "Reclaiming Valuable Metal Wastes," *Pollut. Eng.,* January/February 1972.

8-37. Alter, H., "European Materials Recovery System," *Environ. Sci. Technol.,* Vol. **II**, No. 5, May 1977.

8-38. Ploos Van Amstel, J. and Lee-Framton, J., "Converting Wastes to Raw Materials," *Environ. Sci. Technol.,* Vol. **II**, No. 10, October 1977.

8-39. NSWMA Technical Bulletin, Vol. 7, No. 5, "Resource Recovery from Solid Wastes, a Progress Report," June 1976.

8-40. Quimby, T., "Resource Recovery," *Waste Age,* January/February 1973.

9

Incineration of solid waste

Industry often employs incineration as a final disposal method. Solid waste incineration is seldom handled outside the plant because of cost, so more and more industries are installing their own incineration systems if landfilling is impractical. And, in the future, many of these systems will employ heat recovery devices.

A good waste incinerator for industrial purposes should reduce the weight and volume of a waste material by approximately 95%, but this will vary depending upon the amount of inherent ash in the waste material being burned. Obviously, inorganic materials do not burn and organic materials do burn, so those inorganic materials which are fed into a commercial incinerator will be found largely in the ash. The layman is often amazed to find that iron, steel, aluminum, tin, and other metals do not burn. They do oxidize to an extent, but during a normal incineration process, they are only slightly oxidized and remain with the ash portion to be hauled away and disposed of in some other manner.

The incinerator was not created to burn glass and metals; it was designed to burn organic materials–those materials containing carbon, hydrogen, and oxygen. Unfortunately, those who supply waste to incinerators and the people who operate the incinerators are not chemists. If they utilized principles of chemistry, the whole operation would be greatly simplified. Incinerators are, unfortunately, fed with materials that will never burn, and these materials are found in the ash pit when the residue is removed.

9-1 PRACTICAL AND THEORETICAL CONSIDERATIONS

The chemistry of incineration is very simple. It involves the combustion of carbon and hydrogen. These are the two primary components of virtually all combustible organic waste material. If the carbon and hydrogen are supplied with sufficient oxygen in the form of air and the reaction takes place in a well-stirred reaction system (i.e., under conditions where the air or oxygen has an opportunity to come in contact with all of the carbon and hydrogen in the waste material) of sufficient size, and if the operating temperature of the system is high enough to promote the reaction, then complete combustion of the organic material should occur within the incinerator. Incineration is dependent upon three factors: (1) the residence time of the waste material in the incinerator in contact with air; (2) the amount of mixing which occurs between the air, which is the oxidant, and the waste material; and (3) the temperature at which the reaction takes place.

If we visualize these three basic requirements as being essentially coequal, we can imagine the system as an equilateral triangle, the area of which represents a proper combustion system. If, for any reason, one of these sides is made shorter–either the residence time, the mixing, or the temperature–then it would be necessary to extend the other two sides of the triangle in order to maintain the same area or the same combustion equivalent efficiency. If, for example, the amount of mixing in the incinerator were reduced, then it would be necessary to increase the residence time and perhaps the temperature in order to achieve the same degree of combustion or combustion efficiency. By the same reasoning, if the temperature of the incinerator were reduced, then it would be necessary to increase the turbulence and the residence time in order to achieve an equivalent combustion efficiency.

Each of these factors is significant in the operation of the incinerator. They are very basic principles but, unfortunately, are generally overlooked in the design of good incineration devices. The chemistry of incineration is quite straightforward. It represents the oxidation of the carbon and hydrogen molecules present in most cellulose wastes and in many liquid and gaseous organics.

These equations are as follows:

$$C + O_2 = CO_2 \qquad \text{(Eq. 9-1)}$$

$$H_2 + \frac{1}{2}O_2 = H_2O \qquad \text{(Eq. 9-2)}$$

These reactions are demonstrated in a combined reaction as follows:

$$CH_4 + 2O_2 = CO_2 + 2H_2O$$

If these were the only incineration equations which were required to operate a conventional incinerator, it would be easy to design an incineration system. However, industrial wastes, liquid, solid, or gas do not always permit the latitude of assuming that all wastes are comprised of carbon and hydrogen. Many of the industrial wastes which are handled today contain other combustible materials, such as sulfur, halogen compounds, phosphorous, and inorganic salts. Most of these materials oxidize to one extent or another, depending on the amount of excess air in the combustion reaction, but, after oxidation, they still present a problem. Carbon oxidizes to carbon dioxide, assuming that the reaction is complete; when the incineration process takes place at about 1800°F. If it is below 1800°F, it forms some carbon monoxide, which is, of course, a toxic gas.

Hydrogen, under similar conditions, oxidizes to water vapor which is compatible with the atmosphere. The nitrogen in the air, in many cases, remains unchanged but, under certain conditions, will form various oxides of nitrogen due to a process known as nitrogen fixation. Usually this is found in the form of nitric oxide (NO) since other forms of nitrogen oxides are generally unstable at normal flue gas temperatures. If there is nitrogen in the fuel, it will also form nitric oxide but by a different mechanism. Nitric oxide formation can generally be reduced by minimizing the amount of excess air used in the combustion reaction.

Sulfur in a waste material presents a very definite problem since it will form sulfur dioxide and, under certain conditions, sulfur trioxide. Since the sulfur oxides are considered toxic chemicals if they are present in a significant quantity in the flue gas from an industrial incinerator, they must be scrubbed with a solution that will react with a sulfur oxide and form a sulfite or sulfate which can be removed by precipitation and filtration. Generally, industrial solid wastes contain very little sulfur so this is not normally a problem. But, many liquid industrial wastes do contain sulfur compounds. It is a problem in the burning of coal and fuel oil but not one generally faced in the incineration of waste materials.

More prevalent in the present waste character is the presence of halogen materials. Halogens are fluorine, chlorine, bromine, and iodine. The last two are very seldom seen in industrial wastes. The first two, fluorine and chlorine, are much more common with chlorine being by far the largest troublemaker.

Fluorinated compounds, present in aerosols and various plastics as well as antifreeze compounds, present a distinct problem of removal when generated in an incineration system. Chlorinated compounds present a similar problem. The primary source is polyvinyl chloride materials, which are used for upholstery, clothing, footwear and flooring.

When fluorine and chlorine compounds are burned, the flue gases tend to contain molecular fluorine or chlorine unless some provision is made to hydrolize the halogens in the combustion reaction. This is not a difficult job, but it is one which should be considered in any incineration system. If there is sufficient hydrogen present to hydrolize fluorine to hydrofluoric acid, or chlorine to hydrochloric acid, then this will occur and a relatively high ratio of hydrogen to carbon is required in the waste fuel. For chlorine, it requires about a 4 to 1 ratio of hydrogen to chlorine in order to form hydrochloric acid essentially free of chlorine gas. If sufficient hydrogen is not present in the initial feed to the incinerator, then some free chlorine gas will be formed in the combustion reaction. One method of hydrolizing this free chlorine is to add unburned fuel to the system, or water in the form of steam, at temperatures in excess of 2000°F. These reactions are shown below:

$$CHClCCl_2 + 2O_2 = 2CO_2 + HCl + Cl_2 \qquad \text{(Eq. 9-3)}$$

is the normal reaction of trichloroethylene burned in air, but if excess hydrocarbon is added the reaction is

$$CHClCCl_2 + 3\frac{1}{2}O_2 + CH_4 = 3CO_2 + 3HCl + H_2O \qquad \text{(Eq. 9-4)}$$

or, if steam is used in place of methane

$$CHClCCl_2 + 1\frac{1}{2}O_2 + H_2O_{(v)} = 2CO_2 + 3HCl \qquad \text{(Eq. 9-5)}$$

For all practical purposes, the carbon, hydrogen, sulfur, and halogen reactions are those which will be found in any combustion or incineration system. Phosphorous or other nonmetal oxidation reactions may also be present but are seldom a consideration in either industrial or municipal solid waste incineration systems.

The chemistry of combustion or incineration reactions is important but no more than the kinds of conditions which provide the best possibilities for complete oxidation of the waste material. Solid waste is considerably more difficult to burn than either liquid or gaseous wastes. The fact that the material is in solid form means that the molecules are much closer together; therefore, it is much more difficult to bring each organic molecule of the waste material into contact with oxygen to perform the necessary oxidation reaction. For example, to mix a gaseous waste with air so that it will burn to completion, it is merely necessary to create sufficient turbulence in the reactor so that the incidence of contact between the oxygen in the air and hydrocarbon in the waste will be maximized. A well-stirred gaseous reactor is relatively easy to effect.

In a liquid incinerator, spray nozzles are used to atomize the fuel into the air stream so that a similar condition can be achieved. The finer the atomization, the better the combustion. In a solid waste incinerator, there is an entirely different problem, that of mixing air with a solid waste material. If the solid waste combustible material could be divided into fine particles, it would burn much better than if it were in a bulk state. Many consider compaction to be a great boon to solid waste handling and may be for certain types of solid waste disposal such as landfill, but it is far from ideal when incineration is considered as a final method of treatment.

The more finely divided the solid waste product to be incinerated, the better the incineration process. It presents more surface area for contact with the air or oxygen which is required for combustion. Therefore, shredding, pulverizing, or any other size reduction method which can be used ahead of incineration, and can be used economically, will pay big dividends in the amount of residue which is eventually left in the ash pit of the incinerator.

A second misguided concept of incineration is that large amounts of excess air provide better combustion and, therefore, better incineration efficiency; nothing could be further from the truth. A system which will operate on a low percentage of excess air is by far the most efficient, because it will produce the maximum operating temperature and require the minimum amount of horsepower to supply the air for combustion. In other words, a low excess air system when properly mixed will be by far the cheapest and most efficient system. The problem has been that most low excess air systems would not adequately burn the waste and, therefore, more and more air has been added in order to achieve combustion. Excess air has a tendency to reduce the temperature of the reaction and, increase the amount of carbon monoxide and hydrocarbons in the flue gas, see Figure 9-1.

Most solid wastes have relatively low calorific values compared to their gaseous and liquid counterparts, see Tables 9-1 and 9-2. Since they have a low heat release per pound, the incineration temperature is often below the level required for complete oxidation of the carbon to carbon dioxide. Consequently, auxiliary fuel is often required. Modern incineration systems utilize this auxiliary fuel in a secondary chamber to increase the temperature of the combustion products of the first stage burning in the presence of excess air, so that all carbon and hydrogen is completely oxidized to carbon dioxide and water vapor. This is known as two-stage combustion, see Figure 9-2.

Modern incinerators utilize this two-stage process as it has the advantage of reducing the amount of excess air required in the first stage and adding it to the second stage where good mixing is much more easily achieved. For example, the

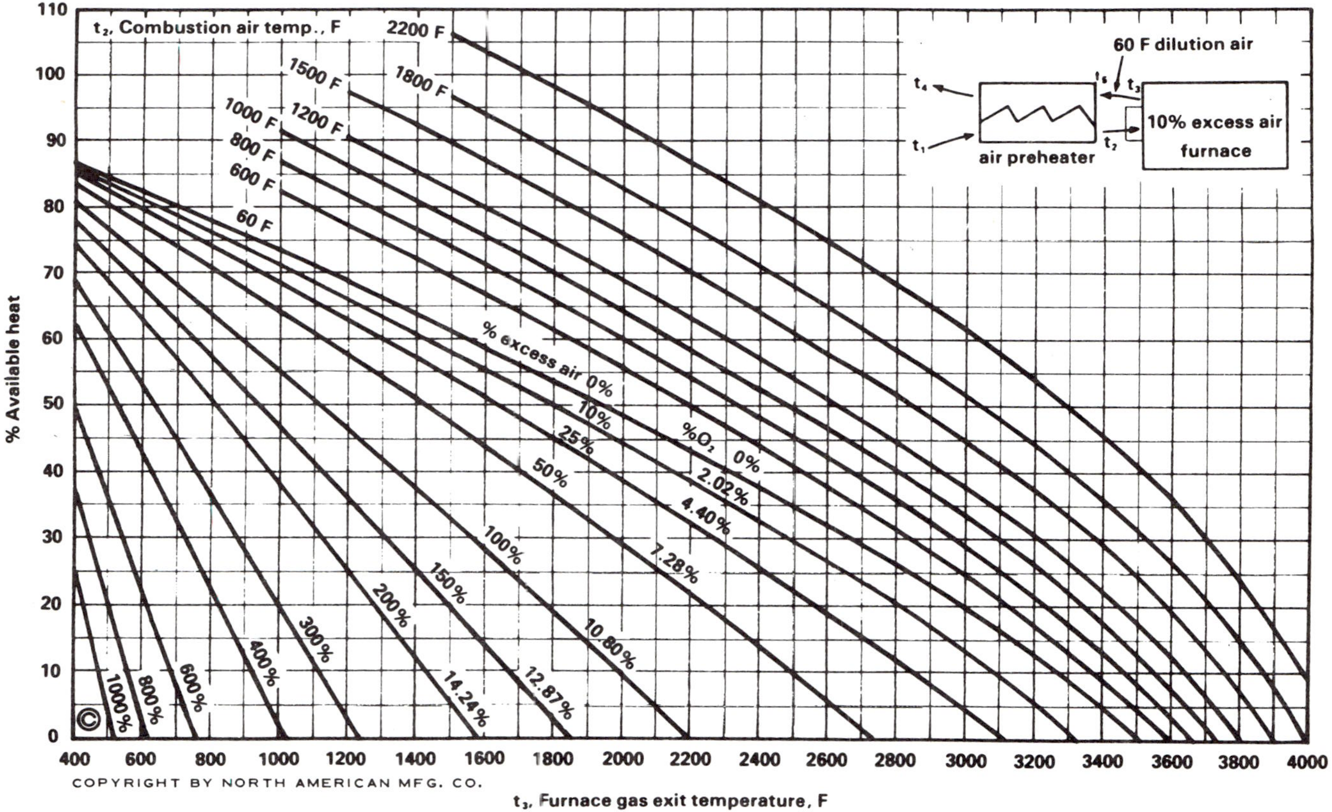

Fig. 9-1. Heat available curve. *Note:* This chart is approximate because it is based on the assumption that the air required per gross Btu is the same for all fuels. *Source: North American Combustion Handbook*, North American Manufacturing Co.

TABLE 9-1. Btu Values of Various Solid and Liquid Wastes

				Content by weight in percentage	
Waste	B.T.U. value/lb. as fired	Wt. in lbs. per cu. ft. (loose)	Wt. in lbs. per cu. ft.	Ash	Moisture
Type 0 Waste	8,500	10		5	10
Type 1 Waste	6,500	10		10	25
Type 2 Waste	4,300	20		7	50
Type 3 Waste	2,500	35		5	70
Type 4 Waste	1,000	55		5	85
Kerosene	18,900		50	.5	0
Benzene	18,210		55	.5	0
Toluene	18,440		52	.5	0
Hydrogen	61,000		.0053	0	0
Acetic acid	6,280		65.8	.5	0
Methyl alcohol	10,250		49.6	0	0
Ethyl alcohol	13,325		49.3	0	0
Turpentine	17,000		53.6	0	0
Naphtha	15,000		41.6	0	0
Newspaper	7,975	7		1.5	6
Brown paper	7,250	7		1.0	6
Magazines	5,250	35		22.5	5
Corrugated paper	7,040	7		5.0	5
Plastic coated paper	7,340	7		2.6	5
Coated milk cartons	11,330	5		1.0	3.5
Citrus rinds	1,700	40		.75	75
Shoe Leather	7,240	20		21.0	7.5
Butyl sole composition	10,900	25		30.0	1
Polyethylene	20,000	40–60	60	0	0
Polyurethane (foamed)	13,000	2	2	0	0
Latex	10,000	15	45	0	0
Rubber waste	9,000–11,000	62–125		20–30	
Carbon	14,093		138	0	0
Wax paraffin	18,621		54–57	0	0
1/3 wax-2/3 paper	11,500	7–10		3	1
Tar or asphalt	17,000	60		1	0
1/3 tar-2/3 paper	11,000	10–20		2	1
Wood sawdust (pine)	9,600	10–12		3	10
Wood sawdust	7,800–8,500	10–12		3	10
Wood bark (fir)	9,500	12–20		3	10
Wood bark	8,000–9,000	12–20		3	10

TABLE 9-1. (Continued)

Waste	B.T.U. value/lb. as fired	Wt. in lbs. per cu. ft. (loose)	Wt. in lbs. per cu. ft.	Content by weight in percentage: Ash	Content by weight in percentage: Moisture
Corn cobs	8,000	10–15		3	5
Rags (silk or wool)	8,400–8,900	10–15		2	5
Rags (linen or cotton)	7,200	10–15		2	5
Animal fats	17,000	50–60			0
Cotton seed hulls	8,600	25–30		2	10
Coffee grounds	10,000	25–30		2	20
Linoleum scrap	11,000	70–100		20–30	1

The above chart shows the various Btu values of materials commonly encountered in incinerator designs. The values given are approximate and may vary based on their exact characteristics or moisture content. *Source:* Incinerator Institute of America.

TABLE 9-2. Heating Values for Gaseous Compounds

Fuel	Symbol	Gross heating value: Btu/cu ft	Gross heating value: Btu/lb	Net heating value: Btu/cu ft	Net heating value: Btu/lb
Hydrogen	H_2	325.0	61,100	275.0	51,623
Carbon monoxide	CO	321.8	4,347	321.8	4,347
Sulfur	S	....	3,982	....	3,982
Methane	CH_4	1013.2	23,879	913.1	21,520
Ethane	C_2H_6	1792.0	22,320	1641	20,432
Propane[a]	C_3H_8	2590.0	21,611	2385	19,944
Normal butane[a]	n-C_4H_{10}	3370.0	21,308	3113	19,680
Acetylene[a]	C_2H_2	1488.0	21,344	1438	20,617
Hydrogen sulfide[a]	H_2S	647.0	7,100	596	6,454

[a]Heating values may be somewhat less for commercial forms of these compounds.
Source: North American Combustion Handbook, North American Manufacturing Co.

first stage of a solid waste incinerator operating on approximately 10% excess air will generate a flue gas which will contain some hydrocarbons, some carbon monoxide, and a very small amount of excess air or oxygen. This flue gas can then be heated to a temperature above 1800°F in a situation where excess air is present in small quantities, thus oxidizing the remaining carbon monoxide to carbon dioxide and oxidizing any remaining hydrocarbons to carbon dioxide and water vapor. A single-stage incineration process does not provide for this because there is never a reheating opportunity after the primary combustion

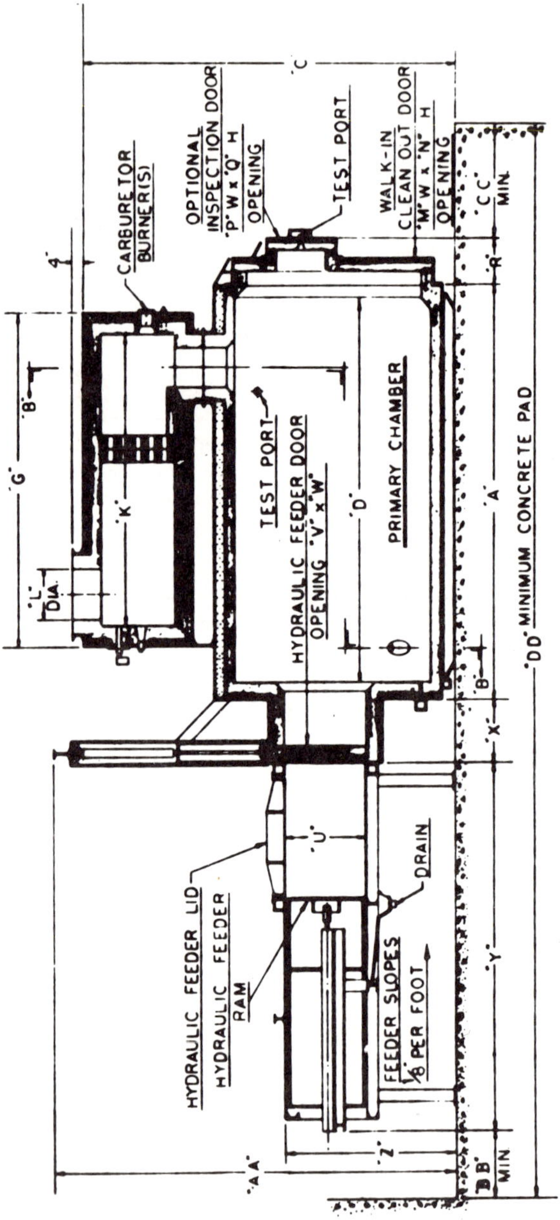

Fig. 9-2. Two chamber incinerator. *Source:* G. L. Simonds Co.

stage. In other words, the single-stage incineration system must achieve everything in the first stage and, in order to do this, auxiliary fuel and excess air must be added in inefficient quantities. Even then the combustion efficiency is relatively low.

9-2 INCINERATOR DESIGN

There are several basic steps that are required for the disposal of solid waste by incineration. These steps are as follows:

1. Separation
2. Preparation
3. Feeding
4. Primary Incineration
5. Secondary Incineration
6. Flue Gas Treatment

These steps are somewhat self-explanatory. The first, separation, involves removal of all materials which can be separated either by gravity, magnetic means, or otherwise, which are not combustible materials. There is no sense in putting bottles and metal into an incinerator; they will not burn. Bottles may melt causing a slag, whereby, the resulting difficulties, aside from the loss of heat, are just not worth the short cut. A good separation system will pay for itself many times over. The separation of ferrous metals is easily achieved by electromagnets, or permanent magnets, at some point in the handling system. Glass and other heavy nonmagnetic materials can usually be separated by gravity, either by the use of high pressure air or centrifugal separation systems. It will not be possible to remove all noncombustible materials from an industrial waste. The purpose of such equipment is to reduce it to the absolute minimum, and rather simple systems can effectively reduce the noncombustible portion to less than 5% by weight of any industrial waste to an incinerator. Much of this material can be recycled so that, at the very least, it can be separately landfilled with considerable ease. There is no value in placing materials which are noncombustible in an incinerator. They create maintenance problems and personnel hazards.

The second step is the possible use of size reduction as a tool. Shredding equipment is expensive, both from a standpoint of first cost and from the standpoint of energy use; nevertheless, anyone who has ever burned a bundle of twigs will agree that they burn much faster than a log of the same diameter as the bundle. The reason is that there is much more surface area for contact between the oxygen in the air and the material to be burned. Unfortunately, such logic often escapes those who are designing incineration systems. While it is not always practical from an energy or first-cost standpoint to prepare solid waste before it enters an incinerator, it does increase the efficiency of the incineration system. A uniform feed of any size is a better feed to an incinerator than a nonuniform feed. It simplifies the mixing of the air with the combustible waste and greatly reduces the amount of time for the incineration reaction. For example, an incinerator at any given volume would probably burn 1-in. diameter, 1-in. long wood chips at a rate three times as fast as it would burn 6-in. diameter by 3-in. long logs. This time comparison may not be exact; nevertheless, it is proportionately correct. Therefore, feed preparation, where it is economically feasible, is desirable. Where it is not economically feasible, then the incinerator must be designed to take a variety of sizes and shapes, but it will not be as efficient as one which is fed a uniform feed.

The third consideration is the feeding of the incinerator. Most industrial incinerators are batch fed. That means they are fed at the whim of the operator. If every incinerator could be operated by a qualified combustion engineer, there would be very little problem with incineration systems. Unfortunately, this is not a practical consideration and, therefore, incinerators must be designed, at least to the best of the designer's ability, to cope with the attitude and ability of the average operator. The operator often looks for the best way possible to reduce his work load. Therefore, he charges the incinerator as full as possible initially, lights it off, and allows it to operate for whatever period of time may be necessary until it is essentially burnt out. Then another charge is applied to the incinerator to keep it going. This cyclical load varies not only the heat load and the ash load but also the particulate load of the flue gas in the stack. This type of incinerator operation usually causes air pollution at the beginning of each cycle. Once the incinerator catches up with what it has been fed, it begins to smooth out and burn properly. If it is a two-stage operation, then the secondary burner can operate at an essentially constant output. If, on the other hand, it is on a batch charge basis, this burner will fluctuate from high fire to low fire quite a bit during the course of a day.

A number of incinerator manufacturers have decided that it is not practical to design a fool-proof incinerator which will cope with the whims of the operator and, therefore, are manufacturing a batch load design which will permit two or three complete batches to burn in a day. At the end of each batch, the incinerator is cooled to some extent and the ash is removed before the incinerator is recharged. From an operating standpoint, this has some good points. From a practical point of view, the incinerator use is perhaps not as efficient as it might be as a continuously charged machine.

There are a number of methods for continuously charging incinerators. This might be achieved through some type of a rotary airlock valve, if the material is relatively fine in nature, or a continuous conveyor for more coarse material. A batch-continuous arrangement is nearly ideal. This is a periodic batch loading on short intervals through an airsealed loading chute or other suitable means. Uniform charging of uniform materials is the best way to operate most incineration systems.

The primary combustion chamber is the place in which the initial burning must start. Once the material is ignited and burning evenly, then the incinerator merely must provide an opportunity for air to mix with the charge. This can be done by using grates or burning on a solid hearth, Figure 9-3. Underfire air and overfire air ports to provide good mixing of air with the waste are essential. The amount of air used in the primary chamber should be stoichiometric or just slightly above. The flue gas from the primary chamber then should pass into a secondary chamber where a reheat burner is utilized, and additional air is injected to complete the combustion reaction. Finally, the flue gas coming from the system may or may not require additional treatment, depending on the waste being handled. If the waste is high in ash, undoubtedly some of this material will be carried into the stack as particulate matter, despite the number of baffles installed in the incinerator. Some of this material will be very fine in nature, even submicron and, if it exceeds air pollution regulations, it must be removed. Removal systems which are discussed in Chapter 14 involve the use

Fig. 9-3. Two chamber incinerator. *Source*: G. L. Simonds Co.

of cyclones, wet scrubbers, electrostatic precipitators, and bag collectors after suitable cooling. All of these systems are possible methods for the removal of particulate matter from industrial incinerator operations.

Industrial solid waste incineration applications will generally involve the incineration of cellulosic type wastes. But, unlike municipal incineration systems, they will more often have to handle certain kinds of chemical waste in the form of either solids or sludges. Therefore, the device which is chosen for the incineration system must satisfy all the types of wastes which will be generated in the industrial facility, and this sometimes leads to considerable complications. Modern incinerators must provide for proper air pollution control as well as complete incineration of the feed material. The only incineration designs which will be discussed in this chapter are those which have the basic capabilities of providing an effluent which is either clean or capable of being cleaned by some mechanical device.

The simplest and most basic type of solid waste incinerator is the stationary grate type of system with a baffled secondary chamber, see Figure 9-4. This is used for applications up to about 2000 lb/hr of refuse. The material is fed to the incinerator through a charging door, either manually or automatically, onto the grate. The grate is located in the primary combustion chamber which is sometimes known as the ignition chamber. Here an ignition burner lights the waste charge, and it begins to burn. Air to burn the waste on the grate is supplied to the incinerator either by a forced draft fan, or by an induced draft fan, which pulls or pushes air through underfire and overfire air ports. The underfire air ports are located below the grate. The overfire air ports are located above the grate. Initial mixing and combustion begins here, but it is hardly complete in most incinerator designs. Flow then passes over a baffle wall and moves vertically downward into the second chamber in an area which is known as the flame port. This change of direction gives the partially burned gases more chance to mix with the combustion air, and, in some cases, auxiliary burners are located in the flame port to maintain combustion temperatures and initiate the secondary burning. These gases then pass into the secondary combustion chamber where the velocity is greatly reduced down to less than 10 ft/sec. Solid particles of ash will drop out in the secondary combustion chamber, and oxidation of the gases created in the primary combustion chamber will be completed. The gases then pass into a stack and subsequently through a spark screen to the atmosphere. Often a third chamber is added to give additional time for settling of particulate matter.

The stationary-grate type of incinerator is used for many commercial and industrial applications. It's generally constructed of brick or metal and lined with standard duty fire brick. It will work well only if the air ports are properly located and the unit is not overcharged. Most designs, however, do not provide for good combustion and, therefore, incineration has gotten a bad name. One

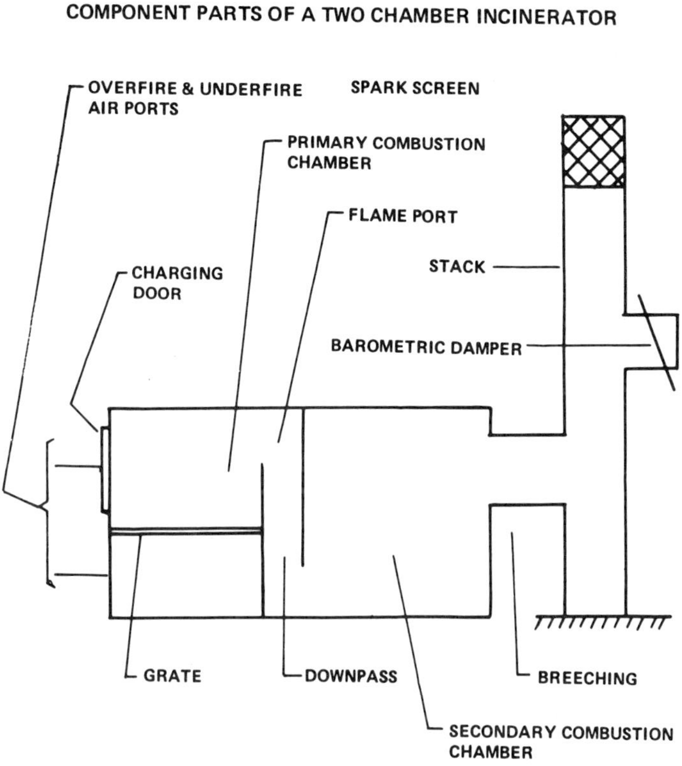

Fig. 9-4. Old design–in line incinerator. ***Source:*** **Cross, F. L. Jr. and G. Noble.**

of the disadvantages of the grate type incinerator is that large unburned particles drop through the grate into the ash pit resulting in large portions of the feed being hauled away before they are completely incinerated. Grates tend to overheat and then to buckle, and refractories require considerable maintenance.

Large incinerators using grates often enploy traveling grates which are essentially endless belts, see Figure 9-5. These have been used for many years as they have been employed in coal fired boilers. Reciprocating grates which move back and forth in horizontal motion have also been used for trash incineration, as have rocking grates which tend to move both up and down and horizontally. Traveling grates, reciprocating grates, and rocking grates are designed for much larger incineration application where it is necessary to move the waste material through the furnace to the ash pit while the combustion is taking place. The disadvantages of any grate type incinerator are usually higher maintenance and a higher unburned portion of the waste in the ash.

Fig. 9-5. Chain grate for incinerator. ***Source:*** **Laclede Stoker Co.**

9-3 TWO-CHAMBER CONTROLED AIR INCINERATOR

In the last ten or fifteen years, incineration systems have been developed which employ better combustion principles and enable burning to take place at more nearly stiochiometric conditions. These incineration systems are often designated as controlled air or starved air systems, see Figures 9-2 and 9-3. The terminology "starved air" is not quite accurate; "controlled air" more nearly describes the operation. These incinerators are generally small in size and will handle up to approximately 2500 lb/hr. They can be charged batchwise or continuously, and they depend on combustion taking place on a solid hearth rather than a grate. These incinerators, also occasionally known as "two-chamber systems," involve a primary chamber into which the waste is fed and ignited by the means of an auxiliary burner on either fuel oil or gas. In this primary chamber, a controlled amount of air for the incineration of the waste is introduced both below and above the waste charge in the primary chamber. The amount of air might be slightly above (controlled air) or below (starved air), the stoichiometric condition. As the incineration takes place in the primary chamber, the gases from this oxidation process pass into a second chamber where they are reheated by a larger burner operating on a conventional fuel, and additional air is introduced so that final combustion of all volatile or organic compounds is achieved. The temperature in this second chamber must be high enough to effect the final combustion of all organics from the system. This concept eliminates smoke and odor from the incineration process by creating a flue gas which is acceptable to the atmosphere.

Velocities in this system must be kept relatively low in order to trap particulate matter within the incinerator. This is usually under 10 ft/sec. There will, however, under most conditions, be some particulate matter in the stack and, depending upon the stack loading, additional air pollution control equipment may be required. Because forced air is used in this type of design, combustion is usually more rapid than in a natural draft system; the amount of air which is used may be slightly less than the stoichiometric amount required to burn the waste, but, in few cases it is more than 5% in excess of the amount required.

The construction of the primary and secondary chambers will vary depending upon the manufacturer, but the cross section is usually circular, although it may be square or rectangular. The circular cross section is much more easily lined with a refractory and has better structural integrity.

The controlled air incinerator as it is supplied by most manufacturers is a modular device. It is designed generally to handle type one waste (Table 9-3). If waste with a high plastic content is to be burned in the incinerator, a special design may be required. This type of incinerator can be loaded manually or by a hydraulic automatic loading device. In many cases, they have been coupled with a compactor which can compact the waste up to two and a half or three times its original volume, but, as mentioned earlier in this chapter, compaction is not in the best interests of incineration and it is not recommended for other than a material handling advantage.

9-4 PIT INCINERATOR

The idea of controlled air incineration has prompted several other devices which have found considerable use in industry. The first of these is the pit type incinerator, see Figure 9-6. This incinerator utilizes a pit as the primary combustion chamber in which the waste is ignited and initial burning started. The pit is blanketed by a curtain of air from a blower and air manifold which supplies the combustion air for the primary burning and also the combustion air for the secondary burning of the gases coming from the primary chamber. Since the air is introduced into the secondary burning area and the hot products of combustion must pass through the air curtain before entering the atmosphere, complete

TABLE 9-3. Classification of Wastes to be Incinerated

Classification of Wastes Type	Description	Principal Components	Approximate Composition % by Weight	Moisture Content %	Incombustible Solids %	B.T.U. Value/lb. of Refuse as Fired	B.T.U. of Aux. Fuel Per Lb. of Waste to be included in Combustion Calculations	Recommended Min. B.T.U./hr. Burner Input per lb. Waste
*0	Trash	High combustible waste, paper, wood cardboard cartons, including up to 10% treated papers, plastic or rubber scraps; commercial and industrial sources	Trash 100%	10%	5%	8500	0	0
*1	Rubbish	Combustible waste, paper, cartons, rags, wood scraps, combustible floor sweepings; domestic, commercial, and industrial sources	Rubbish 80% Garbage 20%	25%	10%	6500	0	0
*2	Refuse	Rubbish and garbage; residential sources	Rubbish 50% Garbage 50%	50%	7%	4300	0	1500

*3	Garbage	Animal and vegetable wastes, restaurants, hotels, markets; institutional, commercial, and club sources	Garbage 65% Rubbish 35%	70%	5%	2500	1500	3000
4	Animal solids and organic wastes	Carcasses, organs, solid organic wastes; hospital, laboratory, abattoirs, animal pounds, and similar sources	100% Animal and Human Tissue	85%	5%	1000	3000	8000 (5000 Primary) (3000 Secondary)
5	Gaseous liquid or semi-liquid wastes	Industrial process wastes	Variable	Dependent on pre-dominant components	Variable according to wastes survey	Variable according to wastes survey	Variable according to wastes survey	Variable according to wastes survey
6	Semi-solid and solid wastes	Combustibles requiring hearth, retort, or grate burning equipment	Variable	Dependent on pre-dominant components	Variable according to wastes survey	Variable according to wastes survey	Variable according to wastes survey	Variable according to wastes survey

*The above figures on moisture content, ash, and Btu as fired have been determined by analysis of many samples. They are recommended for use in computing heat release, burning rate, velocity, and other details of incinerator designs. Any design based on these calculations can accommodate minor variations. *Source:* Incinerator Institute of America.

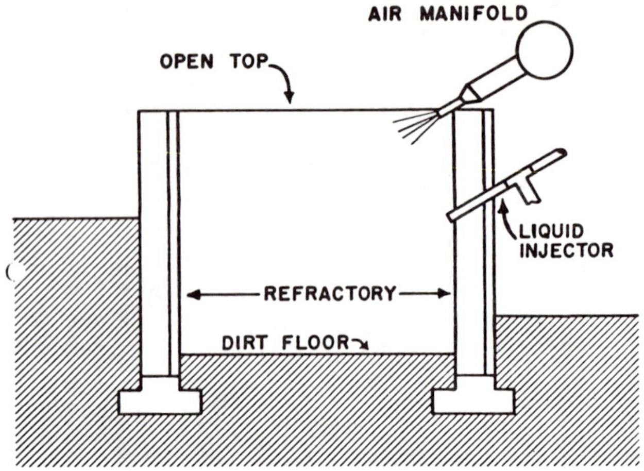

Fig. 9-6. Open pit incinerator. ***Source:*** **E. I. du Pont de Nemours and Company.**

complete combustion is usually achieved with dispatch, but the turbulence of the system and the fact that it has an open top tends to produce a higher particulate loading in the flue gas than does the closed type. This is especially true if the waste consists of materials such as those found in type one waste.

Although spark screens constructed over the pit will reduce these emissions to some extent, this type incinerator has not been widely accepted by air pollution control authorities. It has, however, found a very important use in the land clearing and construction industry. In most cases, air pollution authorities have granted exceptions for their use for the burning of logs and refuse generated during land clearing operations for roads, buildings, etc. Here the pit is not permanent but merely a hole dug in the ground and the manifold and blower are transportable and can be moved from place to place. The blower is driven by a diesel engine. While they are not perfect in terms of particulate emissions, they are definitely better than simple open burning practices which have been used in the past which do not provide for any controlled air mixing with the products of combustion.

9-5 CYCLONIC INCINERATOR

The original pit type incinerator has been enclosed to produce an acceptable in-plant type system known as the cyclonic incinerator, see Figure 9-7. This unit, usually circular in cross section, employs the use of a series of air nozzles in the primary chamber. Solid waste can be charged into the chamber, ignited, and

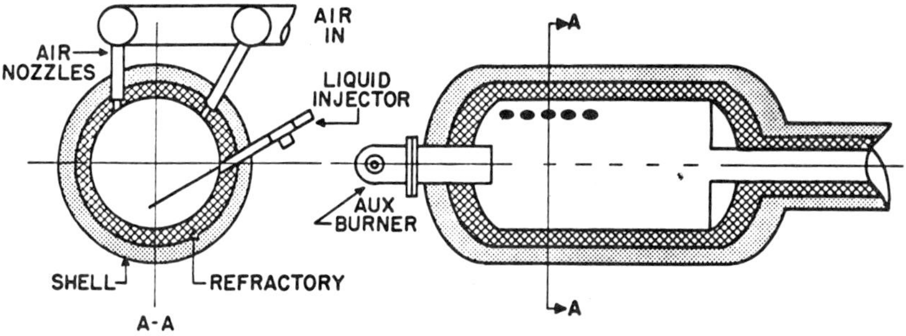

Fig. 9-7. Cyclonic incinerator. ***Source:*** **E. I. du Pont de Nemours and Company.**

burned to completion. The length of the chamber and the diameter of the chamber are a function of both the density of the charge and the type of material.

9-6 MULTIPLE HEARTH INCINERATOR

Many industrial wastes which can be handled by incineration are not truly solid materials; they may be sludges too heavy to atomize in a conventional liquid incineration system and with too low a calorific value to burn properly on a hearth, even with controlled air. Such materials tend to form a cake or an ash blanket over the burning area shortly after ignition and extinguish combustion in this manner. Therefore, they need some type of continuous agitation or a method of getting air to the particles at the bottom of the pile. Several types of incineration systems may work for this type of waste.

The first employs mechanical agitation and is known as a multiple hearth incinerator (Figures 9-8 and 9-9). This is especially designed for low heating value materials which will not create high temperatures in a combustion zone. The unit is dependent upon mechanical rabble arms which continually move the waste material over the surface of the hearth, exposing new areas for proper oxidation of the waste. These units have been used for processing a variety of metallic ores, as well as the incineration of sewage sludge. They generally require large quantities of auxiliary fuel in order to maintain the temperature within the system and, because they are mechanical in nature, generally have a high maintenance factor. They do, however, enable the incineration of low heating value materials which might not otherwise be incinerated. The multiple hearth incinerator consists of a series of four or more hearths, one above the other, each with an opening in the hearth.

The sludge feed to the incinerator is admitted on the top hearth where it is ignited and incineration continues through the first several hearths. The last

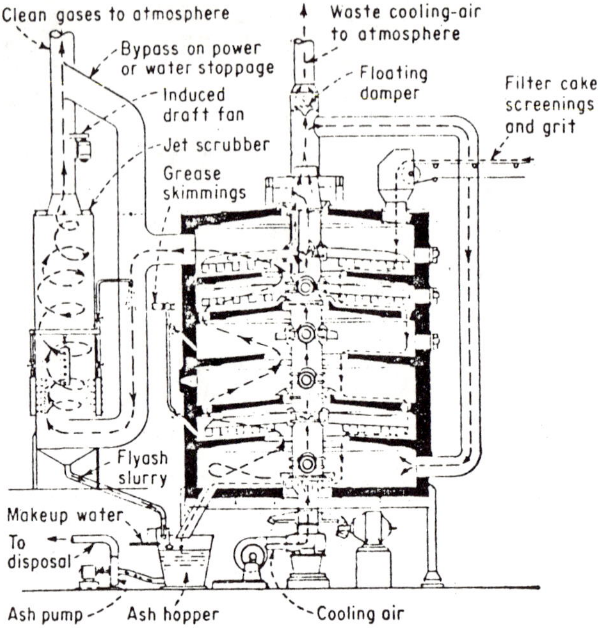

Fig. 9-8. Multiple hearth incinerator diagram. *Source: Chem. Eng.*, **p. 91, August 29, 1966.**

hearth is usually used for ash cooling. The problem with the multiple hearth incinerator, aside from its mechanical shortcomings, is the fact that it may require afterburning and usually will require some type of air pollution control device to remove ash from the stack. It is also a relatively expensive device and will find applications industrially only where the character of the waste demands its use.

9-7 FLUID BED INCINERATOR

A second type of sludge incineration system which is finding wider and wider application in industry is the fluid bed incinerator, Figure 9-10. The fluid bed has been used for a number of years to recover salts from black liquor and also for burning phosphate rock. It has been used to burn refinery sludges containing

Fig. 9-9. Multiple hearth incinerator. ***Source:*** **BSP Division of Envirotech Corp.**

salts and water, Figure 9-11. It consists of a bed of sand or alumina which can be fluidized by hot air. If the combustion air is preheated to ignition temperature and blown upward to fluidize the bed, and the waste is supplied to the top of the bed, it will burn and the products of combustion will pass out through the chamber into a cyclone separator and subsequently into a wet scrubber. It makes an excellent continuous incinerator since the waste sludge can be continually added to the top of the unit and the ash continually removed. Its shortcomings are the air pollution potential which requires considerable abatement equipment and a generally high cost in comparison to other incineration systems. Materials which melt or slag will cause problems with the operation of fluid bed systems; operating temperatures must be carefully controlled.

Like the multiple hearth furnace, the fluid bed incinerator is an expensive device and, therefore, is only economically justified when handling extremely

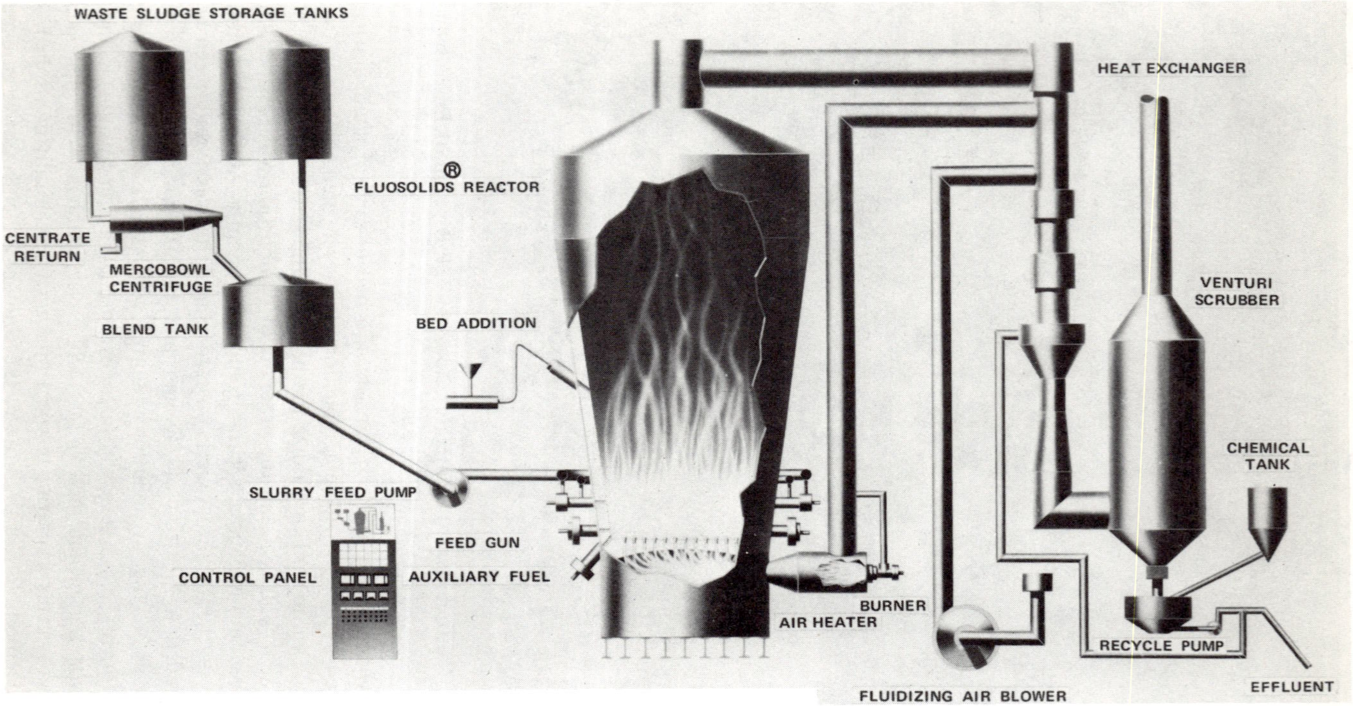

Fig. 9-10. Fluidized bed incinerator for refinery wastes. Fluosolids (a registered trademark of Dorr-Oliver Co.)

Fig. 9-11. Fluidized bed oxidation system for chemical and refinery wastes. *Source:* **Copeland Systems Inc., Oak Brook, Illinois 60521.**

difficult materials. One very apparent advantage of the fluid bed combustion system is that it will generally operate at a lower temperature than some of the other devices available for the incineration of waste material. The optimal operating temperatures are in the range of 750 to 950°C, as compared with temperatures several hundred degrees higher in conventional incineration systems depending upon the material being incinerated. This means that the amount of nitrogen oxides formed will be lower. The fluid bed will decrease the volatilization of highly corrosive alkali salts and trace metals, and it will also inhibit slag and ash formation. The temperature in a fluid bed can easily be controlled by the addition of more excess air, although care must be taken that velocities through the bed do not get high enough to lift the material in the bed into the exit of the unit and through the heat exchanger. Where the fluid bed has been used in phosphate rock processing applications, there is evidence of alkali salts carrying over into the heat exchanger and then condensing.

There also seems to be some evidence that fluid bed combustion systems could greatly reduce or control sulfur oxide emissions by using limestone or dolomite as a material for the bed. The rapid mixing occurring in the fluidized bed would allow for effective capture of the sulfur by the limestone, thus eliminating the need for subsequent scrubbing operations. It is also conceivable that wastes with extremely low heating values could be destroyed in this manner, since coal or some other material with a higher heating value could be added to the bed at the same time. This addition of combustible material would supply the heat for the incineration of the low calorific waste. One of the difficulties, of course, is that the feed material must be crushed or shredded to proper size, usually a maximum of 1/4-in. average diameter.

The fluid bed incinerator really finds its place in the industrial market where difficult to handle and difficult to burn sludges must be disposed of. It is not an all-purpose incinerator, but it is excellent for certain specific problems.

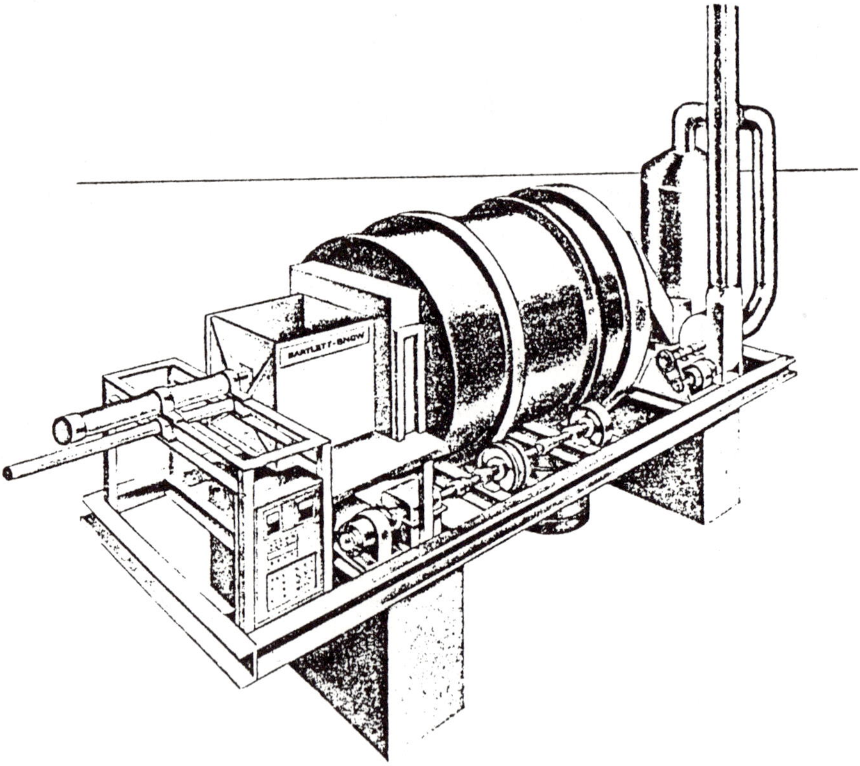

Fig. 9-12. Rotary kiln. *Source:* C. E. Raymond.

9-8 ROTARY KILN INCINERATOR

Another type of solid waste incineration system which has been widely used in industry is the rotary kiln, Figure 9-12. The rotary kiln originally was developed for a number of processing applications, such as drying and calcining. The kiln, which is a large cylinder rotating on steel tires turning on trunnions, is sloped slightly from the feed to the discharge end so that the material being processed will move along the length of the cylinder. For incineration applications, unlike drying applications, the cylinder must be lined with some type of refractory, usually a heavy-duty, hard surfaced fire brick. The waste material is charged at the high end of the kiln and usually ignited by a burner located at the low end of the kiln; however, this is not always the configuration used. Burning equipment may be located at the feed end of the kiln for waste material which is combustible in itself and does not need a great deal of auxiliary heat.

The air for combustion is usually pulled through the kiln by means of an induced draft fan located downstream from the kiln, afterburner, and scrubber system. The rotational speed of the kiln will vary depending upon its size. The forward movement of the material being burned through the kiln will also vary greatly for the same reason. Ignition occurs at the front end of the kiln and combustion progresses until the unburned material or ash is discharged into an ash pit at the low end of the kiln.

The flue gases then pass into an afterburner section where they are reheated to 1500 to 1800°F. The afterburner section is almost essential in a rotary kiln when burning waste because the mixing of the air and the combustible material in the kiln is not sufficient to provide complete combustion entirely within the kiln itself. Consequently, some hydrocarbon material will exit from the kiln even in the presence of excess air and will need reheating to complete the combustion reaction. This is not true in every case, but in the majority of industrial waste kiln applications, it is true.

The combustion, once completed in the afterburner, produces flue gases of carbon dioxide, nitrogen, and water vapor for conventional hydrocarbon materials and such particulate materials as might be generated from the fine ash of the waste. This particulate is usually in sufficient quantity to require some further separation, and, because the gases are at high temperature, it is usually desirable to use a wet scrubber to accomplish this task. The type of wet scrubber will greatly depend upon the size of the particulate matter in the gas. Generally, because of the nature of the combustion operation, these particulates will be fine and a venturi type scrubber will be required. Design pressure drops through the venturi scrubber, of course, will vary depending upon the size of the particulate to be removed, see Figure 9-13.

A new type of kiln developed by the Japanese firm Nittetu Chemical Engi-

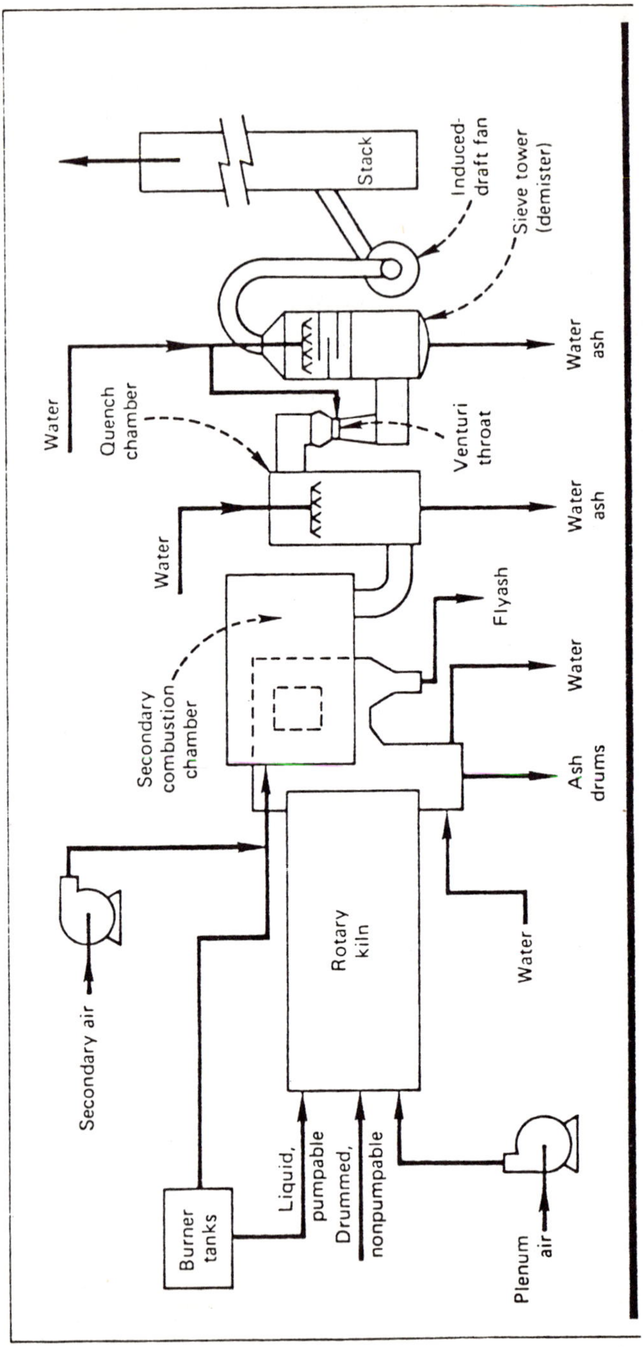

Fig. 9-13. Rotary kiln system. *Source: Chem. Eng.*, Deskbook Issue, p. 117, October 18, 1976.

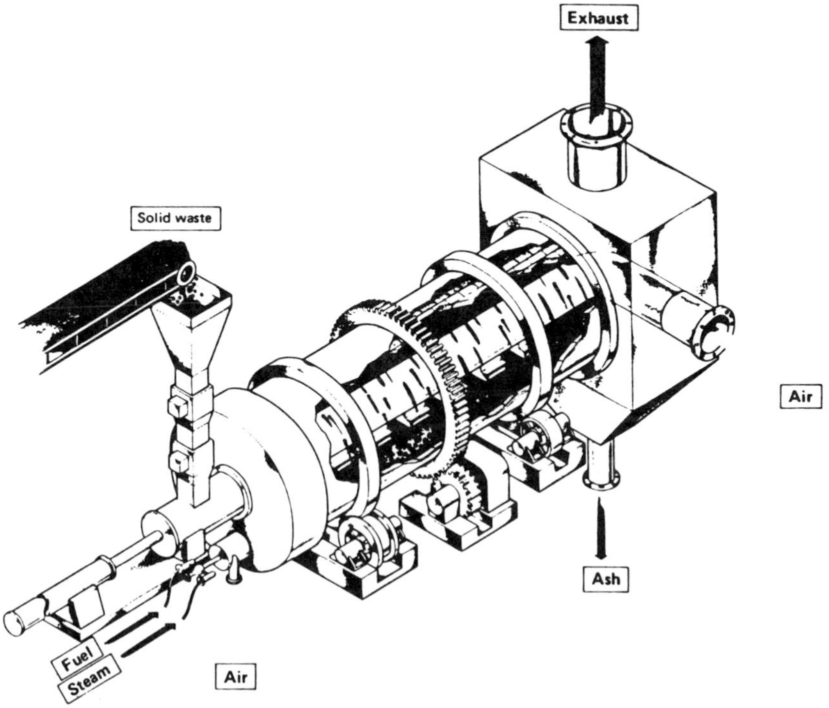

Fig. 9-14. Nittetu VR incinerator. (Courtesy NCE Corp., USA).

neering Co. Ltd. provides a better method of getting air to the waste by means of a central pipe and should prove more efficient, see Figure 9-14.

Kilns have been used throughout the world for the incineration of waste solids and sludges. They are, more or less, a brute-force type of method. Equipment is large, generally requiring a lot of land area for installation, and is expensive in first cost. The operating cost for the kiln system mainly depends upon the amount of fuel required in the afterburner and the horsepower required to pull the flue gases through the venturi scrubber. Kiln type systems have been used in the United States both industrially and for large municipal installations. Their primary advantage is that they provide a moving hearth which is essentially impervious to the high temperatures of operation, and their mechanical features are relatively simple and rugged with good maintenance experience.

9-9 ROTARY HEARTH INCINERATOR

As long as there is need for incineration, there will be people inventing incinerators. Some of these are for broad-based applications, such as the rotary

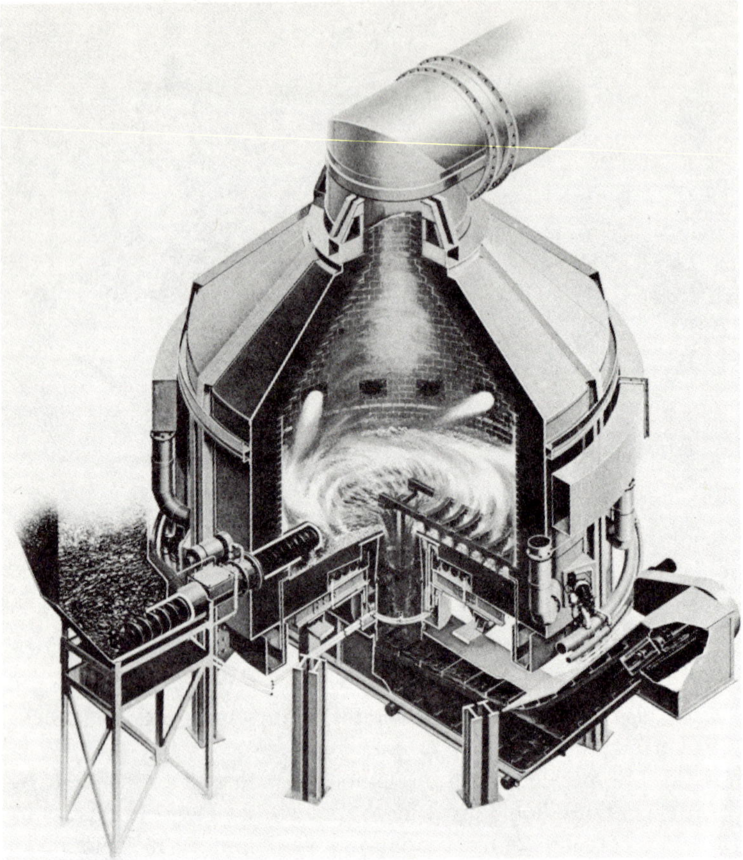

Fig. 9-15. Rotary hearth cyclonic furnace. *Source:* Fluor, Utah.

kiln, and others have been invented for very specific needs, such as the destruction of used automobile tires. The Lucas rotary hearth furnace is such a device, Figure 9-15. While it may have other applications in addition to the destruction of automobile tires, it was designed with this purpose in mind.

The Lucas rotary hearth furnace is somewhat like the cyclonic furnace described earlier in this chapter except that it is vertical instead of horizontal. The cyclonic flow is generated by a series of alternating air nozzles and auxiliary fuel burners generally spaced at equal distances around the periphery of the cyclindrical body of the furnace so that the air and the gas flow tangentially into the combustion chamber as a result of this flow. The cool, dense incoming air is held to the outside wall while the less dense, high temperature products of combustion flow to the center forming a burning vortex several hundred degrees hotter than the gases. This provides good mixing and cooling for the outer wall. The major difference with this type of incinerator is that the hearth actually rotates, somewhat like a slow carousel. The waste is deposited at one point on the hearth and it moves 360° until it comes around to the feed point again. The ash from the waste material moves toward the center, or eye, of the rotary hearth and drops into an ash pit at this point. The top is conically shaped which helps to maintain the vortex as the material is turned.

Of course, this type of design can be used for material other than automobile tires. Sludges can be handled on such a moving hearth and only the manner of feeding the waste into the incinerator is changed. For sludges, however, it is necessary to use ploughs, or some similar device which is water cooled, to continually turn over the waste to expose its surface to the air; otherwise, the material will cake over and cease to burn. Unlike the multiple hearth furnace described earlier where the rabble arms move across a fixed hearth, the Lucas furnace uses a moving hearth and fixed ploughs to accomplish the same task. Also, unlike the multiple hearth furnace, this is a single hearth furnace. It has many of the attributes of the multiple hearth furnace and the fluid bed furnace in that it will handle difficult materials that cannot be burned on a grate or a stationary hearth. It is also very expensive in terms of first cost and because it depends upon mechanical movement of a heavy hearth under high temperature conditions, it would appear to have the possibility of significant maintenance costs. Since this is a relatively new development, information is limited.

9-10 FLASH DRYING AND INCINERATION

Flash drying incineration is described by Figure 9-16. The flash drying system can be used for the incineration of a variety of materials when handled in suspension. Usually the materials which are handled by a flash drying incineration system must be light enough to be air conveyed.

9-11 HEAT RECOVERY

Today, when energy is becoming more expensive and industry is examining every alternative fuel source, the potential for heat recovery from incineration systems is being examined much more closely than it has in the past. Several major municipal projects involving steam generation, such as those at Nashville, St. Louis, and Chicago, have given impetus to industry to examine similar heat recovery on a much smaller scale. Heat recovery may be approached from several different directions when it involves utilizing waste materials from an

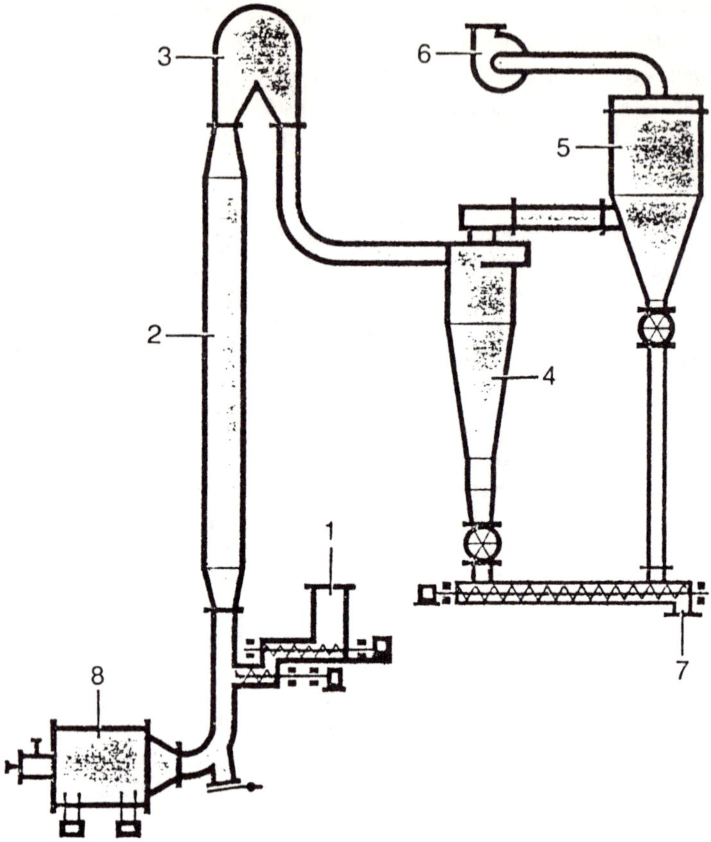

Fig. 9-16. Flash drying system. *Source:* Strong-Scott Co. Flash Dryer for free-flowing products with a *uniform* particle size. The product is metered and fed by screw conveyors directly into the drying duct. The drying gas may be heated by gas-or-oil-fired burners, or by indirect heat exchangers. The dried product is separated from the gas stream in a cyclone separator and bag filter.

1 Feed Inlet
2 Drying Duct
3 Top Bend
4 Cyclone Separator
5 Bag Filter
6 Exhaust Fan
7 Product Discharge
8 Heater

industrial operation and it, of course, depends on the amount and the character of the waste which is available and the present disposal methods. The two approaches are: (1) to put a waste heat boiler or other recovery device on an existing solid waste incinerator; or (2) to adapt existing boilers so that they can take the solid waste generated by the plant. The second option is usually only practical where the plant burns coal and the boiler is fired by a stoker. Here, by properly sizing the waste, it can be mixed with or added on top of the coal and burned as a solid fuel releasing its heat to the boiler to create steam. If, however, the plant has an existing incinerator, then a variety of waste heat boiler designs can be employed to create steam from the high temperature stack gases.

In both cases, the problems inherent in burning industrial trash (i.e., high particulate loading and occasionally acid-generating materials) will be present in the flue gas, but in the case of the coal-refuse fired boiler, scrubbing will probably be required to reduce sulfur dioxide emissions in any case so that this should not be a problem. The waste heat boiler on the existing incinerator may, however, be a source of constant maintenance unless the design has been carefully considered.

When refuse is used as a boiler feed, the desirability of feed preparation and sizing is apparent. And, while it may not be required in a conventional incinerator, it is almost mandatory for this type of operation.

Considering the increased cost of fossil fuels, heat recovery from solid wastes is more than just an interesting possibility.

REFERENCES

9-1. Balakrishnan, S., Williamson, D. E., and Okay, R. W., State of the Art Review on Sludge Incineration Practice, U.S. Dept. of the Interior, FWQA, 1970.

9-2. Unterborg, W., Sherwood, R., and Schneider, G., "Computerized Design and Cost Estimation for Multiple-Hearth Sludge Incinerators," U.S. Environmental Protection Agency, Office of Research and Monitoring, July 1971.

9-3. Battelle Laboratories Report, "Fluidized-Bed Incineration of Selected Carbonaceous Industrial Wastes," for State of Ohio Department of National Resources and U.S. Environmental Protection Agency, March 1972.

9-4. American Oil Co., Mandan, N. D., "Fluid Bed Incineration of Petroleum Refinery Wastes," for Water Quality Office, U.S. Environmental Protection Agency, March 1971.

9-5. Cross, F. L., Jr. and Hesketh, H. E., *Handbook for the Operation and Maintenance of Air Pollution Control Equipment,* Technomic Publishing Co. Westport, Connecticut, 1975.

9-6. Cross, F. L., Jr., *Handbook on Incineration,* Technomic Publishing Co. Westport, Connecticut, 1972.

9-7. Nolan, M. and Marshalla, A., "Incinerator Emissions: Units, Correction and Conversion," *Pollut. Eng.,* May 1973.

9-8. Lucier, T. E., "The Pit Incinerator," *Ind. Water Eng.,* September 1970.

9-9. Monroe, E. S., *The Cyclonic Incinerator,* DuPont.

9-10. Lewis, C. R., Edwards, R. E., and Santoro, M. A., "Incineration of Industrial Wastes," *Chem. Eng.,* October 18, 1976.

9-11. Editorial, "How Trash is Being Turned into Useful Heat," *Environ. Sci. Technol.,* Vol. **10**, No. 9, September 1976.

9-12. Editorial, "Can Plastics be Incinerated Safely," *Environ. Sci. Technol.,* Vol. **5**, No. 8, August 1971.

9-13. Anders, L. M. and Trexler, R. J., "Incineration Cuts Treatment Costs," *Ind. Wastes.* September/October 1974.

9-14. Editorial, "With Room to Spare," *Ind. Gas,* October 1974.

9-15. Editorial, "MHFS Solve a Sludge Disposal Dilemma," *Environ. Sci. Technol.,* Vol. **8**, No. 5, May 1974.

9-16. Kaiser, E. R. and Carotti, A. A., "Incineration of Plastics," A report to the Society of the Plastic Industry, June 30, 1971.

9-17. Cross, F. L., Jr., "Controlled Air Incinerators," *Pollut. Eng.,* December 1973.

9-18. Fife, J. A., "Solid Waste Disposal: Incineration or Pyrolysis," *Environ. Sci. Technol.,* Vol. **7**, No. 4, April 1973.

9-19. Young, R. A. and Lisk, I. O., "Kodak Disposes of Waste in System that Helps Pay its Way," *Pollut. Eng.,* September 1976.

9-20. Young, R. A., "John Deere Makes 45% Per Year Return on Disposal System Investment," *Pollut. Eng.,* October 1976.

9-21. Editorial, "Sludge Incineration," *Environ. Sci. Technol.,* Vol. **10**, No. 12, November 1976.

9-22. Cross, F. L., Jr. and Lawson, J. R., "Incineration of Industrial Wastes," Air Pollution Control Conference, University of Tennessee, 1972.

9-23. Cheremisinoff, P. N. and Young, R. A., "Incineration of Solid Waste," *Pollut. Eng.,* June 1975.

9-24. Ferrel, J. F., "Sludge Incineration," *Pollut. Eng.,* March 1973.

9-25. Dunn, R. L., "The Pros and Cons of Onsite Incineration," *Plant Eng.,* April 20, 1972.

9-26. Ewing, R. C., "Amoco Installs Fluid Bed Incinerator," *Oil Gas J.,* January 22, 1973.

9-27. Becker, K. P. and Wall, C. J., "Fluid Bed Incineration of Wastes," *CEP,* October 1976.

9-28. Editorial, "Fluidized-Bed Combustion: A Better Way to Burn Coal?" *Chem. Eng.,* June 7, 1976.

9-29. Cheremisinoff, P. N. and Rao, K. B., "Fluidized Bed Reactors," *Pollut. Eng.,* September 1972.

9-30. Liao, P. B., "Fluidized-Bed Sludge Incinerator Design," *J. WPCF,* Vol. **46,** No. 8, August 1974.

10

Pyrolysis and wet air oxidation of solid wastes

10-1 DESCRIPTION OF PYROLYSIS

Pyrolysis has been known and used for centuries in the manufacture of charcoal. Recently, it has been used for number of applications in the chemical industry to produce a wide range of products. *Pyrolysis* is described as a chemical change brought about by the action of heat. It is, therefore, essentially a baking process rather than a burning process.

The application of pyrolysis to solid waste dries the feed material, causes the volatile components of the waste to be distilled from the feed material, and results in some chemical reactions. Pyrolysis generally refers to indirect rather than direct heating; although, forms of pyrolysis can also involve combustion. Most pyrolytic solid waste systems today do involve combustion with a definite deficiency of air. In such processes, the burning or pyrolysis with a deficiency of air creates a flue gas which is rich in hydrocarbon materials that may be burned in a secondary combustor or may be cleaned and used as a fuel in some other process.

In addition to energy recovery, the argument for using pyrolysis rather than direct incineration is that it presumably simplifies the air pollution control problem because gases, not solids, are being burned at the end of the system. This provides a much lower particulate entrainment, simplifying the gas cleaning and providing lower emission levels than are produced by incineration systems.

Most commercial pyrolysis systems are designed either for the recovery of heat or the generation of a fuel gas which may be used elsewhere. Pyrolysis has been used for municipal waste with only very limited success. It has, at the present time, very few industrial applications in the waste disposal field and is used in a wide variety of chemical processes to make other materials such as charcoal, carbon black, and activated carbon.

Pyrolysis is often suggested as an alternative to the process of incineration. In the past ten years, a variety of pyrolytic processes have been developed on a semicommercial scale and some even on a commercial scale; most of them have encountered severe problems. It is, however, unfair to condemn the entire area of pyrolysis because of some initial failures. Many successful systems have been born out of failures. Just as man was discouraged many times before the Wright brothers were successful, it could have been concluded that it was impossible to fly and that anyone who tried to do so was foolish. Pyrolysis has a similar history but, fortunately, some of the newer techniques are beginning to show some merit, see Table 10-1.

For example, a company in Birmingham, England, hopes to have a 50,000 metric ton per year scrap tire pyrolysis plant in operation by 1979. It has based its decision to build a full-scale plant on successful trials over the past two years with a 12 metric ton per year plant at the government funded Warren Springs Laboratory in Stevenage. The products from this plant will include oil, char, and scrap steel from the steel belted radial tires. In the process, a disk shredder cuts tires into coarse fragments which are then fed continuously to an oxygen-starved reactor operating at low temperatures in the range of 300 to 700°C. They claim that the oil product will be a high-grade industrial fuel.

Pyrolysis has been defined as the final decomposition of material in the absence or near absence of oxygen under high temperature conditions. It is, in effect, the chemical degradation of organic substances in an oxygen-deficient atmosphere. The primary motivation on pyrolyzing a waste material is to develop a system where a solid waste can be converted into a storable, transportable fuel either as a liquid or gas and to provide a clean, inert solid refuse that can be easily landfilled.

Depending upon the type of waste fuel which is fed into a typical pyrolysis unit, the ultimate product would be a waste gas having a calorific value from 100 to possibly as high as 300 Btu/cu ft, and this gas will consist primarily of hydrogen, methane, and carbon monoxide. This low calorific value fuel could presumably be used in a boiler to generate steam. The second by-product, if the system is designed for its removal, is an oil that is liquid at room temperature

TABLE 10-1. Pyrolysis projects

	Healing Method		Product Distribution			Feed Conditions				Status		
Process	Direct	Indirect	Solid (Btu/lb)	Liquid (Btu/lb)	Gas (Btu/ft^3)	Raw	Size Reduction	Separation	Reactor Temp. (°F)	Research	Pilot Plant (tpd)	Commercial (tpd)
Shaft-vertical												
Union Carbide	•				300	•			3000		5	200
Garret		•	9700	10 500	550		•	•	900		4	200
Torrax Systems, Inc.	•				150	•			3000		75	
Urban Research and Development Corp.					150	•			2600		120	
Ga. Tech.	•		10 000	13 000	200		•		750		25	
Battelle	•				170				1800		2	
Shaft-horizontal												
Kemp Corp.		•	•	•	•		•		1100		5	
Barber-Colman		•			500		•	•	1200		1	
Rotary kiln												
Rust Engineering		•			450				1250			260
Monsanto	•		2500		130		•		1800		35	1000
Devco	•		•		•		•	•	1000		120	1500
Pan American Resources, Inc.		•					•		2000	•		

Fluidized bed										
W. Virginia Univ.		•			450	•	•	1400	•	
Coors	•				150	•	•	1400		1
A.D. Little		•			•	•	•		•	
Others										
Battelle		•						1800	•	
Hercules			•						•	
Bureau of Mines		•			500	•	•	1800	•	
N.Y. University		•						1700	•	
Univ. of Southern California									•	
Anti-Pollution Systems, Inc.									•	
Univ. Calif		•							•	
Wallace-Atkins		•	3000	16 000	500			1600	•	
Resource Sciences		•				•		1800		2

Source: Environ. Sci. Technol., April 1976.

and includes such materials as acetic acid, acetone, and methanol. This liquid by-product is carried off in the gas stream and is later condensed into a usable liquid, see Table 10-2. The amount of liquid by-product may greatly depend upon the nature of the waste fed into the pyrolysis unit. If the charge is high in plastic materials and rubber, then the amount of oil recovery is a higher percentage of the total than would be found in the pyrolysis of conventional refuse.

The final product from the system will be an inert char which is almost pure carbon plus whatever metal or nonorganic materials might have been in the original waste feed. This char is usually a hard, brittle solid product at room temperature and is generated by water quenching after it flows from the pyrolysis unit as a liquid at high temperature. The brittleness of the char will

TABLE 10-2. Pyrolysis products.

Chemical Analysis of Raw Refuse	
	(% by wt)
Moisture	20.00
Carbon	29.83
Hydrogen	3.99
Oxygen	25.69
Nitrogen	0.37
Sulfur	0.12
Ash and metal	20.00
After Pyrolytic Operations	
	(% by wt)
Pre-pyrolysis	
Inorganics removed	9.00
Moisture removed	20.00
Pyrolysis products	
Charcoal	11.25
Liquids	40.80
Gas	7.95
Ash and metal	11.00
Yield and Typical Thermal Values[a]	
Char residue	154–230 lb, 8–13,000 Btu/lb
Tar and pitch	0.5–5 gal.
Light oil	1.2–2 gal. 150,000 Btu/gal.
Gas	11–17,000 cu ft 500 Btu/cu ft
Aqueous liquor	80–133 gal.
Ammonium sulfate	18–25 lb

[a]per ton of municipal refuse.
Based on U.S. Bureau of Mines Data.

greatly depend upon the amount of glass and metal present in the waste material. If they are high, then the residue will have the appearance of black glass, which is apparently a ferro-silicate. If glass and metal are not present, then the pyrolysis waste will be more granular.

10-2 TYPES OF PYROLYSIS SYSTEMS

There are four major pyrolysis systems available today that need to be discussed, even though there are many other pyrolysis systems either at a pilot-plant stage or under prepilot plant discussion and investigation. These four systems seem to predominate:

1. A low temperature flash pyrolysis system developed by Occidental Research.
2. The oxygen-fed, slagging pyrolysis system known as the PUROX system developed by Union Carbide Corporation.
3. Monsanto Enviro-Chem's LANDGUARD system, which is a medium-temperature, slagging-type pyrolysis unit.
4. The TORRAX system of the Carborundum Company, which is an air-fed slagging pyrolysis unit. All of these units operate with substantially different approaches that bear individual discussion.

10-2.1 Flash Pyrolysis

The Occidental flash pyrolysis process, Figure 10-1, has been piloted on a small scale, and now a 200 ton per day pilot plant is being constructed in San Diego, California. It offers specific advantages because it can produce a liquid fuel. However, the liquid produced by this process is very viscous and is corrosive. It is a tar or oil which has a tendency to polymerize when it is stored. It will not mix with fuel oil; however, it can be dispersed in fuel oil as an emulsion which makes it more easily pumpable and therefore usable as a supplementary fuel. The plant claims to produce 7200 to 9600 gal./day of fuel oil with a composition generally as described in Table 10-3.

10-2.2 PUROX Process

Union Carbide is currently operating a 200 ton per day demonstration plant in South Charleston, West Virginia. While there were some early operating problems, they seem to be resolved and continuous operation at or above design capacity has been successfully demonstrated. This system uses a vertical or a shaft type of furnace. The refuse is introduced to the furnace from the top

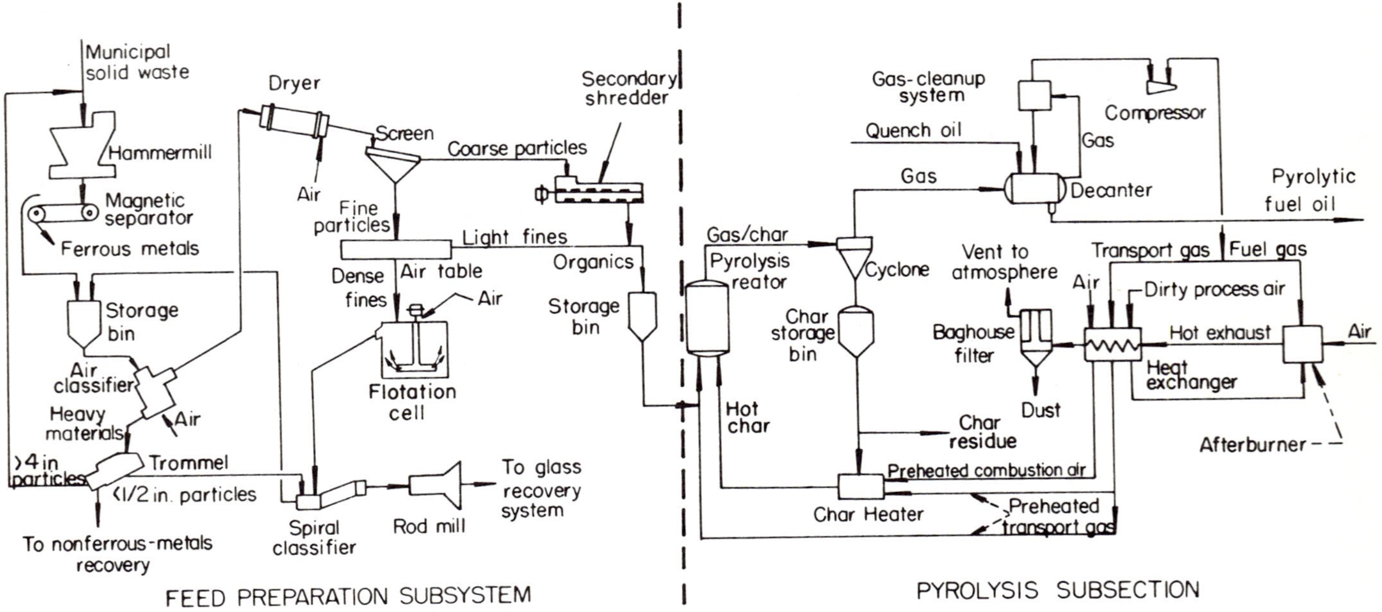

Fig. 10-1. Occidental flash pyrolysis process. *Source: Chem. Eng.*, p. 91, July 19, 1976.

TABLE 10-3. Fuel Oil from Flash Pyrolysis

	Pyrolytic Fuel Oil	No. 6 Fuel Oil
Carbon, wt. %	57.0	85.7
Hydrogen, wt. %	7.7	10.5
Sulfur, wt. %	0.2	0.7–3.5
Chlorine, wt. %	0.3	–
Ash, wt. %	0.5	0.05
Nitrogen, wt. %	1.1	2.0
Oxygen, wt. %	33.2	
Heating value, Btu/lb	10,600	18,200
Specific gravity	1.30	0.98
Density, lb/gal.	10.85	8.18
Heating value, Btu/gal.	114,900	148,840
Pourpoint, °F	90[a]	65–85
Flash point, °F	133[a]	150
Viscosity at 190°F, Ssu	1,150[a]	340
Pumping temperature, °F	160[a]	115
Atomization temperature, °F	240[a]	220

[a]Assumes oil containing 14% moisture.
Source: Chem. Eng., p. 90, July 19, 1976.

through a charging lock. The furnace is filled with refuse, and oxygen is injected into the combustion zone at the bottom of the furnace where it reacts with the carbon char residue from the pyrolysis zone which is just above it. The temperature generated in the hearth is sufficiently high to melt and fuse all noncombustible materials. A molten material continually overflows from the hearth into a water quench tank, where it forms a large granular product, see Figure 10-2.

The hot gases formed by the reaction of the oxygen in the carbon char rise through the descending waste in the middle portion of the vertical shaft furnace. Here the organic materials are pyrolyzed in essentially a reducing atmosphere to yield a gaseous mixture high in carbon monoxide and hydrogen. The typical analysis is approximately 50% carbon monoxide and 50% hydrogen, measured on a dry basis. These hot gas products continue to flow upward, drying the entering refuse in the upper zones of the furnace and also providing excellent heat recovery from the exiting gases. The flue gas temperature averages approximately 200°F which means that the system has a very high thermal efficiency.

The gas mixture exiting the furnace contains water vapor, some oil mist formed by the condensation of high boiling organics present in the waste, and a small amount of fly ash. As the gas leaves the furnace, the oil mist and fly ash solids are removed by a gas cleaning system, and, after cleaning the product, gas is passed through a condenser removing any water and residual oil. The resulting gas is a clean-burinig fuel comparable in many ways to natural gas in its com-

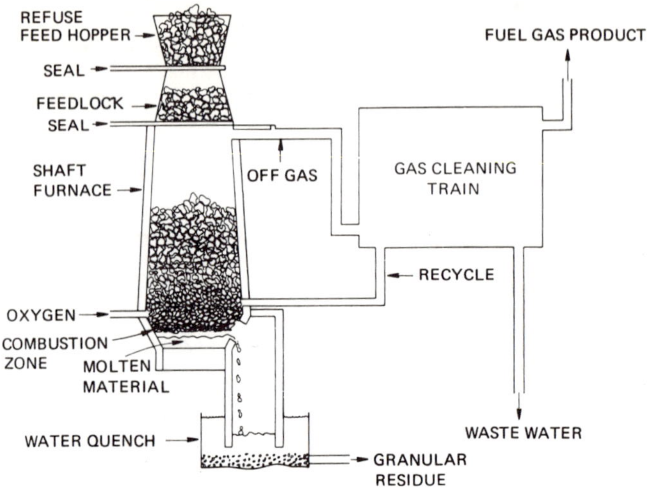

Fig. 10-2. Union Carbide Purox refuse converter. ***Soùrce:*** **Union Carbide Corp., New York.**

bustion characteristics while its heating value is only 300 Btu/cu ft. The gas is essentially sulfur-free and contains only about one-tenth the amount of fly ash normally permissible under air quality standards, see Tables 10-4, and 10-5.

The unique aspect of the PUROX system is its use of oxygen rather than air for a pyrolytic reaction. Since air contains 79% nitrogen and only 21% oxygen by volume, the flue gas volume in the PUROX reactor is only about 20% of that from a pyrolytic system utilizing air as an oxidant. This greatly increases the efficiency of the pyrolysis and reduces the air pollution potential

TABLE 10-4. PUROX System Fuel Gas Composition

Constituents	Volume (%)
CO	47
H_2	33
CO_2	14
CH_4	4
C_2H_x	1
N_2	1
	100
Fuel gas quantity	7 million Btu/ton refuse

Source: Union Carbide Corp. New York.

TABLE 10-5. Comparison of Fuels.

	PUROX System Fuel Gas	Methane	Propane
Lower heating value (BTU/SCF)[a]	286	910	2312
Compression power (KWH/MM BTU)[b]	5.7	1.8	0.6
Combustion air requirements (SCF/MM BTU)[c]	8300	10,500	10,300
Volume of combustion products (SCF/MM BTU)	10,500	11,600	11,200
Heat release/volume of combustion products (BTU/SCF)	95	86	89

[a]Standard cubic foot dry, as measured at 60°F and 1 atmosphere pressure. Heating value calculated at 18°C.
[b]Gas compressed to 35 psig from 1 atm., 100°F with 75% efficiency.
[c]Based on the air needed to convert the fuel to CO_2 and H_2O—no excess air.
Source: Union Carbide Corp. New York.

of the process. The use of oxygen also decreases the amount of solid residue from the shaft furnace to about 3% of the incoming volume. Claims for the efficiency of the system maintain that the PUROX process produces four times as much energy as it consumes; however, this is an energy balance on the furnace itself and does not consider the energy required to manufacture the oxygen used in the process. At the present time, it appears that the PUROX process is the most practical of the current pyrolysis systems, see Figure 10-3.

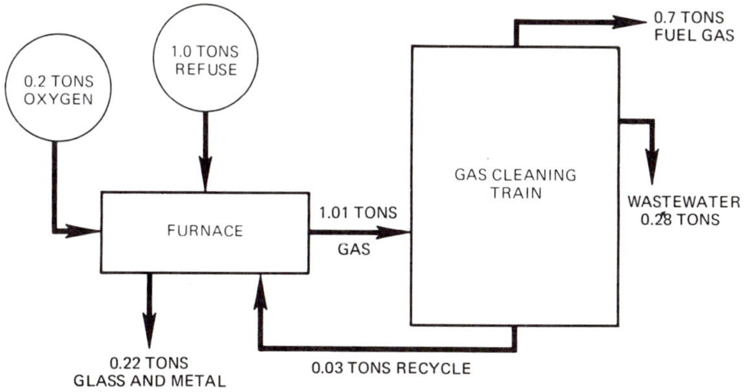

Fig. 10-3. PUROX system inputs and products. ***Source:*** **Union Carbide Corp., New York.**

It must also be considered that Union Carbide Corporation is a manufacturer of oxygen. Whether used industrially or municipally, a PUROX system would require the use of oxygen in some form, either stored as a liquid and purchased from one of the major suppliers or if the pyrolysis system were large enough, an oxygen generation plant would be necessary to operate the pyrolytic unit.

10-2.3 The LANDGUARD System

The Monsanto Enviro-Chem LANDGUARD system, Figure 10-4, has been removed from the market in the last year because of problems with a 1000 ton per day demonstration unit handling the waste of the City of Baltimore. The construction of this unit was predicated on earlier pilot plant work by Monsanto Enviro-Chem Division.

In this system, garbage and trash are brought to the refuse disposal site, shredded, and fed into a single kiln where it is partially burned with air. This is the pyrolysis unit. The by-product is the gas having a heating value of approximately 100 to 120 Btu/cu ft, which is subsequently burned in an after-burner and the heat so generated is employed to generate steam. Since the waste gas is burned on-site, there must be an immediate customer for the by-product steam. And, the fact that the steam generation is dependent upon a continuous flow of waste makes it necessary to have a standby fuel system on the boiler. Such problems, of course, are not insurmountable but certainly are a disadvantage to the process.

The initial operation of the LANDGUARD system commenced early in the summer of 1975, and significant problems were incurred causing many to question its reliability as a long-term solution to the solid waste problem. There were a number of problems involving the feed to the pyrolysis unit which were complicated by numerous mechanical breakdowns. There were also serious slagging problems in the secondary combustion chamber where the waste gas was burned, as well as air pollution problems which could not be surmounted. After virtually two years attempting to bring this system on-stream, Monsanto and the City of Baltimore decided to abandon the pyrolysis project. The equipment which is still on-site has been considered satisfactory to be modified into a straight incineration system. This system will require more air pollution control equipment than the original pyrolysis system, but the operational problems should be considerably less.

10-2.4 ANDCO-TORRAX System

The air-blown slagging pyrolysis unit known as the ANDCO-TORRAX system, Figure 10-5, was developed by Andco Incorporated and The Carborundum Company in a 75 TPD pilot plant during the early 1970's. At the present time,

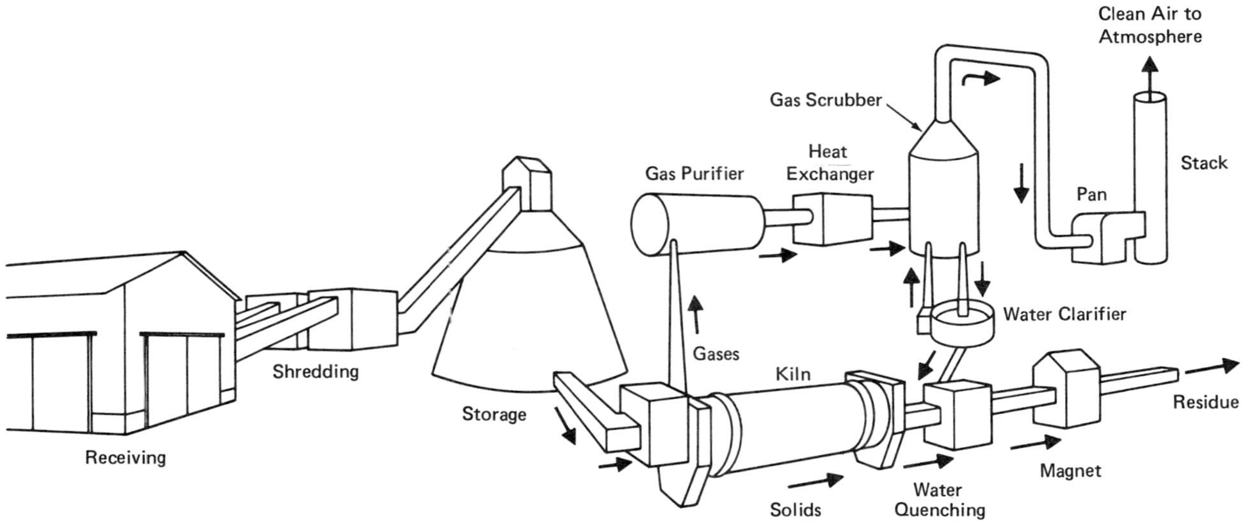

Fig. 10-4. The LANDGUARD system.

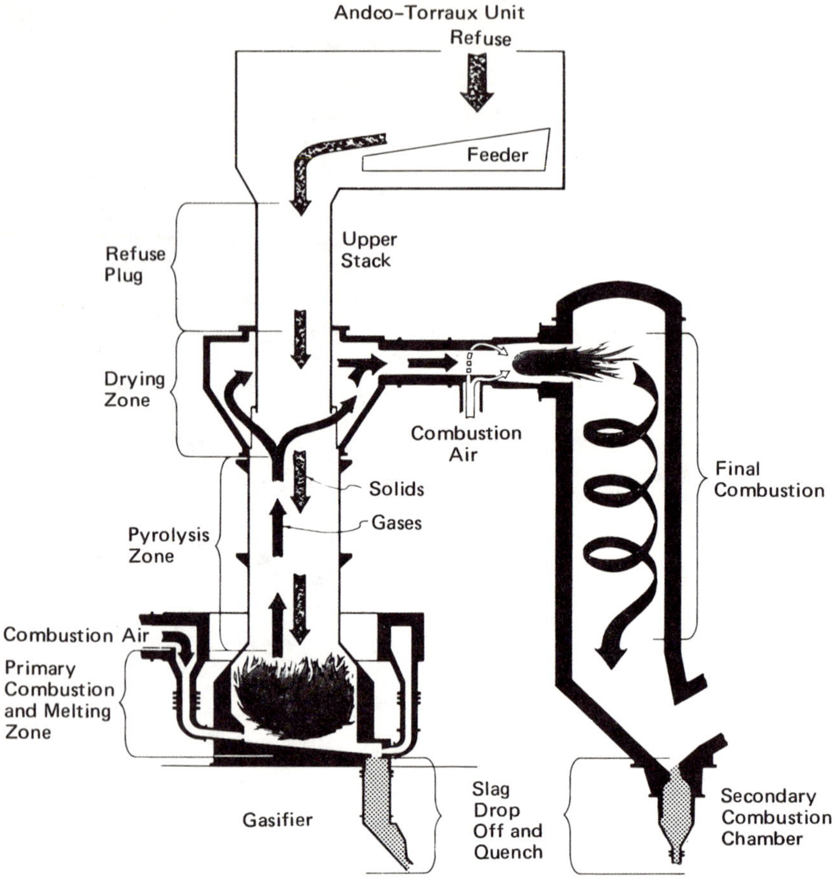

Fig. 10-5. TORRAX system. ***Source: Ind. Water Eng.,*** **July 1970.**

there are four commercial ANDCO-TORRAX plants in Europe, three of which are in operation and the fourth is under construction. These plants vary in capacity between 170 and 400 metric tons per day. The ANDCO-TORRAX countercurrent vertical shaft gasifier is similar to that used in the PUROX process except that it employs preheated air in place of oxygen to achieve slagging pyrolysis. The product pyrolysis gas from the ANDCO-TORRAX gasifier has a low heating value in the range of 100 to 150 Btu/cu ft.

Refuse descends in the ANDCO-TORRAX gasifier by gravity through 3 process zones similar to those in the PUROX process–drying, pyrolysis and combustion/melting. In the ANDCO-TORRAX process, no shredding or sizing of the refuse is required, except to shear oversize, bulky items. Preheated air enters the hearth of the gasifier at temperatures of approximately 1900°F. This air is used to burn the char remaining after pyrolysis to produce heat to melt the inerts and to further pyrolyze the descending refuse. The molten slag produced by the high temperatures in the hearth (up to 3000°F) is continuously drained through a tap hole into a water quench tank to produce a black granular material which is inert and clean. This residue represents approximately 3% of the initial volume of refuse fed to the gasifier. The pyrolysis gases exit the gasifier at 800 to 1000°F and are tangentially admitted into a secondary combustion chamber where they are burned to completion with air to produce an exhaust gas with temperatures of 2200 to 2300°F. Approximately 15% of the hot gases from the secondary combustion chamber are introduced into refractory-filled heat exchangers which heat the primary combustion air up to 1900°F used in the gasifier. These regenerative towers are similar to the stoves used in the steel and glass industries for heat transfer. The remaining portion (85%) of hot gases from the secondary combustion chamber are passed through a waste heat boiler to produce steam. Cleaning of the cooled waste gas is typically effected with a conventional electrostatic precipitator.

10-3 EQUIPMENT FOR PYROLYSIS

The equipment used for pyrolysis systems can generally be divided into three basic types: (1) the vertical shaft which is used by the TORRAX and the PUROX process; (2) the horizontal shaft or rotary kiln, the system employed by the Monsanto LANDGUARD process; and (3) a fluidized bed which is currently only in pilot-plant applications.

Shaft reactors are conceptually the simplest and the lowest in capital cost. The vertical shaft is fed at the top of the reactor, and the waste settles into the reactor by its own weight. Pyrolysis gases generated in the reactor pass upward through the shaft and are removed from the top. Typical feed mechanisms include screw conveyors, rotary devices, rams, and skip hoists. A residue discharge device and gas take-off manifold must be provided on the reactor.

The horizontal-shaft type reactor is basically a rotary kiln, so it must incorporate a feed conveyor system which is usually mechanical in the form of a screw conveyor. Since the feed goes from the high end of the kiln to the low end of the kiln, it is automatically moved within by the rotation and discharged into a quench at the low end. A horizontal reactor which is not of the kiln variety could be used, but the mechanical problems employed in moving the waste through the horizontal reactor are extremely difficult. Both the kiln and the vertical shaft furnace are refractory lined and usually constructed of steel or other metallic material.

The fluidized bed reactor consists of a bed of solid particles such as sand suspended in an upward flowing gas stream. For pyrolysis applications, the solid particles are heated either in the same vessel or in an external vessel with circulation and serve as the heat source for the pyrolysis reactions. A chemical reaction involving the solid particles may take place. The major advantage over other types of reactors is improved heat transfer and temperature control. The primary disadvantages are erosion and carryover problems associated with the solid particles, gas velocity control, and solids transfer and separation problems.

As mentioned before, pyrolysis systems are of two distinct types: direct and indirect heating. In a direct method, heat is supplied to the reaction mixture by partial combustion of the waste and/or supplementary fuel within the pyrolysis reactor. Some oxygen must be supplied, and the reactor product gas contains a significant amount of carbon dioxide and water with the result that the heating value of this gas is reduced. If air is used as the oxygen source, then the environmental problems concerning the formation of NO_x must be considered as well as the further dilution of the pyrolysis gas by large amounts of nitrogen.

With indirect heating methods, the primary heating zone is separated from the pyrolysis vessel. Such a separation can be achieved by a heat conducting barrier or wall or by the transfer of a separate medium, such as sand, between the combustion and the pyrolysis vessels. Wall heat transfer methods are not too practical in large solid-waste applications because of the poor heat transfer and the fact that there is not enough surface area for the transfer of heat in a large system. The use of the separate medium to transfer heat is desirable but can present major problems with regard to solids transfer and separation.

Indirect methods are generally less efficient than direct methods but avoid the problems of excessive carbon dioxide and water formation and reduced heating value and high NO_x and nitrogen content. The indirect methods may have some specific applications in the industrial field but do not appear to have the type of utility that is required for major pyrolysis installations.

10-4 PYROLYSIS PROCESS VARIABLES

Just as in an incineration system, there are several variables which may be used to operate the pyrolysis process: temperature and residence time within the reactor (that is the time under maximum heat). Very little information has been developed to date on these variables. The key reactor control variable is the temperature of pyrolysis. Decomposition of organic materials begins at approximately 360°F. Although most processes operate in the 500 to 3000°F range, at high temperatures the reactor products consist primarily of gas and slag phases where the gas is a low molecular weight hydrocarbon, such as hydrogen, carbon monoxide, and carbon dioxide, and the slag is a fused mass of various solids, usually carbon, ferro-silicate, etc. If the temperature in the reactor is lowered, the gas phase becomes richer in higher molecular weight hydrocarbons, such as methane and ethane; a liquid phase may also be present. Under low temperature conditions, the solid phase also becomes more heterogeneous.

Batch reactor studies with refuse from which glass and metal had been removed indicated that the amount of gas increased as the temperature of the reaction increased, while the amount of residue declined. The liquid fraction remained fairly constant over the temperature range studied. The gas phase composition changed as temperature was increased to produce more hydrogen and less carbon dioxide. The maximum gas calorific value was achieved at a reactor temperature of about 650°C.

The residence time in a pyrolytic reactor is generally defined as the time required to bring the residue to a uniform temperature and, in most systems, this residence time is less than 1 hr, even in the batch reactor. Batch reactor studies have shown that gas yields increase sharply at higher heating rates (i.e, a relatively short time to 800°C) and the hydrocarbon liquid phase seems to be maximized at relatively short heating rates corresponding to about 25 min., while the water phase fell off at the higher heating rates. Gas compositions indicated that high heating rates produce an increase in carbon monoxide and a decrease in carbon dioxide, and the heating value curve reached a maximum at a heating rate corresponding to about 40 min.

The feed preparation to a pyrolysis system will be an important aspect in its subsequent operation. Shredding or size reduction to provide a uniform feed will produce more uniform heating rates and better temperature distribution throughout the reactor and will, therefore, give more uniform results in terms of the composition of the flue gas and the final product. Raw trash or refuse which has not been previously sized or screened will give erratic results, especially in a shaft type furnace. This is due to the fact that such material when charged into a vertical shaft reactor will not be uniform and will have a number of voids within the reactor body. Too fine a distribution of the feed to the reactor will cause the opposite type of problem, i.e., high pressure drops for the air or oxygen and the possibility of greater air pollution.

The process of pyrolysis for handling solid waste material is relatively new, and, therefore, it is very difficult to make definitive statements concerning the desirability of one system over another. Based on the experience with the LANDGUARD process, it is obvious that there is a possibility that some processes now at the pilot-plant stage will fail, although, the possibility that some processes now at the conceptual or experimental stage may become, in the future, much more preferable than the systems that have presently been demonstrated. It is also probable that new pyrolysis systems which have not yet even reached the conceptual stage will be desirable.

Therefore, anyone faced with making an immediate selection of a pyrolysis process for some industrial application will want to have as much background information as possible on the individual process to see whether it is, first, technically feasible and, second, economically feasible. However, even the success of the pilot plant does not seem to guarantee that the full plant operation will not be without serious problems.

10-5 PYROLYSIS ECONOMICS

The economics of the various pyrolysis alternatives will greatly depend on the application and the size of the equipment and the possibility for reuse of the fuel or reclaimed materials which are generated. While economic comparisons for various pyrolysis systems have been published, they are quite suspect becuase of the newness of the process and its lack of continual operation. The net amortized cost estimates, however, generally fall into the range of \$3 to \$10 per ton of solid waste feed.

In general, the very low heating value of the fuel gas produced by any of these processes makes it doubtful that the gas could be transported economically

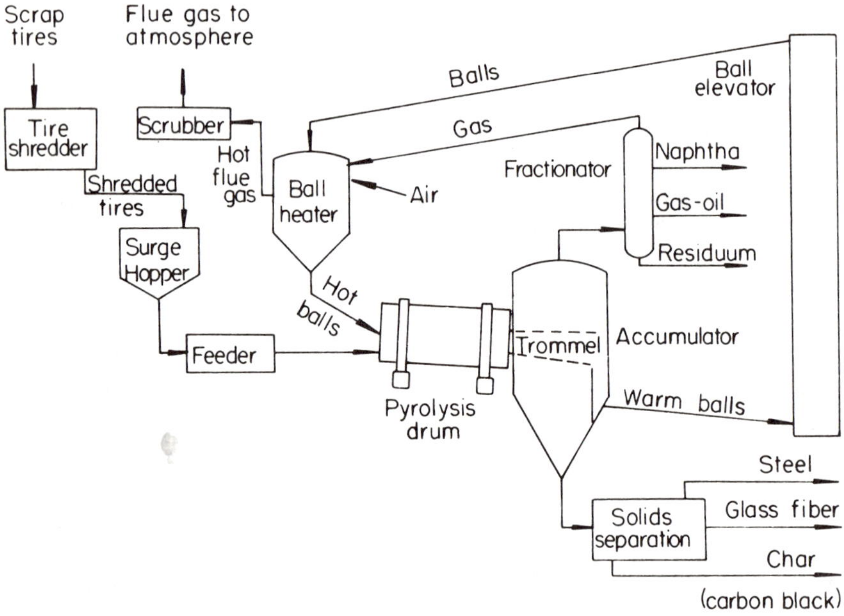

Fig. 10-6. Retorting process to recover materials from tires. ***Source:*** **Ricci, L. J., "Spent-Tire Pyrolysis May Add Bounce to Recycling,"** ***Chem. Eng.*****, p. 52, August 2, 1976.**

for any distance, because of its high volume and the high amount of inerts which must be transported. It certainly is not compatible with normal gas heating systems in either industrial or domestic applications. The gas, however, could be used "on-site" for the generation of steam if, in fact, there is a need for steam. Indirect processes, on the other hand, which do not use air to achieve the pyrolysis can produce gases with sufficient Btu value for transport, but all of these processes are currently in the pilot stage. Such a process is being piloted by Goodyear Tire and Rubber Co. to recover oil–carbon black and steel from old tires, see Figure 10-6.

There does not seem to be any current indication that a reasonable solid by-product can be created from such processes, but pyrolysis processes which produce sufficient volumes of combustible oil certainly will have an advantage if the product can be used as a fuel. In most industrial applications, it would seem that such fuel could be used within the plant rather than as a salable commodity.

10-6 PYROLYSIS APPLICATIONS

Much of the early work on pyrolysis in the United States was by the U.S. Bureau of Mines. A pyrolysis unit built in 1929 was designed to be a research tool and used mainly on specific types of waste material, such as waste tires. For example, from a ton of automobile tires with a heating value of approximately 32.5 million Btu/hr, about 95% of the heat could be recovered in some manner or another as a by-product fuel.

Other work is being performed with various types of plastics. Much of this work has been by Union Carbide Corporation on materials such as high density and low density polyethylene and polypropylene. In these cases, the products are usually a liquid, a grease, or a wax, depending upon the desired end-product, but, in some cases, the material can be converted into gas for use as a fuel or raw material. When only external heat is applied and there is no partial combustion with air or oxygen, the system may be better described as destructive distillation. Where air or oxygen is added to the system, then it is a true pyrolytic operation.

By far, the most "researched" application has been domestic or municipal refuse. The U.S. Bureau of Mines system is shown in Figure 10-7 and the one developed by Urban Research and Development in Figure 10-8.

Some advocates of pyrolysis systems would have one believe that the user is getting something for nothing. The laws of thermodynamics just won't let this happen. Every pound of refuse or trash has a heating value that can be measured. If it is burned properly so that complete combustion is achieved, then this heat release will be very near the ideal for that type of waste. If the

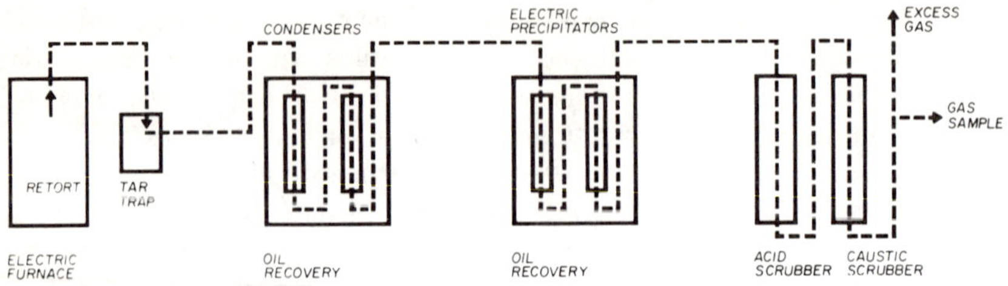

Fig. 10-7. U.S. Bureau of Mines pyrolysis system.

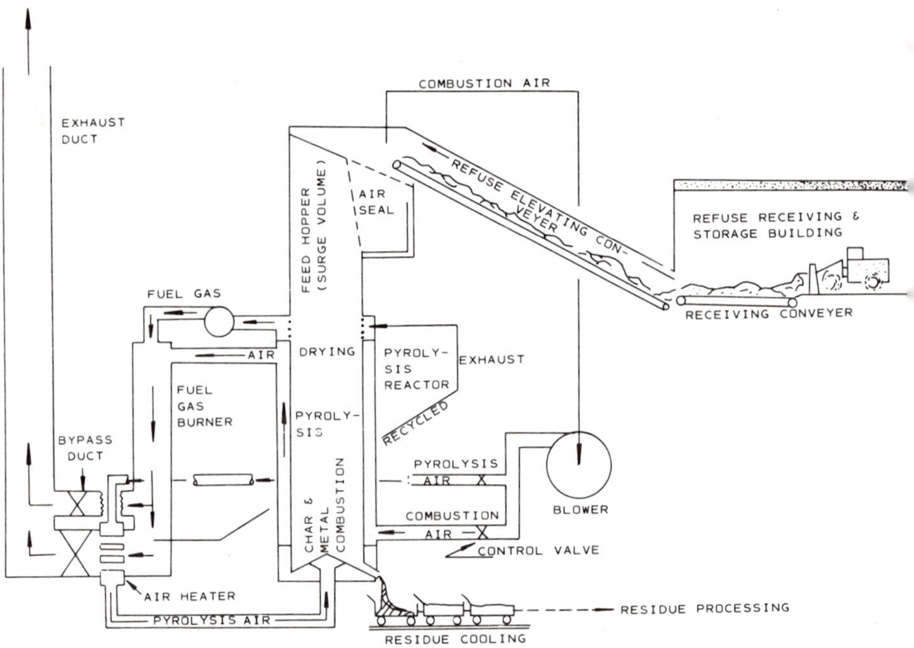

Fig. 10-8. URDC pyrolysis system.

same pound of waste material is placed in a pyrolytic unit, the heat released in the pyrolysis operation, plus the heat released in burning the by-product material, should be equivalent to that released in the incinerator. The question is how to generate the heat. If you wish to save the heat in a gaseous fuel and possibly reduce the attendant air pollution control problem, then it would appear that pyrolysis would have an advantage. Pyrolysis systems are appropriate if the user wishes to develop or generate steam on-site, but this can also be done directly by incineration, recovering the heat from the combustion reaction. Proponents of pyrolysis point out that it is considerably cheaper than the use of a recovery boiler on an incinerator and that it has a much better chance of working, but they have not proved their point of workability to date on the scale where incineration and heat recovery has been proven. This is not to say that pyrolysis would not some day be proven a satisfactory system on a large-scale for both industrial and municipal applications, but presently it is at an early stage in a very interesting scientific and commercial venture.

10-7 WET AIR OXIDATION

Normally, wet air oxidation methods would not be considered appropriate in the discussion of industrial waste disposal because where wet air oxidation has been used in the past, it has been primarily in municipal wastewater applications. But, in the last few years, new approaches have made industrial applications considerably more economical and more interesting.

Wet air oxidation is often known by the name, "Zimmerman process." This process is a rapid oxidation of organic materials in the presence of water at temperatures between 250 and 700°F. It usually requires steam injection to achieve these temperatures. The operating pressures can be as high as 3000 psig or as low as 150 psig. The primary advantage of this process is that no dewatering is required. Sludges having 3 to 6% solids can be treated by this process, and the degree of oxidation depends on the temperature and the pressure and also on the solid content of the entering sludge and the retention time in the processing vessel.

The advantage of the system is the finely oxidized solids in the effluent which are equivalent to those formed in an incineration system, and they are sterile and separable by ordinary gravity or centrifugal means or vacuum filtration.

10-8 APPLICATIONS OF WET AIR OXIDATION

While the wet air oxidation process as supplied by Zimpro, a subsidiary of Sterling Drug, Inc., has been used mainly for municipal sludges, it definitely has a place in the industrial waste disposal effort. The process is currently being used to recover filler in oxidized sludge and recover energy and chemicals from black liquor streams at pulp and paper mills, both in the United States and abroad. It has also demonstrated its ability to destroy toxic wastes in a variety of petrochemical wastewaters and is being used to regenerate carbon and to destroy associated biomass in the biophysical treatment of industrial or joint industrial and municipal waters.

Wet air oxidation has been used industrially for the treatment of a large class of wastewaters that do not readily lend themselves to treatment by other means.

For this class of waste, biological treatment is either impossible or extremely difficult because thé wastes are biologically refractory or are toxic to treatment organisms. Incineration is generally considered uneconomical because the flow is high and the waste is probably dilute. Incineration may also be impractical because of an air pollution problem incurred downstream from the incinerator. It has been used to treat the wastes from the manufacture of acrylonitrile, monosodium gluatamate, polysulfide-based rubbers, metallurgical coke, petrochemicals, and pharmaceuticals. Wet air oxidation may be used as a complete treatment step, or it may be used in conjunction with biological or biophysical polishing steps.

10-9 THE BASIC SYSTEM

Virtually any material that is subject to oxidation, either biologically, chemically, or thermally, can be oxidized at relatively low temperatures in water and in the presence of air. Wet air oxidation should not be confused with submerged combustion in that temperatures in the wet air oxidation process rarely rise above 400 to 600°F and no flame is involved. As shown in Figure 10-9, wet air oxidation starts with a wastewater containing some oxidizable material which is pumped through a heat exchanger where the temperature is increased to a point at which the reaction between the material to be oxidized and the

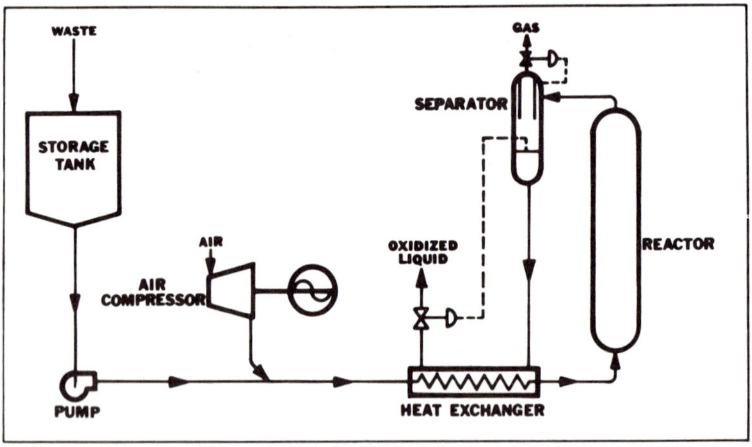

Fig. 10-9. Wet air oxidation flowsheets. *Source:* Pradt, L. A., "Developments in Wet Air Oxidation," *Chem. Eng. Prog.*, 68, 12, 72, 1972 (Updated, 1976).

air will proceed automatically. The temperature is raised by the heat of combustion which is released as the oxidation takes place. Heat is recovered by exchanging the reactor effluent with the incoming waste stream. Spent air, carbon dioxide, and steam are then disengaged from the oxidized effluent in a separator.

As in incineration, the degree of oxidation is primarily a function of the reaction temperature and the residence time. The operating pressure depends on the temperature and, to some extent, the organic concentration. Temperatures range from about 350°F at 350 psig to 610°F at 3000 psig.

During the oxidation process, higher molecular weight compounds are preferentially oxidized to lower molecular weight intermediate products, such as acetic acid, methanol, formaldehyde, and similar materials, that are easily treatable biologically. It is the large, more complex molecules that are troublesome in biological systems. Smaller toxic radicals, such as cyanide, nitrile, and thiocyanate, are also readily oxidized. Various compounds of sulfide and mercaptan are quickly destroyed as well. Troublesome compounds, such as phenol, can also be destroyed by this process.

When speaking of a mixture of components as might be found in an actual industrial waste water, the decomposition or oxidation of specific components is difficult to follow analytically. Figure 10-10 shows data developed from

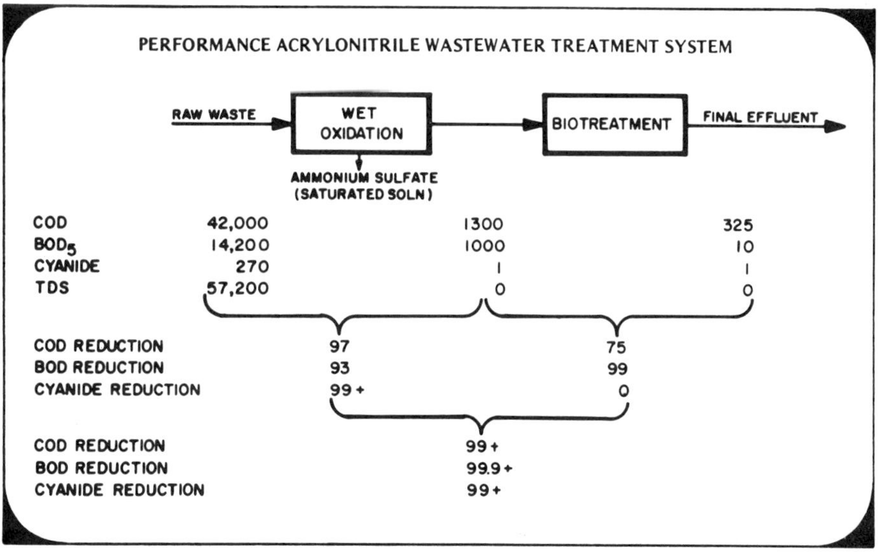

Fig. 10-10. Wet oxidation and biotreatment of acrylonitrile wastewater. *Source: Chem. Eng.*, p. 108, February 16, 1976.

the oxidation of acrylonitrile waste waters. It should be noted that BOD_5 increases substantially as oxidation breaks up toxic and refractory components, in this case, nitrile and polymeric substances. The COD can actually increase as organics resistant to the dichromate-and-acid oxidizing conditions of the standard COD procedure are attacked in the more rigorous conditions of wet oxidation.

It will be noticed that the ratio of BOD to COD is used as an indicator of treatability and plotted on log probability paper, both before and after oxidation. The average increase in BOD/COD from 0.27 to 0.85 is evident in Figure 10-11. The relative amount of treatment done by wet air oxidation to that done biologically will be governed by economics and the character of the waste. Typically, 80 to 95% reduction will be done by wet air oxidation with the remainder to be removed biologically.

The effluent from a wet air oxidation process is quite amenable to further treatment by some type of biological means or, as in the case of toxic waste, a biophysical treatment process in which powdered activated carbon is carried in activated sludge aeration basins.

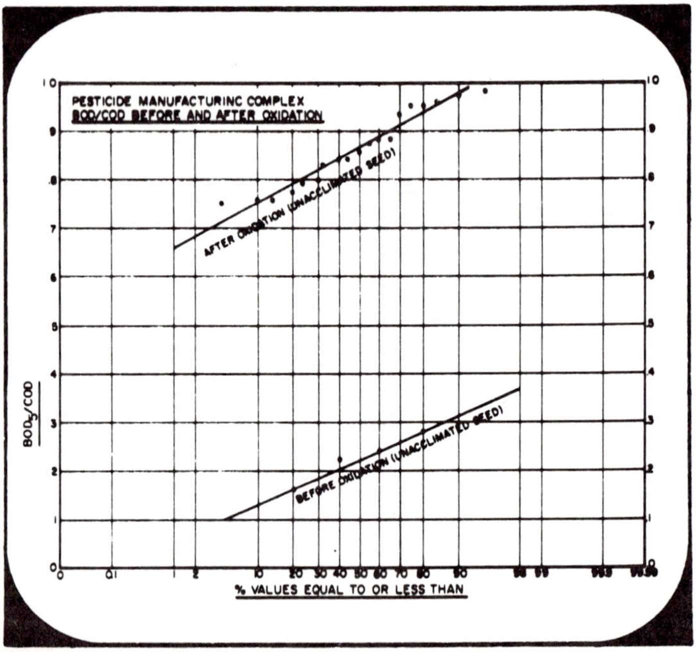

Fig. 10-11. BOD/COD Ratio before and after wet air oxidation. ***Source: Chem. Eng.*, p. 107, February 16, 1976.**

10-10 BIOPHYSICAL TREATMENT SYSTEMS

A basic biophysical flow scheme is shown in Figure 10-12. It may be modified in a number of ways, e.g., two-stage biophysical and one-stage contacting. It differs from a conventional activated sludge process in that powdered activated carbon is carried in the aeration basins at levels in excess of 10,000 to 20,000 mg/liter, and this offers a number of advantages. The carbon particle serves as a heavy nucleus for floculation, thus promoting settling. An active biomass can be carried at substantially higher levels than found in conventional systems, sometimes as high as 10,000 to 15,000 mg/liter. The effect of this is smaller aeration basins and higher sludge age and lower sludge production, and because of the rapid settling and higher sludge age, complete nitrification can usually be achieved without resorting to a multiple stage operation.

The large amounts of carbon present in the system offer another advantage; it serves as the adsorbent for substances that may have survived wet air oxidation. Toxic components are adsorbed and passed out of the system on the surface of the carbon or are desorbed as they become biologically assimilated at the surface. Furthermore, the carbon greatly reduces odors from the aeration basin. The biophysical system is generally as shown in Figure 10-12. The functional aspect of the wet air oxidation process is to regenerate the carbon.

It is also possible to combine wet air oxidation and biophysical treatment, as shown in Figure 10-13. In this case, the raw waste is fed into the wet air oxidation system and a secondary wet air oxidation system is used to regenerate the

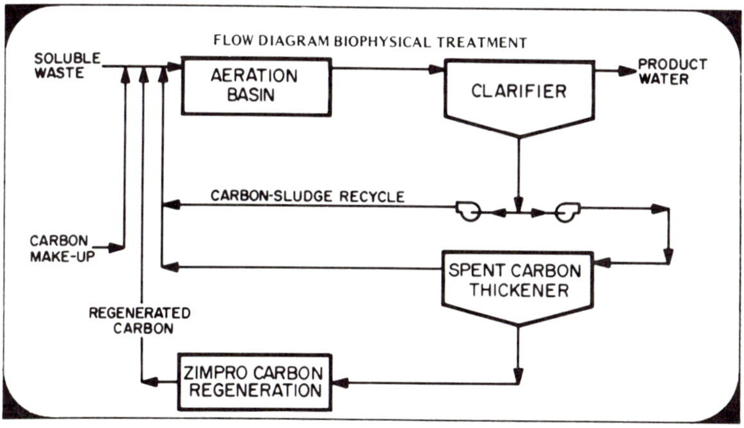

Fig. 10-12. Biophysical treatment with wet oxidation regeneration of carbon. ***Source:*** **Zimpro, Inc.**

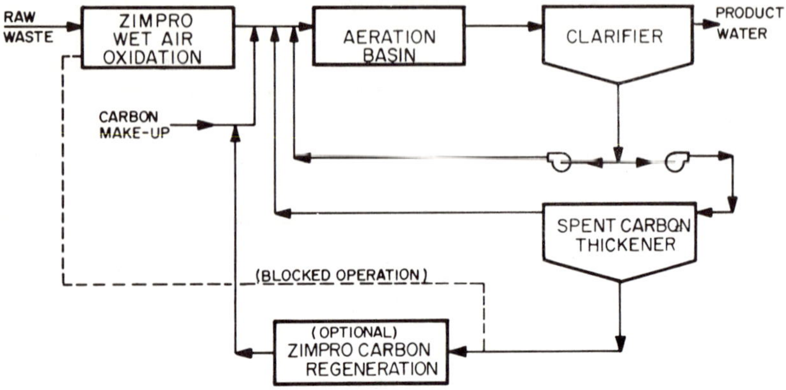

Fig. 10-13. Wet air oxidation followed by biophysical treatment.

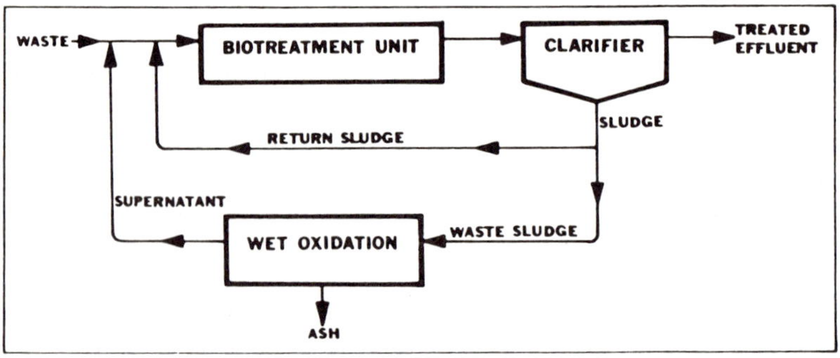

Fig. 10-14. Wet air oxidation of waste sludge. *Source:* Zimpro, Inc.

carbon. Figure 10-10 shows the performance of an acrylonitrile wastewater treatment system using wet air oxidation and biotreatment as steps 1 and 2.

Chemical wastewaters issuing from a series of complex manufacturing processes which were found to be untreatable by conventional biological means can be treated by a combination of wet air oxidation and biological means with a 75% reduction of COD in the first step, as demonstrated by Figure 10-13. These processes serve to demonstrate some of the things which can be achieved by wet air oxidation in the treatment of toxic wastes. Figure 10-14 shows wet air oxidation of waste sludge.

It should be noted at this point that wet air oxidation is applicable only to relatively dilute aqueous solutions of organic solids and, therefore, should be compared to other treatment methods applicable to dilute solid wastewaters.

However, since it is applicable to such materials as toxic waste disposal and pulp mill wastes, it is a viable means of destroying otherwise difficult materials which create problems when incinerated, pyrolyzed, or landfilled.

10-11 PAPER AND PULP MILL APPLICATIONS OF WET AIR OXIDATION

Pulp mills throughout the world need liquor recovery systems which will help combat both the cost increases of disposal and energy shortages. For nonwood fiber and bagasse pulping operations, it is extremely important because the high silica, high fiber, and low solids content of spent black liquor make efficient recovery by conventional evaporation and burning techniques especially difficult. The silica build-up, the lime, and the wet pith disposal are added problems. Wet air oxidation can simultaneously dispose of the pith and the sludge while recovering their heat values and can accept high silica liquors, at the same time meeting environmental standards by oxidizing these dissolved solids in water from 0 to 99% in the presence of oxygen under pressure and at temperatures as low as 315°C. The advantages of the system are that it is completely enclosed, assuring high levels of chemical and energy recovery, and virtually free of smoke and fly ash, associated with conventional incineration.

10-11.1 Spent Black Liquor

In typical wet air oxidation systems, dilute spent black liquor is fed into the system and brought to pressure by a high pressure pump. The wet pith, waste biological sludge, or other hard-to-handle waste streams may be added at this point. The liquor then passes through heat exchangers receiving heat from the outgoing flow and is then mixed with air from the compressors. The air and liquor mixture is injected into a reactor or oxidation tower where the organic material is oxidized and those chemicals which have been used for pulping survive the process and can be recovered in the discharge stream. Spent air, which is saturated with steam, is taken off the top of the reactor and can be used to produce thermal or electrical energy. A typical wet air oxidation recovery plant in a baggase mill is shown in Figure 10-15. This system is applicable to either wood or nonwood type pulping operations; the energy balance for the sodium carbonate black liquor is shown in Figure 10-16.

Wet air oxidation is most applicable to straight alkaline, semi-chemical, or soda liquors, though sodium lignin salts in the black liquor are converted to sodium carbonate for direct return to an alkaline semi-chemical process or for recausticizing. The sodium carbonate will actually be produced as a mixture of sodium carbonate and some sodium bicarbonate. By adding make-up chemi-

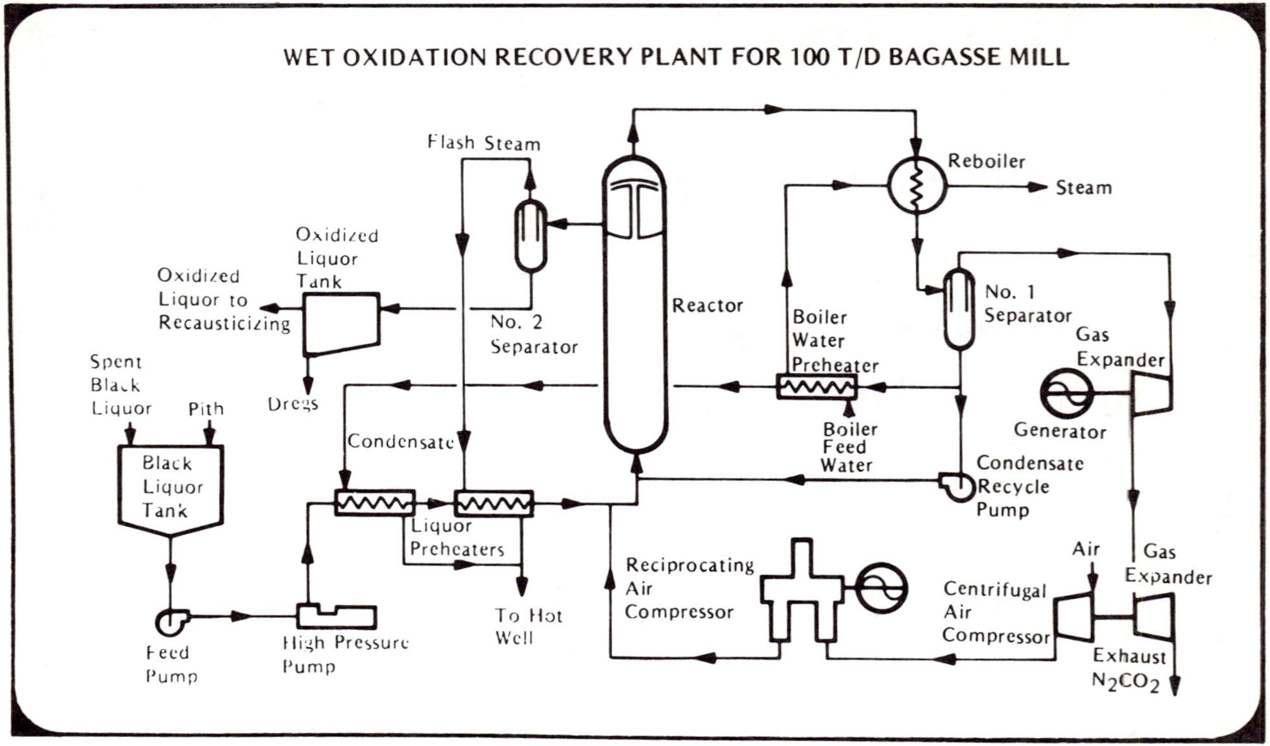

Fig. 10-15. Wet oxidation recovery plant for 100 tons/day bagasse mill. *Source:* Zimpro, Inc.

Energy balance for the wet air oxidation of Na_2CO_3 black liquor. (Basis: 1 gal. liquor, 9.28% solids.)

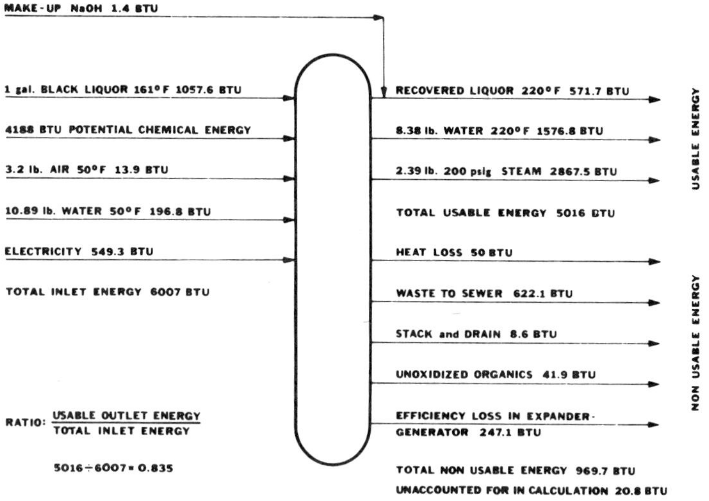

Fig. 10-16. Energy balance for wet oxidation of Na_2CO_3 black liquor. *Source: Chem. Eng. Prog.*, p. 67, May 1976.

cals, as sodium hydroxide, a carbonate hydroxide mixture can be produced suitable for pulping, see Figure 10-17.

The applicability of wet air oxidation to sulfur-containing processes is limited because the sulfur is converted to the sulfate which is difficult to return to a usable form except in a reducing furnace.

10-11.2 Kraft Mill Waste

In a Kraft mill where the recovery boiler is the bottleneck in the process, the excess raw black liquor can be diverted to wet air oxidation. There, the bulk of the organic material will be oxidized leaving an essentially organic-free sodium sulfate liquor that can be mixed with the remainder of the raw black liquor prior to evaporation and conventional burning. The remaining organics will be oxidized, and the sulfur will be reduced. The wet air oxidation unit has removed the excess thermal load and air requirements from the recovery boiler. The potential chemical energy contained in the liquor diverted to the wet air oxidation unit can be recovered. This type of oxidation should be suitable for up to 15% of the total cost of the black liquor flow. Sufficient fuel volume must be contained in the unoxidized liquor to maintain combustion in the recovery boiler.

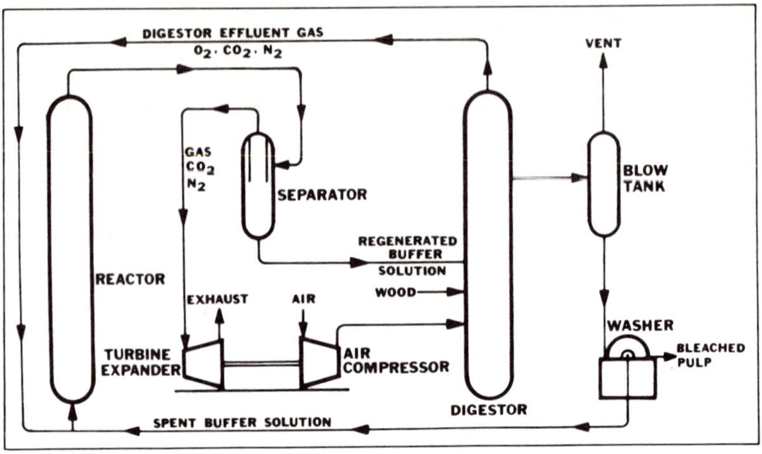

Fig. 10-17. Wet oxidation pulping system. *Source: Chem. Eng. Prog.*, 68, 12, p. 76.

10-11.3 Heat and Energy Recovery

The energy recovery aspects of wet air oxidation are to some degree limited. Oxidation temperatures can not exceed the critical point of water (705.4°F) and, in general, they do not exceed 600°F, so that the production of steam at temperatures of more than 550°F is not practical. On the other hand, low intensity energy, such as hot water or low pressure steam, is abundant and can be made available in many forms.

A low energy steam is important, especially in a paper mill. Some mills use relatively cold water for make-up boiler feed. Heating the water to 220°F before deaeration is a good use of low intensity energy. Other hot water streams can be used for pulp washing or digestor charging. Steam under 200 psig can be used in dryers or for a variety of other applications. Condensate streams from wet air oxidation can also be used for pulp washing or digestor charging. Waste heat from the process air compressor can be utilized by cooling the compressor with make-up boiler feed water rather than conventional cooling water.

The process off-gases from the wet air oxidation unit contain considerable pressure energy. This can be recovered in a turbine or expander. Electrical energy is one possible mode of recovery of the energy from the wet air oxidation system; it is a question, however, where the most economical ratio exists between heat and electrical and mechanical energy in a given plant. Generally, the process off-gases are cooled as much as possible by heat exchange before the gas stream is expanded to recover the mechanical energy. Secondly, this energy can be extracted as mechanical or electrical energy to make the system self-sustaining with no use of outside energy after start-up. All excess energy liberated from the wet air oxidation system will be recovered as heat by heat exchange. Finally, any subsequent energy remaining can be used to heat water or to make steam, see Figures 10-18 and 10-19.

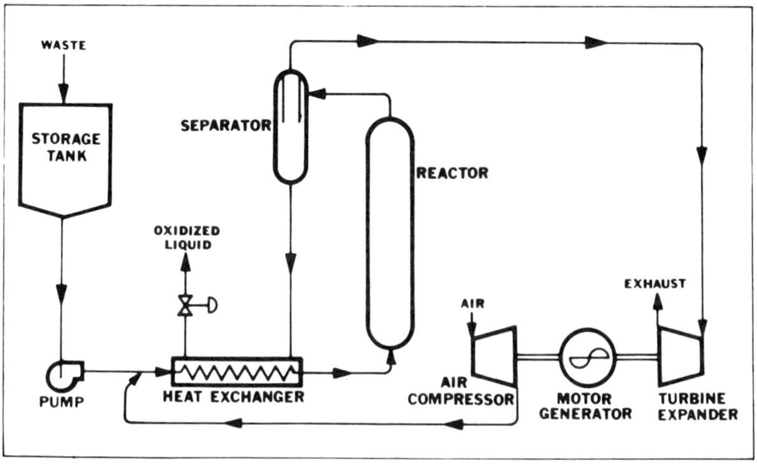

Fig. 10-18. Steam generation flowsheet. *Source: Chem. Eng. Prog.*, 68, 12, p. 74.

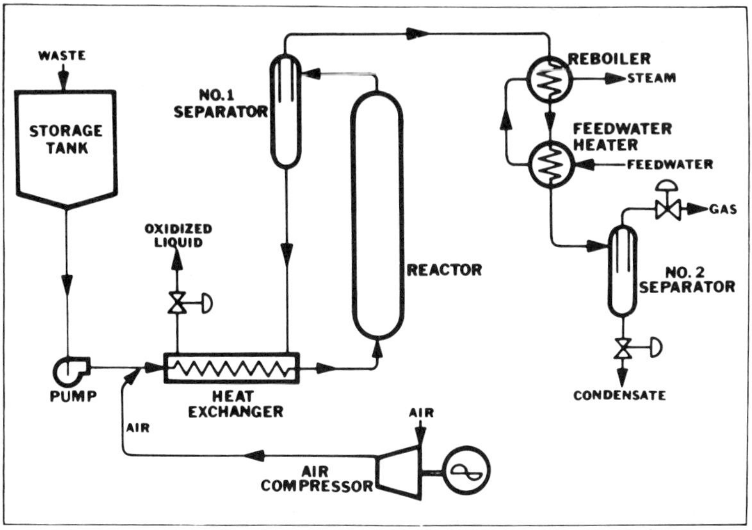

Fig. 10-19. Power generation flowsheet. *Source: Chem. Eng. Prog.*, 68, 12, p. 74.

10-12 ACTIVATED CARBON RECOVERY

Wet air oxidation has been used as an excellent means for the regeneration of activated carbon in liquid phase carbon adsorption systems. Powdered activated carbon alone or in combination with biological treatment provides effective treatment of various wastewaters, but as the carbon becomes saturated with hydrocarbon materials, its activity decreases and it is necessary to regenerate the carbon. The carbon may be regenerated in multiple hearth furnace or similar device once the carbon slurry has been dewatered. However, it is possible to regenerate the carbon in a wet air oxidation system without dewatering and obtain activated carbon equal to virgin carbon with the exception that the very smallest pores are not regenerated.

The properties of regenerated carbon from a physical adsorption process by wet air oxidation are shown in Table 10-6. This table indicates the results of regenerated carbon by the wet oxidation process for two different carbon materials. It can be seen that the relative adsorption efficiencies, surface area, and pore volume are higher for the larger size particles. Adsorption isotherms

TABLE 10-6. Properties of Wet-Oxidation Regenerated Carbon from a Physical Adsorption Process

	Aqua Nuchar A	Darco S 51 FF
Average number of contact-regeneration cycles	6.4	4.2
Loss per cycle, %	4.6	12.7
Relative sorbate efficiencies, %		
Iodine	69	54
Methylene blue	114	135
Erythrosin	122	82
Molasses color	100	98
Surface area, % of parent		
0–10 Å Radius pores	30	46
10–15 Å Radius pores	118	113
15–20 Å Radius pores	102	116
20–30 Å Radius pores	74	112
30–300 Å Radius pores	92	104
300–100,000 Å Radius pores	101	112
BET total, surface area, % of parent	62	86
Pore volume, % of parent		
0–300 Å Pores	65	94
300–100,000 Å Pores	92	150
Total	89	128

Source: Chem. Eng. Prog., 71, 5, 91, 1975.

show that most C_1 - C_4 compounds are not adsorbed from dilute aqueous solutions by even virgin carbon; thus, the compounds most likely to involve the smallest pores are not adsorbed by activated carbon with a full complement of surface, so their adsorption will not be affected by a regenerated carbon material.

10-13 ECONOMICS OF WET AIR OXIDATION

It would be extremely difficult to present any meaningful economic comparison between wet air oxidation and other means of solid waste disposal. As indicated previously, this method of waste treatment has been applied to solids only to the extent that they are transported by liquids and, thereby degradable. It might be assumed, therefore, that industrial solid wastes could be handled in this manner, but no previous work has been done in this field to the author's knowledge. However, wet air oxidation has proven itself a reliable method of handling a wide variety of relatively dilute wastewaters containing solids, so that its ultimate possibility as a treatment regime should not be discounted.

The cost for wet air oxidation systems are generally high as compared with other methods of treatment if considered on their own. However, considered in the context that it gives a clean, compact operation with relatively low temperatures and a stable and controllable combustion process with a possibility of by-product recovery; it does not require any auxiliary equipment to maintain air pollution regulations, and it is compatible with biological and biophysical process. It may, under a number of conditions, be more economical than other options. It is difficult, if not impossible, to provide economic guidelines for the reader without being definitive in terms of the organic loading to the system, as well as the total loading, and the nature of the waste, and the temperature, and the pressure which is required to decompose this waste under a wet air oxidation system. Consequently, no economic data is provided here because of the possibility of misleading the reader. Suffice it to say that the system is generally more expensive than present methods unless evaluated in the complete context of what it accomplishes.

REFERENCES

10-1. Gitchel, W. B., Meidt, J. A., and Burant, Jr., W., "Carbon Regeneration by Wet Air Oxidation," *Chem. Eng. Prog.*, Vol. **71**, No. 5, May 1975.

10-2. Union Carbide Corporation, "Solid Waste Disposal Resource Recovery," Descriptive Brochure on Purox Process.

10-3. Mallan, G. M., "Flash Pyrolysis Turns Refuse into Fuel Oil," *Chem. Eng.*, July 19, 1976.

10-4. Schulz, H. W., "Energy from Municipal Refuse: A Comparison of Ten Processes," *Prof. Eng.*, November 1975.

10-5. Zulver, E., "Baltimore's Pyrolysis Plant—From Concept to Reality," *Resource Recovery*, September/October 1974.

10-6. Takeds, N. and Hiraoka, M., "Combined Process of Pyrolysis and Combustion for Sludge Disposal," *Environ. Sci. Technol.*, Vol. **10**, No. 12, November 1976.

10-7. Eggen, A. and Powell, Jr., O., "Experience in Slagging Pyrolysis Systems," ASME Incinerator Division, October 28, 1971.

10-8. Culham, W. F., "Varied Techniques Reduce Huge Glut of Scrap Tires," *Solid Wastes Management*, February 1975.

10-9. Editorial, "Disposing of Solid Wastes by Pyrolysis," *Environ. Sci. Technol.*, Vol. **9**, No. 2, February 1975.

10-10. Mark, S. D. and Arbeau, P., "Recent Progress in the Pyrolysis of Urban Wastes," Publication of S. A. Heurtey (France).

10-11. Young, R. A., "Treating Refuse by Pyrolysis," *Pollut. Eng.*, March 1975.

10-12. Ricci, L. J., "Spent-Tire Pyrolysis May Add Bounce to Recycling," *Chem. Eng.*, August 2, 1976.

10-13. Kuester, J. L. and Lutes, L., "Fuel and Feedstock from Refuse," *Environ. Sci. Technol.*, Vol. **10**, No. 4, April 1976.

10-14. Fisher, T. F., Kasbohm, M. L., and Riverto, J. R., "The Purox System," *Chem. Eng. Prog.*, October 1976.

10-15. Potts, J. E., "Continuous Pyrolysis of Plastic Wastes," *Ind. Water Eng.*, August 1970.

10-16. Milkovich, J., "What's Happening to Pyrolysis," *Pollut. Eng.*, March/April 1972.

10-17. Editorial, "Wet Air Oxidation Comes of Age," *Environ. Sci. Technol.*, Vol. **9**, No. 4, April 1975.

10-18. Flynn, Jr., B., "Swiss Mill is Trying Wet Air Oxidation to Get Rid of Sludge and Recover Filler," *Paper Trade J.*, May 1, 1976.

10-19. Wilhelmi, A. and Ely, R., "A Two Step Process for Toxic Wastewaters," *Chem. Eng.*, February 16, 1976.

10-20. Flynn, Jr., B. L., "Wet Air Oxidation for Black Liquor Recovery," *Chem. Eng. Prog.*, May 1976.

10-21. Pradt, L. A., "Developments in Wet Air Oxidation," *Chem. Eng. Prog.*, December 1972.

11

Land and ocean disposal of industrial wastes

11-1 INDUSTRIAL WASTES

The earliest solid waste disposal method was dumping on the ground and then burning in place or letting the elements gradually decompose (those that were biodegradable). Items which would not biodegrade would gradually sink into the soil and disappear. Open dumping of municipal refuse was a common means of disposal well into the 1960's. However, the Resource Conservation and Recovery Act of 1976 has virtually eliminated open dumping from a legal point of view; from a practical point of view, however, it still continues.

Aside from the aesthetic aspects of open dumping, there are also a variety of health reasons that have promoted legislative attempts to do away with this type of disposal. An open dump is a harbinger for vermin, for mosquitoes and flies, and offers a hazard from fire.

In terms of municipal waste disposal, there is a combination of trash, garbage, and other refuse, such as plant cuttings, glass, plastics, etc. To dispose of these materials, open dumping has been replaced by a system which is designated "sanitary landfill." Unfortunately, while the word landfill is applicable to most of these sites, the word, sanitary, is not. Theoretically, the approach to sanitary landfill is to haul municipal trash, garbage, and refuse to an area where due to its topographic arrangement, or possibly due to its geology, the waste material may be dumped and covered with a layer of soil on a daily basis. Proper information on the methods of disposing of waste in this manner is found in numerous publications, both governmental and private. The approaches vary widely. The basic premise is to bury the waste and have those materials that will decompose proceed to decompose in the soil; those materials that will not decompose will permanently add to the bulk density of the substrate.

The industrial landfill is different than a municipal landfill for several reasons. First, it will generally handle materials of an entirely different character than the municipal landfill. Industrial waste may be similar to municipal waste in the sense that a large industry may have cafeteria or packing boxes or cartons, crates, or materials similar to municipal wastes, but industrial areas also have building materials and a wide variety of chemical wastes which are very dissimilar to municipal waste. For example, some of the kinds of wastes which industry produces in addition to trash, cartons, wood and other cellulosic materials are: a variety of organic acids, alcohols, and fats; oils of vegetable and animal origin; organic aqueous solutions; materials which contain ammonia, phosphates, and nitrates (usually in solution); biological-treatment sludge; acrylic resin materials; starch and cellulose solutions; amino acids; denatured proteins; a wide variety of acids coming from pickling liquors and plating solutions, including sulfuric, hydrochloric, nitric, hydrofluoric, containing dissolved metals; chromic acid and dichromate mixtures; also, ferric and cupric chloride from etching operations. There are a wide variety of alkalies, such as: caustic soda used for stripping solutions; carbonate and phosphate wash solutions; sodium sulfide and sodium hydrogen sulfide; cyanide; latex suspensions; oil-water emulsions; solvents, such as benzene, xylene, and toluene; alcohols and esters; chlorinated hydrocarbons; waxes; synthetic fabrics; tars and still bottoms; and many, many others. These are the things that an industrial plant must safely dispose of; therefore, an industrial landfill has significantly different problems than municipal landfills.

In February of 1978, the Environmental Protection Agency read into the *Federal Register* proposed rules for solid waste disposal facilities. These regulations are required by the Solid Waste Disposal Act as amended by the Resource Conservation and Recovery Act of 1976. Under Sections 4005 and 4003 of the 1976 Act, all facilities which do not meet these criteria are prohibited. Existing facilities that do not meet these criteria must be closed or upgraded according to a state compliance schedule. After one year from the promulgation of the criteria, the EPA must publish an inventory of all facilities in the United States that do not comply with the criteria. All new facilities must comply with the criteria in Section 4003. Existing facilities that have failed to meet the criteria can be up-graded according to a state established compliance schedule, containing an enforcable sequence of actions leading to compliance within a reasonable time not to exceed 5 years from the date of publication of the inventory. A facility may be placed on this state-established compliance schedule only where there is an EPA-approved state plan and where it has been demonstrated that no alternative which complies with the criteria can be utilized. These criteria will apply to virtually every kind of solid or liquid waste that is put into the ground with the exception of agricultural waste, including manures and crop residues that are returned to the soil as fertilizers or soil conditioners, or to the mining and milling wastes that are to be returned to the mine.

Section 257.3-1 of the Act concerns the location of disposal facilities in environmentally sensitive areas, such as wet lands, flood plains, and permafrost areas, critical habitats of endangered species, and recharge zones of sole-source aquifers. These areas were selected in this criterion because the EPA feels that these areas are natural assets and have beneficial qualities which are not adequately protected by other aspects of the proposed law, see Figure 11-1.

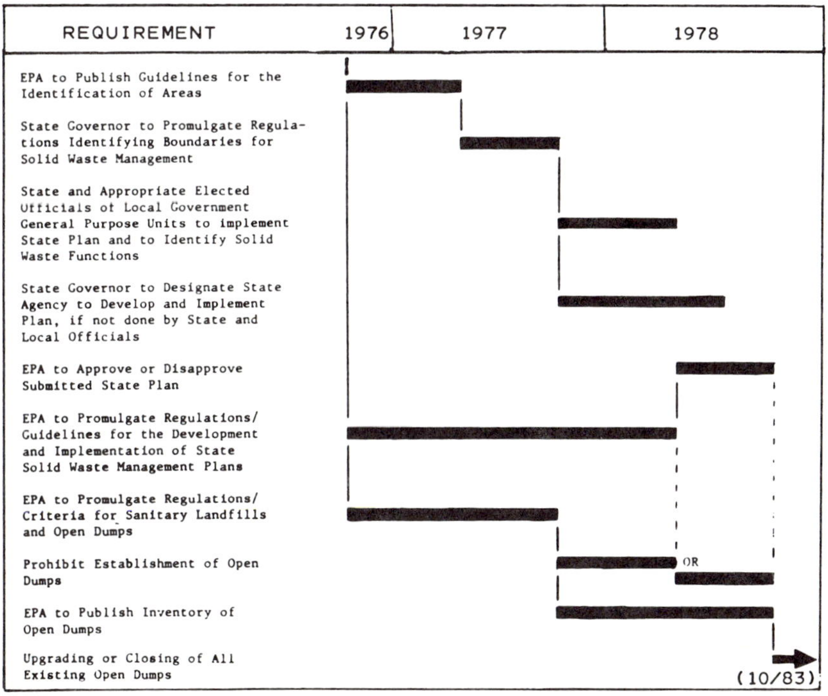

Fig. 11-1. Requirements of Resources and Recovery Act pertaining to solid waste disposal facilities.

11-2 THE INDUSTRIAL LANDFILL

Many industries try to incorporate all of their wastes into a single landfill, but it is very unwise to use the land as a means of disposing of chemical wastes unless particular precautions are taken. Therefore, for the purposes of this chapter, landfills will be considered under two headings: (1) landfills for the disposal of industrial solids which have little or no potential for producing a dangerous leachate; (2) chemical landfills which must handle a wide variety of chemical materials and possibly hazardous materials.

In designing any type of landfill, some of the goals which should be formulated in the design to help the landfill be successful are as follows:

1. Meet all applicable regulations.
2. Protect the physical environment, i.e., ground water, surface water, and air quality.
3. Minimize operational nuisances, i.e., litter, dust, noise, and fires.
4. Minimize dumping time for site users.
5. Minimize the costs, i.e., initial, operational, and total costs.
6. Protect worker and user safety.
7. Provide for maximum use of the land when the site is completed.
8. Maintain an aesthetic site.

When designating a landfill, certain types of data are essential. The first is to know the environmental regulations. Since these are changing every year, it is important to be a student of the environmental changes in order to have a successful landfill. The environmental standards should be examined; the information concerning the operations and monitoring and the nature of plans and license approvals must be understood.

Next, it is necessary to examine the physical characteristics of the proposed site, which will require: a topographic map of the site and the immediate area; surrounding land use patterns; existing drainage patterns; type of vegetation; growth; location of structures roads, and utilities. It is necessary to obtain a property line survey and to understand the existence of any easements or rights-of-way.

The next important aspect of landfill design is to determine the hydrological and climate aspects. Will the surface water run off, if so at what rate, and at what pattern? What type of drainage will be required for on-site ponds, streams, wells, washes, etc? What are the flood plain designations for the area? and What are the evapo-transpiration rates? and what are the wind velocities and directions?

Another important item in the design of a landfill is the characteristics of the wastes. What kind of wastes will be put into the landfill? and what are their loadings in terms of time and also in terms of volume and weight per unit time? Is there a potential of leachate generation or gas generation?

A further consideration in the design of a landfill is the geotechnical data—Where is rock located? and what's the ground water quality? and what is the hydro-geological setting? and what is the regional geology and the physical characteristics of subsoil, such as grain size, permeability, compaction, etc? What is the depth and the types of top soils and subsoils?

After all of these have been determined, it is vital to know something about the transportation system—Where can the site entrance be located? and where are the access roads and the traffic patterns in and out of the area and the access roads to the site? and what are the anticipated arrival times of the loads to the site—are they critical? Do they occur at congested hours or otherwise?

It is also important to know the final use of the area that is being used for a landfill. Is it going to be recovered and used for recreational area or is it to be used for farming or left fallow? A settlement allowance must be considered, vegetation, the requirements for the final drainage pattern, etc.

Last, but not least, a landfill design must include some consideration of the operation of the site. Are there any equipment limitations? How will the landfill be developed? and from what direction? and what hours will it be operated? With this information, a reasonable landfill can be designed.

There is also the consideration in landfill projects that the landfill may be located on the industrial site or at some remote location. If industrial waste material is being disposed of by contract disposal, then, obviously, the landfill will be a remote landfill in almost every case. The problems associated with contract disposal are covered in another section of this book. For the purposes of this chapter, industrial landfill will be considered as wholly-owned by the company using it as a disposal site and that it may be either remote or a part of the industry complex.

The remote site has several advantages and several disadvantages. From an aesthetic point of view, the landfill should not be adjacent to the industrial facility; this will improve the overall attractiveness of the plant site. The disadvantage, of course, is that the waste material must be hauled from the point where it is created to the point where it must be disposed. Another advantage of using a remote site is that the topography and geology of the area can be selected to handle a landfill operation; whereas, on-site disposal means that land which has been selected for plant construction must be used, and it may not be totally acceptable. On-site disposal also has the advantage of better control and shorter hauling distances.

The industrial landfill, which is not to be used for chemical or hazardous wastes, is very similar to a municipal landfill. It should be a well-planned and engineered installation. It may differ from a municipal landfill in that it does not contain materials which will rapidly decompose (such as garbage) and which create an odor and vermin problem. Nevertheless, if the material is to be put into the land, it should be properly prepared and properly covered.

Chapter 8 deals with the size preparation of various types of wastes which may be landfilled or incinerated. Generally, landfill is less sensitive to the size and uniformity of the waste material than is incineration. Consequently, landfill is used for a variety of materials of different shapes and sizes. Nevertheless, landfill is most economically utilized when the material to be placed in the landfill is relatively uniform in size and has been shredded or ground into individual pieces having a normal cross section of 1 in. or less. Large items disposed of in a landfill create voids and make compaction of the land under and over the object much more difficult.

11-2.1 Designing the Landfill

It is necessary to do a complete planning job before a landfill site is selected or before any waste is placed in the landfill. Certain considerations are as follows:

1. The amount of waste to be disposed of.
2. The types of waste.
3. The topography of the land.
4. The geology and hydrology of the landfill site.
5. The climate.
6. The availability of fill soil.
7. The costs associated with operating the landfill properly.
8. The cost of the land and its possibility for other uses.

It is obvious that the amount of waste must be considered in the operation of any landfill, but it is surprising how often landfills run out of land long before they were calculated to do so. In the selection of the landfill site, the user must contemplate the number of cubic yards of waste that will be disposed of on an annual basis and make sure that the volume of the projected landfill will accept this volume for whatever period of years the owner intends to utilize the site, assuming adequate compaction and proper cover after each filling operation.

Industrial landfills, unlike municipal landfills which are normally selected for ten-to-twenty-year operation, may not be necessary for long periods; nevertheless, it is an important and initial consideration.

The types of waste that are going into the landfill, exclusive of chemical and toxic wastes which will be covered elsewhere, should also be considered. If any of the materials can be leached by rain water to create a material that can eventually reach the water source in the area causing a contamination problem, then special types of protection must be considered. In general, industrial waste which is not considered "chemical waste" is not faced with this type of consideration. It should be obvious that no industrial waste, chemical in nature or which would form any type of dangerous leachate, should be placed into a general landfill. A general landfill is one in which no special waterproof barrier or membrane is used between the landfill and the outside area. Such an arrangement is usually suitable where the material in the landfill is easily decomposed or will not leach in any way.

The topography of the landfill is extremely important. If the landfill is located on flat land, then it will obviously be necessary to dig away the soil in order to create space or volume for the waste. However, where a landfill can be located in a canyon or ravine, then it may be quite suitable for landfilling because a considerable volume of waste material can be deposited in this type of topography without the removal of large amounts of soil. If, however, this canyon or ravine has a stream or underground springs, it would not be a good selection. Any flowing water in the area must be intercepted before it reaches the landfill.

It is also important to consider the geology and hydrology of the site. Surface drainage features are an important consideration so that waters can not erode and interfere with the operation of the fill. Water which seeps into a fill operation can cause collapse of the landfill by washing away the cover, ultimately causing the waste to float to the top. Landfill operations should never be considered when the water table is close to the surface, for obvious reasons. This means that shore and marsh lands are bad locations for projected landfills, either municipal or industrial. There is also the ecological consideration of the use of wetlands, swamp, or tidal basins because of the change of wildlife species which occurs when there is alteration to such wetlands.

The climate of a landfill site is also quite important. If it is in an area of heavy rainfall, then it certainly must be capable of good drainage. If it is in an area of high wind and dry climate, then it may be very difficult to protect the overburden since the soil will be dry and will blow away before the site can be adequately covered; it also creates a serious dusting problem.

Landfills in an extremely cold climate or areas which are below freezing for a high percentage of the year may be difficult, if not impossible, to operate, since frozen soil cannot be easily moved nor compacted for cover operations.

If the rainfall in a given landfill site is normally heavy, then there must be means of removing the water from the site either through the percolation characteristics of the soil or by adequate drain-off provisions. In situations where an undesirable leachate will be created and cannot be avoided, there are really only two possibilities. One is the use of a membrane to separate the leachate from the other ground water, and the other is to locate the landfill and route the drainage from the landfill in such a manner that the collection and treatment of the leachate will be no problem.

One of the most important factors in the design of a landfill is the soil type. After gathering data on the solid waste characteristics, the first data needed by the landfill operator is soil data developed by the U. S. Soil Conservation Service which gives information on the upper 6 ft. of the earth. Once the soil type in a given area is known, the boundaries of the soil type are indicated, and a map is overlaid onto an aerial photography map with other important topographical points noted. Soil survey maps are not sufficient in themselves in locating a landfill, but they are a good indicator. The Soil Conservation Service has data available on the soil types of many but not all areas of the United States. If the data is available in the region where a landfill site is being considered, land with a potential for solid waste disposal can be located by determining the Soil Conservation Service limitations of the particular soil for landfilling.

There are several criteria for evaluating the suitability of soils, including the depth to the ground water, the ability of the soil to attenuate leachate, etc. A chart showing the soil limitations ratings for sanitary landfills is shown in Table 11-1. The type of cover soil which is used for the landfill is also important. The U. S. Department of Agriculture has published a Soil Classification Chart with boundaries showing the limitations of soil suitable for cover as suggested by George Noble in his *Sanitary Landfill Handbook;* it is shown in Figure 11-2.

The costs of landfilling can vary widely depending upon the factors which have been outlined above. In general, industrial solid wastes cost from $6 to $10 per ton to landfill at the present time.

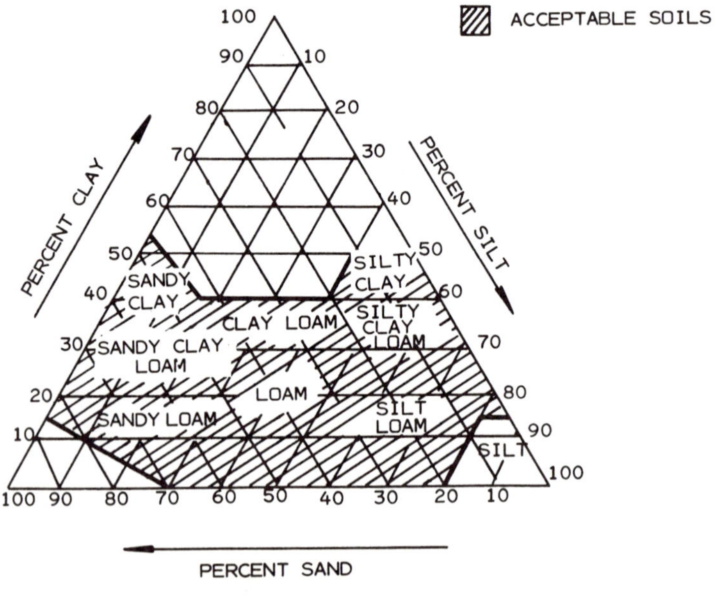

Fig. 11-2. Soil textural classification chart.

TABLE 11-1. Soil Limitation Ratings for Trench-Type Sanitary Landfills.

		Degree of Soil Limitation		
Item Affecting Use		**Slight**[b]	**Moderate**[b]	**Severe**
Depth to seasonal high water table		Not class determining if more than 72 in.		Less than 72 in.
Soil drainage class		Excessively drained, somewhat excessively drained well drained, and some[c] moderately well drained	Somewhat poorly drained and some[c] moderately well drained	Poorly drained and very poorly drained
Flooding		None	Rare	Occasional or frequent
Permeability[d]		Less than 2.0 in./hr	Less than 2.0 in./hr	More than 2.0 in./hr
Slope		0–15 pct	15–25 pct	More than 25 pct
Soil texture[e] (dominant to a depth of 60 in.)		Sandy loam, loam, silt loam, sandy clay loam	Silty clay loam[f] clay loam, sandy clay, loamy sand	Silty clay, clay, muck, peat, gravel, sand
Depth to bedrock	Hard	More than 72 in.	More than 72 in.	Less than 72 in.
	Rippable	More than 60 in.	Less than 60 in.	Less than 60 in.
Stoniness class[g]		0 and 1	2	3, 4, and 5
Rockiness class[g]		0	0	1, 2, 3, 4, and 5

Chart is based on soil depth (5–6 ft) commonly investigated in making soil surveys.
[b]If probability is high that the soil material to a depth of 10–15 ft will not alter a rating of *slight* or *moderate,* indicate this by an appropriate footnote, such as "Probably *slight* to a depth of 12 ft," or "Probably *moderate* to a depth of 12 ft."
[c]Soil drainage classes do not correlate exactly with depth to seasonal water table; the overlap of moderately well drained soils into two limitation classes allows some of the wetter moderately well drained soils (mostly in the Northeast) to be given a limitation rating of *moderate.*
[d]Reflects ability of soil to retard movement of leachate from the landfills; may not reflect a limitation in arid and semiarid areas.
[e]Reflects ease of digging and moving (workability) and trafficability in the immediate area of the trench where there may not be surfaced roads.
[f]Soils high in expansive clays may need to be given a limitation rating of *severe.*
[g]For class definitions, see the Soil Conservation Service *Soil Survey Manual.*
Source: Solid Wastes Management, 56, August 1977.

11-2.2 Methods of Sanitary Landfill

There are several methods which are used for landfilling both industrial and municipal waste; they have slight variations. In general, the refuse is delivered to the site, spread to a given thickness, compacted within a small area, and then covered with soil. The structure which is created as the result of building a layer of refuse encased in a layer of soil is called a cell. When the final elevation of the landfill has been reached, a 3- to 4-ft layer of cover soil is placed over the top of these cells and compacted to complete the layer. The three basic methods which are available for landfill operations are the trench method, the area method, and a combination of the two, often called a ramp method.

Since one objective of any landfill operation is to get the greatest possible volume from it, the refuse and the cover soil should be compacted to as great an extent as possible. In general, compacted density of the refuse must be in the range of 800 lb/cu yd or more. However, industrial wastes will vary widely from this figure, and where construction materials or wastes are compacted, much higher densities are obtained. However, much lower densities apply to rubber, plastic, and other resilient materials.

The design of the individual cell may vary in municipal applications from 6 to 8 ft thick to, in some cases, as high as 30 ft, but this will greatly depend upon the nature of the waste and the available cover soil.

The trench method is utilized when there is no water table problem (Figure 11-3) i.e., where it is possible to dig a large trench or hole in the ground into which the refuse is dumped and the soil which has been removed from the trench is used as the cover. In areas where the water table is near the surface, it

Fig. 11-3. Trench method–sanitary landfill. ***Source:*** **Pavoni, J. L., Heer, J. E., Jr., and Hagerty, D. J.,** ***Solid Waste Management,*** **p. 197, Van Nostrand Reinhold Co., New York, 1973.**

is not possible to use this method. Then the area method of landfilling can be used, Figure 11-4. Here, the refuse is spread and compacted on an existing ground surface, and cover soil is spread and compacted over the waste to achieve the completed cell design. The ramp method (Figure 11-5) is a combination of the two where the refuse is spread and compacted on a slope and then covered with soil which has been obtained by excavating in front of the working space. Of these three, the area method is usually the most expensive because it requires the hauling of soil to the landfill site; whereas, the other two methods utilize soil immediately available at the site.

Often the problem in the operation of a suitable landfill is the limitation of the available equipment. Scrappers, loaders, dozers, drag lines, back hoes, trucks, and graders are used to open the landfill, to place the waste material in the landfill, and to cover it with suitable soil. The number and the type of vehicles needed in the landfill depends primarily on the amount of waste handled and the traveling distance for cover materials. Figure 11-6 shows the number of vehicles needed at landfills for various daily waste quantities without any need to transport cover materials from a long distance.

The excavation and transportation of cover materials will increase the number of pieces of equipment needed. In addition, new equipment for handling the waste and cover material and support equipment is also required for maintenance and repair, dust control, fire control, and litter. Most industrial landfills have plentiful cover material available nearby and can be handled by the usual loader,

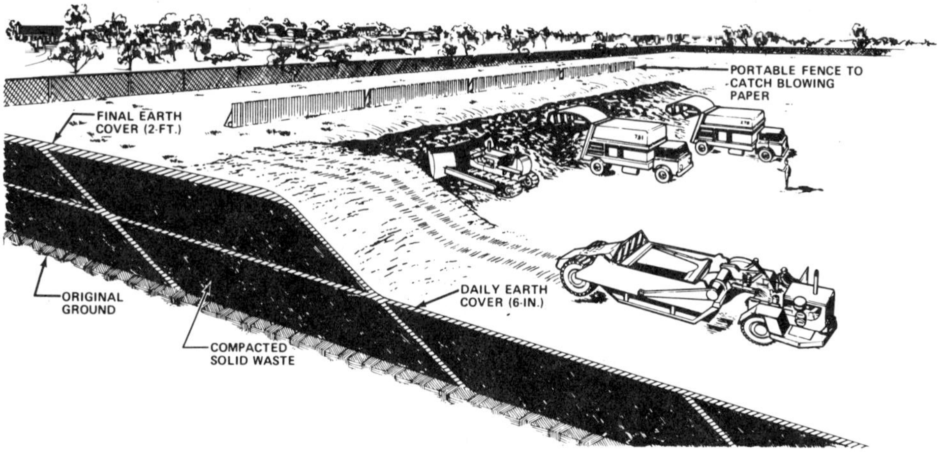

Fig. 11-4. Area method–sanitary landfill. *Source:* Pavoni, J. L., Heer, J. E., Jr., and Hagerty, D. J., *Solid Waste Management,* Van Nostrand Reinhold Co., New York, p. 198, 1973.

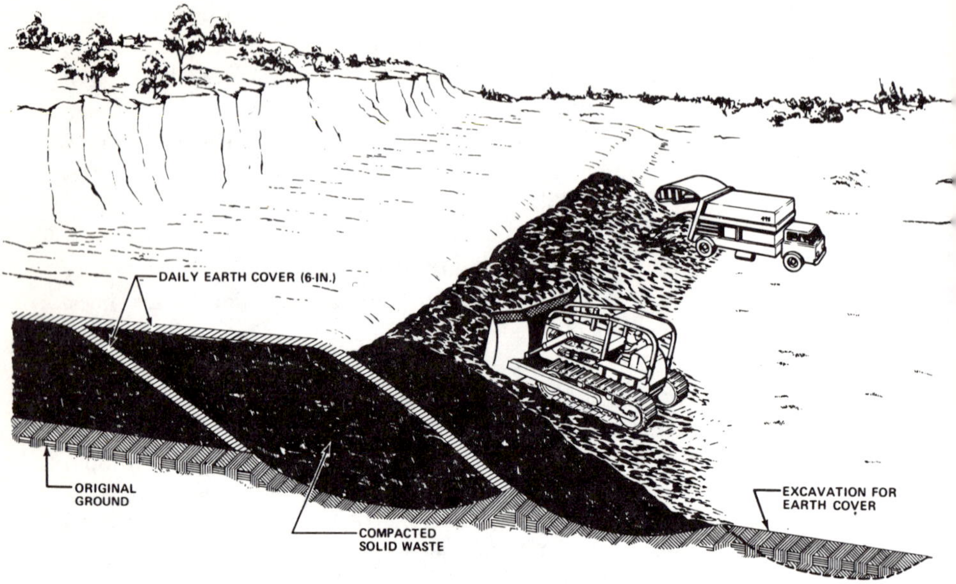

Fig. 11-5. Ramp method–sanitary method. ***Source:*** **Pavoni, J. L., Heer, J. E., Jr., and Hagerty,** ***Solid Waste Management,*** **Van Nostrand Reinhold Co., New York, p. 199, 1973.**

dozer, and compactor. But, if the cover material is quite distant, that is more than 600 ft away, additional pieces of equipment may be required. Rubber tired scrapers are very efficient for excavating and transporting soil for cover when the cover soil is greater than about 1000 ft from the working trace. They are fast and maneuverable and can move a lot of earth in a short period of time, especially when assisted by a bulldozer. Drag lines are also effective earth movers but only applicable when the area to be reached is relatively close by and, therefore, not suitable for transporting cover material. Back hoes can be used for excavation, but they are limited in scope. Dump trucks are used, of course, for bringing the material to the landfill and depositing it in the cell.

The landfill design also must consider the problem of gas formation. Landfills are initially an aerobic decomposition device. Air is compacted into the landfill when it is initially created and when the water content of the wastes in the fill is below 40%, such aerobic decomposition occurs slowly; but, if the water content increases to about 80%, rapid aerobic decomposition occurs, and both water and carbon dioxide will be produced by the microorganisms in the landfill. If, however, the landfill exists for a number of years and is sealed or covered by considerable rainfall, ultimately it becomes anaerobic, which means that the

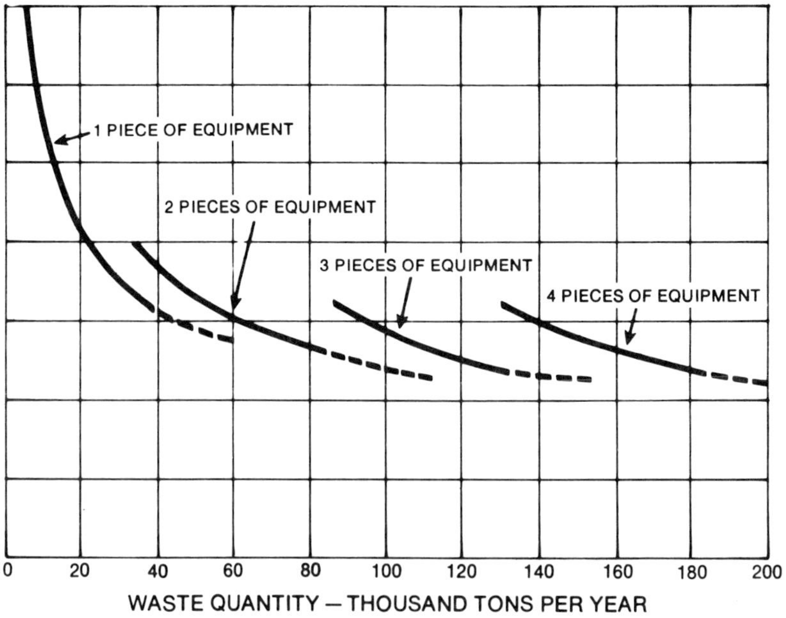

Fig. 11-6. Equipment requirements–sanitary landfill.

microorganisms now decomposing the waste work without oxygen and, as such, can create methane gas. Methane, which is the primary constituent of natural gas, presents an explosion hazard if it is present in sufficient concentrations. Its explosive range is from 5 to 15% by volume in air. In order to eliminate the fire and explosion hazard from the landfill, it is necessary to provide some means of removing such gases from the landfill. This is most commonly done with a gravel-type vent. This is a gravel-filled trench extending the full depth of the landfill in a number of different locations, which allows for the venting of methane gas caused by anaerobic decomposition. The trench should be wide anough so that it is not easily clogged by infiltrating fine materials from the soil and it must also drain surface water either naturally or artificially so that the water does not form a seal against gas release, see Figure 11-7.

Many older landfills did not provide for this gas release, and it was found that methane gas had formed and was coming up through voids in the landfill. In such cases, vent pipes were installed in various refuse cells to allow for the escape of these gases. In the design of a new landfill, an impermeable layer should be provided over the fill, and the collector pipes should be inserted through the layer.

Wherever impermeable barriers, such as clay or other artificial liners are used for landfills, they will prevent migration of gases through the landfill and

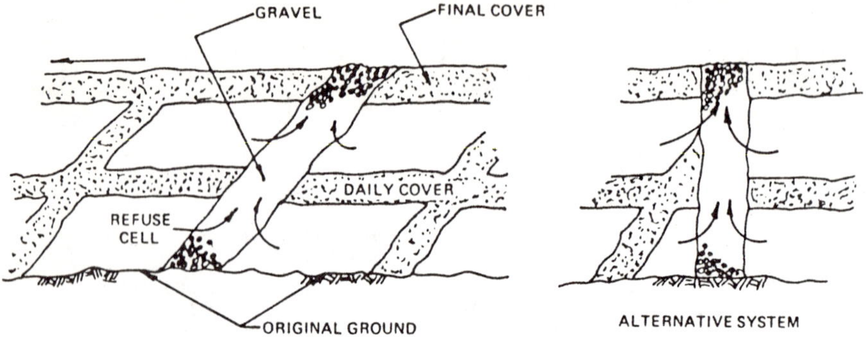

a

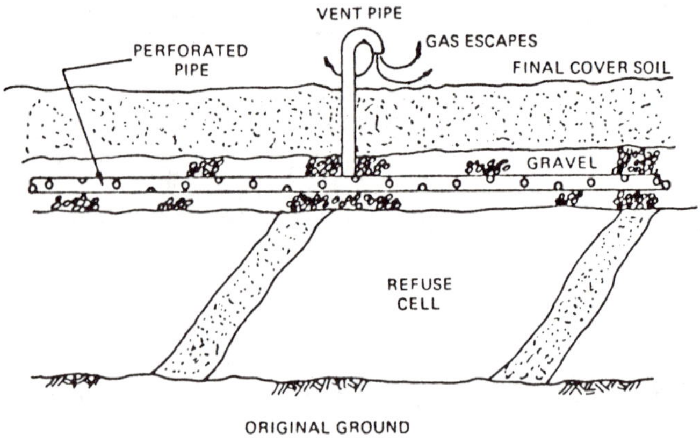

b

Fig. 11-7. Gravel and pipe vents for gas removal. ***Source:*** **Pavoni, J. L., Heer, J. E., Jr., and Hagerty, D. J.,** ***Solid Waste Management,*** **Van Nostrand Reinhold Co., New York, p. 205, 1973.**

then pipes or trenches can be used to bring the methane gas to the surface. Combustion at the surface of the landfill through a flare or other type of incinerator has been used but, in most cases, is not required unless large quantities are formed.

In general, a sanitary landfill requires lots of planning. It requires expensive, special equipment and sound consideration for environmental protection. While it has been used in the past and will be widely used in the future, it has shortcomings to the industrial user in that he must provide more and more land to accept his waste.

11-2.3 Gas Formation in a Landfill

The biological activity in a landfill follows a particular pattern. Solid waste is initially decomposed aerobically and, as the oxygen supply is excluded and exhausted, anaerobic microorganisms predominate producing odorless and colorless methane gas. The temperature within the landfill increases to the high mesophillic-low thermophillic range (60 to 150°F) due to microbial activity, the products of this aerobic decomposition are carbon dioxide, water, nitrates, and nitrites. Anaerobic decomposition will characteristically produce methane, carbon dioxide, water, organic acids, nitrogen, ammonia, sulfides of iron and manganese, and hydrogen.

As water moves through the solid waste, it will produce a leachate containing dissolved and suspended waste constituents along with microbial products. This may either be a ground water or an infiltrating surface water source. The leachate will leave the fill at the ground surface as a spring and will percolate through the soil and rocks underlying and surrounding the waste.

The gases as generated in a landfill are a natural product of the solid waste decomposition. The quantity and composition of the gas depends on the type of solid waste which is decomposing. If it contains large amounts of easily degradable organic material, it will produce more gas than a waste which is largely ash and construction debris which is normally found in industrial landfills. The rate of the gas production depends solely upon the microbial decomposition of the solid waste and, as decomposition ceases, gas production also ceases. If, however, the decomposition is carried to completion, each pound of refuse containing 25% inert material is capable of producing up to 6.6 cu ft of gas. Methane and carbon dioxide are the major constituents, but hydrogen sulfide, ammonia, hydrogen, and nitrogen are also generated; these must be removed from the landfill in a safe manner and either captured for some other use or flared.

11-2.4 Chemical Landfill

It is difficult to get industry to agree to what a chemical landfill should be and what materials should go into it. In this chapter, landfill technology has been divided into two areas: (1) industrial landfill, containing nontoxic, nonchemical type wastes, i.e., materials which are similar to municipal wastes but not identical; (2) those materials which we consider to be chemical landfill wastes.

Today's industry will certainly burn those wastes which appear to offer significant heating value in some type of incinerator and use a waste heat boiler to reclaim steam, hot water, or some other form of heat for their processes. There are, however, in most chemical plants, numerous materials which might be combustible but do not lend themselves to clean incineration. Such materials are usually solid in nature or, if not solid, they are so viscous or elementally composed that incineration would be very difficult if not impossible. Many of these materials take the form of sludges or semisolids, i.e., combinations of solid and liquid materials which are not pumpable and atomizable. Many of these materials are not combustible at all, containing small amounts of organic materials mixed with inorganic catalysts, filter media such as diatomaceous earth, or chemically contaminated materials such as slag, limestone, etc.

Many of the chemical waste materials which cannot be otherwise destroyed react with water; ignite on contact with air (pyrophoric); produce, as indicated before in this chapter, a number of undesirable leachates; give off undesirable odors; or react with other wastes which might also go into the landfill.

Often, these very heavy and nompumpable wastes are difficult to transport and are packaged in a steel or fiber drum, because it is the most convenient container at the site where the waste is created. For example, heavy still bottoms coming from a distillation column may not be pumpable and are barely fluid at the operating temperature of the column. When drained from the column into a drum or into a fiber pak, they are then available for removal. To design a suitable incinerator for such large batches of material in bulk would be a very expensive proposition, plus the fact that the air pollution potential for wastes such as this in bulk quantities is very high, and pollution abatement is very difficult from a technical point of view.

Some chemical plants have passed on such drums and fiber paks to a contract disposal organization willing to take the material, perhaps hoping that it would never be seen again. Unfortunately, the out-of-sight, out-of-mind theorem did not always work. In many cases, accidents occurred in the handling of the waste because the material was inadequately landfilled by inexperienced contractors, and the company later was identified as the supplier of the waste material. It gradually became evident that chemical companies creating chemical wastes to be disposed of by chemical landfill, usually must provide the facilities themselves, monitor these facilities with their own people, and operate these facilities with their own people. Today many major chemical producers recognize that they must provide their own chemical landfills, or at least insure in some way that they are utilizing valid and technically sound commercial organizations to take their waste.

The advent of the truly safe chemical landfill involves a commitment by management that such materials are impounded so that wastes in a variety of physical states can be satisfactorily landfilled or buried in an area without the least possibility of contaminants reaching the ground water and without the significant possibility of fire or explosion. This must be done in such a way that these wastes can either remain for an unlimited period without adverse an effect upon the environment or that they will biologically decompose into products which will no longer be a threat to the environment. Obviously, those materials which are landfilled in containers such as fiber paks or drums are in the first category, but they are landfilled in these containers because virtually no other method of treatment is worth consideration. Any materials which will eventually biologically decompose to products which will no longer contaminate the environment should be landfilled in bulk.

A case in point is a major chemical company which had quantities of arsenic sulfide for disposal. Arsenic sulfide is a crystalline material which has little commercial use unless it can be returned to the parent metal, and this is very costly. Disposal of large quantities of sulfide is virtually impossible in bulk in a landfill without a resultant objectionable leachate, and the permissible quantities of arsenic in water are extremely low, on the order of several parts per million. Therefore, the only practical disposal of such waste, since it is not incinerable, is to landfill, but the landfill must in effect be dedicated to perpetuity since the material will not decompose and will not change in physical structure.

Drums filled with arsenic sulfide were, therefore, landfilled in a concrete type vault which would remain there virtually forever with a chemical barrier that would prevent any leachate from the vault and with a surface designation of what was buried there. Obviously, such materials cannot be buried in an area where the land can be reclaimed or where eventually buildings or some other structure would be built; yet, there is no other suitable way of disposal other than reclamation and reuse of the material, or ocean disposal. Perhaps this is an extreme case but it illustrates the difficult problems that one faces with a chemical landfill.

Considerations in the design of a chemical landfill are as follows:

1. It should have a large volume, not only for today's wastes but for at least several years' storage.
2. It should be close to the area where the wastes are being created to reduce the transportation, supervision, security, and operating problems.
3. The landfill should be designed so that it can control the leachate and is in an area where excessive rain or flooding will not adversely affect its operation.
4. A chemical landfill should be designed for the protection of the people who work in the area as well as those who work in the plant.
5. The landfill should be divided in such a way that it is safe from fires and especially property loss through fire damage and also so that the odor problem is minimal.
6. Waste which might be interreactive must be kept in different areas of the landfill to prevent a hazard resulting through leakage from one area to another.
7. The possibility of gas generation from biological oxidation must be considered by blending the waste in such a way with the earth that gas generation would be minimized and avenues permitting the gas to escape uniformly over the entire working face of the fill must be provided.

The chemical landfill wastes usually can be best handled through a trench type landfill where the soil is removed, mixed with the chemical waste and then placed back into the void. Because it is a chemical waste, there is much more concern about the leachate. In most older chemical waste landfills, little consideration had been given to the geological problem, while admittedly a good deal of thought had been put into the hydrological problems. There often was no permanent and absolute seal between chemical wastes and its ultimate receptor, the groundwater. There are three possibilities for sealing the landfill. The first and the least expensive is possible in certain parts of the United States because of the soil characteristics but not in others. This is to use an impermeable clay lining for the landfill. Areas such as south central Louisiana and central South Carolina have excellent clay soil which is essentially impermeable and which will not allow leachates from the chemical landfill to reach a relatively high water table. The second possibility is to use an asphaltic membrane. This has been tried by several major companies in the Northeast and has proven to be quite successful. This asphaltic membrane, similar to a roadway barrier, is used to line the trench prior to the introduction of the chemical waste material. The third, and perhaps the most expensive approach, is to use a plastic liner such as might be used for lagooning liquid wastes. This is usually a 30-mil thick sheet of PVC material which is used to enclose the entire landfill. Seams between sheets are sealed at the time of installation. Such liners provide for the complete containment of any leachate which then can be piped to a single retaining pond for further treatment before discharge to the environment.

The chemical landfill must be designed so that leachate or drain water flows into some type of duct or pipe which carries this water to an impoundment basin for further treatment. Also, the chemical landfill should be designed with surface water drains to carry off excess water before it reaches the material in the chemical landfill. Table 11-2 shows contamination possibilities for various chemical wastes.

Union Carbide Corporation employed a wet system for the removal of leachate in a chemical landfill as early as 1969. Using a diked area, they designed one dike that would leak in such a way that the water in the landfill could pass out through the leaky dike into a leachate container pond which was bounded by an impervious dike. The collected leachate was then transported to the plant's waste water treatment unit for stabilization, Figures 11-8, 11-9, and 7-3.

Chemical wastes should normally be blended with an average of one volume of earth for each volume of waste, except in the case where drums or similar solid waste containers are used. Some drums are precrushed to open them so that the materials can biodegrade but, if materials cannot biodegrade, there is little purpose in opening the drum.

It would be easy to say that all chemical waste could go to a chemical landfill, but obviously this is not a reasonable proposition. Each chemical waste

TABLE 11-2. Relative Distances of Contamination for Various Industrial Wastes

Pollutant	Observed Distance of Travel	Time of Travel
Tar residues	197 ft	
Picric acid	several miles	
Picric acid	3 miles	4–6 yr
Mn, Fe, hardness	2000 ft	
Miscellaneous chemicals	3–5 miles	
Chromate	1000 ft	3 yr
Phenol	1800 ft	
Phenol	150 ft	
Chlorides	200 ft	24 hr
Gasoline	2 miles	
Chemical weedkiller	20 miles	6 mo
Radioactivity	–	5 days

Source: General Electric, "Solid Waste Management Technology Assessment."

must be thoroughly analyzed and a decision on the method of disposal of the waste made by responsible technical people before admitting it to the landfill, incineration, or any other type of disposal. The possibility of toxic leachate should also be considered, and a suitable treatment method for this leachate should be available before the material is placed into the landfill. A classification of these wastes, as established by Union Carbide Corporation, bases a special hazard rating in the form of each known chemical component in the areas of health, flammability, stability, or reactivity; a number from one to four is entered in the label under each heading to indicate the relative hazard.

In all general landfills as well as chemical landfills, there should be monitoring. This monitoring can be done by the use of test wells which should be located outside the landfill and at varying depths around the area. The number of sampling points is controlled by the expected variability for each parameter and the degree of accuracy desired in the monitoring. The sample point distribu-

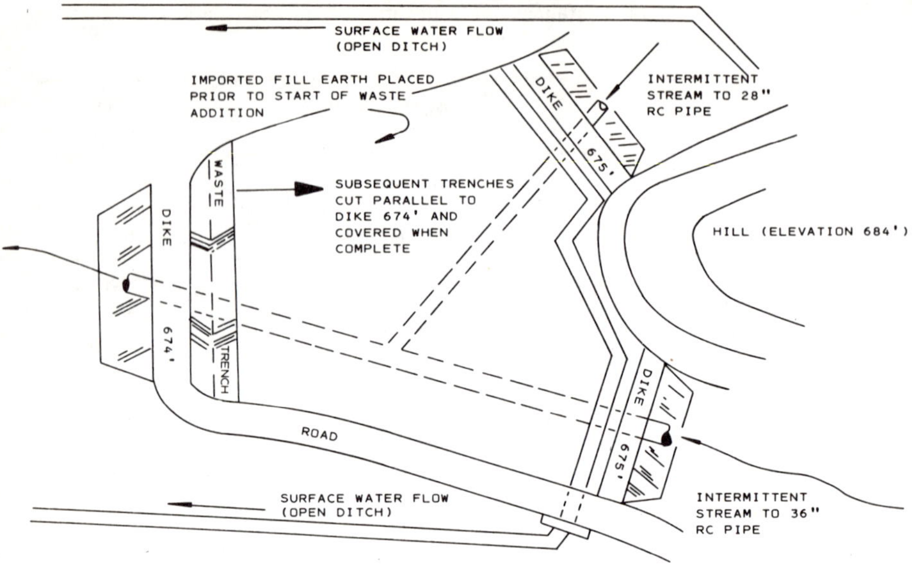

Fig. 11-8. Chemical landfill–plan view. ***Source:*** **Slover, E., Union Carbide Corp., November 5, 1975.**

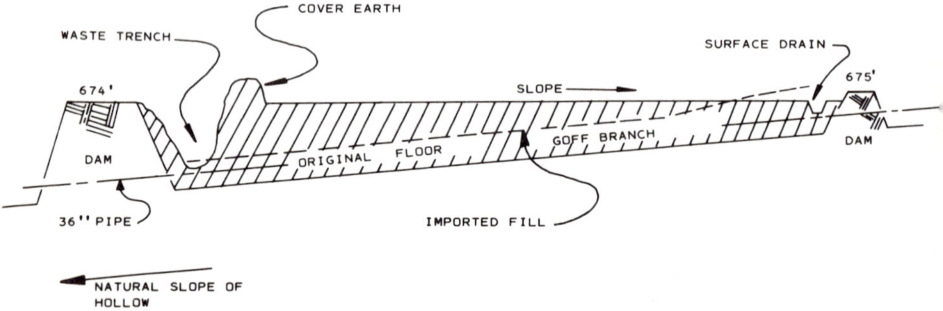

Fig. 11-9. Chemical landfill–impoundment design. ***Source:*** **Slover, E., Union Carbide Corp., October 13, 1975.**

tion is set by geologic, hydrologic, and chemical complexities likely to be encountered in the area. In unconfined chemical landfills, the test wells should be located in the direction of groundwater flow and at depths between the groundwater conditions and the subsurface water location. In confined chemical landfills, these wells should be located outside of the membrane or impervious enclosure in such a way that they would immediately detect leakage from the membrane enclosed area.

11-2.5 Closing of the Landfill

One of the important aspects of landfill operation is the fact that it is not permanent; it will some day be full and will have to be closed. The importance of properly closing a landfill cannot be overemphasized, whether it is industrial or municipal. It should be done aesthetically and in a way to control erosion and run-off and to minimize environmental problems not only for the immediate future but also to control potential problems. If it is a chemical landfill, undoubtedly leachate control and water treatment will have to continue for many years, even though the landfill is not in operation; so the dikes and run-off provisions which were made initially must be kept intact and, in some cases, will have to be excavated and repaired when evidence indicates that leachate is getting outside of determined barriers. Test wells must continue to be operated and monitored on a regular basis. The vermin control measures, if vermin are present, will have to be carried out for some years after the closing of the landfill.

The landfill should be properly landscaped to make it aesthetically pleasing to the surrounding area. Even though it may not be possible to build on this area, it should be properly screened by landscaping. The landscaping should be selected to control dust and erosion; therefore, attention must be given to the soil type, the plant hardiness, and the effectiveness of plant growth for satisfying the goals and objectives. The soil type is important to plant growth because each soil has different capabilities to support various types of plants. Therefore, soil should be analyzed to determine its characteristics to affect plant growth and to provide the proper cover. The material which is placed over the top of the landfill should be a relatively impermeable clay designed to cap the site and reduce infiltration of water into the landfill. Development of the surface water run-off drainage plan after the capping is also important. Top soil should be applied over the clay in order to encourage the growth of the landscaping material.

A chemical landfill should be adequately marked. In effect, especially if drums of material have been placed in the landfill, it is a landfill for perpetuity and must be maintained by the company that established it with adequate safeguards so that excavation cannot take place in a foreseeable number of years.

The records which were established during the landfill's life should be maintained so that they can always be referred to in the future. It is imperative that future generations know specifically what is in a chemical or industrial landfill so that consideration of future use will be proper. If an artificial liner has been used, it is important that this liner never be ruptured by drilling through the landfill and through the liner since this will immediately affect the surrounding groundwater. If there is a possibility of forming explosive gases in the landfill after it is closed, then the same types of gas removal devices which were planned for the active landfill should be kept in working order for the future.

11-2.6 Chemical Fixation

Many materials which are currently being disposed of by neutralization, incineration, or chemical landfill lend themselves to a pretreatment known as chemical fixation. Several companies have established service organizations based on their capabilities of mixing various types of chemical wastes with inert materials which tend to fix the solubility of the wastes in a dry aggregate or solid material which can be used for landfill in a way that leachate is minimized or totally eliminated. Chemical fixation primarily applies to acid-type wastes which may be found in steel mill pickling operations, metal plating, limestone wet scrubber sludges, and oil refinery wastes. They contain heavy metals which have very limited permissible levels in groundwater and which, unless otherwise handled, might create severe leachate and treatment problems.

In general, chemical fixation utilizes lime and fly ash in various percentage admixtures which react with the chemical waste producing a solid material which is essentially leachate free. Such processes have been highly successful in certain types of acid-heavy metal wastes and not so successful in others.

One of the processes is the Poz-O-Tec process as developed by IU Conversion Systems, Inc. of Philadelphia. The basic chemistry of the Poz-O-Tec process is a combination of several simultaneous reactions. First among these is a rapid reaction taking from 48 to 72 hr which involves certain alkaline earth compounds which may be present in a scrubber sludge or added as needed. These reactions produce a fiberous gypsum that reduces permeability and increases structural properties. In addition, there are cementitious reactions between alumina/silica in the fly ash, the lime compounds and the sulfur oxide salts. The water content of this mixture is extremely important.

The stabilization of sulfur oxide sludge and fly ash involves large quantities of calcium sulfite or sulfate which result when lime is used in the treatment of sulfur oxide, acids, or waste gases. The sulfites and sulfates are beneficial to the cementitious reactions and allow them to proceed rapidly. Careful study of the sludge and fly ash, both of which vary greatly in chemical and physical properties, is essential to produce desired results.

The engineering properties of the final product are such that it is possible to place the stabilized fill material in a landfill without the addition of an aggregate. This does not mean that an aggregate cannot be used because such materials, bottom ash from coal burning power plants, by-products slags, culm, or mine refuse materials, as well as conventional aggregates, have been used. Work has also been carried out with these mixtures to encapsulate the potentially toxic materials, such as crushed battery cases, containing high concentration of heavy metals. The process locks up soluble ions of many of the deleterious trace elements and substantially reduces their solubility.

The resulting material from a stabilization process like Poz-O-Tec has a very low permeability on the order of 10^{-6} cm/sec. This low permeability provides

TABLE 11-3. Engineering Properties of Poz-O-Tec Structural Flyash-Sludge Compositions

Age of Stabilized Mixture–Cured at 73°F (days)	Compressive Strength (psi)	Subgrade Modulus (lb/cu in.)	California Bearing Ratio (%)	Permeability (cm/sec × 10^{-6})
3	40	620	22	1.9
7	80.5	–	63	–
14	140	2300	97	0.8

excellent protection against leaching of undesirable components. In addition, the permeability improves with age so if the mass of sludge which has been fixed in this manner is essentially impermeable, leaching can only occur from the surface, and the degree of leaching depends upon the solubility of the mass.

Such stabilized material has other specific benefits due to their engineering properties. A product can be made to withstand heavy loads, such as foundations of bridges or a road base. It has a bearing capability of heavy sand or gravel and its light weight makes it extremely attractive for embankments, especially when constructed over poor ground such as alluvial clay or silt where excessive weight could produce settlement beyond allowable limits. Because of its high load-bearing properties, light weight and self-hardening characteristics, the product makes an ideal construction raft for building development over poor ground. It reduces settlement to a minimum and the hardened, stabilized fly ash forms a monolithic structure which can bridge weak spots in the subgrade and thus reduce differential settlement.

Some of its engineering properties are shown in Table 11-3.

Other potential uses are land reclamation and embankment construction, dams and impermeable liners, pavement construction, roadway, airfield and parking areas, aggregate substitute in Portland cement, concrete and bituminous concrete. It has been used with other waste materials and sulfur dioxide sludges. It has been applied to industrial and municipal wastewater sludges. When applied to the latter, the resulting product has been used for road embankments. In industrial sludges that contain high quantities of metal ions, the process is somewhat more experimental and leachate tests should be made in order to determine whether subsequent problems will develop.

11-3 BIOLOGICAL METHODS

While biological methods of waste degradation are prevalent in the industrial sector for liquid wastes, there has been very little done towards developing biological systems to destroy solid wastes. The landfill itself, of course, is a biological system, sometimes being aerobic and other times being anaerobic, but it has been considered here as a separate disposal method and is not included under biological systems.

The biological systems available for solid waste management tools are composting and anaerobic digestion. Neither of these has been used widely by industry, but it is worthwhile reporting some of the techniques since they do offer possibilities to the industrial sector and may be of interest to the future.

In order to biodegrade any material, it must be organic in nature. And, for that reason, materials which will not degrade in a landfill nor burn in an incinerator are not suitable for these processes and must be removed either at the beginning or at the end of the process. Since industry produces large volumes of waste of mixed materials, certain industries would have more interest in biological systems than others. Specifically, these would be industries involved in forest products, agriculture, and food products.

Biological systems, in order to be economical, have to be quite large. Therefore, for industry to consider a biological method of waste degradation, they must have a large biodegradable waste source. Only under these circumstances does a biological system begin to make sense to an industry.

11-3.1 Composting

Composting as a process has been around for a long time. It, of course, was much better understood by the agricultural industry since it has been used by this industry for literally hundred of years, both as a scientific method for creating a material that could be used as fertilizer and soil conditioner; and possibly, in many instances, composting just happened to provide these benefits without any scientific knowledge of the benefits on the part of the farmer.

Anyone who has a backyard compost pile understands the value of compost. In most cases, the would-be gardener or farmer throws organic refuse onto a pile throughout the summer and fall and allows it to decay during the winter months so that in the springtime, the center of the pile delivers a rich loamy material resembling soil. This can be fed back into the garden adding needed nutrients and, at the same time, conditioning the soil in the garden for future planting. This home-type compost operation is essentially the same process that industrial composting operations utilize except that it is on a much smaller and slower scale.

Composting of industrial and municipal wastes involves an accelerated operation which will produce a similar compost product in a much shorter period of time and in a much greater quantity. Since wastewater sludge is often used as an activator for the compost, it is usually a much more odorous proposition than the backyard compost pile. In this respect, it has found some difficulties in acceptance both in the municipal and in the industrial community.

Since materials that are not organic in nature do not participate in the composting process, it is desirable to remove them prior to composting. Removal procedures are much as described in another chapter. Metal is removed by gravity or magnetic means and so are glass and other nonbiodegradable materials. The remaining meterial then can be composted directly. Composting works on material of a uniform size much better than it does on random pieces of organic waste; therefore, shredding is usually a precursor to the composting operation. This does not mean that fine shredding is necessary, but materials of irregular cross section, like sugarcane, should be broken up before they are placed into the composting process.

Moisture is a very important component of the composting process and should be between 50 and 60%. Also, the carbon to nitrogen ratio should be relatively low, approximately 35 to 1. Rubbish and garbage may have a carbon to nitrogen ratio as high as 80 to 1 or more; composting of this type of material without the addition of some material rich in nitrogen is bound to fail. It is for this reason that sewage sludge or raw sewage is often added to the composting operation in order to make it proceed at a rapid pace.

Composting is actually a form of fermentation. The temperatures in the compost should be at 140 to 170°F. These temperatures can be obtained in two ways: heat generated by bacterial activity; adding heat.

It is also vitally important that the material be turned or agitated regularly to provide for aerobic degradation. If the material is not turned, anaerobic bacteria will increase and eventually putrifaction will occur. This results in odor problems which have been the downfall of many composting operations.

With the proper materials in the composting process, temperatures reach 160 to 170°F in less than a week. In another three weeks, oxygen becomes limited turning the material into an anaerobic system unless it is constantly agitated to allow more air to reach the organic material. As the material is turned, temperatures begin to come down towards ambient but they will usually stay at 130 to 140°F which represents moderate thermophilic microbial development. As the oxygen is gradually used up again, the temperature will continue to drop to about 110°F, providing the desired mesophilic climate for the final stage at the end of which the temperature will fall to ambient. This temperature drop signifies the end of a proper composting process and a well-decomposed, stable end product.

In a well-designed composting system, the highest temperatures are produced at the initial stage with the temperatures gradually decreasing towards the end. This high initial temperature is also important to assure that all pathogens in the raw materials are destroyed. The minimum temperature for such purposes has been set at 140°F.

In a continuous composting process, such as the Dano process, there is little problem in getting atmospheric air to the compost material. In other types of processes, a blower is normally required. Efficient composting requires good forced aeration, and this means that different quantities and qualities of warm, moist air are needed throughout the process. Direct or indirect sunlight is toxic and should be excluded.

Because composting is essentially a fermentation operation, it is important that the carbon dioxide concentration of the air in the composting process is high. It should vary from stage to stage but should be about 1 to 20%, which is considerably higher than atmospheric air. The highest concentration is normally in the early stages where it is needed to form carbonic acid which dissolves non water soluble microbial nutrients in the raw materials. This and the high temperature in the initial stage are definitely prerequisites for rapid composting.

The history of composting is perhaps more interesting than its present, especially in the United States. Today substantial quantities of compost are being produced and utilized in Europe. The history of mechanized composting operation goes back to the early 1920's when Sir Albert Howard developed his Indore process which was used in India. Later Beccari patented a similar process in Italy. Both processes, however, were of questionable value for two reasons: first, they both employed anaerobic degradation which made them very slow and, secondly, there was no great demand for the final product after it was produced.

In the 1930's, however, there were several attempts to mechanize and speed up the process by using aerobic decomposition. One of the best known of these, the Dano process, appeared in Denmark; a similar process, the Vuilafvoer Maatschappij process was developed in Holland. Subsequently, Eweson received a patent for a similar process in the United States, and later in 1949, the Frazer process was similarly developed in the United States. In the early 1950's, a number of people began to see composting as a possible answer to municipal solid waste problems. However, no one really understood how aerobic composting worked, and a number of municipalities were sold on the idea without sufficient research into the process to make it successful. There seemed, at the time, to be a limited knowledge of the biology which was required to make the units operate properly. Since the mechanical details had not been clearly developed, many of the early plants were failures for one reason or another.

Research was needed in order to develop the entire basis for the composting process so that rates of aerobic decomposition could be determined and, therefore, the size of the equipment required to dispose of a given waste rate, and finally to establish the economics of the process versus the amount of material to be handled. Also, very little was known of the problems of controlling disease-carrying pests, such as flies, rats, and mice and, for that matter, what type of treatment was necessary with wastewater sludge to really dispose of pathogenic materials. Also, not much was known about the value of the ultimate product in terms of its use as a fertilizer or soil conditioner. For this reason, the U.S. Public Health Service and others started research programs which were aimed at solving some of these fundamental problems. The accomplishments of these programs determined the parameters under which aerobic composting could be carried out.

Windrow type composting in the open was shown to be a simple procedure. It was determined that sepcial inoculums, usually considered to be secret materials by compost promoters, were unnecessary and, in many cases, were worthless, and the recycling of the finished composting of the feed also was of little value.

It was further determined that a simple, consistent turning of the material was an effective way of maintaining aeration, and that forced aeration was not at all needed. It was also determined that composting could be carried out by both aerobic and anaerobic means and that the aerobic decomposition resumed once the material was turned.

Perhaps most importantly, the process was found to be applicable to a great variety of agricultural, industrial, and municipal wastes, and that proper control and operation of the composting process not only produced a product which was acceptable as a low grade fertilizer and soil conditioner but also that such aspects as fly control and disease destruction occurred in the process.

It must be remembered that the compost created is not a high grade fertilizer such as can be created by the combination of various chemicals, such as, potash, nitrogen, ammonia, and urea. The finished compost has a low percentage of releasable nitrogen and phosphorus. Its main advantage is in its moisture retaining and soil conditioning properties.

Perhaps the most important single thing which was determined during this period was that the finished product was not going to be an easily marketable material. Many cities went into the business of compost production in the 1950's with the idea that they could sell the compost to gardeners for use in their soil. However, the material did not really offer a marketable product and, for this reason, many of the so-called self-sustaining composting operations failed miserably.

It should be stated, at this point, that the technological aspects of composting are rather simple and straightforward. A full-scale mechanized plant is technologically feasible. It naturally involves mechanical problems, but these can be surmounted. The biological aspects of a composting system have been and can be achieved in large-scale operation, and, while they are complex, the answers to most of these problems are known. The aesthetic problems of a compost plant cannot be solved any more quickly than the aesthetics of a wastewater treatment plant or a public incinerator. Composting operations are not a traditional backyard type of operation. Once they reach industrial or municipal proportions, there will be some odor and there certainly will be the need for large storage facilities for both the incoming wastes and the final product.

The finished compost product has several possible uses. One is its use as a fertilizer and soil conditioner. The second is its use as a landfill material. It can greatly reduce the volume of solids which would normally be placed in a landfill if placed in a landfill before composting. This reduction in volume by composting, however, is usually not greater than 2 to 1, but the material produced by a composting operation is much more uniform than the normal landfill feed and, for such reasons, will probably give landfill savings of up to 3 to 1 or greater.

The chemical makeup of municipal compost (it would be impossible to accurately state the chemical makeup of industrial compost even if there were one or two composting plants on which to base the information) generally is less than 1% nitrogen and from 0.05 to 0.2% phosphate plus 0.2 to 0.3% potassium. Consequently, the compost itself is not much of a fertilizer, and it may not be sold in the United States as a fertilizer. Its major advantage, however, is not in its fertilizer capabilities but in its ability to improve the physical characteristics of the soil, to hold moisture, and to support a biological system from which nitrogen can be released over a period of time. The failure of composting operations to sell or even give away its product here in the United States is primarily due to the fact that it has very low nutrient value and that despite its soil conditioning value, it is really not a product suitable for U.S. agriculture.

Many might ask, then, why this process has been so successful in Europe and in the Far East countries such as Japan. The answer lies in the type of farming that is carried out in these countries in comparison with the farming in the United States. Here most farms that produce food for market are operated on a large scale. Commercial fertilizers, such as ammonia and urea, are much more economical to purchase and much more economical to distribute over the land surface than a heavy, bulky material such as compost. In other countries, the compost is being sold to the specialty farmer who operates a very small farm with a great deal of manual labor, and the material thus sold is suited to the market.

The problem of composting, especially in the United States, is more of a marketing and economics problem than it is a technological feasibility problem, and, from that standpoint, industrial wastes can be processed into a compost if they are readily biodegradable. On the other hand, the cost of doing so would certainly not be worth the effort for most industries unless they had a limited landfill which they were trying to extend by several years or unless their landfill was to be reclaimed at a future date as a building site and a uniform material was more desirable than a random material for the fill.

Composting processes can be divided into three basic types: (1) the windrow which is nothing more than a pile or windrow of waste material that can be in the open or under cover of a shed; (2) a completely mechanized industrial plant; and (3) a mixing process with soil and finished compost.

The first method requires a considerable land area plus a limited amount of physical processing; therefore, it is the least expensive. However, it is also very slow and, consequently, the product will not be available for some period of time. The second process requires a very complex mechanical system but, nevertheless, offers more rapid production than the first. The third also requires appreciable land area and is essentially a semilandfill operation.

From an industrial point of view, the only system which would seem to be practical is the mechanical system, Figure 11-10. This involves a separation

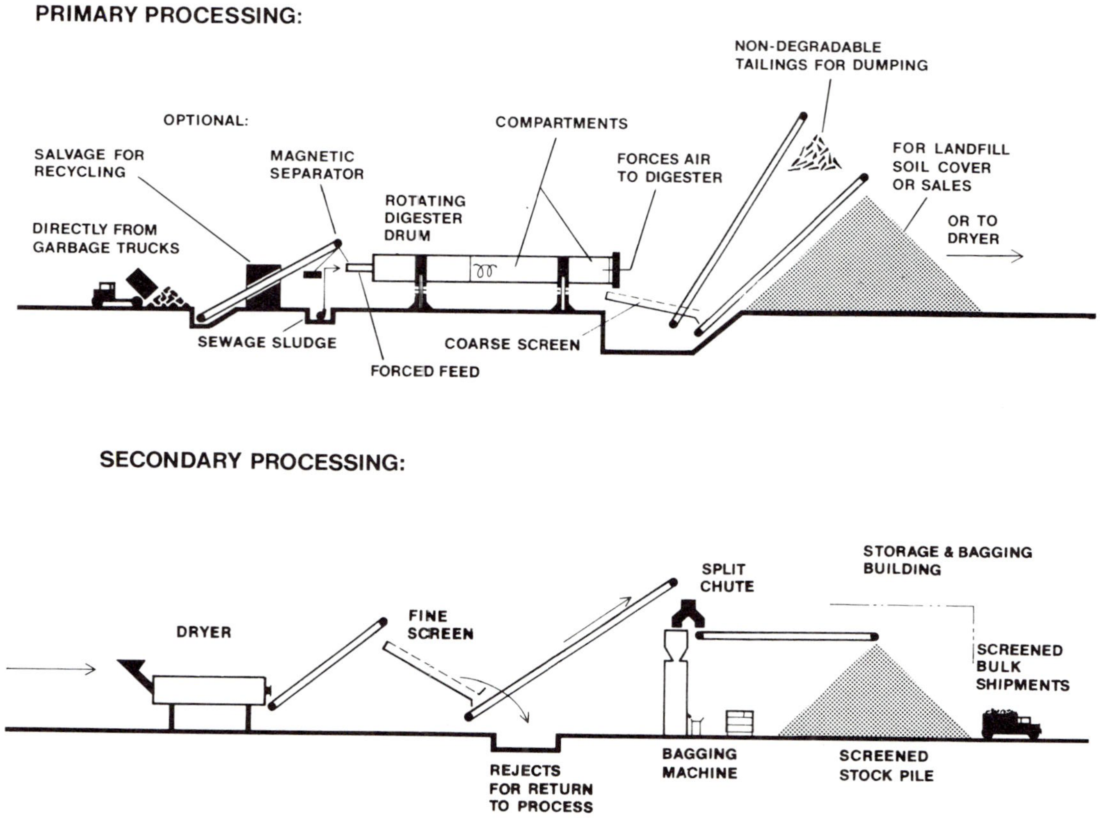

Fig. 11-10. Mechanical Composting System. *Source: Pollution Engineering.*

function where the crude refuse is dumped onto a belt conveyor where metallic particles and glass particles are removed by means of magnetic and gravity separation. The next step is the biodegradation section of the system, which can occur in a rotating cylinder similar to a kiln or on a series of screens or plates similar to a multiple hearth furnace. The degradation by aerobic bacteria actually occurs on the surface and, in general, the holdup of the system must be as much as five days' supply of refuse depending upon the degree of treatment required. The reason for this is, of course, the rate which the bacteria can attack the waste material. Consequently, all such systems have to be relatively large, and the maximum capacity is usually limited to approximately 200 tons per day of product, although smaller amounts can be handled effectively, if not economically.

The wastes are generally delivered to the rotating cylinder in a moist condition, which may be achieved by the addition of water or, in some cases, wastewater sludge. The cylinder rotates slowly, much like a cement kiln, and the waste is moved from one end to the other where it is thoroughly mixed and granulated by abrasion. Air is added at low pressure in controlled amounts throughout the length of the cylinder, giving the aerobic conditions necessary for rapid decomposition of the waste with minimum odor and other problems. Normal operating temperatures up to 140°F are developed by the decomposition. The final material from the rotating cylinder is then delivered to a conveyor belt and screened where the compost is separated from more durable materials such as splinters of glass or bottles. Any uncomposted material screened out in the final treatment area may be recycled back into the feed for further processing.

Commercial operations which have tried to make composts for fertilizer application have found that compost can be produced for costs of approximately $40 per ton, not including the collection costs of the waste material. Consequently, it is high compared to landfill or other means of disposal and perhaps even high in comparison to incineration if subsequent treatment of flue gases is not required.

Once again, it should be stated that composting has not been particularly successful in the United States. A list of plants that have been started and finished and volumes of written material on composting are available from the Environmental Protection Agency.

11-3.2 Aerobic and Anaerobic Digestion of Solid Waste

Very few words will be spent on this area of industrial waste disposal because there is a serious question as to whether it is really applicable. Some research work has been done on the development of digesters which will produce methane from the anaerobic digestion of solid wastes. For this reason, it is mentioned in this section as a possibility.

For all practical purposes, when discussing composting, we have discussed aerobic digestion of solid wastes. There are several other systems which have been used that might more nearly represent aerobic digestion. One of these was installed at Altoona, Pennsylvania, and is known as the Fairfield Hardy process. This system composts garbage with combustible rubbish in a rotary digester after the refuse is pulped in a wet pulper. The unit is similar to a large clarifier having a rake or agitator which moves around the periphery of the unit. The feed is brought in at the outside of the circular digester and is slowly moved to the center where the finished compost is eventually discharged. Composting time is about 5 days with operating temperatures as high as 170°F. Air is supplied to the digester by blowers and distributed throughout the pulp by pipes. This could be called aerobic digestion or composting.

Anaerobic digestion, due to the absence of air and due to its requirement of a different type of bacteria and consequently a different type of degradation process, produces methane in some quantity depending upon the material in the feed, just as methane is produced in anaerobic wastewater digestive processes. The difficult aspect of the anaerobic process is determining the best method for its acceleration since it is inherently slow. About the only method that can be used effectively is an artificial increase in temperature up to a point where the bacteria will grow rapidly but not to a point where sterilization will occur. This can generally be done by the use of low pressure steam.

Several companies have, on an experimental basis, created an anaerobic digester which can produce methane with the long range intention that our waste products may ultimately become a source for fuel gas. The conversion rate of the product is quite low, and the amount of steam required to supply the external heat is high so that the economics are rather poor. Until more definitive data are available on this process, it can merely be mentioned as a suggestion to industry for future consideration.

11-4 LAND DISPOSAL OF LIQUID WASTES

Dilute liquid wastewaters, which are biologically active, can be disposed of by using the wastewater as an irrigation source, assuming that the location of the liquid disposal is properly located so that an active run-off does not go into nearby streams and that the material being used for irrigation is aesthetically acceptable, i.e., that it has no severe odor problems.

Most spray irrigation of industrial wastewater is naturally done in rural areas since this is where the irrigation is most needed. It has been used, however, for public parks and golf courses near urban areas where the quality of the wastewater was acceptable.

Land disposal of waste reduces the volume of water required in areas where irrigation is necessary and recycles water which may be very valuable. Functionally, the land acts as a large trickling filter to reduce BOD and COD contents to acceptable levels while at the same time irrigating crops.

Wastewater from a plant system is usually put into a lagoon where it is given some aeration by mechanical floating aerators; then, it is pumped to sprinkler systems which are located across the area to be irrigated in such a way as to give essentially complete coverage. This is usually done with a center pivot type spray system which consists of a large boom with many spray nozzles which rotate around the center pivot covering a large circular area or, for that matter, many circular areas.

Naturally, the type of waste which can be disposed of in this manner is limited. Waste which contains dissolved metals or high levels of certain inorganic materials are not acceptable but wastes from canners, meat packers, nutritional product manufacturers, beer and beverage companies which primarily have BOD and COD discharge limitations can be disposed of in this manner. With the water and usable nutrients of the waste, more crops can be grown in a given area than with conventional dry land farming. The yields can be expected to be higher and, if the waste has significant nutrient value, commercial fertilizers will not be necessary. Normally, this method of disposal is used only on crops which are not directly consumed by humans.

Certain industrial biological wastes can create odor problems. Normally these installations are located away from living areas; thus, any odor problems will be slight. The timing of the application, the weather conditions, are important in terms of odor control.

There has been virtually no problem with this type of disposal where properly used relative to local groundwater. The soil filters solids and provides for biological decomposition of the wastewater before the water reaches the natural water table.

Perhaps the biggest advantage to land disposal is its relative cost compared to biological and physical-chemical methods. Wastewater treatment plants are often overloaded; so if an industrial waste can be treated by land disposal, it will free up capacity in a municipal treatment plant and also provide a much less expensive method of treatment. While it is possible to go directly from primary treatment to land applications, it may not be desirable depending upon the BOD and COD level, but many localities in the United States are doing this at the present time.

One of the shortcomings of liquid waste land disposal is that it is not functional at low temperatures. Obviously, below freezing, the spray will freeze on the soil and at temperatures below 45°F, greatly reducing the biological activity. Therefore, land disposal is more practical in warmer climates or during the portion of the year when ambient temperatures are higher.

Also, pH can be a problem in waste disposal. If the waste is extremely alkaline or extremely acid, it is obviously going to affect the pH of the soil so that only certain types of crops, if any at all, can be grown there without some type of neutralization. It is usually easier to neutralize a low or high pH in a waste stabilization pond than after it has been sprayed on the soil. A typical wastewater analysis for an Iowa packing company plant is shown in Table 11-4. The company, located in Sioux Center, Iowa, meets all state and federal regulations for wastewater disposal for 1985. It provides a tertiary treatment capability without an expensive, power-consuming waste treatment plant; and it improves crop yields for farm land. It provides approximately 20 MM gal. of water annually which is used in the processing of hogs. A complete report of this research is published as, "Water Conservation and Waste Load Reduction in a Modern Hog Packing Plant," Region VII, Environmental Protection Agency, Kansas City, Missouri. The effluent from two slaughtering operations moves towards two 1-¼ acre anaerobic lagoons after passing through a solids separator that skims off the floating wastes. The waste from the anaerobic lagoon passes over a bed of iron filings to reduce hydrogen sulfide. Grease from the rendering operation is added to the top of the lagoon as a cap to keep it anaerobic. Two holding ponds receive the effluent from the anaerobic lagoons (see Table 11-5).

Six monitoring stations are provided around the fields where the wastewater disposal is carried out, and three of these stations are wells which go into the water-bearing sand aquifer some 35 ft below the surface. No significant

TABLE 11-4. Typical Analysis of Land Disposal Waste from Meat Packing Operation

Parameter	March 1974 (ppm)	March 1974 (lb/day)
COD	14,568	788
BOD	6000	324
Suspended solids (SS)	None	None
Total dissolved solids	4402	238
Nonvolatile TDS	1000	165
NH_3 as N	25	1
Organic N	147	8
NO_3 as N	2	0.1
Phosphorus	4	0.2
Calcium	275	15
Magnesium	28	2
Potassium	60	3
Sodium	200	11
Iron	100	5
Manganese	4	0.2
Chloride	1662	90
Sulfate	50	3

TABLE 11-5. Design Data for Lagoon and Irrigation Plot

Acreage provided = 15.4 acres
Irrigation rate = 55 gpm at 60 psi

	Loading Rate	Acreage Req.	Volume
Nitrogen loading			3000 lb N/yr
Design loading	300 lb N/A/yr	10.0	
Max. recommended loading	350 lb N/A/yr	8.5	
BOD loading (40% removal in lagoon)			61,000 lb/yr
Design loading	21.5 lb/A/day	10.0	
Max. recommended loading	50.0 lb/A/day	4.3	
Hydraulic loading (40-wk basis)			51,000 gal./wk
Design loading	2 in./wk	10.0	
Max. recommended loading	1.6 in./wk	1.2	
Lagoon capacity (40-wk basis)			
Minimum aerated capacity			150,000 gal.
12-wk capacity			600,000 gal.
Operating capacity			750,000 gal.
Emergency capacity			850,000 gal.

changes have been observed in groundwater quality since the beginning of the operation in early 1974.

Land disposal of semisolid orgnaic materials which are also biodegradable has also been successfully tried over the years. Sludge from petroleum refining operations has been plowed into soil with excellent results. Usually the material decomposes completely over a three-year period. The major concern, of course, with the use of land as the means of disposal is that land itself is costly and such utilization must also be logistically practical for industry. Urban industries would find this approach expensive because of transportation needs; whereas, rural industry has this opportunity if they own sufficient land area to carry out the process.

11-5 OCEAN DISPOSAL

A logical place for the disposal of many industrial wastes has been the seemingly unlimited capacities of the Pacific Ocean, Atlantic Ocean, and the Gulf of Mexico. Ocean disposal has always had the great advantage of the landfill without requiring fill for cover. It is presumed the deeper the disposal site, the more effective the disposal.

One of the reasons that the oceans (off our coasts) and the Gulf of Mexico have been used for both industrial and municipal disposal sites is that very few regulations have been written to restrict this type of waste disposal. As early as 1899 under the Rivers and Harbors Act, the Corps of Engineers in New York designated several sites in the New York Bight where dredging and construction debris could be dumped. Then, in 1924, sewage sludge was permitted to be dumped in the same sites; subsequently, waste acids in 1948 and later, toxic wastes in 1965. These are but a few examples of controlled ocean disposal.

There are literally hundreds and hundreds of cases where disposal was neither regulated or even documented, so it is impossible to estimate how many tons of industrial and municipal wastes have been dumped into our oceans and the Gulf of Mexico over the years. It is known, however, that it has been increasing and with no particular pattern. As a matter of fact, in the past no one seemed to be particularly concerned whether this dumping had any great effect on the ecology of the oceans, whether it was causing pollution, killing fish, or actually promoting their growth. After all, the oceans and the Gulf of Mexico are very large bodies of water, and the amount of waste disposed in them seemed to be insignificant. One of the unfortunate aspects of human rationale is that we never really concern ourselves with a problem until it is too late. Perhaps in the last ten years we have come further towards appreciating the possibilities of indiscriminate pollution and waste disposal than we had in the previous 190 years of our country's existence.

It must be remembered that in addition to the materials which are dumped into disposal sites in the oceans and the Gulf, the rivers of this country and, for that matter, the rivers of the world, carry all types of organic and inorganic wastes into these bodies of water, creating a difficulty in measuring the actual amount of contamination which enters coastal waters in any particular area and more difficulty measuring its overall effect.

The Marine Protection and Sanctuaries Act of 1972 was passed to reflect the concern that studies by the National Marine Fisheries Service had made which demonstrated the degradation of water quality in areas that were used for sewage sludge dumping. This law does not prohibit ocean disposal, but it does place strict regulations on the types of waste which may be released and where they may be released. Currently, the number of waste disposal sites is being reduced, and only 14 active sites are in use for industrial and municipal wastes.

Additionally, an international agreement, which came out of the International Convention on the Prevention of Marine Pollution by the Dumping of Wastes, became a public law in March of 1974. This agreement prohibits ocean outfalls from exceeding certain levels in trace contaminants as identified in the ocean dumping criteria. This international convention specified that certain materials could not be released at all, and other materials could only be released in trace quantities. For example, it is permissible to dump organohalogens into certain ocean dumping sites, but the trace concentration in the mixing zone should not provide more than 1 part in 10,000. Before it is dumped, waste cannot contain more than 1 part in 100 (i.e., one hundredth) of a lethal concentration.

A number of studies have been made concerning the effect of dumping industrial wastes into the ocean. This involves construction wastes, chemicals, acids, bases, and all types of municipal garbage and trash. To date, none of the studies has been entirely conclusive that such dissemination of wastes into the oceans or the Gulf of Mexico actually causes ecological problems. But then, such a study is very difficult to define and more difficult to carry out. It is difficult to make a determination based on a short period of time in a particular area. Ecological systems change over many years, and it may be that the ultimate damage is indeterminate until it is far too late. The question of ocean dumping is essentially two-fold. First, its economics must be considered versus other methods of disposal, and, second, its ecological implications must be accurately evaluated in relatively small test areas before conclusions permit larger and more extensive ocean dumping. A number of studies are in progress concerning disposal sites off the east coast in the New York, Philadelphia, and Florida areas. Also, shipboard incineration in the Gulf of Mexico is under study sponsored by the EPA.

Not all the laws governing ocean dumping are sound. Public Law 92-532 suggests that ocean disposal sites should be located beyond the continental shelf wherever possible; however, sites located beyond the continental shelf are much more difficult to monitor and documentation cannot be made early enough to find alternative disposal methods if they are required. Also, material dumped in the deep ocean has a very slow decomposition rate. Less information is available concerning deep ocean disposal, and much more effort would be required to accurately research the problem. Wastes dumped off the continental shelf may not resurface for as much as a thousand years, and this could create problems for future generations. Deep water disposal has the advantage that it leaves no immediate trace, but first ocean dumping must be evaluated to know whether ecosystems are being changed for future generations.

For the industry considering ocean disposal, there are a number of contract organizations which will take a variety of wastes (usually those which are liquid in nature), put them aboard a barge, take them to an approved disposal site, and dump then accordingly. These companies must obtain permits for ocean dumping, although it is suspected that many indiscriminate and unregulated dumping operations take place. Insofar as the Corps of Engineers and the U.S. Coast Guard are able to patrol our coasts, however, ocean dumping is regulated. There are a number of different types of ocean dumping permits as follows:

1. The *research permit.* This permit may be issued for the research of a particular waste impact upon the marine environment. The permit allows mate-

rials other than those prohibited by law to be released, and an evaluation of the scientific merit of the operation must be significant enough to outweigh the potential damage of the release. A detailed project statement and an environmental impact statement of the project must be provided before the permit is issued, and a public notice and hearing are usually held. A permit can be issued for no longer than 18 months.

2. This type of permit is a *general permit.* It is issued generally for small amounts of nontoxic materials. A general permit usually specifies the type and quantity of waste to be discharged and the approved site for the discharge. It is subject to notice and public hearing, and it is a one-time affair. There are no renewals of general permits.

3. Another type of permit is the *special permit.* This permit is issued for the disposal of waste materials which could not be considered under a general permit but do meet the ocean dumping criteria. Such materials as dredge spoils which are not polluted would fit this category. Special permits usually have a fixed expiration date with a three-year maximum, and they specify the exact quantities and locations where the material will be dumped. They are renewable.

4. The next type of permit is the *interim permit.* This permit is used where materials which exceed the established criteria must be dumped or abandoned. Its purpose is to prevent or strictly regulate the disposal of material which may damage the marine environment. Any material identified as exceeding the criteria for trace constituents may be released under an interim permit. This permit can be issued for a maximum of one year and cannot be renewed, but a new interim permit can be issued when it appears wise to issue one. Many industrial wastes and sewage sludges are examples of material which would be handled under an interim permit.

5. Finally, there are *emergency permits* which may be issued after consultation with the Department of State and other appropriate persons or agencies. This is for materials which pose an unacceptable human health risk and would admit no other feasible solution for their disposal. It is not applied to routine industrial waste discharges. It might apply to radioactive materials or highly toxic substances. This permit cannot be renewed.

The ocean dumping criteria permit the dumping of certain types of industrial wastes and prohibit the dumping of other types. Prohibited materials are biological, chemical, or radiological warfare agents; high level radioactive wastes; and persistent inert, synthetic or natural floating materials that could be washed ashore.

Materials which are prohibited in other than trace quantities usually would be the heavy metals such as mercury and cadmium, organohalogens, and such materials as oil and greases which float on the surface of the water if released in large quantities.

Other materials which come under strict regulation or require special care

are the element or compounds of such materials as arsenic, beryllium, chromium, copper, lead, nickel, selenium, vanadium, and zinc; organosilicon compounds; inorganic processing wastes including cyanides, fluorides, and titanium dioxide wastes; petrochemicals; organic chemicals and organic processing wastes such as solvents, amines, phenols, phthalic esters, plastics and plastic intermediates and other by-products. Also, biocides, not elsewhere prohibited, and including carbamate compounds, herbicides, insecticides, and organophosphorous compounds. Also regulated are oxygen-consuming or biodegradable organic material and radioactive wastes which are otherwise not prohibited for dumping; materials on the toxic or hazardous substance list; flammable materials such as gasoline, carbon disulfide, and toluene; and any materials which might represent hazards to navigation such as large construction debris which would float; or large quantities of any material, acids, and alkalies; materials in containers; and materials containing living organisms.

The equipment required for barging quantities of industrial wastes to disposal sites is of such specialty and cost that most industrial or chemical companies would not consider owning and operating their own system but rather contract with independent companies who would then obtain the necessary permits and convey the material to the disposal site. One of the disadvantages of this method of disposal is it usually means double handling. Material must be shipped by truck or rail to the disposal contractor's facility unless the industrial plant requiring the disposal is located on the waterway where barges can be brought adjacent to the property and their tanks filled directly from the plant operation.

Today wastewater sludge is by far the largest waste which is disposed of by ocean dumping. However, industrial wastes are a close second, and it would be difficult to characterize industrial wastes by quantity or disposal location. Pickling acids and other materials of similar nature, available in large quantities, are the most likely industrial wastes disposed of today. The disposal of solid wastes by ocean dumping, aside from construction debris and municipal wastewater sludge, is usually neither as economical as landfill nor is it as readily permissible. Certain types of materials, such as low-level radioactive wastes and other toxic chemicals and munitions, are contained in drums or in concrete vaults and dropped into deep areas of the ocean. In the future, the disposal of such materials will again have to follow the International Open Dumping Convention regulations which include a list of prohibited substances. The cost of ocean dumping ranges widely depending upon the material being handled. It ranges from \$3 to \$5/per ton for liqud waste to \$20/per ton or more for bulky waste.

Another type of ocean disposal is ocean incineration. In 1967, D. Sobinger, A German engineer and specialist in combustion processes, became interested in the use of combustion to destroy industrial waste. He subsequently started a company named SBB which purchased a coastal tanker and converted it to a sea-going incinerator ship. He increased his fleet to two ships and began burning chlorinated industrial waste, making HC1 which subsequently condenses and goes into the sea where it is diluted millions of times. Another company, Ocean Combustion Services, a subsidiary of West German shipping company, outfitted the Vulcanus with incinerators which can accommodate 4200 metric tons of chlorinated wastes per year. This ship became operational in 1972.

Since the first test burning of chlorinated hydrocarbons on the Vulcanus in May of 1975, the EPA has granted a special permit to the Shell Chemical Company to burn various chlorinated organics under a 2-½ year permit up to 50,000 metric tons of chemical waste by-products. The wastes are organochlorine materials from five process waste streams generated in the manufacture of epichlorohydrin, epoxy resins, glycerin, and vinyl chloride. The data from the early incineration of sea applications indicated that these wastes were compatible with the provisions of the Marine Protection Research and Sanctuaries Act, Public Law 92-532.

Similar tests for qualification of the products of combustion were made for the ship Matthias III in the North Sea in 1977.

At the present time, these wastes are being incinerated in a special area in the Gulf of Mexico some 190 nautical miles south of Cameron, Louisiana and 170 nautical miles south-southeast of Galveston, Texas. The EPA has stated that this site in the Gulf of Mexico is good for five years and it filed an impact statement on July of 1976. Only the Vulcanus and the Matthias III are qualified to burn such wastes in this area, and the EPA has decided that the operations of the Vulcanus and the Matthias III are equivalent.

If, as the EPA contends, the complete incineration of organic chlorine materials yields water, carbon dioxide, hydrogen chloride, and nitrogen, then the ocean because of its high chloride content is adapted to receive large amounts of hydrogen chloride without serious imbalance of chlorine content. Also, because the ocean has a built-in bicarbonate buffering system, the ocean is able to deal with any acidity produced by the HCl in order to maintain its normal pH of approximately 8. No significant effects on the ocean were found below the plume of either ship, and there were no significant changes in the pH and the organic chloride content of the ocean in either case even when ocean samples were subjected to gas chromatographic separation and concentration which were followed by mass spectrophotometric analysis. The level of organic material lies below or at the level of sensitivity of the analytical system.

Before burning, the chemical waste materials are analyzed for their calorific value and chlorine content, the resulting pH and trace elements, such as heavy metals, and the presence of certain organic compounds. Also, a bioassay of the wastes is run. The approximate capacity of the Vulcanus burning in the Gulf of Mexico is 25 tons/hr, for the Matthias III it is 28 tons/hr.

The International Governmental Maritime Consultive Organization, which is a specialized agency of the United Nations, has under study the regulations and safety of such ships, and they are considering the possible designation of a burning lane across the ocean. The economics of such an operation can favor a sea lane for burning across the Atlantic, and these problems will be addressed by the Safety and Regulations Committee. It will be necessary for this committee to identify all of the incinerator ships which will be used in this manner, to study safety considerations, to designate a sea route for such incineration ships, and to set special provisions for the prevention of marine pollution from such incineration.

The Vulcanus has two incinerators with 3 nozzles in each; whereas, the Matthias III has only one incinerator and 20 burners around the base of the furnace. The Matthias III also is capable of coincineration of solid wastes and wastes stored in drums. On each ship, the burners are liquid air atomizing type with air and liquid fed through a firing nozzle through concentric tubes. The 20 liquid waste burners on the Matthias III are arranged symmetrically, evenly spaced around the base of the incinerator. The burner flame, which can be seen through the burner mounting ports, was observed to be a bright yellow to white, free of dark areas, and to be stable and nonsputtering.

The storage facilities for the various organochlorine wastes are below decks in tanks that are contained inside a double hull to lessen the possibility of ocean contamination in the event of a collision. The incineration temperature is maintained above 1200°C at all times and safely shuts down if this temperature is not achieved. Flame temperature is measured by an optical pyrometer and wall temperatures in the incinerator.

The Vulcanus has two combustion chambers; three burners are located roughly symmetrically around the bottom of each furnace. The Vulcanus has rotating cup fuel injection that provides vortex turbulence and the distribution of feed throughout the entire chamber.

The EPA decided that ocean incineration is an acceptable, if not a desirable, alternative to the disposal of such chemical wastes on land. However, the process is new and requires further investigation.

Burning costs for combustible materials are \$50 to \$60 per ton and for materials needing auxiliary fuel up to \$80/ton. This is considerably more than ocean dumping, but it is also a more acceptable disposal method.

11-6 DISPOSAL OF FLUE GAS CLEANING WASTES

One of the major problems today with the requirement for removing sulfur from flue gases is the disposal of the solid waste created by either dual alkali or lime-limestone treatment of these waste materials. There is serious concern about the disposal of calcium sulfite because it is, to some extent, soluble. For landfill disposal, dewatered or treated scrubber wastes are hauled to the disposal site or waste slurries are pumped to the site for dewatering or treatment there. The dewatered wastes are placed and compacted, although sanitary landfill procedures are not necessary. Exposure to rainfall should be minimum so as to reduce leaching. The disposal site should be designed so that a minimum of water run-off occurs; some run-off will probably be generated around the site so that trenches for collection and recycle or treatment should be provided. These materials must have enough physical stability to be hauled by earth-moving equipment for landfill disposal, which means dewatering the material to at least 50% solids and perhaps even more. There are several methods for improving dewatering. One is to add dry solids such as in the chemical fixation procedure described in this chapter; the second is to improve the dewatering equipment; the third is to control the crystal size of the calcium sulfite; and the fourth is oxidation to obtain gypsum, which is relatively insoluble.

The third approach which involves the increase in the size of the calcium sulfite hemihydrate crystals has involved the development of a process that controls crystal size through the use of mechanical equipment in conjunction with lime-limestone scrubbing by the use of special reactor outside of the scrubber loop. Increasing the reaction product crystal size is also a purpose of the oxidation of these wastes into gypsum, since gypsum crystals are typically much larger and thicker than sulfite crystals and therefore settle quicker and trap less water.

One of the considerations which has been given to the disposal of these wastes is the concept of coal mine disposal. This was motivated by the existence of railroad transportation between the coal mine and the power plant and the need for material to fill the void left by the mining of the coal, especially in strip mines. However, definitive information on such disposal has not yet been established. Others have also considered ocean disposal of various types of flue gas desulfurization residues with increasing awareness of the adverse effects of much of this type of activity.

An examination of the physical and chemical characteristics of flue gas desulfurization wastes identified the following environmental concerns relating to ocean disposal. First, ocean floor sedimentation—the particle sizes of most untreated flue gas desulfurization wastes are much smaller than the coarse grained sand particles of the ocean floor which are most conducive to the marine life found on the continental shelf; therefore, deposition of these wastes in large quantities may have an undesirable effect in the relatively shallow continental shelf area, particularly if disposal results in paving the ocean floor. Also, solids suspended in the water column indicate a problem in the settling of these wastes towards the ocean floor as well as resuspension of solids from the ocean floor which would expose various types of marine life to elevated concentrations of suspended sediments, and this effect could be serious depending upon the organisms involved. High sulfite concentrations present a potential problem of sulfite toxicity to marine life and potential chemical oxygen demand if oxidation to sulfate occurs. The toxicity depends on the degree of ingestion by the marine life, and the COD depends on the kinetics of both sulfite dissolution and the sulfite oxidation to sulfate. Finally, there are trace contaminates in the waste which include mercury, zinc, selenium, cadmium, and nickel. Without significant dilution of the waste with sea water, they are not acceptable for ocean disposal.

REFERENCES

11-1. Pavoni, J. L., Heer, Jr., J. E., and Hagerty, D. J., *Solid Waste Management,* Van Nostrand Reinhold Co., 1973.

11-2. Ross, R. D., *Industrial Waste Disposal,* Reinhold Book Co., 1968.

11-3. Reindle, J., "landfill course", 10 Lessons, *Solid Waste Management,* 1977-1978.

11-4. National Solid Waste Management Association Technical Bulletin, Vol. 8, No. 2, February 1977.

11-5. Mead, B. E. and Wilkie, W. B., "Leachate Prevention and Control From Sanitary Landfills," *Waste Age,* March/April 1972.

11-6. Hainley, J. D. and Slover, E. E., "Industrial Chemical Waste Landfill," *Waste Age,* November/December 1971.

11-7. Slover, E. E., A Case History: Implementing a Chemical Waste Landfill—NSWMA. Fourth Congress on Waste Management, Atlanta, Georgia, November 12-14, 1975.

11-8. Editorial, "Diversified Hauler Develops System for Hauling Both Solid and Liquid Wastes," *Solid Wastes Management,* January 1978.

11-9. Schomaker, N. B., "Construction Techniques for Sanitary Landfills," *Waste Age,* March/April 1972.

11-10. NSWMA Communication, "EPA Solid Waste Disposal Criteria," February 1978.

11-11. Editorial, "Stabilizing Waste Materials for Landfills," *Environ. Sci. Technol.,* Vol. **II**, No. 5, May 1977.

11-12. Editorial, "How Hot is Ocean Incineration," *Environ. Sci. Technol.,* Vol. **9**, No. 5, May 1975.

11-13. Ricci, L. J., "Offshore Incineration Gets Limited U.S. Backing," *Chem. Eng.,* January 5, 1976.

11-14. Barrett, B. R., "Controlling the Entrance of Toxic Pollutants into U.S. Waters," *Environ. Sci. Technol.*, Vol. **12**, No. 2, February 1978.

11-15. Editorial, "First U.S. Deep Sea Incineration Conducted in Gulf of Mexico," *Solid Waste Management*, December 1974.

11-16. Eweson, E., "Making Composting Work," *Pollut. Eng.*, June 1973.

11-17. Jones, J. W., "Disposal of Flue Gas Cleaning Wastes," *Chem. Eng.*, February 14, 1977.

11-18. Editorial, "Ocean Incineration Anew," *Environ. Sci. Technol.*, Vol. **II**, No. 3, March 1977.

11-19. Walker, W. H., "Monitoring Toxic Chemicals in Land Disposal Sites," *Pollut. Eng.*, September 1974.

11-20. Lord, W. H., "Disposal of Sludge From Flue Gas Desulfurization," *Pollut. Eng.*, June 1976.

11-21. Flower, F. B., "The Refuse Landfill," *The Executive Housekeeper*, February 1974.

11-22. Black, R. J., "A Review of Sanitary Landfilling Practices in the United States," (Reprint from Proceedings–Third International Congress, International Research Group on Refuse Disposal), Trento, Italy, May 1965.

11-23. Schroering, J. B., "Sanitary Landfilling–Applications and Limitations," *Pollut. Eng.*, November 1969.

11-24. Editorial, "Truckloads of Landfill From Waste Sludge," *Chem. Week*, January 26, 1972.

11-25. Hainley, J. D., "Unique Landfill Process Tried," *Water Wastes Eng.*, January 1972.

11-26. Hersey, I., "Fertility and Waste–Can We Control the Cycle?," *Eng. Opportunities*, July 1969.

11-27. Routson, R. C. and Wildung, R. E., "What Happens in Soil–Disposal of Wastes," *Ind. Water Eng.*, October 1970.

11-28. Farvolden, R. N. and Hughes, G. M., "Sanitary Landfill Design," *Ind. Water Eng.*, August 1969.

11-29. Young, R. A., "Company Sprays Away Wastewater Problem and Makes Hay," *Pollut. Eng.*, March 1977.

11-30. De Vries, J., "Soil Filtration of Wastewater Effluent and the Mechanism of Pore Clogging," *J. WPCF*, April 1972.

11-31. WPCF Policy Statement, "Use of Land for Wastewater Treatment and Disposal," *J. WPCF*, December 1973.

11-32. Flower, W. A., "Earth Used to Treat Pulp Mill Wastes," *Factory,* November 1967.

11-33. Dotson, G., Dean, R., Kenner, B., and Cooke, W., "Land Spreading and Non-Polluting Method of Disposing of Oily Wastes," U.S. Dept. of Interior, July 1970.

11-34. Editorial, "Land Disposal of Wastewater," *J. WPCF,* July 1973.

11-35. Editorial, "Ultimate Disposal: No Final Answer," *Environ. Sci. Technol.,* Vol. 7, No. 7., July 1973.

11-36. Michel, H., Gilbert, P., and Creed, H., "Land Disposal–What's the Realistic View?," *Water Wastes Eng.,* June 1974.

11-37. Perna, A. J., "International Solid and Liquid Waste Handling and Disposal," *Pollut. Eng.,* Yearbook and Product Reference Guide.

11-38. I. U. Conversion Systems Paper on Poz-O-Tec Process.

11-39. Editorial, "Composting Municipal Solid Waste," *Environ. Sci. Technol.,* Vol. 5, No. 11, 1971.

11-40. Hart, S. W., "Solid Waste Management/Composting European Activity and American Potential," Report SW-2C, U.S. HEW, 1968.

12

Contract disposal of industrial waste

One method of disposing of industrial liquid and solid waste is to contract for their removal from the site where they are created. Everyone is familiar with the municipal rubbish truck which goes down the streets and hauls away garbage and trash. This concept on an industrial basis is essentially the same except that the industry must usually pay for the removal of these materials since this is not generally included in their taxes. An industry may contract for the removal of all of its liquid and solid wastes or just selected wastes, depending upon the facilities for waste disposal which it has at its manufacturing sites, and also depending upon an economic evaluation of the advantages of contract disposal versus "in house" disposal.

12-1 TYPES OF CONTRACT DISPOSAL

The possibilities for industrial waste disposal by contract methods are numerous. First, consider those which are under the jurisdiction of the municipality or community in which the industrial property resides. The opportunity for an industry to dispose of its solid waste in municipal facilities does exist in the United States, and the fee schedule for such disposal varies quite widely. Many solid industrial wastes do not vary significantly from domestic wastes and can be landfilled or incinerated along with domestic waste. The cost for such disposal can either be part of the tax base or it may be separately charged on a tonnage basis. Industrial wastes which consist of wood, paper, cardboard, and a nominal amount of plastic materials are essentially identical to domestic wastes and create no problems in existing landfills and incinerators.

Industrial wastes are not always constant, and occasionally their composition may create a problem in their disposal. Most municipal incinerators are designed to handle materials having heating values up to 6000 or 7000 Btu/lb, and they can tolerate a nominal amount of plastic waste, in some cases as high as 20%. Many of these plastic wastes have heating values betweeen 15,000 and 20,000 Btu/lb. It readily can be seen that a large percentage of combustible plastics thrown into an incinerator at any given time can cause temperatures within the incinerator far beyond its design capabilities, thus, destroying existing incinerator systems. Also, because of the limited amount of air available in the incinerator for the destruction of plastics, clouds of black smoke can be generated constituting an air pollution problem.

Therefore, if municipal incineration systems are to be used for industrial waste disposal, industry will have to be responsible for the materials which it sends to the incinerator. The same is obviously true of any landfill. If industry includes materials which are not satisfactory for landfill and which will cause groundwater contamination problems, it will only be a limited time until local authorities determine that industry is at fault and will no longer take its waste.

Many industries, however, do not have this option open to them. The municipalities, if there are municipalities in the area of the industrial plant, refuse to take the waste. They say to industry, "This is your problem. Dispose of it in the best manner that you can." Under these circumstances, there are usually private contractors available who are willing to take the industrial waste and dispose of it in some manner or another. In some cases, they may contract with the municipality at some price per ton for the disposal of the waste in either an incinerator or landfill; more often, they are either owners of their own disposal facilities or they contract with other private owners for this disposal. These contractors run the gamut from being efficient, capable, and honest to those with totally opposite characteristics.

For years, industry in the United States has been faced with the problem of solving the industrial waste disposal dilemma. Industries that tried to safely dispose of their wastes were penalized economically because many others gave their wastes, without second thought, to private contractors and could have cared less what was done with them. This, fortunately, is not the case today. Industry, regardless of who handles its waste, must ultimately be responsible for the disposal of the waste materials. Court decisions have severely punished companies who have handed their wastes to an irresponsible outside contractor. Therefore, the days of not caring where the waste goes are gone.

States have established specific jurisdiction for solid waste disposal as well as liquid waste disposal. Many states have outlawed chemical landfills. Others are closing down undesirable solid waste landfills or making them meet certain prescribed regulations. All of these things are tending to upgrade the waste disposal industry. This industry represents literally billions of dollars in total revenue each year. The National Solid Waste Management Association, headquartered in Washington, D. C., has several thousand members, all of which are involved in the disposal of solid waste or its reclamation in one way or another.

Contract waste disposal operations can be divided into the following categories:

1. A municipal disposal facility which contracts with industry for the disposal of its wastes.
2. A quasi-governmental facility in which the industry participates, much like a cooperative.
3. A private disposal contractor who does not own his own disposal facility but contracts again for the disposal of the industrial waste.
4. A contract disposal organization which has it own disposal facilities and disposes of the industrial waste on its own site.

Each of these bears some individual discussion.

12-1.1 Municipal Facility

The first type is merely a city or community accommodation to industry in allowing industry to utilize its facilities for disposal of certain types of industrial wastes. This may be solid waste which can be deposited in a community landfill or burned in a community incinerator (providing there is capacity available to handle it), or it may be a city sewage system which permits the dumping of liquid waste of certain character into existing sanitary sewers for further treatment at the municipal city sewage plant. Obviously, both of these types of systems require close monitoring, not only by the industry but by the community.

Violations of regulations concerning the character of the waste which is given to the community may result in very serious repercussions. For example, suppose that a food processing company were permitted to dump its waste into a city sanitary sewer system as long as these wastes were of a particular character and their BOD was not any greater than a specific level. The food processing company does this for a number of years and, at some point, decides to add a new operation which involves the manufacture of its own plastic containers for the food which it sells. This operation includes the use of solvents or possibly even a material such as phenol. The operators are not technical and therefore feel that wastewater is wastewater and inadvertently wash large quantities of solvent or phenol-containing water into the same sanitary sewer with the rest of their wastes. The result is that the city wastewater treatment plant may be in serious trouble. Many of these materials act as a bactericide, killing useful bacteria and cause the whole wastewater treatment operation to malfunction. This is one of the continual concerns with allowing industrial waste to flow through municipal sewage treatment facilities.

The same type of situation might occur in the handling of wastes sent by industry to a landfill. Unless the industry and the municipality monitor the type of wastes, it is possible that soluble or leachable materials might be deposited in the community landfill, and most community landfills do not have or require leachate monitoring or control. For example, assume that an industrial customer of the municipal landfill would include a number of fiber drums of some soluble inorganic chemical, perhaps as simple as sodium chloride, and this was not discovered by the landfill operators until after the material had been placed in the landfill and covered with soil. The salt would gradually leach out into surrounding streams or wells and could cause enough contamination to present a short-term problem.

The municipality would also have to be concerned about the type of industrial waste which went to an incinerator, because wastes of high calorific value, above 7000 Btu/lb, may cause physical destruction, over a period of time, which increases the operating costs for the community. In general, the municipal disposal facility, if available to industry, is a very desirable arrangement because the cost for disposal is usually lower than other types of contract disposal. The cost for solid waste disposal in either landfill or an existing incinerator varies so widely throughout the Country that it would be meaningless to give any figures on which to base a comparison. Also, in most cases, a true contractual relationship with the municipal operators is not required since they usually charge on a volume or weight basis and pick up on a regular schedule or on specific request.

For liquid waste going into a treatment plant, the charge is often based on the total oxygen demand of the waste, which is the combination of the biochemical and chemical oxygen demand requirements above a certain base level. Municipal treatment plants usually charge a flat rate on the basis of actual or estimated gallonage from an industrial facility as long as the total oxygen demand is below 200 ppm and then apply a surcharge for higher total oxygen demand requirements above this base. Therefore, industry theoretically can dilute the waste but pay on a higher gallonage basis or leave it concentrated and pay on the basis of total oxygen demand.

Municipal wastewater treatment plants do not like to accept materials with exceptionally high BOD and COD requirements from industry and try to discourage industry from using the municipal plant by high tarrifs and surcharges. On the other hand, industry's alternative is to build their own biological system or use some other form of treatment which requires a sizable capital outlay plus operating cost. Federal pretreatment standards apply to many types of industrial wastes. (See also Section 7-1.2.)

Since the conditions for municipal handling of industrial waste vary so widely throughout the country, it would be unwise to be specific in terms of economics with this information; however, it can be generalized. Municipal handling of industrial solids and liquid wastes will be cheaper than other approaches which are available on a contract disposal basis and will probably be subject to less restrictions than other methods. In return, industry should realize that it is probably getting a good deal and should do everything possible to maintain the waste character which has been authorized for acceptance by the municipality. Once the municipality begins to have problems with the industrial waste, they will probably refuse to accept it, and then industry must find an alternative.

12-1.2 Quasi-Governmental Facility

The second system, that of having a quasi-governmental organization develop a cooperative for the disposal of industrial wastes, has been tried in only a few places. One of the situations where it has been quite successful is the Gulf Coast Waste Disposal Authority just outside of Houston, Texas. Several years ago this authority built a plant with public funds to solve some of private industry's dilemma along the ship channel which stretches from Galveston to Houston. This first plant is essentially a biological treatment plant used to handle aqueous wastes from a variety of industries in the Pasadena, Texas, area. Various industries with varying problems, all of which could be handled in a biological plant, were asked to join the cooperative which was operated under the quasi-governmental organization. The Gulf Coast Waste Disposal Authority obtained its financing for the construction of the plant through municipal bonds. The companies participating in this venture will pay off the bonds, plus the operating costs of the plant on a contract basis, i. e., so many dollars per unit volume or unit weight of waste which is brought into the plant. This first experiment seems to have worked quite well. The Gulf Coast Waste Disposal Authority is now interested in building additional plants for more complicated and more difficult disposal problems.

A quasi-governmental facility has some significant advantages when compared with other types of industrial waste disposal, but it also has some rather serious disadvantages. Perhaps the most difficult aspect of a quasi-governmental arrangement is trying to match the waste producers with the facility which is to be constructed. In order to set up a cooperative type of waste disposal facility, it is necessary to have firm commitments from the waste producers, not only as to the type of waste which they intend to send to this facility but also the quantity of waste and the number of years which they are willing to commit to the waste disposal facility for the disposal operation. It would be impossible to build a multimillion dollar facility for the disposal of wastes, on a cooperative basis, using public money, unless there was some specific evidence that the bond issue which was used to raise the money could be paid off in a reasonable period of time. Without industry's commitment to use that facility for a given number of years, at least the term of the bond issue, no bond issue could be obtained and the facility could not be constructed.

Those quasi-governmental installations which are currently successful require that the industrial partners take a firm stance that they intend to use the facility for a period of years. In the chemical industry this is especially difficult since products arrive upon and leave the market often in very short periods of time, perhaps in much shorter periods of time than it would be practical to finance such a facility and, consequently, the companies involved may have made commitments for the future which may not be in some respects in their long range best corporate interest.

In this type of facility, there are very few cases where industry is not as interested in the success of the waste disposal facility as they are in their own operation because they are, in essence, partners with the government in the operation of the facility. If they should ship some material without proper analysis which was not included in their initial contract and if it should damage the operation of the facility, it would, in fact, be hurting that industry as well as many other industries in the area. Consequently, there are perhaps less operational problems in this type of facility than there are in others. In almost every case, the cooperating industries have designated the types of waste which they will send to the facility, and the facility has been designed specifically for these wastes. This does not mean that other industries may not join the cooperative, but their waste must be of a character which can be accepted by the facility for treatment.

The type of contract which industry will have with a quasi-governmental organization will vary depending upon how the money was raised for the facility. If it is a true cooperative, then the industries involved may have to guarantee the bonds which are issued for public sale, although this is not always the case. In any event, the contract which industry signs will delineate the initial price per unit weight or volume for the disposal of those specific wastes which they have designated they will send to the facility. Since it is a nonprofit operation, the costs of operating the facility are essentially divided among the participants either directly or by increasing the cost of the waste treated.

This type of facility does not usually provide its own transportation to the disposal site. This is either handled by the industrial partners with their own vehicles or pipelines or is contracted to special organizations or common carriers.

Since the quasi-governmental system is relatively new and has been tried only in several areas, there is meager economic information available. It would seem, however, if operated correctly, it could be more economical than private disposal on-site by an industry and also more economical than private contract disposal by others. It must be kept in mind, however, that the local governments and industries who attempt to build and operate such a facility are dealing with public money and any indication to the public that this benefits private interests more than it does the public sector is bound to create severe problems, either initially when plans to build the facility are first offered, or later after the facility is in operation. Governments which have been successful in developing such a program are to be complimented for their foresight in a very difficult and sensitive area.

12-1.3 Private Disposal Hauler

The third type of disposal operation involves the private contractor or hauler who takes the waste from the industrial plant and disposes of it through a third party. Of all the possibilities, this is, perhaps, the least desirable because it provides for no continuity of responsibility. This type of contractor is usually a hauler or truck line operator. His primary interest is to take waste material away from the industrial plant to a third party. In many cases, he collects a fee on both ends, especially if he can find something in the waste which is recycleable. This was especially true in years past when cheap solvents, such as xylene and toluene, were being landfilled by many such organizations, a dangerous practice in itself. Often, these haulers would find that there were reasonable quantities of expensive solvents mixed in with the cheap solvents, such solvents as methyl ethyl ketone could be reclaimed for a reasonable price. These haulers would then locate a reprocesser and take the material to the reprocesser; instead of paying three or four cents a gallon to put it in a landfill, they would receive five or six cents a gallon for the value of the solvent. In most cases, the originating industry didn't know where the material was going. If, for example, this material had been contaminated with a hazardous substance, the ultimate processer who was reclaiming the solvent would never have known it and may have subjected his personnel to a very hazardous situation.

Such middlemen often use landfills but have also often used ocean dumping. Now that ocean dumping is becoming a bad word and public scrutiny is focused in that direction, we will probably see less of this type of operation.

The private contract hauler with no disposal site of his own was most often the culprit in cases of illicit dumping or disposal. He may have claimed that he was taking the material to a state-operated or approved landfill when often he had a friend with a farm where he dumped the waste, or he dumped the waste on public or private land without the owner's knowledge, which may have been illegal and decidedly dangerous.

A large eastern company had been disposing of liquid wastes through a contractor of this type for some time. They had no repercussions so they felt that the waste was being properly handled, but apparently made no effort to find out precisely where it was going until one day a number of its drums of relatively dangerous wastes were found semiburied on a farm within the state. The state authorities immediately put pressure on the company (since its name was on the drums) to describe the disposal process. The company had to admit that it really didn't know where the waste was going. The waste had been given to a contractor and he was supposed to dispose of it. At the time of this discovery, the contractor had skipped town and the industry which generated the waste had to dig up all the drums and transport them at its own cost to another facility where it was required to pay considerably more for the disposal. Fortunately, in this case, no one was hurt, but had there been repercussions which involved the health or the life of individuals, the cost could have been much greater.

This does not mean that there are not honest middlemen or contractors which take wastes from an industrial plant and see that they are disposed of properly by a third party, but it does suggest that this third party approach is perhaps not as good as the others suggested.

In the development of a contract with such an organization, industry should want to know precisely what the hauler is doing with the waste, and the contract should delineate this with respect to recovery, if such is to be accomplished, and with respect to ultimate disposal. If it is to be in a landfill, then the landfill should be designated in the contract or the possible landfills that would be used. The industry producing the waste should find out whether these landfills are, in fact, suitable for the waste which they are giving the disposal contractor. If the wastes are chemical materials, is it a properly acceptable chemical landfill which provides for leachate collection and treatment? If the wastes are going to an incinerator, are they being burned in accordance with pollution regulations within the state? If the wastes are going out of the state, then to what state are they being taken? Are they being carted in proper vehicles in accordance with the Department of Transportation regulations? Is the material properly marked? If it is being disposed of in the ocean, industry should be aware of the method of transportation, the disposal site and whether it meets current laws and possible future regulations concerning marine disposal. Industry must not only think of the present disposal problem but also any adverse publicity which may arise from improper disposal by such a contractor which will ultimately revert back to the industry itself.

12-1.4 Contract Disposal Organization

The fourth possibility is the use of a contractor who has his own disposal facilities. These facilities may be designed to handle the wide variety of industrial wastes or they may be limited to several categories, such as biologically treatable aqueous wastes, solvents, waste oil, and landfill material. The advantage of dealing with such a contractor is that you can inspect his facilities and determine whether he really has the capability of handling the waste material which you intend to give him. He will generally charge more than the middleman type of contractor but, then, he must, because he owns a facility and meets local and state regulations in the disposal of whatever waste that you give him. Consequently, his costs are going to be higher because he must do a better job.

Such companies may use a single disposal method, such as incineration, and may only take incinerable materials, or they may have a wide variety of treatment methods which can handle virtually any type of wastes which would be generated by industry, including hazardous wastes. Some may not have final disposal capabilities on their treatment site areas and may contract for this by either subsequent land disposal or an approved landfill, or ocean dumping by means of an approved ocean handling company.

Such contract disposal companies vary from very small companies to organizations with hundreds of millions of dollars per year in revenue. They may be staffed with competent, technologically oriented people, or they may not be, and this is one of the things industry has to determine before it gives its waste to such an organization. Many such companies are set up strictly for reclaiming rather than final processing. This is especially true in the oil reprocessing business.

The good contract disposal organization will usually require that a complete analysis of the waste which it handles is provided first by the industry supplying the waste and, secondly, they will have facilities to make complete analyses themselves. Many of these organizations have very sophisticated laboratories, and if an industry does not know specifically what their waste contains, the contract disposal organization will analyze it for them and give them a price for the disposal.

Not only will they analyze the sample of the waste prior to giving a disposal price but, to be sure that they are receiving the waste for which they have contracted, they will continually analyze each shipment of waste which comes into their plant to insure that it is in accordance with the initial sample. In some cases, it may be necessary for them to ask for additional revenue if the character of the waste is significantly different than the initial material and, in some cases, it may be necessary for them to refuse to take a shipment if there is a substantial difference between the sample and the contracted waste which was to have been shipped to them. It may also be necessary for them to refuse a shipment of a specific waste which is under contract if they are repairing or having difficulty with those disposal facilities which will be used for the destruction of the waste. The extent to which any contract disposal organization is responsible for proper disposal of its customer's waste is a function of the contract it has with the generator and the laws and regulations which govern its operation.

In examining a contract disposal organization, the industrial producer should look for the same things as he would look for if he were building his own facility. He should be intimately involved in the type of waste disposal which is being used to destroy his waste. For example, if incineration is being used, he should know what type of incineration system the contract operator is using and at what temperature this system is operating so that he can not only satisfy himself but governmental authorities if he should be requested to do so that the waste is being handled and disposed of properly. If the waste is being treated chemically, the same would be true. Most contract disposal facility operators, if not now, will soon be subject to stringent regulations, especially if they are handling hazardous wastes, which most of them are. Industry will not have to be as introspective in the future as they are now because of these regulations.

The type of contract which should be developed with a contract disposal organization is discussed further under the legal aspects in this chapter, but, in general, industry wants to know as much detail as possible about the handling of its waste and, in return, it should be willing to pay a premium for its disposal.

12-2 Transportation

Sometimes contract-disposal companies have their own transportation system. That is, they will come and pick up waste at your plant, take it to their facility for processing and charge both transportation and a processing charge. Others will utilize existing common carriers for the hauling of the waste from your plant to their facility where they will perform the disposal operation. The advantage is on the side of the company which can provide the total service because there is no split responsibility. When the material leaves your property, it should become the property of the company which has contracted for its disposal, and presuming you have given the disposal company complete information concerning the hazards of the handling and disposal of this waste, they should be willing to give you reasonable indemnification for anything which happens subsequently. This does not mean, however, that you are totally absolved from any problems that may occur after the waste has been delivered to the processor or in the process of transportation. It is your waste and, if it ultimately causes a problem because of a spill on the road or because of contamination of groundwater or air pollution at the processor's site, you may be considered totally or partially responsible by the court. Therefore, it can be said that no idemnification, no matter how it is written, is truly a total indemnification of the producer. You should never believe any contract disposal organization which gives such indemnification as sacrosanct.

12-3 ORGANIZATIONS

There are a number of large organizations in the United States which are active in the contract disposal business but which operate through smaller subsidiary organizations, either totally or partially owned. Such companies have established network waste-disposal companies in a number of states by purchasing small private contractors, giving them additional funds to expand their business and taking a portion of their profit. These companies are quite sophisticated in their methods, both business and technological, and, in general, have improved the cause of safe and sane waste disposal throughout the United States. Such multi-million dollar organizations, as Browning Ferris, Inc., Waste Management, Inc., and S.C.A. are indicative that the waste disposal business is a profitable venture.

The EPA has listed in its publication EPA/530/SW-146, dated March 1975, a number of companies which are presently in the business of accepting and disposing of various hazardous wastes. This list is not an endorsement by the Office of Solid Waste Management, nor does it vouch for the environmental adequacy of their operations. But it does indicate the number of companies which are available for waste management operations, most of them private in nature, and gives a brief summary of their capabilities and their areas of interest. This list will be revised quarterly by the EPA.

12-4 LEGAL ASPECTS

As stated above, one of the more difficult aspects of contract waste disposal is how to stay out of trouble. Lawyers disagree on the exposure of any company using contract waste disposal services. For this major reason, some very large U. S. companies have essentially become self-sufficient in the waste disposal area because they feel that contract disposal may result in serious legal implications if improperly handled. On the other hand, just as many large companies seem to be satisfied that handing their waste to competent disposal operators is a satisfactory approach, though perhaps not as conservative as handling it "in house."

Several things should be emphasized in this regard. The first is that there is no such thing as complete indemnification. If a waste from an industrial plant causes property damage, illness, or death in its subsequent transportation and disposal, the producing company will undoubtedly become involved in some type of legal action, and may or may not have a judgement against it, despite the fact that the handling company may have given it a certificate of indemnification and have satisfied the insurance requirements of the producing organization. Second, the best insurance against such miscalculations is to contract with a company whose capabilities, technologies, equipment, and facilities meet the necessary standards for proper disposal.

In other words, any industry contracting to dispose of the waste should know where it's going, how it's being transported, how it's being treated, and what the final disposal of the waste is going to be. The industry should be satisfied that such methods are totally acceptable before such a contract is signed. Third, the industry should insist that the company contracting to take the waste out of the plant not contract to resell all or any part of the waste unless previously approved by the waste generating organization. This keeps the waste from going someplace unknown.

Observing these three rules will solve most of the legal problems attendant with contract disposal. If these rules are ignored, all types of problems can occur.

12-5 ECONOMICS OF CONTRACT DISPOSAL

The economics of the disposal of various industrial wastes by contract methods vary widely from locality to locality and section of the country to section of the country. One cost in the disposal of waste is transportation. The farther a waste must be hauled to a disposal site, the more expensive the entire process or contract is going to be. Transportation costs are calculated on a CWT (hundred weight) basis with a special commodity rate. The longer the haul, the greater the cost. This averages 56¢/CWT/100 miles for waste materials.

Obviously, the simplest and cheapest form of disposal is sanitary landfill. If the material is acceptable to a sanitary landfill and does not present a leachate problem, then this has to be the least expensive method for disposal. In a chemical waste landfill which is especially designed to prevent leaching, the cost will be slightly higher than it will be for rubbish-type solid waste, but it will still be the least expensive method for disposal. Chemical waste averages $20/ton.

The second least expensive method would be biological systems which will usually take aqueous wastes having a BOD or a COD concentration up to approximately 1000 ppm. This type of contract disposal normally involves large trickling filters or similar devices requiring quite a bit of land area plus inexpensive equipment and treatment methods. Costs are about 1.5¢/lb.

The next least expensive type of disposal will be that requiring chemical reaction. This is generally true for acids and bases, or solutions containing heavy metals, where reaction with a neutralizing agent will be required and some means of precipitating and removing the heavy metals will be necessitated. Costs vary but are generally in the 2¢ to 3.2¢/lb. range.

Finally, generally the most expensive method for contract waste disposal will be for those materials requiring ultimate disposal by incineration. Incineration requires a significant equipment investment on the part of the contract disposal organization and often also requires fuel consumption. If incineration effluents must be scrubbed to remove materials such as SO_2 and HCl, then neutralizing costs must be added to the price.

The following table generally indicates current disposal prices for various types of wastes on a contract basis in the eastern part of the United States at the time of this writing. As chemical landfills are closed, which seems to be the present trend, in fact in some states they are not even permitted, then the more sophisticated methods of disposal will have to be used and undoubtedly these prices will increase.

Solid Waste	
Trash landfill	\$3–\$16/cu yd
Chemical landfill	1¢–3¢/lb
Liquid Waste	
Biological treatment	1.5¢–3¢/lb
Chemical treatment	2–3.2¢/lb
Incineration	4¢–8¢/lb + chemical costs for scrubbing (some as high as 20¢/lb)
Deepwell	7¢–15¢/gal.

REFERENCES

12-1. Williamson, Peter, "Structural Integrity in a Centralized Disposal Facility," 76th National Meeting AIChE, March 1974.

12-2. Nemerow, N. L., "Cooperative Treatment of Industrial Wastes," *Ind. Water Eng.*, June 1969.

12-3. Editorial, "Central Waste Handling at Bay Port," *Ind. Water Eng.*, May 1969.

12-4. Lombard, W. K., "The Design and Operation of a Private Waste Disposal Plant near Sarnia, Ontario," 24th Annual Purdue Industrial Waste Conference, Purdue University, May 1969.

12-5. Faatz, A. C., Jr., "Centralized Waste Disposal Facility is Economical," *Oil Gas J.*, August 11, 1969.

12-6. Williamson, Peter, "Processing of Industrial Wastes at Regional Treatment Facilities," ASME Incinerator Conference, September 9, 1971.

12-7. Alsentzer, H. A., "Treatment of Industrial Waste at Regional Facilities," National Paint, Varnish and Lacquer Association, November 8, 1971.

12-8. Martin, R., "World's First Pollution Control Plant," *Petro./Chem. Eng.*, August 1970.

13

Disposal of hazardous materials

13-1 WHAT WASTES ARE HAZARDOUS?

The reason for a separate discussion of the disposal of hazardous material is not so much because of it being a special group or category of wastes which could be classified as hazardous but the very fact that waste disposal involves the handling and final disposal of a wide variety of materials about which very little may be known. The fact that there is very little known about the waste materials which are handled on a daily basis by industrial plants and contract organizations means that they are, in fact, hazardous. Any time it is necessary to handle materials about which there is little published information, there is a hazardous situation. Or, any time someone with limited knowledge of a particular material must process it, there is a hazardous situation. A bomb demolition expert goes through many rigorous days of training before he is allowed to defuse his first bomb. The lion tamer at the circus has spent years learning his profession, and men who jump from planes to fight forest fires have had months of training for their assignment. All of these occupations or vocations are hazardous. Yet, someone with limited education is permitted to haul away chemical wastes and dispose of them in a presumably safe manner so that they cannot get back into the environment. No one bothers to tell him whether these wastes are toxic, flammable, explosive, corrosive or many other things which he really should know in order to do his job properly. There is also a good chance that even if he did know the hazards of the waste material that he was handling, he wouldn't know how to protect himself, other people, or the environment both during and after its disposal.

This, therefore, is what really makes wastes hazardous. It is easy to single out a few bad actors and say that this waste is toxic and that waste is a suspected carcinogen, or that another waste will burn the eyes and the skin, and yet another waste will contaminate groundwater. But sometimes the hazards are much more subtle than this. Sometimes the hazards do not exist until two or more chemicals are mixed together, possibly at the disposal site, which creates a third chemical which is toxic or even has lethal capabilities.

13-2 DEFINITION AND CLASSES

To examine what is meant by a hazardous waste, a definition or a series of conditions are necessary to identify the hazardous waste as opposed to the nonhazardous waste. This requires the answers to the questions, who, what, where, why, and when?

The first question should deal with who or what is affected or might be affected by the hazard. Injury or harm can be sustained by a variety of things: humans, animal life, vegetation, property, or the general environment. We should not only be concerned with human health and property but for all other living organisms and the environment. The environment is somewhat of a broad-based term which, if taken literally, includes human, animal, and vegetable life. If we look at it selfishly, any change in the environment may, in the long run, affect our way of living.

The second question which must be asked is, Why is the material hazardous? For example, does it contain toxic substances which might be ingested or absorbed by humans, animals, or plants leading to their ultimate destruction. The material might be flammable; it might be explosive or highly reactive. It might cause irritation or sensitization. It might be corrosive or radioactive. It might involve bioconcentration or even have genetic change potential. But, there must be a reason why it is considered hazardous.

Finally, it is necessary to determine when the material is hazardous. Some materials are not hazardous until their threshold of exposure has been exceeded. For example, minute quantities of ozone are created during a thunder and lightning storm; yet, it has a threshold value of 0.1 ppm, indicating that it is an extremely toxic substance. Everyone knows that sulfuric acid is very corrosive, but, in very weak concentrations, it will not burn the skin; however, a drop of concentrated sulfuric on the skin would be very painful and leave an open sore. Therefore, the hazard is not a function of the chemical formula but more of the concentration. It is important, therefore, to know when things are hazardous and when they are not. Some materials are quite safe to handle and store, but they may have a significant hazard associated with them when mixed with other materials. For example, sodium sulfide is not a particularly hazardous material in itself. It may be handled as a generally nonhazardous substance. But, if a spent sodium sulfide solution should be mixed with a strong acid at a disposal site, the result would be hydrogen sulfide gas which is extremely toxic and, in relatively low concentrations, is lethal. Other materials may be hazardous and become more hazardous with time. Ethyl ether, which has been used for years as an anesthetic and a solvent, is hazardous because of its flammability. However, when stored for a long time, it becomes more hazardous because of its tendency to form peroxides which are extremely explosive. The waste ether might auto-ignite or detonate without warning.

It is, therefore, not enough to just say waste is hazardous. It is necessary to define why it is hazardous, situations where it may be hazardous, and conditions when it is hazardous, such as, concentration, form and degree. With this information in mind, it is possible to find some of the areas or aspects which make the waste hazardous in the first place, and this should lead to possible solutions, see Table 13-1.

Bioconcentrative Wastes. Wastes which contain lead, cadmium, mercury, or polychlorinated biphenyls (PCB's) are not readily expelled by living organisms and are considered to be bioconcentrative. Such wastes would normally come from industrial operations involving the use of these materials. Normally, tight control is held over such materials, but the waste may be offered to public contractors in an acid containing these metals. Disposal of such wastes can be handled by precipitation of the heavy metal and reclamation or disposal in a designated landfill. Under proper conditions, PCB's may be incinerated.

Toxic Wastes. Toxic wastes are those that may affect living organisms either by ingestion through the food chain, respiratory system, or through the surface of skin. Degrees of toxicity, of course, vary. Some forms of toxic wastes are not initially toxic, or at least to all appearances, but may become toxic or may manifest their toxicity after repeated exposure with the ultimate destruction of the organism. The proper handling and disposal of such wastes is extremely critical.

TABLE 13-1. Hazardous Wastes Type

Miscellaneous Inorganics

Ammonium chromate
Ammonium dichromate
Antimony pentafluoride
Antimony trifluoride
Arsenic trichloride
Arsenic trioxide
Cadmium (alloys)
Cadmium chloride
Cadmium cyanide
Cadmium nitrate
Cadmium oxide
Cadmium phosphate
Cadmium potassium cyanide
Cadmium (powdered)
Cadmium sulfate
Calcium arsenate
Calcium arsenite
Calcium cyanides
Chromic acid
Copper arsenate
Copper cyanides
Cyanide (ion)
Decaborane
Diborane
Hexaborane
Hydrazine
Hydrazine azide
Lead arsenate
Lead arsenite
Lead azide
Lead cyanide
Magnesium arsenite
Manganese arsenate
Mercuric chloride
Mercuric cyanide
Mercuric diammonium chloride
Mercuric nitrate
Mercuric sulfate
Mercury
Nickel carbonyl
Nickel cyanide
Pentaborane-9
Potassium dichromate
Selenium
Silver azide
Silver cyanide
Sodium arsenate
Sodium arsenite
Sodium bichromate
Sodium chromate
Sodium cyanide
Sodium monofluoroacetate
Tetraborane
Thailium compounds
Zinc arsenate
Zinc arsenite
Zinc cyanide

Halogens and Interhalogens

Bromine pentafluoride
Chlorine
Chlorine pentafluoride
Chlorine trifluoride
Fluorine
Perchloryl fluoride

Miscellaneous Organics

Acrolein
Alkyl leads
Carcinogens (in general)
Chloropicrin
Copper acetylide
Copper chlorotertrazole
Cyanuric triazide
Diazodinitrophenol (DDNP)
Dimethyl sulfate
Dinitrobenzene
Dinitro cresols
Dinitrophenol
Dinitrotoluene
Dipentaerythritol hexanitrate (DPEHN)
cinate (LDNR)
Lead styphnate
Lewisite (2-chloroethyenyl dichloroarsine)
Mannitol hexanitrate
Nitroaniline
Nitrocellulose
Nitrogen mustards 2,2′,2″ trichlorotriethylamine)
Nitroglycerin
Organic mercury compounds
Pentachlorophenol
Picric acid
Potassium dinitrobenzfuroxan (KDNBF)
Silver acetylide
Silver tetrazene
Tear gas (CN) (chloroacetophenone)
Tear gas (CS) 2-chlorobenzylidene malononitrile)
Tetrazene
VX (ethoxy-methyl phosphoryl N.N. dipropoxy-(2-2), thiocholine)

Organic Halogen Compounds

Aldrin
Chlorinate aromatics
Chlordane
Copper acetoarsenite
2,4-D (2,4-dichlorophenoxyacetic acid)
DDD
DDT
Demeton
Dieldrin
Endrin
Ethylene bromide
Fluorides (organic)
Guthion

TABLE 13-1. Continued

Pentaborane-11	GB (propoxy (2)-methylphosophoryl fluoride)	Heptachlor
Perchloric acid (to 72%)	Gelatinized nitro-cellulose (PNC)	Lindane
Phosgene (carbonyl chloride)	Glycol dinitrate	Methyl bromide
Potassium arsenite	Gold fulminate	Methyl chloride
Potassium chromate	Lead 2,4-dinitroresor-	Methyl parathion
Potassium cyanide		Parathion
		Polychlorinated biphenyls (PCB)

Source: Office of Solid Waste Management Programs. Report to Congress; disposal of hazardous wastes. Environmental Protection Agency Publication No. SW-115. Washington, U.S. Government Printing Office, p. 110, 1974.

Flammable Wastes. Flammable wastes are those considered flammable under the National Fire Protection Association, Category #4.

Explosive Wastes. Explosive wastes are any materials which would have the possibility of explosive detonation, with or without an external source of ignition, when stored in a nonpressurized container.

Reactive Wastes. Reactive wastes would be those considered to be reactive under the National Fire Protection Association, Category #4.

Irritating or Sensitizing Wastes. These are wastes which may come under the toxic waste heading but might not be toxic in the sense that they will cause permanent damage. They may, however, be irritating to the eyes, the skin, or the respiratory system when in contact, or they may have a sensitization potential so that increased contact with the wastes will cause more dramatic and more serious irritation.

Corrosive Wastes. Corrosive wastes are those that will damage human organisms or property by causing rapid oxidation of the material within their contact or will react chemically with these organisms or property.

Radioactive Wastes. A radioactive waste is one that emits neutrons, alpha, beta, or gamma radiation due to nuclear decay. Radioactive wastes will not be considered in this discussion on hazardous wastes since present regulatory controls established by ERDA essentially eliminate the possibility of uncontrolled radioactive industrial waste. This does not mean that it is not a hazardous waste, indeed it might be exceedingly hazardous, but normal detection methods would not be sufficient to determine its presence unless the waste was suspected of being radioactive.

Other Hazardous Materials. Certain materials might be considered hazardous because they will affect genetic changes in living organisms or because they may have some long-range etiological effects. Environmentalists have been accusing industrial wastes of causing such damage for a number of years. However, such things are often hard to prove. For example, black lung disease, prevalent in miners, is an etiological effect of an airborne material in coal mines. And, the higher incidence of respiratory problems in cities such as London, which at one time burned high sulfur coal, is probably also a direct etiological effect of the sulfur dioxide in the atmosphere. The genetic effect of the use of the drug thalidomide was malformed children. We rely on the genetic effect of certain insecticides to change certain types of troublesome insects in such a way that the species will disappear. There is the possibility of hazardous wastes causing similar etiological and genetic effects if they reach the atmosphere.

Today in industry there is a great deal of confusion between the Toxic Substances Control Act and the EPA's guidelines and regulations towards the disposal of hazardous material. These are two distinct and different parcels of legislation, the latter is controlled currently by the Resource Conservation and Recovery Act.

The Toxic Substances Control Act (TOSCA) specifies 15 classes of known or suspected toxic substances which are currently being reviewed concerning their potential risk to the public. Those materials currently under review are acrylonitirole, benzene, lead, polybrominated biphenyls, inorganic phosphates, cadmium, asbestos, 2-3 dibromopropylphosphate, hexachloralbenzene, trichloralethylene, mercury, benzadine, arsenic, polynuclear aromatics, and vinylidene chloride.

As other items are added to the control list, the inventory published by the EPA will include data on molecular structure and the location of manufacturing facilities for each material now used in the U.S. Material for which production has been halted as of July 1, 1974, or later, may go on this list; otherwise, substances produced before but not in the succeeding three years will be subject to premarket notification to the EPA along with the new substances.

13-3 HAZARDOUS WASTE MANAGEMENT GUIDELINES AND REGULATIONS

The Resource Conservation and Recovery Act of 1976, Public Law 94-580, attempted to take on the very difficult problem of hazardous waste management from a legal or governmental point of view and involved itself with almost every aspect of hazardous waste management. As this book is being written, the final guidelines on hazardous waste management have not been completed and a good bit of the information required by the public and those handling hazardous waste has still to be developed at the state level. But, during the Fall of 1977, the EPA conducted a number of public meetings on these subjects with the intention that final regulations would be complete by May 1978, and once the final regulations are published, generators and transporters and hazardous waste facility operators will have 90 days in which to notify either the EPA or the state administrator of their compliance. The various sections of these regulations are as follows:

3001 Identification and Listing of Hazardous Waste
3002 Generators of Hazardous Wastes
3003 Transporters of Hazardous Wastes
3004 Owners and Operators of Hazardous Waste Storage and Disposal Facilities
3005 Permits for Treatment, Storage and Disposal
3006 Guidelines for State Programs
3010 Preliminary Notification

To expand upon these section titles, a brief outline of what they contain is perhaps in order.

3001 Identification and Listing of Hazardous Wastes

This section will give the criteria for the identification and listing of hazardous waste, that is, those things which make the waste hazardous. It will identify the characteristics and list specific hazardous wastes which could cause or significantly contribute to a present or potential hazard. It will also deal with the prospective content of various wastes. Such things as household refuse, small waste generators, mining and milling wastes will not be included. It will give a definition of what a waste material is and give its characteristics, such as flammability, corrosiveness, toxicity, etc. Presumably, this section will list substances and processes which produce hazardous wastes.

It will further deal with the definition of wastes which are to be destroyed or buried and wastes where a resource is to be recovered.

3002 Standards Applicable to Generators of Hazardous Wastes

This section deals with what the generators of hazardous wastes will have to do regarding record keeping, labeling, containerization, information about the waste material and how to develop a manifest system and a reporting system for these wastes. The record keeping, safe handling and shipment information must be developed in such a way that a permanent record can be kept for a period of at least 3 years. Section 3002 also provides a method of enforcement of such record keeping. This section has unresolved problems regarding labeling, safe handling and shipment of the material because it has not yet been determined whether the EPA or the Department of Transportation will have final responsibility. It appears at the present time that the Department of Transportation regulations will be paramount in determining how materials shall be labeled and shipped. The Department of Transportation, Materials and Transportation Bureau, in the *Federal Register,* Monday, August 2, 1977, indicated that they are already into the problem of tightening up their regulations for shipment and packaging of various hazardous wastes. The difficulty remains, however, that most state agencies in developing their own hazardous waste program are going to have to walk a tightrope between the EPA's guidelines and the Department of Transportation's regulations. It is the interface between the hazardous waste management guidelines and the DOT regulations which is the most critical, and this will be the manifest system. The manifest will be a tracking document to track any shipment of a generator of hazardous wastes to the transporter to the waste facility and, if necessary, back to the generator. The EPA permanent, 3-year record will be developed from this manifest and will be a verification of shipment and also a basis for enforcement, so record keeping is extremely important.

The small generator classification of wastes has not yet been defined adequately and could be a problem. Also, shipment to resource recovery facilities under the present Resource Conservation and Recovery Act does not require record keeping, labeling or manifest, although it probably will in the future.

3003 Standards Applicable to Transporters of Hazardous Wastes

This section deals with transporters who move the wastes from the generator to the waste disposal facility, and this will include record keeping, proper labeling and compliance with the manifest system, and the responsibility to see that all hazardous waste goes to the designated permitted facility. Obviously, these transporters will have to work in close coordination with the regulations of the Department of Transportation. This means that the transporter, to the greatest extent, will be responsible for the acceptance of a certified manifest before picking up the hazardous waste and be responsible to see that the materials which he carries are properly labeled and packaged. He will also be required to see that there are no incompatible wastes and will have to meet all DOT loading and stowage requirements. He will also have to see that the material is removed from the vehicle in the proper manner and certify where it is deposited. The marking and placarding of vehicles will have to be in accordance with DOT regulations, although development of placards for various hazardous wastes has not yet been determined, and the transporter will be hard pressed to develop close integration of DOT and EPA regulations unless some other resolution is made.

3004 Standards Applicable to Owners and Operators of Hazardous Waste Treatment Storage and Disposal Facilities

This section primarily authorizes the EPA to promulgate regulations pertaining to the records of wastes and the manner in which they are handled, reporting monitoring at various facilities, inspection of the facility, the location of hazardous waste disposal facilities, the design, operation and construction of these facilities, and the training of their personnel, the ownership and the continuity of operations and financial responsibilities, contingency plans and their performance standards which are required to protect the environment and human health.

This would include such considerations as the protection of groundwater so that it would not have to be specially treated to make it suitable for drinking purposes and the protection of ambient air not only in the disposal facility and loading areas but in the general areas surrounding any waste treatment plant. It would include air and groundwater leachate monitoring requirements which implies that all surface water run-off from active areas would have to be collected and confined to a single point source and that all landfills would have a leachate collection system or natural conditions capable of a 500-year containment or the equivalent. It would also imply that incinerators must achieve combustion efficiencies of 99% or greater and operate above 1000°C for 2 seconds retention or the equivalent. Obviously, this is a sticky portion of the regulations because it must deal with such abstracts as financial responsibility and whether it is really legal to require zero discharge or, for that matter, even practical.

3005 Permits for Treatment, Storage or Disposal of Hazardous Wastes

This section requires each owner or operator of a hazardous waste treatment storage or disposal facility to obtain a permit, and the EPA is authorized to promulgate regulations specifying the content of the applications, how the permits will be issued, the time limitations for completion, or modifications necessary to bring a facility into compliance with the Act and the procedure for the revocation of permits.

This whole area of permits is one which will be very difficult to resolve, at least initially, since the hazardous waste disposal facilities currently available in the United States are few and their methods, in many cases, are questionable so that interim permits will undoubtedly have to be granted in order to have some place to put the hazardous waste which is being generated at the present time. Ultimately, however, it appears that a permit system can be made workable. There is also the question of whether an Environmental Impact Statement will have to be prepared for every facility. It would only seem fair that such facilities would have to have Environmental Impact Statements if waste producers have to have Environmental Impact Statements for the construction of a plant which creates the hazardous waste.

3006 Guidelines for Authorized State Hazardous Waste Programs

This section authorizes the EPA to promulgate guidelines to assist the states in state programs, describe the procedural requirements for states applying for authorization, and describe the interface of the EPA's over-sight of the state program as well as the required provisions for withdrawal of their authorization. It will certainly have to deal with legislative authority and the method of developing the criteria and a permit mechanism for storage, treatment and disposal; a manifest system; an identification of the resources; a delineation of the responsibilities of the various agencies; who will have primary jurisdiction; and provide an opportunity for the public to participate in the over-all program. It will also have to cope with the problems needed for protection of public health and the environment and will have to deal with the discrimination between waste produced within or outside any state. It will undoubtedly be a regulatory nightmare to complete, and there will have to be an authorization plan for bringing all elements of the program into compliance with the requirements which say full authorization by October 21, 1980. It will have to deal with the relation of the hazardous waste generator and the transport and waste management industries and also deal with the proper balance of economic and environmental impacts for the disposal of certain wastes.

3010 Preliminary Notification

This section will give the EPA quthority to promulgate regulations to require notification by hazardous waste handlers within 90 days of promulgation of regulations under Section 3001 covering criteria for identification or listing. In that notification, a hazardous waste handler must state the location, general description of activity and the identified or listed hazardous waste handled which must be filed with the EPA or with states having authorized hazardous waste programs under Section 3006.

The purpose of the above discussion is to illustrate the complexity of the entire hazardous waste disposal problem now that hazardous wastes will come under

Federal Government regulations. On the one side, it will be necessary for the generator, the transporter, and the hazardous waste disposal facility operator to coordinate their activities as some have done and others have not in the past. On the other side, it will require the EPA, the states, and the Department of Transportation, and both the federal and state governments, to develop regulations which will protect the life, health, and property of its citizens while at the same time remaining within the bounds of economic feasibility. State governments must consider the possibility that they may have to operate hazardous waste disposal facilities. Unless they can get private operators to effect proper disposal under economically beneficial conditions, hazardous waste management will become a municipal or governmental responsibility which will have to be paid for with tax dollars.

Since many states at the present time are developing hazardous waste legislation and perhaps will have developed such legislation by the time this book is in print, it would be foolish to cite any more than potential content of these regulations. However, specific information on any state can be received from the National Solid Waste Management Association, 1120 Connecticut Avenue, N.W., Suite 930, Washington, D.C. 20036.

13-4 THE HANDLING OF HAZARDOUS WASTE MATERIAL

Once the "why's" of the hazard have been defined, then it is necessary to devise a system which will permit safe handling of the hazardous waste between its point of creation and its ultimate disposal. Actually, the problem of handling a hazardous waste is perhaps more difficult than the disposal problem. This is because there are many more options in the handling of such materials than there are for their ultimate disposal. Waste materials can be packaged in drums, bottles, cartons, cans, trucks, tanks, cylinders, etc. The variety of possible containers is almost unlimited. Obviously, the transportation of hazardous waste virtually eliminates uncontained gaseous wastes. Most gaseous wastes which are hazardous in nature must be treated on the site and, therefore, transportation is not involved with the exception that it may have to be piped from one section of the plant to another for its disposal. Liquid wastes, on the other hand, are very often transported to disposal sites, and, the bulk of hazardous materials handled in the U.S. today are those in liquid form. There are some hazardous solids but the preponderance of solid waste materials, industrial, and otherwise, is not hazardous.

The handling of hazardous wastes will, of course, depend upon the type and degree of hazard. All waste materials should be carried in containers which are immune to the corrosive aspects of the waste. People handling these waste containers should have proper personal protection. This might involve special suits of clothing or only the use of special eyeglasses and gloves. In any event, if the material is severely corrosive, the individuals handling the material should be protected from any physical harm which might be caused in the event of a spill or leak in one of the containers.

Highly flammable or explosive materials should be handled in special areas that are equipped with explosion-proof electrical fixtures and spark-proof machinery with special provisions for the elimination of all static discharge.

Toxic wastes, or bioconcentrative materials, or materials which are suspected of being carcinogens, should be transferred in sealed containers from the point of generation to the disposal site; and, where it is necessary on occasion for workers to be exposed to these materials or their vapors, they should have proper respirators as well as complete body cover so that the materials cannot come in contact with the skin, eyes, or the respiratory tract. This is also true where irritating or sensitizing chemicals may be used. Even though the degree of protection may not be quite as important, respiratory protective agents definitely should be utilized.

13-4.1 Shipment of Hazardous Materials

Congress passed the Hazardous Materials Transportation Act of 1974 which gave the Secretary of Transportation the authority to regulate the shipment of such materials. This automatically has been delegated to the Materials Transportation Bureau of the Department of Transportation which has jurisdiction over almost all moving of hazardous materials in interstate commerce whether by road, rail, air, water, or pipeline. The Secretary's authority also covers materials considered to pose an unreasonable risk, all carriers and shippers, container or package makers, storage and unloading, safety aspects, and commerce affecting interstate commerce, or his representative, can hold hearings, issue orders and injunctions and move for civil or criminal penalties.

Industry accordingly organized the Hazardous Materials Advisory Committee (HMAC) in 1974, also headquartered in Washington, which deals with all components of the transportation community and works toward achieving practical, efficient laws and regulations for the safe movement of hazardous material and bringing together truckers, airlines, pipeline operators, waterborne carriers, insurers, and others concerned with these shipments. This organization holds seminars telling industry how to comply with federal hazardous material regulations with the focus on the latest rules. It teaches what is required of shippers and carriers; it provides workshop and problem-solving sessions; it also trains those in the program to train others in the company pursuant to the requirements of Public Law 93-633 so that industry is helping to make the handling and transportation of hazardous wastes a much safer proposition. Industrial wastes generators, handlers, and disposers should not be without the proper information to deal with hazardous materials since it is available from these sources.

Hazardous waste materials should always be transported in the largest feasible container unless there is a serious chance of a violent reaction or explosion; then, perhaps, the smallest feasible container makes more sense. The most difficult disposal problem today is dealing with the hundreds of relatively small containers that we often use to transport our waste. Laboratory chemicals which are no longer useful are often discarded in their original containers, usually small glass bottles. These laboratory containers could be virtually anything, from acids and bases, either strong or weak, solvents, very reactive materials, organic dyes, even explosives, such as iodides and peroxides. Unlabelled, they constitute one of the most serious hazardous waste disposal problems in the United States today. Even when they are properly labelled, they constitute a serious problem because many uninformed people feel that they can be dumped into a landfill or an incinerator with complete impunity because the quantities are so small nothing serious could result. Anyone who has ever seen metallic sodium thrown into water quickly realizes that a very small amount of sodium makes a big flame. Imagine a bottle of metallic sodium or potassium, which is equally reactive, thrown into a chemical landfill where flammable solvents might be present. The resulting fire could shut the landfill for weeks or months, and someone could be seriously hurt in the bargain.

To properly dispose of such wastes, each bottle or container must have some process established for its ultimate demise. Obviously, this could become very expensive. Such materials as acids, bases, and certain organic materials could be combined at the time they are discarded in a way that their disposal process would be the same. Then, the number of individual steps would be greatly reduced.

If larger containers are used, such as drums, which are perhaps the most common receptable for industrial waste liquid material, there is still a serious disposal problem. One drum absolutely full holds only 55 gal. of material, and each drum is filled with a hazardous waste that becomes a potential hazard itself. If the drum is made of suitable material which will not corrode from the waste which it contains, it may yet have to contend with atmospheric corrosion before the ultimate disposal of its contents if it is stored outdoors for any period of time. In some circumstances, a drum will begin to leak its contents onto the ground terminating any thought of proper disposal. Each drum in a truckload of drums full of waste hazardous material is a potential hazard. Each drum has the capability of leaking by itself and creating a separate hazardous situation both to the people handling the material as well as to those transporting and receiving it, not to mention the general public.

The best approach to any type of liquid waste disposal, hazardous or otherwise, if feasible, is to use a tank truck or railroad tank car. These vehicles were designed for such purposes and come equipped with a variety of corrosion-proof linings and can be ordered for handling almost any type of corrosive hazardous waste. Rubber-lined, glass-lined, and plastic-lined tank trucks and tank cars are available in most areas. If a tank truck leaks, of course, it is a rather serious situation; if a railroad car or a tank truck is in an accident where the vessel is broken in the accident, then a very serious situation results. However, the incidence of failure or accident using tank trucks and tank cars is much less than it is when using drums for carrying waste materials.

Solid waste materials can be carried in a variety of containers, usually by truck or rail car. If they are hazardous, then the options are somewhat limited. Waste cyanides from heat treating operations could be considered to be a hazardous waste. They are toxic to a limited degree, but, of greater concern, they can react with acids to produce hydrogen cyanide gas, which is deadly. Such materials should be shipped in the largest feasible closed containers which are impervious to moisture or to any material which may be outside the container. Often, this is best effected by enclosing the solid waste in some type of impervious plastic envelope and then placing this envelope in the drum. "Fiber pack" drums have often been used for waste filter cake which is hazardous in nature, especially if the waste material is organic and lends itself to incineration, the advantage being that the entire package can be burned as a unit. This is often ideal from a final disposal standpoint but not too acceptable from a transportation safety standpoint. These fiber packs are somewhat vulnerable to mechanical damage by fork lift trucks or other conveying devices. They cannot be left outdoors for any extended period of time because the fiber eventually deteriorates with water contact, exposing or leaching out the contents. Often such fiber drums shipped inside steel drums make a satisfactory double container arrangement. The steel drums, of course, must be the type with removable tops so that the fiber packs can be removed prior to incineration. Large returnable bins, or other types of sealed containers, might be preferable for certain types of wastes. Each waste will have its own character and will demand a separate type of container for transport.

Often, solid waste materials have only one acceptable mode of disposal, landfill. If they cannot be landfilled in bulk because of their hazardous nature, which most likely will be toxicity, then they can be buried in drums or fiber packs if they are immediately covered with a sufficient quantity of earth in a proper "dedicated" landfill where no leaching of the material by groundwater can take place.

13-4.2 Identification

One of the biggest problems in handling a hazardous waste material, especially when it is being shipped off the plant site where it was created, is in its proper identification. In other words, what is it? What are its hazardous characteristics? What should be done with it at the disposal site? And, what should be done with it in the event of an accident along the road or railroad during the course of its transportation? The ICC has been concerned with this for a long time and has certain regulations where these carriers come under interstate jurisdiction. However, there are literally thousands of carriers which do not come under interstate jurisdiction and are subject only to the relatively meager laws that states have provided. This is not to say that states are not becoming concerned about short-haul carriage of hazardous wastes. They are beginning to put together legislative packages to govern the hauling and the handling of hazardous waste. In most cases, this begins with a manifest, or bill of lading, which tells not only the carrier but also the receiver, what the waste is, and also its degree of hazard or danger. While this bill of lading is not foolproof, it does give the carrier and the receiver of the goods some written warning concerning the hazards of the material and is significantly better than present practices which essentially provide no such warning.

The National Fire Protection Association (NFPA) has developed a hazard identification system which most companies dealing with hazardous and flammable waste currently use on all containers which hold the material within the plant and those which go outside the plant. The NFPA code symbol is diamond shapped and has four sections: one deals with the health aspects of the material and is coded blue in degree of seriousness numbered from 0 to 4; the second area deals with flammability and is similarly numbered; the third deals with the reactivity and is similarly numbered; and the fourth square indicates whether water is compatible or not with the material being carried or stored in the container, see Figure 13-1.

The placarding of transport vehicles under Section 3003 of the Hazardous Waste Management Guidelines has not yet been developed but will not necessarily obviate the NFPA code. It will only apply to vehicles and will be controlled by the DOT.

13-4.3 Storage and Disposal

Disposal of a hazardous waste is not unlike the disposal of a nonhazardous waste of similar physical and chemical properties. Actually, the act of disposal is quite simple and straightforward. The major hazard involves the handling of material from the time it is brought to the disposal site until the time that it is ultimately disposed of. First, looking at the problem of on-site storage of hazardous wastes, if the wastes are highly flammable, they should be stored in accordance with rules and regulations for flammable materials. This includes that the storage tanks should be remotely located from buildings and other facilities; that they should have breather type vents (the discharge of the breather lines preferably run through activated carbon beds); That any pumps or agitators or other electric equipment in the immediate area be of Class 1, Group D, Div. II, Explosive Proof Construction; and that the tanks be diked in accordance with

Type of Possible Injury (Blue)		Susceptibility of Materials to Burning (Red)		Susceptibility of Release of Energy (Yellow)	
Signal		Signal		Signal	
4	Materials which on very short exposure could cause death or major residual injury even though prompt medical treatment were given.	4	Materials which will rapidily or completely vaporize at atmospheric pressure and normal ambient temperature, or which are readily dispersed in air and which will burn readily.	4	Materials which in themselves are readily capable of detonation or of explosive decomposition or reaction at normal temperatures and pressures.
3	Materials which on short exposure could cause serious temporary or residual injury even though prompt medical treatment were given.	3	Liquids and solids that can be ignited under almost all ambient temperature conditions.	3	Materials which in themselves are capable of detonation or explosive reaction but require a strong initiating source or which must be heated under confinement before initiation or which react explosively with water.
2	Materials which on tense or continued exposure could cause temporary incapacitation or possible residual injury unless prompt medical treatment is given.	2	Materials that must be moderately heated or exposed to relatively high ambient temperatures before ignition can occur.	2	Materials which in themselves are normally unstable and readily undergo violent chemical changes but do not detonate. Also materials which may react violently with water or which may form potentially explosive mixtures with water.
1	Materials which on exposure would cause irritation but only minor residual injury even if no treatment is given.	1	Materials that must be preheated before ignition can occur.	1	Materials which in themselves are normally stable, but which can become unstable at elevated temperatures and pressures or which may react with water with some release of energy but not violently.
0	Materials which on exposure under fire conditions would offer no hazard beyond that of ordinary combustible material.	0	Materials that will not burn.	0	Materials which in themselves are normally stable, even under fire exposure conditions, and which are not reactive with water.

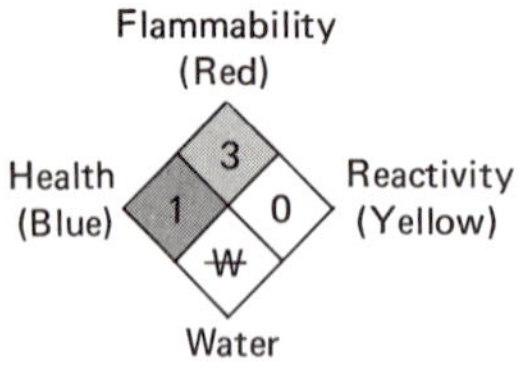

NFPA Hazard
Identification System

Degree of Hazard
4 Severe
3 Special
2 Moderate
1 Minor
0 No Hazard

Fig. 13-1. NFPA hazard identification code. ***Source:*** **NFPA.**

petroleum refinery practice. Large quantities of extremely flammable material should not be stored at the disposal site any longer than absolutely necessary. Usually flammable material is incinerated, and it should be incinerated with the greatest dispatch.

Incineration may also be applicable to explosive materials or highly reactive organic materials. These materials should not be stored, if at all possible, at the disposal site. They should be incinerated promptly upon receipt. While this may not be possible in every case, any storage at the disposal site provided for explosive materials should be remote from the rest of the facility and properly bunkered and diked to prevent the spread of a fire in the event of an explosion.

Materials which are reactive and form lethal gases upon reaction should be carefully separated from each other at the disposal site so that no chance of inadvertent mixing of the two can occur. Acids and inorganic cyanides and sulfides should be located on opposite sides of the disposal site in a way that an inadvertent mixing cannot happen.

Toxic or irritating wastes should be stored in tanks or containers which suit the corrosive properties of these wastes; breather vents should be trapped with activated carbon. Such wastes should never be stored in open lagoons or ponds.

While drum storage at disposal sites is inevitable, it would certainly be desirable to have drums stored on a concrete pad which has provision for adequate drainage into a sump which can be analyzed and treated before the runoff can get to the groundwater. It would be ideal to have all drums stored under roof so that they would not be open to the weather. Fiber pack drums must be stored in this manner, or they may ultimately result in contamination of the ground and the water table.

13-5 FINAL DISPOSAL

The final disposal of hazardous wastes is essentially the same as the disposal of any nonhazardous wastes. The available options are limited, however. The first and probably the most desirable, yet not the least costly, is incineration. Incineration provides for complete destruction of hazardous wastes if the system is properly designed and the ultimate temperature is high enough. A consideration of the combustion characteristics of the waste must be given and, if the waste does not have a high enough calorific value of its own, then auxiliary fuel must be used.

Halogenated wastes will produce halogen acids which must be scrubbed before admitting the final effluent to the atmosphere. Wastes containing sulfur must also be scrubbed before release. But, in all cases, incineration at final temperatures above 1800°F is recommended for complete destruction of the waste. Residence times in the incinerator will vary depending upon the specific material. All organic materials are applicable for incineration. Inorganic-organic mixtures will present difficulties because of the final effluent which must be removed by high energy scrubbing. In the case of an inorganic-organic mixture, it is usually desirable to incinerate the organic material if it is hazardous rather than using some other approach.

The second possibility is the chemical landfill. Chemical landfills should have a design such that any leachate from the landfill can be trapped in some type of sump or lagoon, to be further treated before release. The design of the chemical landfill will greatly depend upon the geology of the disposal site. An impervious clay soil is available in certain parts of the United States, which is suitable as a barrier between the chemical landfill and the groundwater. In most areas, some type of artificial barrier or membrane will have to be constructed between the soil and the contents of the landfill to be totally satisfactory. Such a landfill should be monitored at the perimeters with wells which are used for testing the possible contamination of the groundwater. This procedure is covered generally in an earlier chapter but is much more important in the case of hazardous chemical waste.

Two types of hazardous wastes are deposited in chemical landfills. One is a waste which will normally biodegrade, thus removing the present and potential hazard. The other type of waste is one which is impossible to degrade in any manner. This type of waste placed in a chemical landfill will never disappear. Such things as oxides or sulfides of heavy metals which are highly insoluble in water and which do not present a severe leachate problem may be deposited in a well-designed chemical landfill or they may find their final resting place in a dedicated landfill which presumably can never be opened once it is closed. A dedicated landfill could consist of a large concrete vault into which drums or other containers of this type of material are placed and then covered with concrete. The question concerning the use of a dedicated landfill is whether this is not a very poor use of the land and whether in generations to follow it may not come back to haunt us. There are, nevertheless, certain types of inorganic waste which, if they are not recycled, cannot be destroyed either biologically or by incineration; therefore, they must find a resting place for perpetuity. This has been true, of course, for radioactive wastes, especially those with high levels of radiation and long decay periods. Such materials, until now, have been placed in underground tanks or vaults to remain there for hundreds, or even thousands, of years. Earthquakes or geological changes in the earth's crust at that particular location might cause serious problems. In those cases, sites have been selected to obviate such circumstances.

A third possibility for the disposal of hazardous wastes is the use of the deep well. Deep wells have been used for this purpose in Louisiana, Texas, and in Michigan. Generally, it is a short-sighted approach to the problem. While the waste is safe as long as the deep well is sound, it offers all kinds of complications if the well begins to leak into nearby aquifers. Contamination of groundwater under such circumstances may not be a present danger but is always a future consideration. State and federal laws likely will soon prevent the use of such wells for the disposal of certain hazardous wastes.

Biological systems may be used for the disposal of certain types of hazardous wastes which are in diluted form, if they are organic, and if they lend themselves to biological degradation. Care should be given to see that refractory types of materials are not present in quantity.

Another possibility for the destruction or disposal of hazardous wastes is offshore disposal. This method has not been proven positive or negative with respect to hazardous wastes. Certain radioactive materials have been encapsulated in concrete and dropped to deep sections of the ocean, presumably to remain there forever. If the casing, for one reason or another, should break open, the dilution factor of the ocean would be such that it would render most hazardous wastes ultimately harmless. However, there might be a serious effect in the immediate vicinity for some time after the rupture occurred. It is not a final solution to the problem unless it is known that the hazardous waste will decompose in the sea water within a reasonable period of time. Studies now under way to determine the effects of dumping both municipal and industrial wastes into the sea should give us guidelines concerning the types of wastes and quantities of wastes we can expect to dispose of in this manner.

13-6 PCB'S

One class of hazardous waste that requires special mention are the polychlorinated biphenyls better known as PCBs. The U.S. EPA has already halted the manufacture of PCB's and has proposed zero discharge standards for all PCB materials in relation to ambient water quality with criterion of 1ppt. One ppb had been established by several states as an interim discharge standard.

Currently, Polychlorinated biphenyls are found primarily in industrial capacitors and transformers, heat transfer fluids, and in some types of hydraulic and binding oils. The disposal of these materials has become a significant problem because it is difficult to find a safe, environmentally acceptable method of eliminating them without creating a secondary liability. Polychlorinated biphenyls are what is known as refractory wastes. The physical characteristics of these materials, their inherent high stability and strong ultraviolet absorption, all help in determining whether or not significant quantities are present. The accepted identification technique for low level concentrations involves the use of an electron capture gas chromatograph and is performed only with difficulty.

The major contamination problem with PCBs is spills, which are handled under Section 311 of the FWPCA until other more acceptable methods may be developed. It is impossible to ever remove trace levels of PCBs in soil or absorbent materials such as concrete. The important thing is to isolate the locations of such spills where PCBs are used either as a heat transfer oil or as a capacitor or transformer coolant.

Polychlorinated biphenyls in waste streams containing less than 1 ppm can be reduced to concentrations below the 1 ppb level either by carbon adsorption and regeneration, destruction by dechlorination using ultraviolet assisted ozone, and adsorption on ion exchange resins and other materials. But, by far the most acceptable method of disposing of high level concentrations of PCB wastes is high temperature incineration. A 2 sec dwell time at 2000°F with 3% excess oxygen or a 1.5 sec dwell time at 2700°F and 2% excess oxygen in the stack appears to be sufficient for destruction. Incinerators, of course, must be equipped to remove the HCl formed during incineration.

These PCB materials may also be containerized and sent to a secure chemical waste landfill. Such a landfill provides a long-term protection of the quality of surface and subsurface water, but it does not eliminate the future possibility of contamination if the landfill is ever damaged by earthquake or other holocaust.

13-7 HAZARDOUS WASTE MANAGEMENT AND THE PUBLIC

One of the significant problems with hazardous waste management is the area of public reaction. The term hazardous waste immediately implies a threat or danger to the public. Therefore, if a private operator or governmental agency intends to establish a hazardous landfill or a hazardous waste incinerator in a given area and, if this is publicized, the public in that area will undoubtedly be up in arms about the project, little realizing that such hazardous wastes have been disposed of in much more hazardous ways in the past. But, since the public was not aware that these hazardous wastes were being disposed of in an unsuitable landfill, they were never particularly concerned. Now that hazardous waste management has become a matter of governmental regulation, it also has become a matter of public information and, consequently, a matter of public concern. All of industry, the generator and the disposer, will have to become acutely aware that the public relations aspect of the problem may be as important as the disposal and handling methods, at least for the public.

It seems that the only recourse is education. If a little information creates a panic, then a well-planned information campaign may help avert one. It will become incumbent upon the generator, the transporter, the disposer, and the governmental agencies to coordinate their efforts to develop their plans in detail with solid technical back-up data before announcing a disposal site and the method of disposal. The public will want to be assured and reassured that they are not in danger. This is not to say that such measures in the past were not well planned but, in many cases, public acceptance was assumed rather than public rejection. The worst public reaction should be assumed and sound plans laid to respond to it.

A case in point was the disposal of "agent orange" by the Air Force five years after the end of the Vietnam War. Millions of gallons of defoliant used in Vietnam were stored in drums at two sites, one in the continental United States and one on Johnston Island in the Pacific. While the drums gradually rusted away, some of them began leaking a supposedly hazardous and toxic material containing small amounts of dioxins (considered to be capable of causing animal mutations). The federal government and state governments argued about the safety of disposing of these materials within the continental United States. While safe and acceptable methods were available to destroy these chlorinated hydrocarbons, they were rejected out of fear and the concern for public reaction; finally, these materials were disposed of at sea by incineration, not because the ultimate disposal method was any better, but because it was less obvious and it seemed to represent a lesser risk of public reaction. This does not condemn the disposal method but only the delay period, which was significantly more dangerous than land-based incineration at facilities which were known to have the suitable capabilities. It would at least have prevented leakage of those drums which deteriorated over the interim period. This is not cited as a criticism of governmental inaction as much as it is an example of the type of pressure which public concern can bring to bear on hazardous waste or toxic waste disposal.

Perhaps the most important area of hazardous waste disposal is the contingency plan. There will be serious errors in the future just as there have been in the past. Similar accidents may happen as the one which occurred in Seveso, Italy, in 1976 where a chemical factory inadvertently spread dioxins over a small town apparently permanently contaminating an area and reportedly adversely affecting the health of thousands. Spills from hazardous and toxic substances, both, inside plants, on public highways, and in waste disposal facilities, will never be completely eliminated, and these will cause injuries and death. But, those organizations with a firm contingency plan for containing the problem and reclaiming the land will survive and those who do not have contingency plans will create such public and governmental uproar that they may not continue in operation. Any company dealing in hazardous wastes must in the future not only comply with all the government regulations, but must take it upon themselves to insist that their personnel meet even higher standards.

To reiterate the emphasis of this chapter, the disposal of hazardous wastes is not so much a difficult disposal problem as it is a difficult handling problem. In most cases, hazardous wastes can be disposed of safely if all people involved in the handling and disposal of wastes are acutely aware of what it is, how it should be handled, and the disposal or treatment regimen that must be used for safe final disposal. It is always desirable to dispose of hazardous wastes at the point nearest to where they are created, rather than to give them to a third party for disposal, but it is recognized that this is not always possible. When transportation to an ultimate point of disposal is necessary, then complete information should follow the containers describing the hazards in the simplest of terms so that those who handle the material are safe from the hazards of the waste.

REFERENCES

13-1. Morton, W. I., "Safety Techniques for Workers Handling Hazardous Materials," *Chem. Eng.*, Deskbook Issue, October 18, 1976.

13-2. Cheremisinoff, P. N. and Holcomb, W. F., "Management of Hazardous and Toxic Wastes," *Pollut. Eng.*, April 1976.

13-3. Lindsey, A. W., "Ultimate Disposal of Spilled Hazardous Material," *Chem. Eng.*, October 17, 1975.

13-4. Koppenaal, D. W. and Manahan, S. E., "Hazardous Chemicals from Coal Conversion Processes," *Environ. Sci. Technol.*, Vol. **10**, No. 12, November 1976.

13-5. Editorial, "U.S. Toxic Substances Control Law. Real Impact Fuzzy Until 1980," *Chem. Eng.*, November 22, 1976.

13-6. U.S. EPA, "State Program Implementation Guide: Hazardous Waste Transportation Control," EPA/530/SW-512, 1976.

13-7. McConnaughey, W. E., Welsh, M. E., Ladey, R. J., and Goldman, R. M., "Hazardous Materials Transportation," *CEP*, Vol. **66**, No. 2, February 1970.

13-8. Editorial, "Oregon Traces Flow of Hazardous Materials," *Solid Wastes Management*, March 1975.

13-9. Cheremisinoff, P. N., "Disposal of Hazardous Wastes: Treat or Truck," *Pollut. Eng.*, May 1975.

13-10. "Pesticides and Pesticide Containers," *Fed. Reg.*, Vol. 39, No. 85, Part IV, U.S. Environmental Protection Agency.

13-11. Hazardous Waste Management Facilities in the United States. EPA/530/SW-46.

13-12. U.S. EPA, Summary Materials for Public Meetings on Hazardous Waste Management Guidelines/Regulations Pursuant to Subtitle C—Resource and Recovery Act of 1976, (P.L. 94-580), 1977.

13-13. Fuller, J. G., "The Poison that Fell from the Sky," *Readers Digest*, August 1977.

13-14. Price, D. R., "Considerations in Planning the Management of Hazardous Waste Materials," *Pollut. Eng.*, February 1977.

13-15. Scurlock, A. C., Lindsey, A. W., Fields, T., and Huber, D. R., "Incineration in Hazardous Waste Management," EPA/530/SW-141, 1975.

13-16. Landfill Disposal of Hazardous Wastes. EPA/530/SW-146.

13-17. Russell, D. L., "PCB's: The Problem and What Must be Done," *Pollut. Eng.*, August 1977.

13-18. Editorial, "How to Ship Hazardous Material," *Environ. Sci. Technol.*, Vol. **11**, No. 4., April 1977.

13-19. State Register of Minnesota, Vol. 2, No. 11, pp. 521-616, September 19, 1977.

13-20. *Fed. Regist.*, Vol. **42**, No. 162, pp. 42206-42209, August 22, 1977.

14

Air pollution considerations in industrial plant operations

14-1 THE AIR POLLUTION PROBLEM

Because of the wide range of information available on air pollution in general and the many books which have been written about particular aspects of air pollution and air pollution abatement, it is not the purpose of the author to provide a long and detailed disertation about the various aspects of the air pollution problem but merely to relate it to the problem of industrial waste disposal.

Every industry has a combination of wastes which must be coped with by devising a method or group of methods for their disposal. Some of these wastes are solids, some are liquids, and some are gases, and each type of waste requires a different treatment regimin and a different type of equipment. Often in the disposal of one type of waste, we create another pollution problem. In the incineration of solid waste, there is often an effluent which has the potential to cause air pollution; consequently, it must be treated for the removal of solid materials or other gases which constitute the pollution problem.

The industrial plant has a significant stake in the solution of air pollution problems because usually these problems occur within the plant. While the community may demand that industry clean up its air because of pollution outside of the plant boundaries, the greatest amount of contamination usually occurs within the plant increasing maintenance on buildings, equipment, and machinery and also increasing the health hazard to its employees.

The major sources of air pollution are:

1. Transportation
2. Domestic heating
3. Electric power generation
4. Refuse burning
5. Industrial fuel burning and process emissions
6. Natural causes, such as, forest fires, windstorms, etc.

It would be almost impossible to assign an accurate percentage to these classifications because positive documentation does not exist. However, it is reasonable to assume that transportation of all types, where internal combustion and turbine engines are used, contributes approximately 60% of the total annual emissions, and that electric power generation adds another 10 to 15%, with domestic heating about 10%. This leaves 25% to be assigned to the areas of refuse burning and industrial fuel burning and process emissions. Refuse burning accounts for a small percentage on the order of 3 to 5%. While it is more opinion than fact, it appears that industrial operations are not much greater than 20% of the total weight discharge of air pollutants to the atmosphere, even though industrial operations are often blamed much more. One of the reasons for the public reactions against industrial air pollution is that it is usually obvious, although often the unseen air pollutant is far more of a health and safety problem.

14-2 TYPES OF POLLUTANTS

Natural or clean air is considered to be approximately 20.9% oxygen, 78% nitrogen, 0.9% argon, and the remainder is carbon dioxide, hydrogen, and other trace elements. This air can be contaminated by two basic types of pollutants: gaseous pollutants and particulates.

14-2.1 Gaseous Pollutants

Any gaseous material found in the atmosphere in concentrations higher than those normally found in air is a gaseous pollutant; whether it is dangerous or not really depends upon the concentration. These gaseous pollutants are generally various types of hydrocarbon materials, carbon dioxide, carbon monoxide, the oxides of sulfur, the oxides of nitrogen, chlorine, and other halogens and compounds of these halogens, ozone, and photochemical oxidants. The photochemical oxidant is not really an industrial pollutant because it is not manufactured within the industrial complex. It is manufactured in the air by a combination or interaction of hydrocarbons, nitrogen dioxide, ozone, and sunlight, as described in Figure 14-1. It, in effect, results from industrial or other types of emissions, primarily internal combustion engines, which release hydrocarbons and nitric oxide.

Gaseous pollutants, such as those outlined above, have certain legal maximum concentrations, either measured in parts per million, which is a volumetric basis, or on a weight basis measured in either micrograms or milligrams per cubic meter. The weight basis is gaining wider usage because it is a positive measurement of the amount of pollutant in a given volume of air. The concentration of the pollutant, however, is not the only thing to be taken into consideration. The duration of the concentration of the pollutant in terms of time is equally important. It is possible to tolerate a high concentration level of the particular

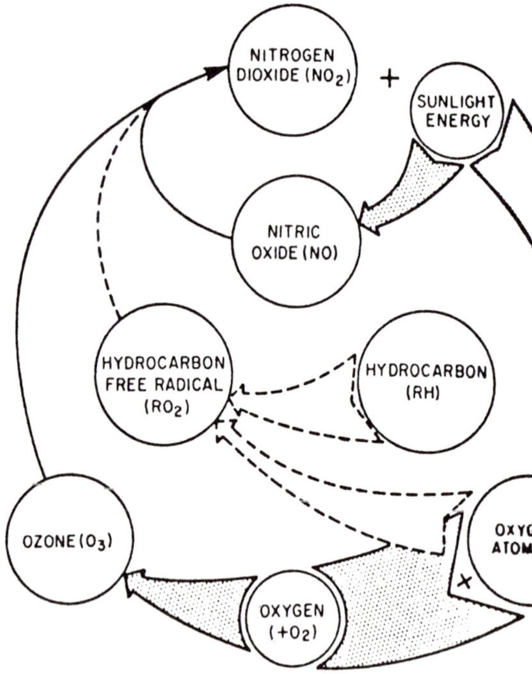

Fig. 14-1. Interaction of hydrocarbons with NO_2 in the presence of O_2 and sunlight. ***Source: Air Pollution and Industry,*** **Ross, R. D., ed., Van Nostrand Reinhold Co., New York, p. 240, 1972.**

pollutant for a short time, but on a continued basis, it may represent damage to health and to property.

Therefore, standards for gaseous pollutant levels are based on 24-hr exposure periods with a maximum concentration during that period. For example, the long-term primary ambient air quality standard for sulfur oxide is 80 $\mu g/m^3$, which is 0.03 ppm. This is an annual arithmetic mean. The short-term primary ambient air quality standard for sulfur oxide in a maximum 24-hr period is a concentration of 365 $\mu g/m^3$, which is 0.14 ppm; this is not to be exceeded more than once per year. Similarly, the long-term ambient air quality standard for carbon monoxide is a maximum 8-hr average concentration of 10 mg/m^3 (9 ppm), which is not to be exceeded more than once a year. For hydrocarbons, it is generally desirable to keep the level below 200 $\mu g/m^3$, which is 0.3 ppm and, for nitrogen oxides, a single primary ambient air quality standard of 100 $\mu g/m^3$ (0.05 ppm).

These figures are meaningless in the truest sense to the industrial plant operator because they are ambient air quality standards which are measured at ground level and not at the top of the stack or the point of emission. At the point of emission, the permissable concentrations may be considerably higher. Some of the aspects of the Clean Air Amendments of 1977 tend to control the actual emissions in terms of weight and not in terms of air quality standards. Rather than give pages of regulatory information in this chapter, since such regulatory information varies widely from state to state, it is suggested that you contact your State Department of Environmental Protection concerning specific standards which must be maintained in your area.

14-2.2 Particulate Pollutants

Particulates are, of course, solid or liquid particles which because of their size and specific gravity become airborne either in an industrial gas stream or are blown by the wind. Particulates could be inert materials, such as finely divided soil or crushed stone; they could be chemically reactive materials, such as calcium carbide or cement dust or lime; or they might even be hazardous or toxic materials, such as powdered lead, zinc, or asbestos. The effect of these materials on health and on plant equipment (in terms of erosion or corrosion), on animals or vegetation may be determined by refering to the EPA's criteria documents. These documents give the cause and effect relationship of various pollutants at different concentrations. Particulates are not always solid particles; they may be fine liquid mists or aerosols. They can be classified by size, but the more common types of particulates are divided into 4 groups:

1. Ash particles, primarily noncombustible components of fuels containing oxides, sulfates, chlorides, carbonates, nitrates, silicates, and many trace metals

2. Carbon
3. Sulfuric acid
4. Tar particles containing hydrogen, organic acids and bases, phenols, etc.

The typical, annual, average particulate level in nonurban areas ranges from 10 to 60 $\mu g/m^3$.

Particulates, just as gaseous pollutants, have an effect on human health, but, in the case of the particulate, the size rather than the quantity of the particles is the governing factor. This is because filtration mechanisms of the respiratory tract remove particles above 5 to 10 μ in diameter and, of course, there is a wide difference in the toxicity of various particulates. Pennsylvania, New York, and New Jersey moved in 1975 to an average annual particulate ambient air quality standard of 65 to 70 $\mu g/m^3$, which represented a 45 to 55% reduction of particulate levels over the estimated 1970 exposure levels in the same region. Air quality standards vary depending upon the area but are in the same general range of 50 to 75 $\mu g/m^3$.

In the measurements of particulates, the most common is a process weight chart, which is a weight rate of emission, although some areas still use a weight per unit volume, i.e., grains per cubic feet or micrograms per cubic meter.

Concentration standards are normally corrected to some base line, such as atmospheric pressure and dry gas at 70°F. This is because atmospheric conditions will vary, and, consequently, weight rates could vary widely if expressed on a volumetric basis. Incinerator standards are usually based on a flue gas containing 12% carbon dioxide. This is to prevent circumvention of the standards though dilution of the particulate material in the stack gases by the addition of air. For example, if a process were tested and the flue gas were shown to contain 1 grain/cu ft and the standard was 0.6 grain/cu ft, the standard could be met by adding enough excess air to reduce the concentrations to below the standard. This, of course, does not change the actual amount of particulate present one iota, merely its concentration; therefore, in this example, doubling the amount of air would bring an incinerator or other device within the legal range. But, if the law requires adjustment to a base condition, such as the 12% CO_2 regulation for incinerators, this type of circumvention is not possible.

Much effort has been expended to use process weight curves for all particulate sources, but this has not been totally successful. The EPA attempted to use a process weight chart for incinerators when national emission standards for this source category were considered, but they were not included in the law. The process weight curve for incinerators is shown in Figure 14-2.

By far the most important air pollution consideration for the industrial plant operator is particulate emissions and the visible aspect of the emissions, usually as related to opacity. While opacity can be related to particulate levels, there is no exact correlation because of particle size, and, if opacity is judged by the

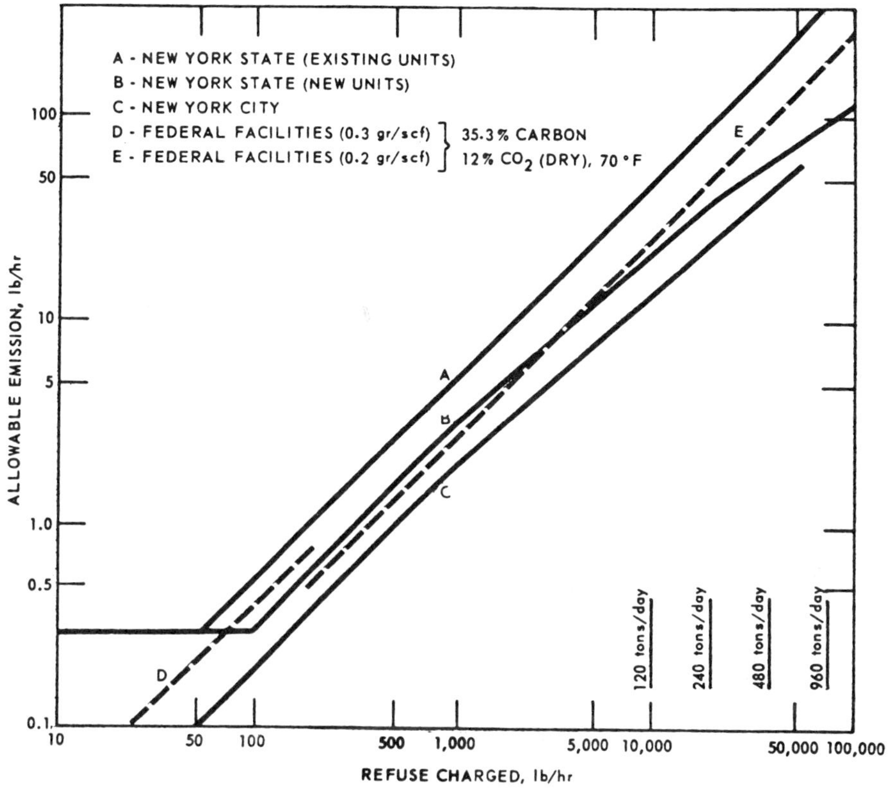

Fig. 14-2. Process weight curve for incinerators (Ref. 14-4).

human eye, it is subjective rather than objective. Opacity measuring equipment has seen developed in the last few years which can produce fairly accurate and repeatable results, but errors caused by environmental lighting and background contrasts still cause problems. Field studies by the EPA show that qualified observers are able to consistently read opacities of black and white plumes with positive errors not exceeding 7.5% opacity based on sets of 25 consecutive readings.

Opacity is affected by three basic atmospheric effects. The first is altitude. Due to the sunlight scattering, visual opacity increases as the sun's altitude increases; therefore, the opacity of a particulate emission would be greater in the summertime than in the winter, and greater at noon than it would be in the afternoon. The second opacity effect is azimuth for the same reasons that the sun's altitude affects opacity. The third factor is the stack diameter. Visual opacity increases as the stack diameter increases even though the grain loading

TABLE 14-1. Industrial Process Emission Levels for Clear Stacks (Ref 14-5).

Industrial Process Classification	Electrostatic Precipitators, Fabric Filters Grains/ACF at Stack Exit Gas Temperature, °F	Estimated Humidity[a] Vol. %, H_2O	Wet Scrubbers, Grains/SCF Dry[a] (SCF @ 70°F and 14.7 Psia)
Utilities and industrial power plant fuel-fired boilers			
Coal-pulverized	0.020 @ 260–320	5	0.029–0.031
Coal-cyclone	0.010 @ 260–320	5	0.014–0.016
Coal-stoker	0.050 @ 350–450	5	0.080–0.090
Oil	0.003 @ 300–400	8	0.0047–0.0053
Wood and bark	0.050 @ 400	10	0.090
Bagasse	0.040 @ 400	10	0.072
Fluid coke	0.015 @ 300–450	Dry	0.022–0.026
Pulp and paper			
Kraft recovery			
Boiler	0.020 @ 275–350	30	0.040–0.043
Soda recovery			
Boiler	0.020 @ 275–350	30	0.040–0.043
Lime sludge kiln	0.020 @ 400	30	0.046
Rock products (Kilns)			
Cement–Dry	0.015 @ 450–600	3	0.026–0.030
Wet	0.015 @ 450–600	25	0.034–0.040
Gypsum	0.020 @ 500	25	0.048
Alumina	0.020 @ 400	20	0.040
Lime	0.020 @ 500–600	8	0.039–0.050
Bauxite	0.020 @ 400–450	8	0.035–0.037
Mag. oxide	0.010 @ 550	20	0.024
Steel			
Basic oxygen			
Furnace	0.010 @ 450	20	0.021
Open hearth	0.010–0.015 @ 450–600	8	0.023–0.027
Electric furnace	0.015 @ 400–600	5	0.026–0.032
Sintering	0.025 @ 300	10	0.040
Ore roasters	0.02 @ 400–500	10	0.036–0.040
Cupola	0.015–0.02 @ 250–400	20	0.029–0.036
Pyrites roaster	0.02 @ 400–500	10	0.036–0.040
Taconite roaster	0.02 @ 300	10	0.032
Hot scarfing	0.01–0.015 @ 150–250	25–40	0.019–0.025
Mining and metallurgical (Non-ferrous)			
Zinc roaster (Ore)	0.010 @ 450	5	0.018
Zinc smelter (Melting)	0.010 @ 400	5	0.017
Copper roaster	0.010 @ 500	5	0.019
Copper reverberatory furnace	0.015 @ 500	5	0.030
Copper converter	0.010 @ 500	5	0.019

TABLE 14-1. Continued

Industrial Process Classification	Electrostatic Precipitators, Fabric Filters Grains/ACF at Stack Exit Gas Temperature, °F	Estimated Humidity[a] Vol. %, H_2O	Wet Scrubbers, Grains/SCF Dry[a] (SCF @ 70°F and 14.7 Psia)
Aluminum–Hall process	0.025–0.075 @ 300	5	0.075
Soderberg process	0.001–0.003 @ 200	25	0.003
Ilmenite dryer	0.020 @ 300	25	0.038
TiO_2 process	0.010 @ 300	5	0.015
Molybdenum roaster	0.010 @ 300	5	0.015
Ore beneficiation	0.020 @ 400	5	0.034
Miscellaneous			
Refinery catalyst regenerator	0.015 @ 475	15–20	0.032
Muncipal incinerators	0.015 @ 500	20	0.034
Apartment incinerators	0.02 @ 350	20	0.038
Spray drying	0.01 @ 400	25	0.022
Precious metal refining	0.01 @ 400	5	0.017

*Based on estimated water vapor content of hot gases

remains constant. This is borne out by the fact that observers trained to read opacity on small stacks have difficulty with larger ones.

Many states have smoke schools to train regulatory and industrial operators to visually determine the opacity of black and white smoke plumes from stacks, and almost each stack will have to be measured to give an opacity reading with a corresponding particulate loading which will meet regulations.

Generally, air pollution regulations state permissible opacity levels as percent opacity with 100% being black and 0% white. Most regulations will allow up to 20% opacity for short periods but essentially zero opacity during normal operating conditions. This opacity may also be measured by the Ringelmann chart, which is an opacity scale with clear air at 0.

One of the problems of opacity measurement is that the optimum particle size for scattering light and producing a dark plume is 0.1 to 1.0 micron; so that even a high collection efficiency in terms of the total weight of particulate removed may allow a visible plume—many small submicron particles are contained in the plume. Such particles remain suspended in the atmosphere for long periods of time and during inversions accumulate to cause severe visibility reduction and soiling of buildings and materials. These small particles are also potentially injurious to human health because they more easily enter the body. Table 14-1 shows industrial process emission levels generally expected to produce visually clear or near-clear stacks with the exception of condensed moisture plumes.

14-2.3 Odors

This third type of air pollutant is really a gaseous pollutant except that it manifests itself in a form which is evident to the human olfactory system. The gas containing the odor may or may not be visible, may or may not be toxic, and, in many cases, is only aesthetically unpleasant. On the other hand, odors can also be of such a nature that they cause nausea and vomiting without any long or short range toxic effects.

Odor control is a complex air pollution control problem, but it can be solved even though the cost may be high. Odors constitute one of the most complicated and objectionable problems in air pollution today. Monitoring information is of questionable value because of the wide variety of trace materials that need to be measured. The human nose is still the present standard for detecting and determining the odor intensity in the ambient air and stack effluents.

Existing odor control regulations consist of a variety of partially successful measures including:

1. Nuisance-type restrictions based on ambient air detection of odors.
2. Process restrictions for certain known odor-producing sources.
3. Control equipment requirements for specific source requirements.

Odors are usually measured by a 6 to 8 member panel where subjective opinions concerning the odor are paramount. This is described in ASTM Procedure D 1391-57. There are a wide variety of odorous compounds from the standpoint of their chemical formula. The materials listed in Table 14-2 may be detectable as odors in concentrations as low as 1 ppb. Some of the typical odor threshold values are presented in Table 14-3.

If the odor is due to a single substance and found in high concentrations, then the compound may be measured by direct analytical techniques. But, if the material is a single compound or a mixture of compounds at very low concentration, it may be necessary to use a subjective method for analysis. There are a variety of odor control methods which shall be discussed later in this chapter.

TABLE 14-2. Odorous Compounds

Type of Compound	Chemical Formula
Unsaturated	$R_1R_2C = C\ R_3R_4$
Oxygenated	$ROH\ R_1OR_2\ \ R_1R_2C = 0$
Halogenated	RF, RCl, RB_r
Nitrogenous	RHN_2, R_1R_2, NH
Sulfurous	$RSH, R_1SR_2, R_1R_2C = S$

TABLE 14-3. Odor Threshold Values

Compound	Threshold Value (ppm)
Carbon tetrachloride	100
Ethanol	10
Acetic acid	1
Styrene	0.1
Ethyl mercaptan	0.001
Trimethylamine	0.0002

14-3 DETERMINATION OF THE PROBLEM

The first step in the evaluation of any industrial process air pollution problem is to establish the quantity of emissions. This then may be compared with performance standards and with emission regulations, especially the EPA's national performance standards.

Emissions should be estimated for the determination of product loss, the estimation of the performance of the air pollution control equipment, and to determine the amount of emission. They may be determined by material balance, the various types of emission factors, or source testing. The material balance is perhaps the most common method used by industry to determine process losses, but it may also be used to estimate air pollution emissions. Test measurements must be made on the basis of steady state operation in order to be reasonably viable.

The Environmental Protection Agency's publication AP-42 gives a number of emission factors whereby the amount of the material fed to the process is multiplied by several emission factors to provide an estimate of the emissions. These were developed by summarizing or averaging source tests and reference material on process emissions and are used quite extensively in developing state implementation plans and are good whenever an average value is satisfactory. Table 14-4 shows emission factors for auto body incineration, both with and without an afterburner.

By far the best method for determining emissions is by source testing. Source testing utilizes a probe and a collection train maintained under vacuum which removes the particulate material from the process and collects it so that it may be analyzed and weighed. Figure 14-3 shows the EPA method test train which is used in sampling and Figure 14-4 the ASTM sampling train which is quite similar. Federal regulations now require that the EPA Method Five technique is used, so it is wise to test with this type of system preferentially to be sure that you are within regulations.

In addition to determining the amount of particulate or gaseous pollutant, it is extremely important to determine the particle size where particles are present.

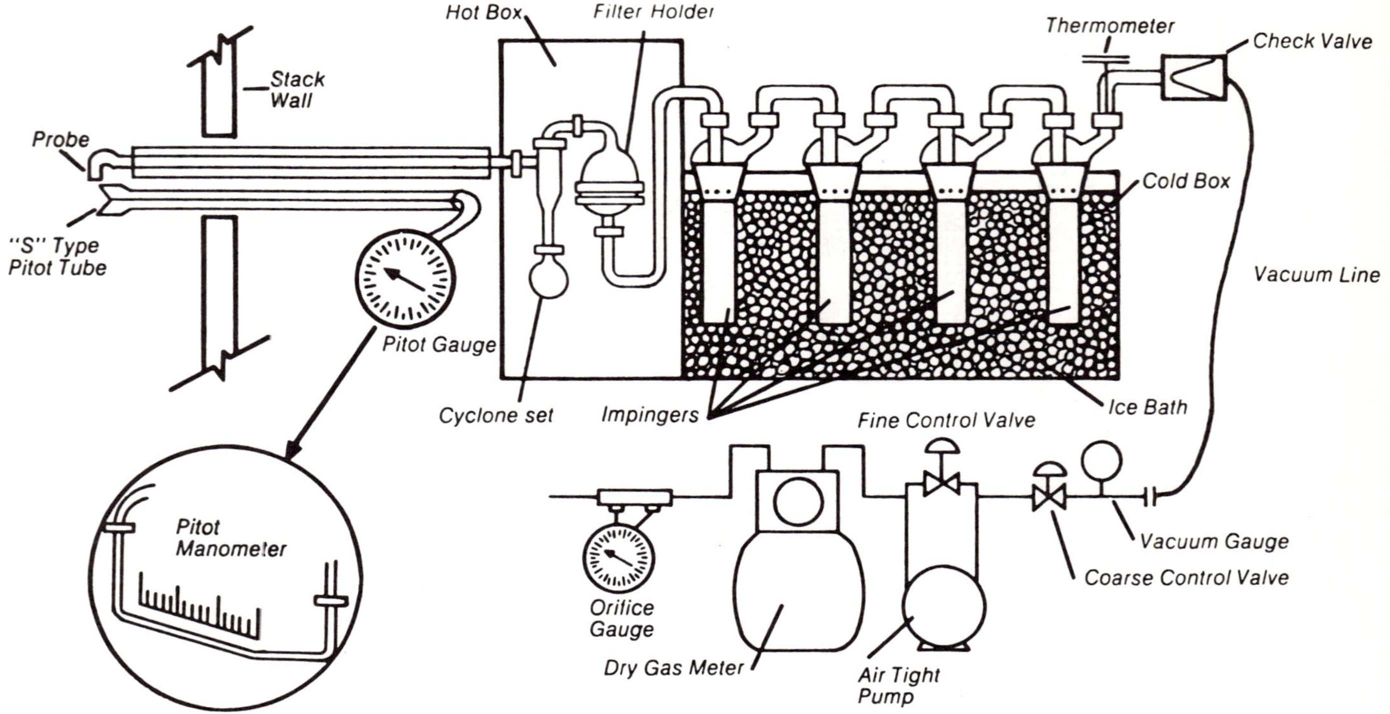

Fig. 14-3. Environmental Protection Agency sampling train (Ref. 14-8).

TABLE 14-4. Emission Factors for Auto Body Incineration (Ref 14-7)

Pollutants	Uncontrolled		With Afterburner	
	lb/car	kg/car	lb/car	kg/car
Particulates	2	0.9	1.5	0.68
Carbon monoxide	2.5	1.1	Neg	Neg
Hydrocarbons (CH_4)	0.5	0.23	Neg	Neg
Nitrogen oxides (NO_2)	0.1	0.05	0.02	0.01
Aldehydes (HCOH)	0.2	0.09	0.06	0.03
Organic acids (acetic)	0.21	0.10	0.07	0.03

[a]Based on 250 lb (113 kg) of combustible material on stripped car body.

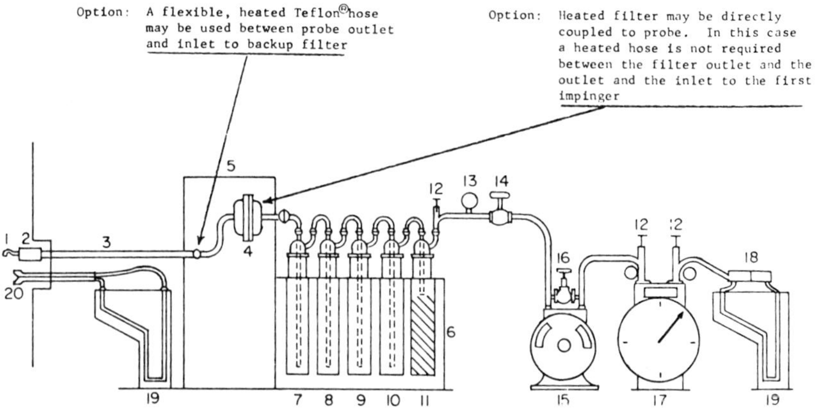

1. Nozzle
2. Thimble or flat filter
3. Stainless steel probe
4. Backup filter holder
5. Heated box for filter
6. Ice bath for impingers
7. Modified impinger, dry
8. Modified impinger with 100 ml water
9. Modified impinger with 100 ml water
10. Modified impinger, dry
11. Silica gel trap
12. Thermometer
13. Vacuum gauge
14. Flow control valve
15. Pump
16. Flow control valve
17. Dry test meter
18. Calibrated orifice
19. Manometer
20. "S" type pilot tube

Fig. 14-4. ASTM sampling train.

The current state of the art for determining particle size is by the use of in-stack, impactor-type collectors. There are a variety of various types of in-stack particle size collectors which will pick up particle sizes as small as 0.1 μ. The particle size distribution is important as various classes of collectors remove various size particles.

Today there are literally dozens of measurement and detection instruments available as standard hardware for measuring a variety of gaseous concentrations. There are instruments to measure nitrogen oxide and sulfur oxides, many of the halogen gases, and other types of pollutants. However, accurate determination of paritculates still remains a more difficult and more complicated measurement problem. Emission testing is a complete scientific area in itself and one in which a number of companies have expertise. It is, therefore, usually desirable for industrial plant operators to engage one of these companies to make the necessary emission measurements prior to installing any pollution control equipment unless they have highly competent and qualified people within their own organization to make such emission tests. Generally, they would be forced to demonstrate to local, state, or federal authorities that air pollution regulations are being met.

14-4 EQUIPMENT FOR PARTICULATE REMOVAL

Until recently, the problem of particulate removal from gas streams has been more of an art than it has been a science. Perhaps today it is 50% art, or at least based more on empirical data than on actual definitive calculations. The problem in the removal of particulates from a gas stream is that there are two, and sometimes as many as five, major variables. These variables which affect any type of particulate removal equipment are:

1. The loading in grains per cubic foot or milligrams per cubic meter.
2. The size distribution of the particles in terms of the number of particles of any given micron size or within any given micron range.
3. The temperature of the gases to be handled.
4. The volume of the gases to be handled.
5. The resistivity of the particles to electrostatic attraction.

The success of any particulate removal system is not only dependent upon the primary removal device. It is also dependent upon: properly sized ducts to minimize friction losses; a properly designed stack to maximize dispersion and minimize ground level concentrations; the fan which is used either to force the flow through the particulate removal device or pull the volume to be handled through the removal device; hoods and inlet structures that optimize the particle capture velocity; materials of construction that are both lightweight and temperature and corrosion resistant; and the materials handling system which is used to collect or dispose of the dust which has been captured. The problem is not a single piece of equipment but a combination of pieces of equipment working in harmony with one another to give the best possible conditions for the removal of the particulate.

Since all of the components are important, the system must be properly designed or the system will not work correctly. However, the most important component of the system is the collector itself, and the main categories of collectors may be classified as:

1. Mechanical collectors
2. Fabric filters
3. Electrostatic precipitators
4. Wet scrubbing devices

The basic features of the various collectors are outlined in Table 14-5. All of these collectors involve a source of dirty gas passing through a gas cleaning device with a fan to move the gas through the total process and a stack to discharge the clean gas to the atmosphere. The fan is usually located downstream of the gas cleaning device to protect it from the erosive and possibly corrosive or unbalancing effects of the entrained particulate material. The size of the collector and its cost usually are directly proportionate to the gas quantity to be cleaned on a volumetric basis. The collection efficiency of the collector is calculated simply by subtracting the weight of the particulate out of the device from the weight of the particulate into the device, divided by the weight of the particulate into the device times 100.

The size of the particulate into the device has a great effect on the collection efficiency. Fine dust is harder to collect than coarse dust, and most particulate dusts are a mixture of both fine and coarse. The proportion of the fine dust will have an important bearing on the collector efficiency and, therefore, will be a major factor in the selection of the type of collector needed to meet the collection efficiency requirements. Heavier concentrations of dust are usually easier to collect than low concentrations.

14-4.1 Mechanical Collectors

Mechanical collection of dust was perhaps the first and the most rudimentary design employed, and for many years, it was the only design employed for the removal of solid particles from air or flue gas streams. There are as many different designs as there are models of automobiles produced in the United States, each with a slight advantage or disadvantage compared with the other. Currently, none of them can adequately remove all of the dust particles to a level of legal acceptance except in a few cases where all of the particles are very large. The basic types of mechanical collectors are the gravity settling chamber, the inertia baffle type separator, and the high efficiency centrifugal type unit.

Gravity settling chambers, as shown in Figure 14-5, utilize the weight of the dust at low velocities to separate itself from the air-stream. Velocities through such a unit are low and settling velocities range from 60 to 600 ft/min. The

TABLE 14–5. Fundamental Operating Characteristics of Commercial Particulate Collectors

Basic Type	Specific Type	Basic Operating Force	Basic Measure and Unit of Capacity	Typical Capacity in cfm of Gas per Basic Measure of Capacity	Typical Required Plan Area for Collector to Clean 100,000 cfm[c]	Typical Energy Required and % Collection Efficiency on Fine Dust[d]		COST IN $1000 FOR A COLLECTOR TO CLEAN 100,000 CFM: Cost to Purchase fob Shop and Install		Annual Cost for Maintenance and Power		Effect On Collection of Operation at Over or Under Capacity[e]	
					sq.ft.	inches w.g.	%	fob shop[a]	install	maintenance	power[b]	over	under
	settling chamber	gravity	casing volume (cu ft)	20	2600	0.2	25	10	3	1	0.5	down	up
Mechanical collectors	baffle	intertia	inlet area (sq ft)	1200–2600	300	0.5	40	13	4	2	1.3	up	down
	high-efficiency cyclones	centrifugal	inlet area (sq ft)	3000–3600	125	4	80	20	5	2	10	up	down
	manual cleaning	filter cake filtration	fabric area (sq ft)	1–4	1000	4	99	35	20	7	10	down	no change
Fabric filters	automatic shaker cleaning	filter cake filtration	fabric area (sq ft)	1–4	1000	5	99	60	30	10	13	down	no change
	automatic reverse jet cleaning	felt fabric filtration	fabric area (sq ft)	3–8	600	8	99	80	35	15	20	no change	no change
	impingement baffle	liquid capture	baffle cross-sectional area (sq ft)	500–600	300	4	80	30	15	7	10	up	down
Wet scrubbers	packed tower	liquid capture	bed cross-sectional area (sq ft)	500–700	250	7	90	40	20	9	17	down	up

	venturi	liquid capture	throat area (sq ft)	7,000 to 30,000	100	30	99	50	50	11	75	up	down
Electric precipitators	single field	electric	collectrode area (sq ft)	5	270	0.5	95	75	55	4	1.3	down	up
	multiple field	electric	collectrode area (sq ft)	3	500	0.5	99	100	70	5	1.3	down	up

[a]Fob shop costs, installation and maintenance costs are based on 1970 mild steel construction costs, and do not include auxiliary equipment such as supporting structure, connecting flues, thermal insulation, foundations, stacks, etc. They are subject to substantial variation due to the specific requirements of each installation including geographic location. Use only for general comparison purposes.
[b]Energy costs are based on electric power at $0.015/kwh.
[c]Fine dust is considered to be 70% by weight minus 10 μ.
[d]Some collectors can be modified to maintain collection efficiency at over or under capacity.
[e]Plan areas do not include connecting flues.
Source: *Air Pollution and Industry*.

Note: Costs shown are relative based on years and values shown and must be made current for estimating purposes.

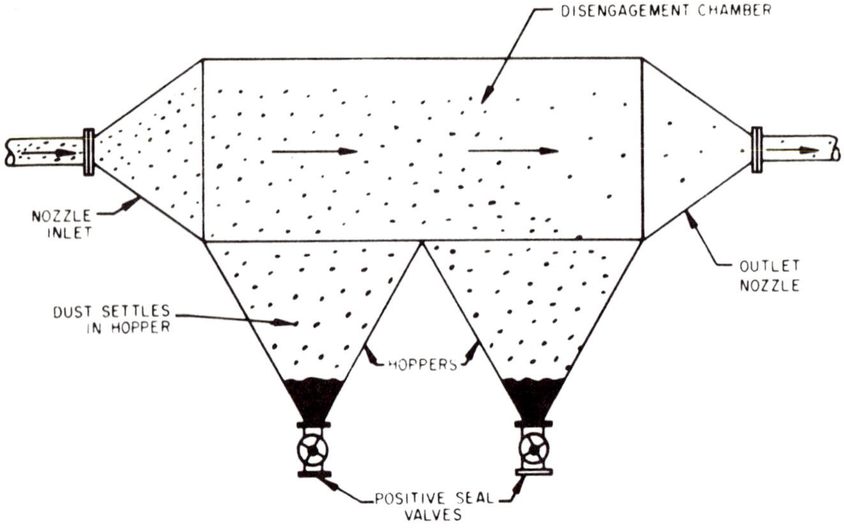

Fig. 14-5. Gravity settling chamber. ***Source:*** ***Air Pollut. & Ind.*** **Ref. 14-12.**

units are large, but the initial cost is generally quite low and the collection efficiency is low. The efficiency of such systems increases as the volumetric load decreases because the lower the velocity, the better the separation. Improvement on the gravity settling chamber may be made by adding baffles on the inlet and the outlet of the gas stream, which tend to impart a sudden high velocity turn to the dust-laden air. The low-specific-gravity gas easily makes the sharp turn, and the heavier dust driven by inertia is unable to make the sharp turn and is restrained below the baffle until it is captured by the dust slot. The dust is conveyed at a lower velocity as the dust slot expands over the collection hopper into which it settles. Units, as shown in Figure 14-6, have collection efficiencies better than that of the settling chamber but not sufficient to meet most emission control requirements. It can be used as a settling chamber or as a precleancr to other types of systems.

The third basic type of collector is a centrifugal collection, commonly called a cyclone. This type of collector separates particulate matter from the carrier gas by using the velocity of an inlet stream to throw the heavier particles to the outside of the collector, creating a descending outer vortex and an ascending inner vortex, both confined within the cyclone. The rapidly rotating descending vortex holds the heavier particulate material against the outside walls of the cyclone by centrifugal force, and it descends into a hopper at the bottom. The ascending inner vortex is cleaned gas which rises through the center and through a baffle to the clean gas outlet.

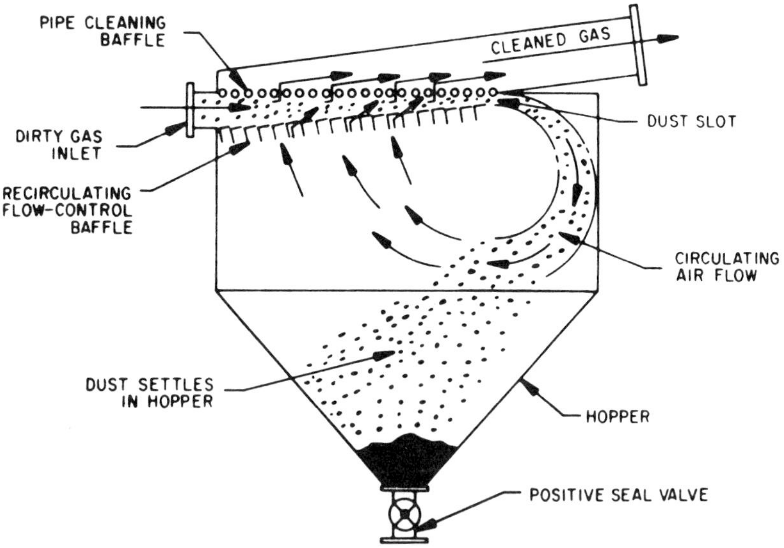

Fig. 14-6. Recirculating baffle collector. *Source: Air Pollut. & Ind.* **Ref. 14-12.**

This type of unit is very, very simple and reliable and has a high efficiency compared to the other types of mechanical collectors and has been widely used for many years for separating solids from gas. The design of the unit must be such that the entrainment of the particulate in the gas stream is minimized, but it offers the advantage that it is generally smaller in size than gravity or baffle type separators and can handle a much higher velocity and, therefore, a greater flow than other types of collectors. Most cyclones operate at an inlet velocity of 3000 to 4000 ft/min. The energy requirement is proportional to the square of the inlet velocity. When handling atmospheric air, the draft loss will generally be in the range of 4 to 9 in. w.c. when measured As a difference of static pressure between the inlet and the outlet.

The design of each cyclone collector represents a compromise among three factors: collecting efficiency, draft loss, and size. In general, higher collection efficiencies require higher draft losses or more expenditure of energy and a larger size which increases the installed cost of both. A typical single cyclone type collector is shown in Figure 14-7.

A number of designs have used a small diameter vane axial type of cyclone collector wherein the gas stream is separated from the particulate content by the use of multiple cyclones arranged in a common shell, having a common outlet and a common inlet and a common dust storage hopper. This type of arrangement was devised because the designer felt that small diameter cyclones are generally more efficient than large diameter ones, but, in some cases, large

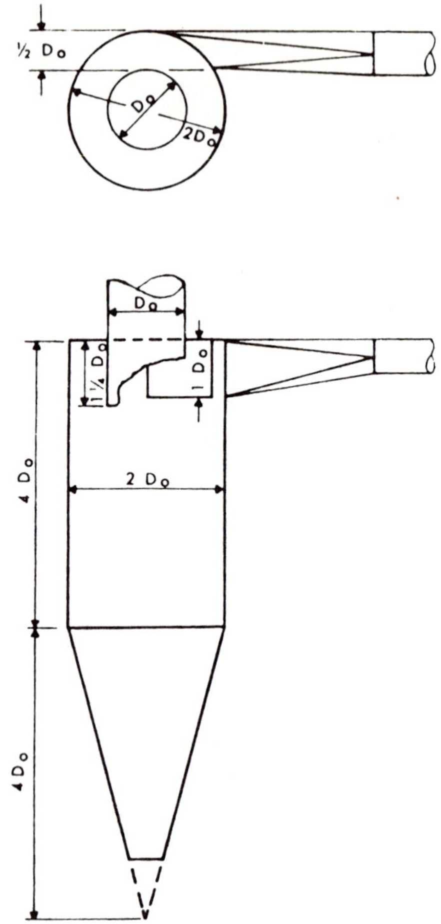

Fig. 14–7. Single cyclone dimensions. ***Source: Air Pollut. & Ind.*** **Ref. 14–12.**

diameter cyclones are more efficient than small diameter ones. This is usually a factor of the draft loss and the ratios of the dimensions of the cyclone.

Large flows are often handled by the large diameter cyclone clusters having an involute inlet. Figure 14–8 shows a number of involute cyclone arrangements. The major problem with this type of an arrangement (often used on large coal-fired boilers) is the variation in pressure drop from one to the other. If this is not balanced out by the size of the ductwork and if the velocities in the system are not equivalent, then more of the dirty gas will pass into the cyclones nearest the inlet, and these will have lower efficiency than those furthest away from

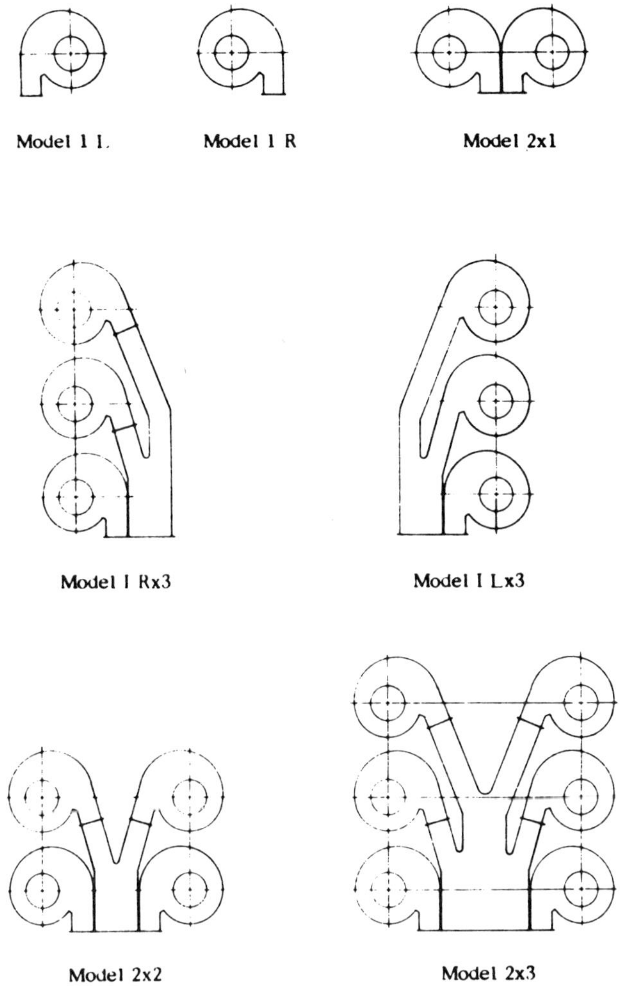

Fig. 14-18. Arrangement of involute cyclone collectors. ***Source:*** ***Air Pollut. & Ind.*****, Ref. 14-12.** ***Courtesy:*** **Belco Pollution Control Corp.**

the inlet. The object, of course, with this type of cyclone is to develop a unit which will fit into a common shell and have a number of separation devices.

The cyclone has been improved by the addition of a fines eductor at the top of the unit as shown in Figure 14-9. This eductor tends to pull off the fines at the top of the cyclone or the cyclone cylinder and continuously remove them because they are trapped in the upper eddies and never pass to the bottom of the cyclone. These are removed from the cylinder and transported by a spiral by-

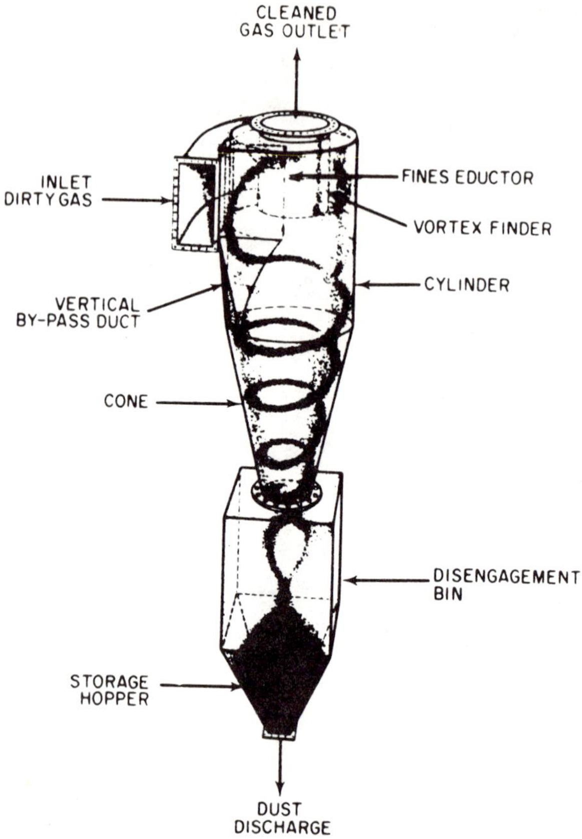

Fig. 14-9. Flow pattern in mechanical cyclone with vertical fines eductor. *Source: Air Pollut. & Ind.*, **Ref. 14-12.** (*Courtesy:* **Belco Pollution Control Corp.**)

pass duct to a fines injector where they are combined with the coarser dust particles at a level where the cylinder and the cone meet. This feature improves collector efficiency without increasing draft loss by the addition of the upper eddy which helps to initially trap the fines and assist the removal of the fines from the top of the cyclinder with their injection into the coarser dust which increases their capture and reduced reentrainment. This type of unit operates at about the same draft loss as other cyclone separators.

The advantages of the cyclone collector are generally low cost and a relatively simple operation which requires little maintenance. The disadvantage of this type of collector is generally lower efficiencies, especially on fine dust (usually below 90%) and the fact that the unit is subject to severe erosion if the dust has properties which will erode the material of the cyclone's construction and possible corrosion if there are corrosive materials present when the cyclone is operated below the dewpoint of the gas.

The cyclone separator may also be operated in a wet configuration. This is generally a large diameter type in which the capture and retention of the dust is increased by spraying a liquid, usually water, into the inlet. This spray is captured as the dust wets the walls of the cyclone. It is also often used to keep the temperature of the cyclone within the structural design limits of the materials of construction when cleaning very hot gases.

Ultra-high efficiency cyclone collectors are often made with very small diameters or very tall cyclone cylindrical sections with relatively small inlets, and these are capable of high efficiencies at the added cost of size, capital investment, and draft loss.

The cyclone collector is used today as a precleaner for fabric filters, wet scrubbers, and electrostatic precipitators. It may be used effectively for such applications as sawdust or wood chips in a small carpentry or saw mill application, but, because it will not capture the fine dust, it is not suitable as a final cleaning device. It finds wide use on coal fired boilers and, until now, has been used as a final collector on many large industrial boilers. However, because of its inefficiency, it will no longer suffice. It will, however, remove enough of the particulate to reduce the size and cost of the secondary collector or at least help the secondary collector achieve the desired collection efficiency.

The materials of construction of most cyclone collectors are carbon steel, although extremely erosive dust may require construction of hardened steel sections, and extremely corrosive dust may involve the use of other materials. For dust at ambient temperatures, fiberglass-reinforced plastic materials are beginning to find application.

14-4.2 Secondary Collectors

The secondary collector for any particulate material must be a venturi scrubber, a fabric filter, or an electrostatic precipitator. These are the principal collectors capable of meeting the current atmospheric standards. The efficiency of these collectors for fine particles is generally as demonstrated in Figure 14-10. There was a time that air pollution control was primarily interested in removing particles above 50 μ in diameter. Now, as removal efficiency requirements become more stringent, the need is to remove particles below 1 μ in diameter; each of the secondary type of collection devices are capable, with proper design, of removing such particles from a flue gas.

14-4.2.1 Electrostatic Precipitators The electrostatic precipitator is a secondary particulate removal device which is capable of handling particles in the submicron range with collection efficiencies approaching 100%. While the precip-

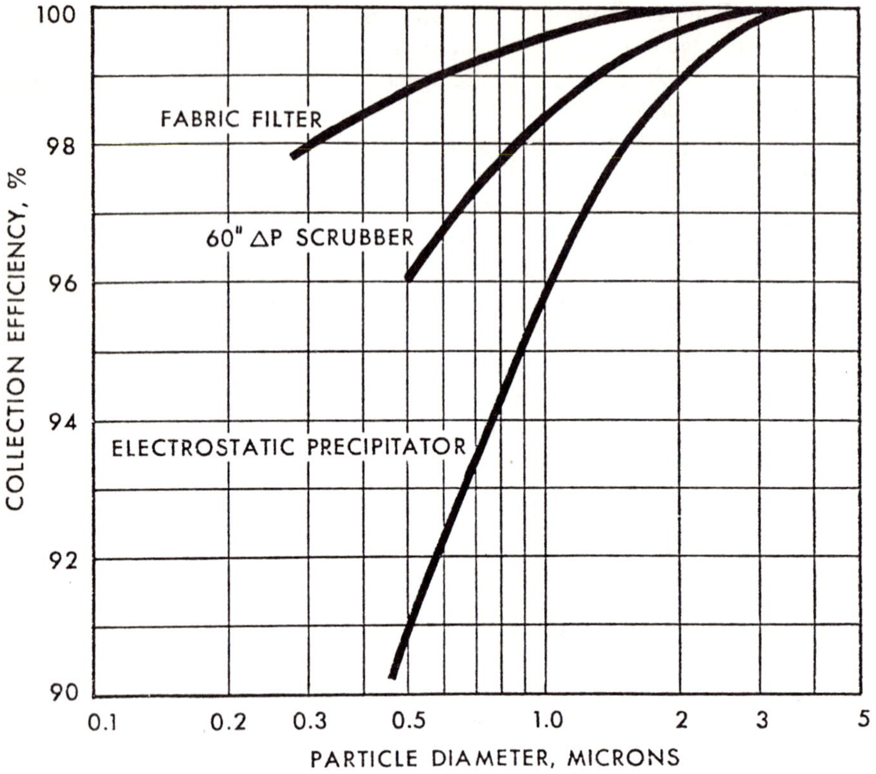

Fig. 14-10. Efficiency of particulate collectors. *Source:* **Arthur D. Little, Inc.**

itator's size is usually dictated by the gas flow through it, at least in terms of its crosssection, the depth of the unit is dependent primarily upon the resistivity of the dust which it must collect. The principal advantages of the electrostatic precipitator are that it has relatively low power requirements compared to other types of collectors, has a high collection efficiency for submicron particles, and is not particularly sensitive to the exhaust stream temperature. It is sensitive to temperature in the sense that higher temperatures usually promote greater collection efficiencies up to the structural limits of steel, which is the standard material of construction of the electrostatic precipitator. The disadvantages of the electrostatic precipitator are that it is very large in physical size and therefore high in initial cost; it is not suitable for materials which are flammable or explosive; it does not work at all well on nonconducting materials; and it, of course, does not remove any gaseous contaminants. It can, however, remove both solid particulate and liquid particulate matter.

The electrostatic precipitator separates entrained particulate matter from the gas stream by charging the dust to a negative voltage of about 50,000 V and then precipitating it onto grounded collection electrodes and then eventually dropping the agglomerated dust into a hopper. Despite the high voltages used, the energy consumption is very low and the draft loss across the electrostatic precipitator is usually on the order of 0.5 in. w.c. which is the lowest of all high efficiency collectors. In the commercial type of electrostatic precipitator, the gas passes through narrow vertical passages formed by parallel rows of grounded collecting electrodes which are in the form of plates. In between these rows of plates are electrically insulated high voltage wires which are spaced precisely on the center lines of each gas passage thereby causing the dirty gas to pass between the high voltage wires and the grounded plates. Typical gas passages on a large commercial precipitator are about 9 in. in width, 30 ft high and 27 ft long. The distance between the high voltage discharge electrode wires and the grounded collecting plates is therefore one-half of the gas passage width or 4-1/2 in. Each plate panel is approximately 6 ft long and is assembled from roll-formed sheet steel. A typical electrostatic precipitator design is shown in Figure 14-11.

There is a current flow between the wires and the plates. This is the result of a massive migration of negatively charged gas ions away from the negatively charged wire towards the relatively positively charged collecting plate. Figure 14-12a shows the 6 essential steps of the Cottrell process, so named because the inventor of the original electrostatic precipitator in 1910 was Frederick Cottrell. Here it can be seen that the negative ion gas flow moves toward the collecting plates and that the dust is negatively charged and precipitated onto the surface of the plate where it builds up to some predetermined thickness. Mechanical rappers shake the dust from the surface of the collecting plates into a hopper at the bottom of the precipitator. The potential difference between the wires and the plates is sufficient to cause billions of electrons to fly off the wires and bombard the gas molecules in between the collecting plates, close to the wires. The force of this bombardment forms both positive and negative gas ions which respond to the powerful electrostatic force by moving toward the oppositely charged electrodes. The positive gas ions return to the negative wire and regain their lost electrons. Negative gas ions move toward the grounded plate which is relatively positive, thereby flooding the gas passage with billions of negative ions. Most current flow within an electrostatic precipitator is due to the flow of negative gas ions from wires to plates.

As the dust particles pass through the gas passages, they collide with the negative gas ions and are thus charged negatively. Because the gas ions are thousands of times smaller than even submicron dust and because of their great number, they charge almost any dust which is in between the plates. The negatively charged dust particles then attach themselves to the grounded or positive

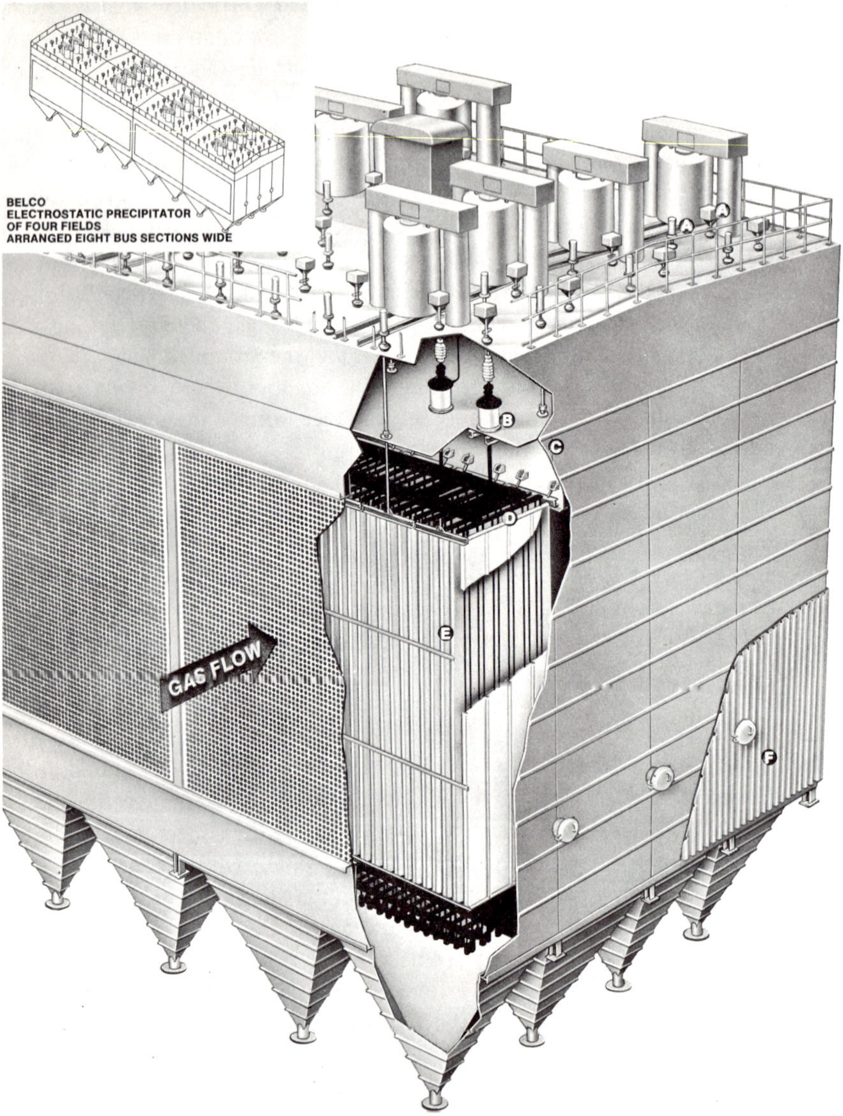

Fig. 14-11. Electrostatic precipitator. *Source:* Belco Pollution Control Corp.

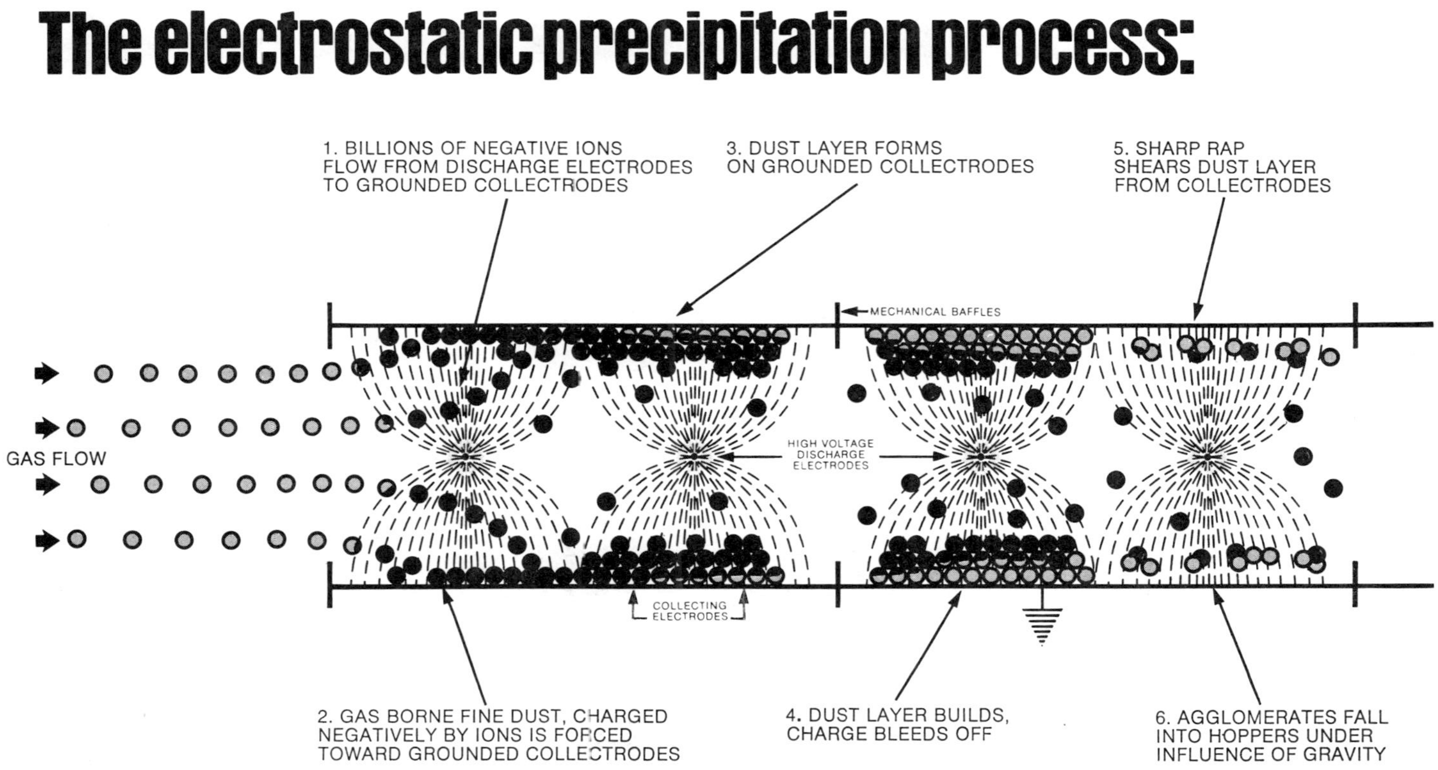

Fig. 14–12a. **The Cottrell process.** ***Source:*** **Belco Pollution Control Corp. and** ***Air Pollut. Ind.***

plates and are held there by powerful electric forces. As the dust layer builds on the plates, the negative charge gradually bleeds from the dust to the grounded plates. The dust layer thickens and the charge on the recently precipitated dust must be conducted through the layer of that dust which is already on the collectrodes; this is termed "dust resistivity." Precipitators are successful on dust whose resistivity is in the range of 10^7 to 10^{11} ohms/cm.

After a layer of dust between 1/16 and 1/4-in. has been built up on the collectrodes, the dust particles next to the collectrodes have lost much of their charge to the grounded plate. The electrical attraction has been significantly weakened, and the recently arrived dust particles still hold much of their charge since they are electrically insulated from the plate by a dust layer. At this point, a sharp rap on the edge of the plate causes the dust layer to shear away from the collectrode. The force of the rapping has to be reasonably restrained so as to allow the layer to be shattered into relatively large agglomerations of dust particles and not to create a fine dust which will be reentrained in the gas stream. The large agglomerations of dust then fall into the hoppers by gravity, just as in a gravity settling chamber. Most electrostatic precipitators are operated at gas velocities in the range of 3 ft/sec to allow ample settling time. There is no change of voltage during the rapping or dust removal cycle so that any dust which is reentrained may be recharged and reprecipitated; this is especially true of the finer dust particles. The discharge electrode wires are usually constructed of round 12-gauge steel spring wires which are reinforced top and bottom to insure good electrical contact and resist mechanical and electrical erosion. An independent rapping system is used to keep the high voltage discharge electrode system continuously clean.

Since the electrostatic precipitator operates on direct current, a high-voltage silicon diode power pack has been developed for the severe load conditions to be encountered. These power packs should have substantial overcapacity both as to voltage and to current so that in actual practice the precipitator can operate within a wide range of resistivity materials and can be maintained without having a continuous electric arc formed between the wire and the collectrode at one or more points within the precipitator. These arc-over voltages change from second to second as operating conditions vary, so modern power packs are equipped with a system of voltage regulation which continually monitors the high voltage just short of the arcing condition.

The high voltage circuit for an electrostatic precipitator is shown on Figure 14-12b and consists of the power supply, a saturable reactor to regulate power, a high voltage transformer to increase the voltage, a rectifier to change the alternating current to direct current, and precipitator electrode systems.

The rapping system on an electrostatic precipitator is extremely important. Rapping is done by both electric and pneumatic devices, or combinations of both, or in some types of precipitators by a mechanical device. However, most

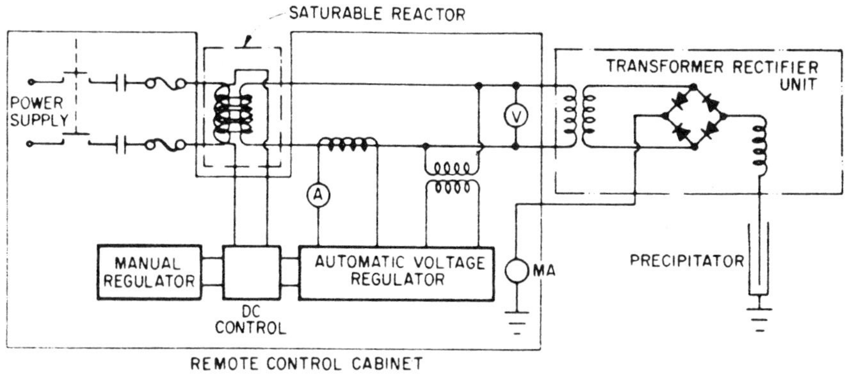

Fig. 14-12b. High voltage circuit for electrostatic precipitator. ***Source:*** **Belco Pollution Control Corp. and** ***Air Pollut. Ind.***

modern precipitators are equipped with two interdependent, electro-mechanical rapping systems, one for keeping the high voltage discharge electrode fields continuously clean and the other for sequential rapping of the collecting plates. Both are designed for automatic rapping while the high voltage remains on and collection continues. Modern rapping systems eliminate puffing and minimize reentrainment by the sequential rapping of small, shock-insulated collectrode modules and by regulating the rapping intensity to match the dust adhesion and the interval of the rapping to match the dust loading.

Rigid Frame Design Over the years, the companies which have manufacture and used electrostatic precipitators in the United States have preferred the suspended wire design where the discharge electrodes are merely a wire weighted at the bottom for tension and located precisely on the centerline between the collecting plates. In Europe, however, a slightly different design was developed which utilized a rigid frame for the discharge electrode. This rigid frame is tubular and has a number of discharge electrode wires stretched between the top and the bottom of the frame. The frame is located at the midpoint between the collecting plates. The collecting plates are quite similar to the U. S. design. They use a roll-form stiffening which reduces dust entrainment during rapping by creating a baffle. The basic difference in the rigid frame design is in the discharge electrode and in the method of rapping. Electrode rapping systems are usually mechanical, using a hammer and anvil rapping principle. The electrode rappers are driven from a single mechanical eccentric unit.

Functionally, the two types of precipitators are essentially equivalent in efficiency; however, the European design, although generally somewhat more expensive, claims to be more rugged and, consequently, has a better maintenance

TABLE 14-6. Efficiency of Multifield Precipitators

% Collection Efficiency	Number of Fields	
	U.S.	Europe
95	1 or 2	1
96–98	2 or 3	2
99	3 or 4	2 or 3
99+	4 or 5	3

record. The primary advantage over the suspended wire design is that occasionally the wires break and the weights at the bottom then fall into the hopper. Some designs have obviated this problem, but the repair of the wires is, nevertheless, an important consideration.

Multifield Electrostatic Precipitators for High Efficiency An electrostatic precipitator may only have one field, but, generally, for most high efficiency applications it has a series of fields in the direction of the gas flow. The number of fields required to achieve any specific collection efficiency is a compromise between the high cost of many short fields and the lower cost of a single long one and also the type of construction of the unit. Table 14-6 shows the collection efficiencies by field number of both U. S. and European designs.

The materials of construction of an electrostatic precipitator are usually carbon steel unless something in the gas dictates that this construction should be otherwise. They are very seldom shop fabricated. While the plates and some of the casing may be fabricated in subsections, it is usually such a large device that field erection is desirable. Small electrostatic precipitators, except for very special applications, are usually not economically viable. The electrostatic precipitator is used for large gas flows and relatively low particulate loading. Its primary application throughout the world has been on coal fired boiler systems.

The resistivity of the dust expressed in ohms per centimeter is a factor which generally determines the size and the number of fields that the precipitator will require to achieve a certain efficiency. Each dust has a different resistivity and this is particularly true of the ash which comes from coal fired boilers. The design of an electrostatic precipitator is a science up to this point and then the resistivity of the dust makes it an art. Empirical data on the attraction of various dusts of different resistivities is almost essential in order to properly design a unit and guarantee a removal efficiency.

Acid Mist Precipitator A special type of electrostatic precipitator is used occasionally in the manufacture of sulfuric acid to collect sulfuric acid mist. In this precipitator, all parts exposed to the gas stream are fabricated of lead.

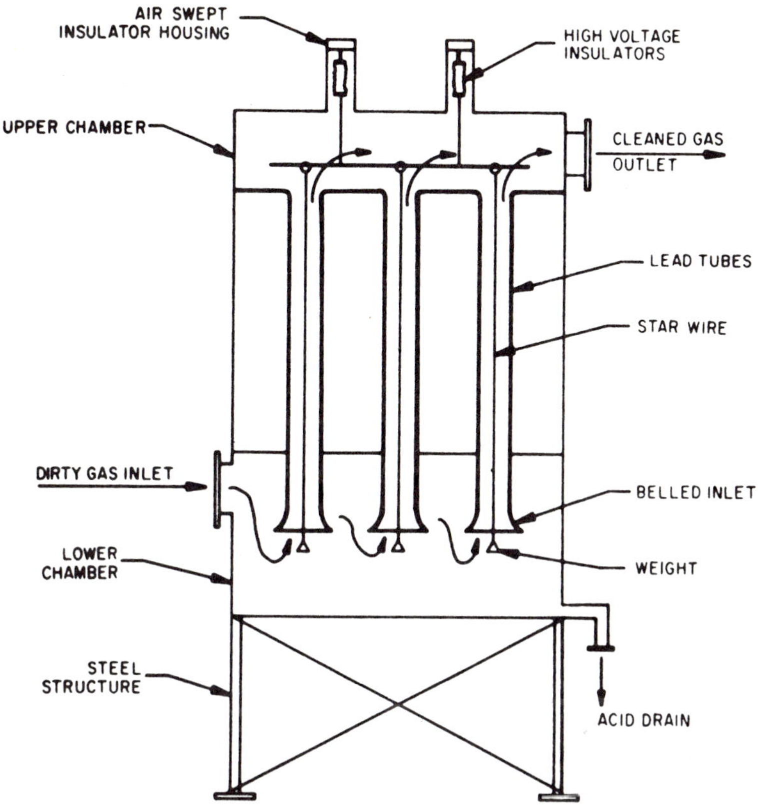

Fig. 14-13. Acid mist precipitator. *Source: Air Pollut. & Ind.* **Ref. 14-12.**

The basic design shown in Figure 14-13 includes an upper cylindrical chamber and a lower cylindrical chamber. The two chambers are connected by 10-in. diameter lead tubes. Star shaped, weighted, tensioned high voltage lead wires with steel cores hang in the center lines of the tubes whose inner surfaces constitute electrodes. The dirty gas inlet is at the bottom, and the gas is passed upward through the lead tubes into the upper chamber and then to the stack. The acid mist is charged negatively and is then precipitated as it passes up between the wires and the tubes. Since it is a mist, it forms a liquid film which drips into the lower chamber and out through an acid drain. Wash sprays are often installed in the upper chamber to flush out precipitated particulates. Acid mist precipitators can be quite large and are often installed in parallel units.

Wet Precipitators Water can be used in precipitation operations either intermittently or continuously. It is difficult to say whether such units are wet precipitators or electrostatic scrubbers. They have the advantage of acting as a particulate collector and absorber at the same time. Some units use water sprays at the inlet and subsequently on the parallel plates to remove solid build-up. Others employ a pipe and wire arrangement similar to the acid mist precipitator. Wet precipitators are advantageous in handling highly resistive particles or those which would stick to plates such as tars and oils. These units have low power requirements compared with a venturi scrubber, which is the principal competitive device.

14-4.2.2 Fabric Filters

Perhaps the most efficient type of particulate removal equipment is the fabric filter which uses a woven or a felted fabric to filter out the fine particles from an air or flue gas stream. The fabric filter is often called a bag house because of its design. It consists of a number of vertical, tubular filter bags. The dust either enters the chamber in which the filter elements are suspended and is collected on the outside of the bag or enters the inside of the tubular bag and is collected there. In both types of systems, there is a means of removing the dust from the bags intermittently and dropping it into a hopper at the bottom of the bag house.

Types of Bag Houses There are four types of bag houses:

1. Mechanical shaker
2. Reverse flow
3. Pulse jet
4. Ring jet

These four types refer to the method of cleaning the bag and not necessarily to the flow pattern through the unit. For example, the reverse flow bag house may collect dust on either the outside or the inside of the bags; this is similarly true of the mechanical shaker bag house; however, the reverse pulse jet normally collects dust on the outside of the bags. Figures 14-14, 14-15, and 14-16 show the schematic arrangements of these three types of systems. The ring jet cleaning system collects dust on the inside of the bag and uses a traveling air ring to dislodge the dust, Figure 14-17.

The bag house itself is a large steel hopper-like structure enclosing a number of bags. These bags are made in the form of a sleeve and are usually 4 to 12-in. in diameter. They are supported internally by a wire screen or cage when cleaning is from the inside, and the dust is collected on the outside. One end of the bag protrudes through a tube sheet which may either be at the top or the bottom of the bag, and the other end of the bag is closed off in such a way as to force the flow of air through the side walls of the bag and into the plenum. At

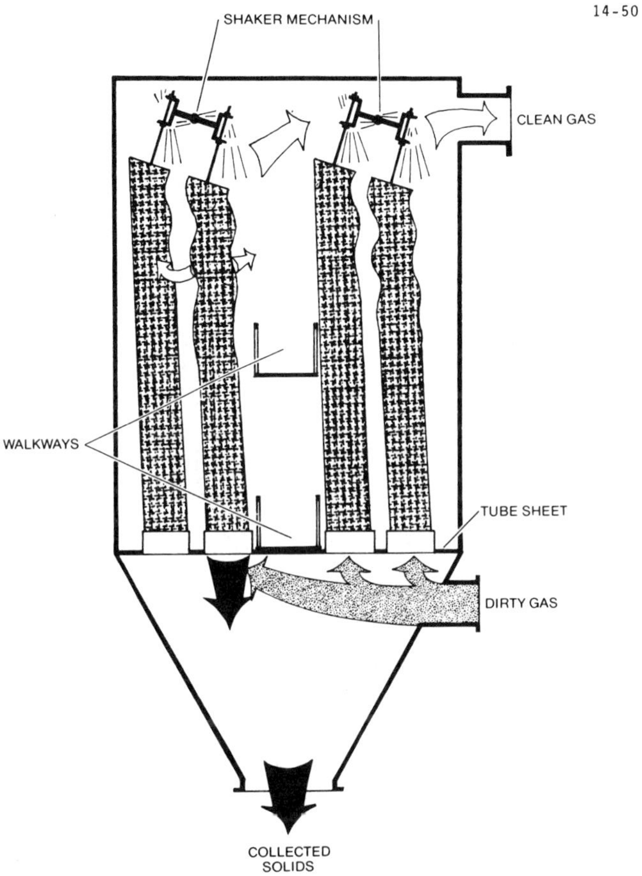

Fig. 14-14. Mechanical shaker baghouse. ***Source:*** **Standard Havens, Inc.**

the bottom of the bag house is a large hopper for the collection of the dust or a series of hoppers usually supplied with some type of removal device such as a screw conveyor or rotary valve which will periodically remove the dust from the hopper. The cleaning of the bags is carried out on an intermittent basis, usually a section at a time; modern bag houses have automatically programmed electrical systems to accomplish this. The frequency of the cleaning operation is usually a function of the dust loading.

There are intermittent bag houses which cannot be cleaned while on-line and thus are generally limited to low dust loadings or infrequent operation. Such units are generally small and are found in small applications with low dust loading, such as a wood-working shop. The continuous, automatic bag house, on the

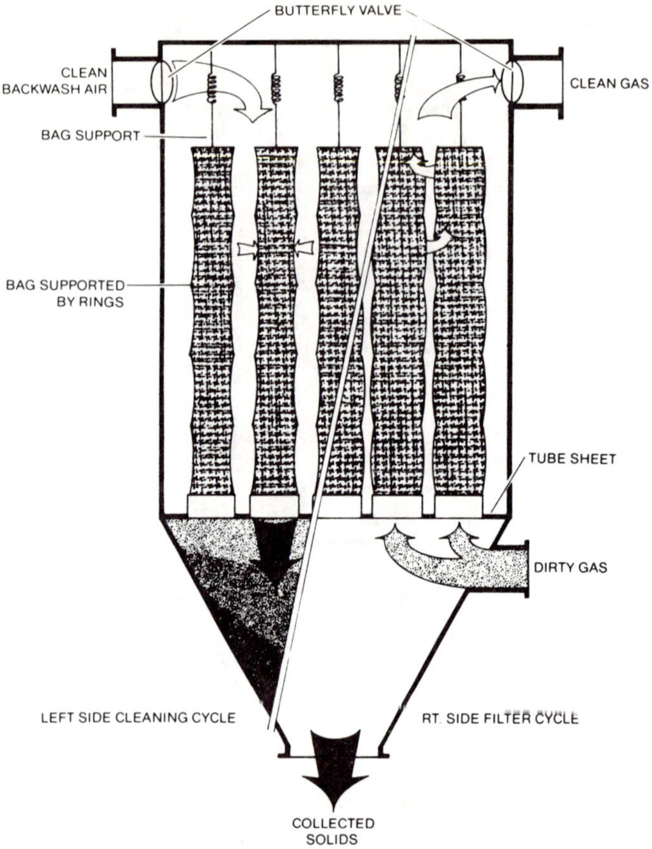

Fig. 14-15. Reverse flow baghouse. ***Source:*** **Standard Havens, Inc.**

other hand, is considerably more expensive but able to operate 24 hours a day and can handle high dust loading.

The mechanical shaker type bag house, as shown in Figure 14-14, has the bag supported by a structural framework which is free to oscillate when driven by a small electric motor. Periodically, on a timed basis, the damper isolates a compartment of the shaker bag house so that no air flows. The bags in the compartment are then shaken for approximately 1 min during which time the collected dust drops from the bag into the hopper for subsequent removal.

The reverse flow bag house, as shown in Figure 14-15, is equipped with an auxiliary fan that forces air through the bags in the direction opposite to filtration. This backwash action collapses or expands the bag and fractures the dust cake. When the bag is reinflated, by being brought back on-line, the fractured

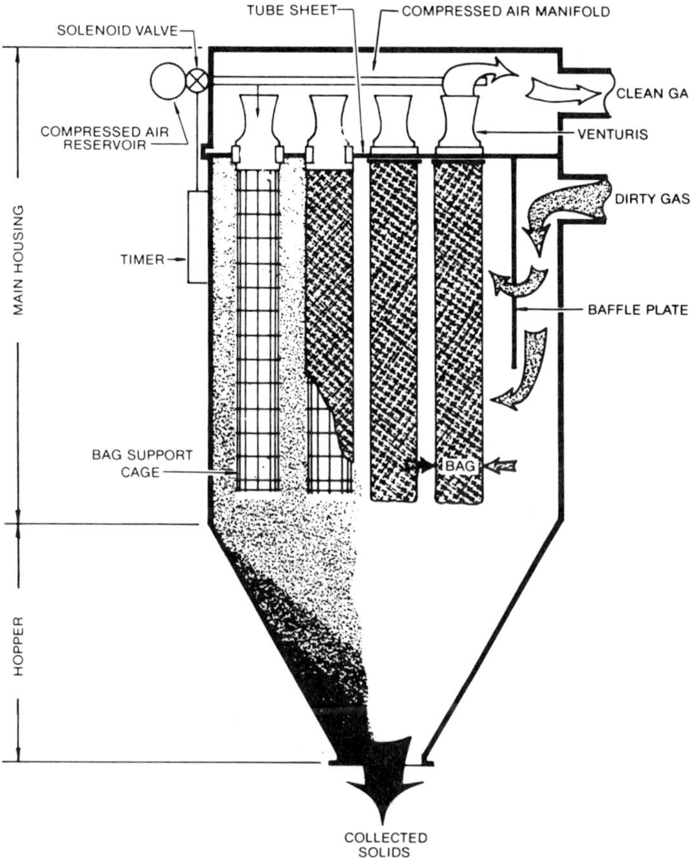

Fig. 14-16. Pulse jet baghouse. ***Source:*** **Standard Havens, Inc.**

dust cake is dislodged into a hopper. If the unit operates under suction, then the main fan is located on the clean side of the bag house.

The reverse pulse jet bag house, shown in Figure 14-16, incorporates a small venturi at the top or open end of the bag which is fed by a jet of high pressure air. The air line to each venturi is controlled by a solenoid valve which is actuated by a timer. When it is time to clean a specific bag or section of bags, the high pressure air jet comes on pulling additional air into the top of the bag and expanding it in such a way as to break the cake from the outside of the bag.

Another type of bag collector, seen more often in Europe than in the United States, is the envelope-type system. Rather than a round bag, the filtering surface is a relatively flat envelope supported on a screen. The advantage of this type of bag house is a smaller space requirement, but often the dust bridges or

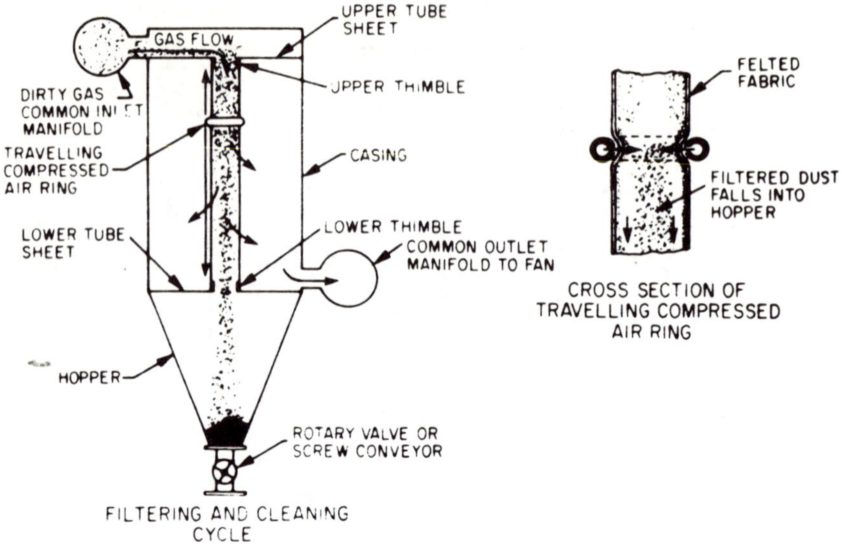

Fig. 14-17. Ring jet baghouse. *Source: Air Pollut. & Ind.* Ref. 14-12.

plugs the spaces between the envelopes. These are usually used for low-flow applications.

Air-to-Cloth Ratio. The fundamental criterion used in applying any bag house to any application is the air-to-cloth ratio, defined as the ratio of the air flow line to the net cloth area. The air flow should be expressed in actual cfm and the net cloth area in square feet. The air-to-cloth ratio is equal to the superficial face velocity of the air as it passes through the cloth. Shaker and reverse air bag houses normally operate at a face velocity or air-to-cloth ratio of 1-to-1 to 3-to-1. Reverse pulse bag houses can operate at a much higher velocity with air-to-cloth ratios of 3-to-1 to 6-to-1, although many are now operating in the range of 10 or 12 to 1. This, however, is more a function of the specific application and is generally determined for a specific type of dust on an empirical basis. The lower ratio shaker and reverse air bag houses are generally equipped with extra compartments. In this arrangement, there is always an extra section of the bag house to accommodate cleaning while the other sections handle the full air flow under normal operating conditions. Generally, because the pulse jet system operates at higher air-to-cloth ratios, it is smaller and has a tendency to be less expensive. However, the use of compressed air is expensive and the operating costs of the pulse jet are usually higher than the reverse air cleaning type of system. Filter drag, or resistance to airflow, is important in a bag house design. When the filter is cleaned initially and comes back on-line, there is only a minimal pressure drop at the start of the filter cycle. Then, as the cake builds up on the filter and the pressure drop increases because filtration is taking place through the dust layer, the operation of the bag house becomes more efficient. As the cake continues to build, higher pressures result and the bag house needs to be cleaned again. This drag cycle is important to note since the pressure drop varies with the square of volumetric airflow.

Filter Media. In the bag house filter installation, particulates in the exhaust gas stream are collected on a fabric filter which permits the clean gas to pass through to the atmosphere. The filter media is made of woven material which has threads 450 to 550 μ in diameter and spaced from 75 to 250 μ apart. These form a mat which allows some of the particles to pass until the filter cake starts to build up on the media. The large aperatures between the threads are criss-crossed with fine fibers in the range of 3 to 8 μ in diameter. Particles smaller than the voids in the fabric will be trapped by impingement on these fine fibers. The choice of the filter fabric will usually depend upon the following factors:

1. The temperature of the gas stream
2. The physical and chemical characteristics of the particles to be collected
3. The chemical composition of the carrier gas
4. The moisture content of the carrier gas
5. The mechanical strength of the fabric in relation to the other requirements

Table 14-7 shows a variety of materials which have been used for filter fabrics and their resistance to temperature, acids and alkalis, and abrasion. For example, the fiberglass bag, which is perhaps the best choice for high temperature boiler applications, has a relatively poor flexing resistance; so, after it has been flexed a number of times, it often fails. The single most important maintenance cost in bag collectors is the replacement of the bags themselves.

Advantages and Disadvantages of the Bag House Design. For large and difficult applications, the bag house is a good choice as long as the temperature isn't too high. The bag house design has the advantage that it operates on a dry basis as compared with a scrubbing system; it is efficient on submicron sized particles; it has a generally high efficiency, 98% or above; and it operates on a very low pressure drop which results in low power requirements. Some of its disadvantages are that it may have clogging problems in gas streams which are near the dewpoint; it cannot be considered at temperatures above 600°F and probably cannot be wisely considered at somewhat lower temperatures; it has a relatively large space requirement; and it is expensive to maintain. On a first-cost basis, the bag house and electrostatic precipitator for large installations are fairly closely matched with the bag house often being slightly cheaper but having a higher pressure drop.

TABLE 14-7. Filter Fabrics

Fabric	Melting Temperature	Maximum Continuous Operating Temperature*	Acid Resistance	Alkali Resistance	Flex Abrasion
Cotton	Decomposes at 302°F	180°F	Poor	Very good	Very good
Wool	Chars at 572°F	200°F	Very good	Poor	Fair to Good
Nylon 6, 6[a]	480°F	200°F	Fair	Excellent	Excellent
HT-1[a]	Chars at 700°F	400°F	More resistant than nylon, inferior to dacron & orlon.	Not as resistant as nylon, superior to dacron & orlon.	Good
Dacron[a]	482°F	275°F	Good to most mineral acids. Dissolves partially in concentrated H_2SO_4.	Good to weak alkali. Fair in strong alkali.	Very good
Orlon[a]	482°F Softens	260°F	Good to excellent in mineral acids.	Fair to good in weak alkalis.	Good
Creslan[b]	475°F Softens	250°F	Good in mineral acids.	Good in weak alkalis.	Good to very good
Dynel[c]	325°F Softens	160°F	Little effect even in high concentration.	Little effect even in high concentration.	Fair to good
Polypropylene	333°F	200°F	Excellent	Excellent	Excellent

Teflon[a]	Decomposes at 750°F	500°F, emits toxic gas at 450°F	Inert except to fluorine.	Inert except to chlorine, tri-fluoride and molten Alkaline metals.	Fair
Fiberglas	1470°F	550°F	Fair to good	Fair to good	Fair
Filtron[d]	505°F Softens	270°F	Good to excellent.	Good	Good to very good
Nomex[a]	675°F	425°F	Good	Excellent at low temperatures	Excellent

[a]Du Pont Reg. Trademark
[b]American Cyanamid Reg. Trademark
[c]Union Carbide Reg. Tradmark
[d]W.W. Criswell Trademark
*Temperatures recommended by Industrial Gas Cleaning Institute
Source: Standard Havens, Inc.

14-4.2.3 Wet Scrubbers

There are literally hundreds of designs of wet scrubbers available in the United States today which are advertised for the removal of particulate materials from gas streams. The size and initial cost generally are much lower than that for an electrostatic precipitator or bag house; however, the operating cost because of the pressure drop is usually much higher.

Of the many types of wet scrubbers available for particulate removal, the only one which usually can really be considered suitable for submicron particle removal is the Venturi-type of unit. Some of the scrubbers which fit into this category are not truly Venturi scrubbers at all but merely orifices; however, in general, they function essentially the same. The Venturi-type of scrubber mixes the gas stream with the scrubbing liquid just ahead of a restricted area, which could either be an orifice or a Venturi. The true Venturi with a 15° diverging section has the best pressure recovery, but any type of restriction increases the velocity of the dust particles in the gas mixture and this high velocity enhances the capture of the dust by impingement upon the water droplets. As the droplets accelerate from zero to gas velocity in the expanding section of the Venturi, the noncaptured dust overtakes the droplets at decreasing relative velocity. Capture of fine dust by fine droplets through the mechanism of diffusion is increased as relative velocities of dust and droplet approach zero. This mixture is then fed into a centrifugal type separator with velocities further reduced, and the dust and water particles are removed from the bottom as a slurry, see Figure 14-18. There are literally dozens of different Venturi designs but for fine particle removal, pressure drops will be anywhere from 20 to 75 in. w.c., that is, it will take this type of pressure drop to place the Venturi scrubber on a par with the electrostatic precipitator and the fabric filter with respect to efficiency.

Other types of low energy scrubbers which utilize water or another scrubbing liquid to remove particles from a gas stream will remove them in direct porportion to the amount of energy expended in the scrubber.

Ionizing Wet Scrubber. An ionizing wet scrubber utilizes electrostatic particle charge in combination with a wet packed crossflow section has been proven effective for very small particles where Venturi pressure drops would be inordinately high. It can simultaneously remove corrosive materials and submicron particles. It utilizes a high voltage ionizer section to charge particulates in the gas stream before wet scrubbing. The charged particles are attracted to or attached to a neutral surface and then eventually washed out in the scrubbing liquor.

14-4.2.4 Mist Eliminator

There are two basic types of drop or mist removal devices: an impingement separator and a unit that operates on the principle of Brownian movement or diffusion.

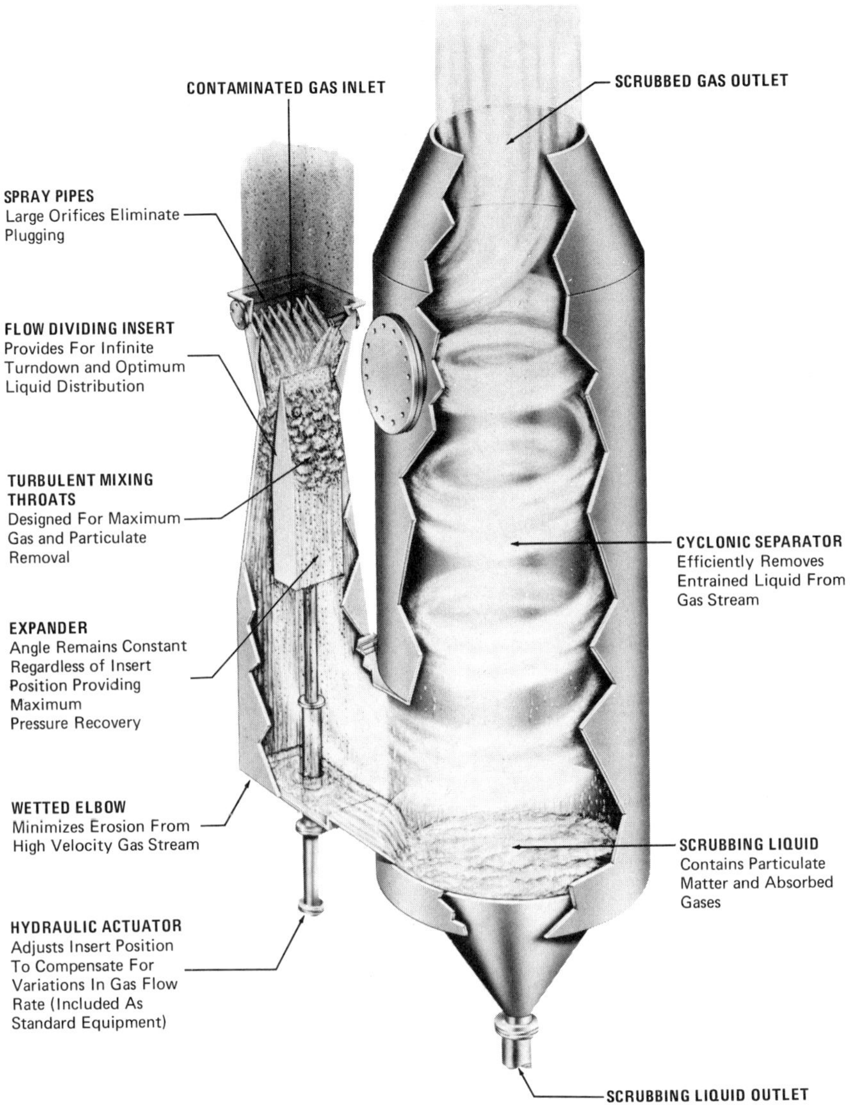

Fig. 14-18. Venturi scrubber. ***Source:*** **FMC Corp., Itasca, Illinois.**

The impingement type separator removes droplets or mist from a wet gas stream by their impingement on a surface. Impingement separators comprise dozens of different designs, but their basic configuration is either a mesh of wire or some other similar material or a series of plates designed to change the direction of a gas stream several times. Impingement type separators will remove droplets down to approximately 10 μ in diameter at about a 99% efficiency level. Velocities through impingement type separators are usually maintained between 5 and 30 ft/sec, and they are generally arranged in the gas stream horizontally so that the gas flows vertically up through the separator, although this is not always the case. Thus, the droplets removed fall back into the entering gas stream and some reentrainment occurs; however, if the unit is operating correctly, there is a certain amount of coalescing which takes place forming larger droplets which are heavier and tend to drop back through the gas stream to the collection point. The advantage of this type of mist separator is that it has a very low pressure drop on the order of 1-2 in. w.c.

The other type of mist separator is the fiber bed unit made of fiberglass, polyester, or Teflon fibers which are very fine and are packed within the bed to high density. This type of mist separator removes droplets by Brownian diffusion. The droplets entering the fiber bed will tend to follow the tiny fibers until coalescing occurs near the bottom of the unit, see Figure 14-19. This type of mist eliminator is capable of removing droplets or mists down to 0.3 μ at efficiencies up to 99%. It differs from the impingement type separator not only

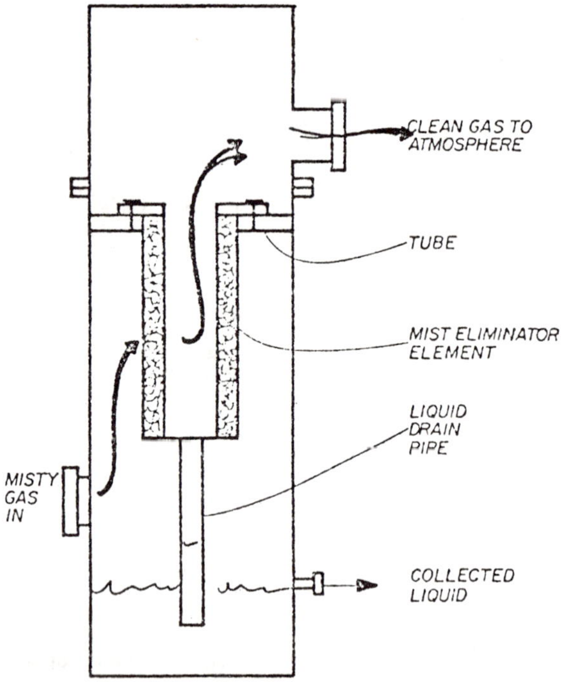

Fig. 14-19. Fiber bed mist eliminator.

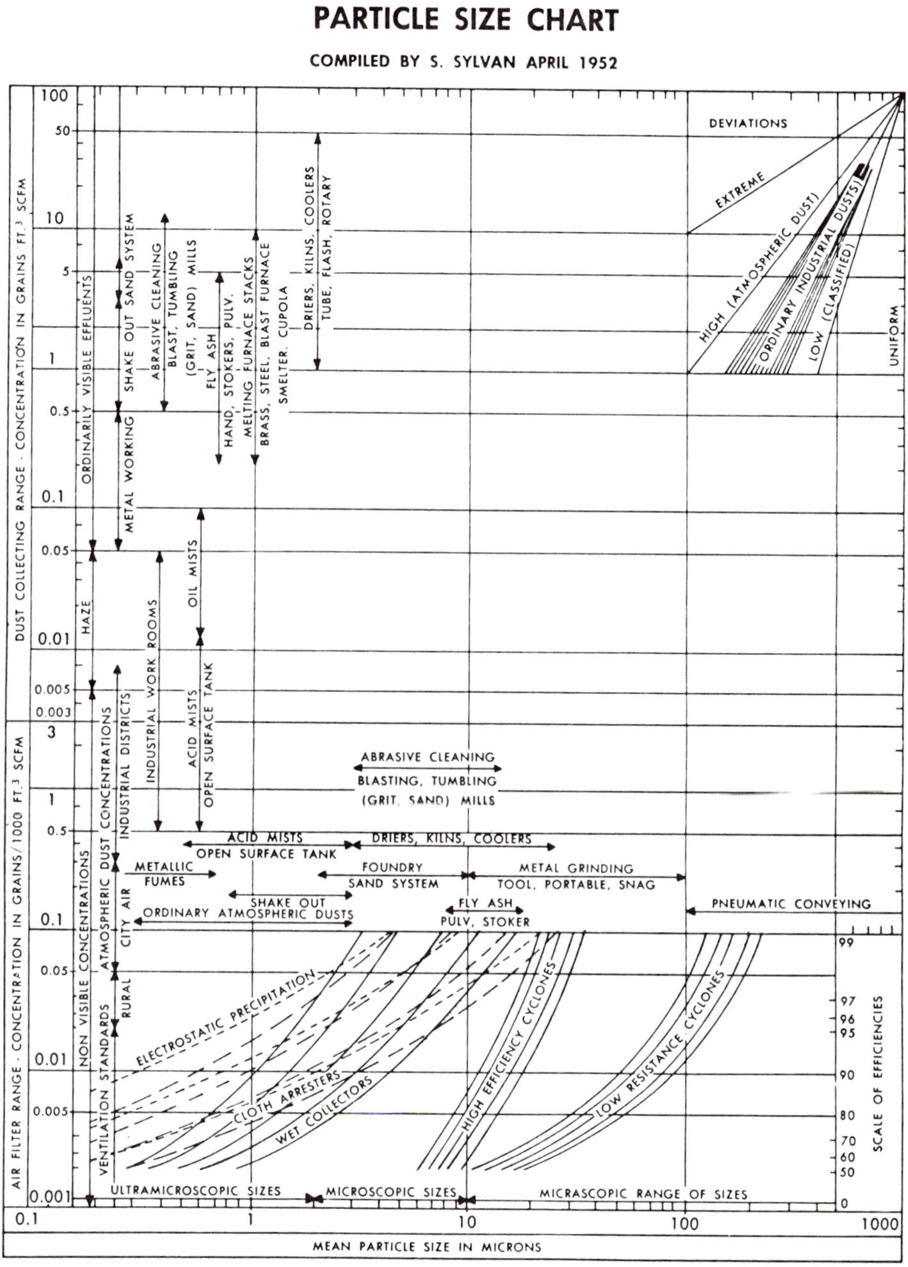

Fig. 14-20. Sylvan chart. *Source:* American Air Filter Co., Inc., Copyright 1952.

in its method of operation but in its pressure drop. The pressure drop through a fiber bed mist eliminator will be from 6 to 16 in. w.c., depending upon the flow and the efficiency required. This type of system was developed for sulfuric acid plants and is an accepted part of the control technology on this type of plant.

The mist eliminator of the fiber bed type is not generally recommended for streams containing solids. However, if the solids are soluble in water or some other solvent and if they are in relatively low concentrations in the gas stream, then the mist eliminator may be used for this purpose. A back wash nozzle affixed to the top of the mist eliminator element can be used to periodically flush the solids from the fiber bed, and, in this sense, the unit becomes a wet scrubber with the fiber bed acting as the medium for scrubbing. Such systems have been very successful in the ammonium nitrate manufacturing industry.

14-4.2.5 The Sylvan Chart

The Sylvan Chart developed by American Air Filter Company some years ago still remains a very good means of estimation of both particle size and grain loading for various applications. It is shown in Figure 14-20. It is, of course, not a 100% accurate estimation of particle size and grain loading but gives the industrial plant operator an idea of the size and loading of dust which he must handle for various standard applications.

14-5 SULFUR DIOXIDE REMOVAL

Perhaps the most auspicious pollutant, and one that has gotten the most extensive design concern, is sulfur dioxide. Virtually all coal burning power boilers, except those operating on very low sulfur Western coal, produce sulfur dioxide levels which are above the acceptable limits in most states; consequently, a means of removal of the sulfur dioxide is necessary. While bag houses or electrostatic precipitators may be used to control the particulate emissions, the sulfur dioxide can only be removed by chemical reaction with either lime, caustic, soda ash, or some similar basic material. A list of the major flue gas desulfurization processes is shown in Table 14-8.

While there are a number of variations on these processes, they essentially involve scrubbing a high sulfur dioxide gas, usually on the order of 0.3%, with some type of alkaline material. In the case of the limestone, the reaction produces a mixture of calcium sulfate and calcium sulfite, see Figure 14-21. In the dual alkali system, the reagents are sodium carbonate and calcium hydroxide, and the sludge at the end of the process is calcium sulfite. In the magnesium oxide system, magnesium sulfate is the end product, and in the case of the Wellman Lord process, a similar soluble bisulfite is used as an intermediate.

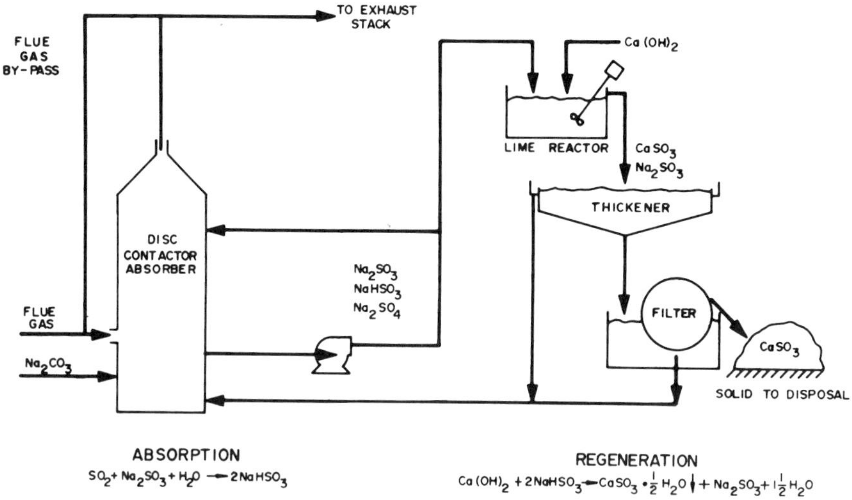

Fig. 14-21. FMC dual alkali process. ***Source:*** **FMC Corp., Itasca, Illinois.**

All of these processes are quite expensive and essentially add a complete chemical plant to the tail end of a boiler system. But, because of sulfur dioxide restrictions, it will be necessary to desulfurize the flue gas or to consider the possibility of desulfurizing the fuel before it is burned. While there are proponents of both approaches, at the present time the desulfurization of the flue gas has by far the greater number of installations. Table 14-9 shows a summary of control alternatives for sulfur dioxide. Aside from using alternative fuels or coal liquification with solvent refining, there does not seem to be any other attractive means for removing sulfur dioxide.

Sulfur dioxide, of course, does not only come from power boilers; it often comes from sulfuric acid plants. The acid plant tail gas contains enough sulfur dioxide to cause serious air pollution problems. An ammonia scrubbing system for the removal of sulfur dioxide as a tail gas with the creation of ammonium sulfate as a product was developed by Davy Powergas and is shown in Figure 14-22. These plants normally operate at a conversion efficiency of 97 to 98%; however, atmospheric emissions from plants of this type contain in addition to oxygen and nitrogen, 2000 to 4000 ppm of sulfur dioxide, 1 to 5 mg/cu ft of sulfuric acid mist, and 1 to 5 mg/cu ft of sulfur trioxide. Recent federal and state environmental agencies, in an effort to improve the ambient air quality, have instituted emission regulations for existing as well as new sulfuric acid plants which require the plant to operate in excess of 99.25% conversion efficiency.

Molecular Sieves. Sulfur dioxide may also be removed by use of a molecular sieve. A molecular sieve is actually an adsorbent material, generally activated

TABLE 14-8. SO_2 Removal Processes

Process	Classification/ Operating Principles	SO_2 Particulate Efficiency	Application	Advantages	Disadvantages
Lime/limestone scrubbing	Nonregenerable process/ wet absorption in scrubber by slurry; insoluble sulfites, and sulfates disposed of as waste.	Up to 90 percent SO_2 removal/99 percent fly ash removal by most scrubbers.	Old or new power plant; coal- or oil-fired.	Cheapest of existing processes; elimination of particulate control requirement.	Waste and water pollution problems; reheat of scrubber exit gases needed; supply and handling of large volumes of reactant may be problems.
Double alkali process	Nonregenerable process/ wet absorption in scrubber; reactants and soluble; reaction products precipitated and removed from recycled reactant solution outside of scrubber; most common reactant sodium sulfite.	High efficiency >90 percent SO_2 removal/high particulate removal as above.	As above with potentially lower cost and greater ease of operation favoring some inroads into smaller plants.	Potentially cheaper and more reliable than lime/limestone system.	Similar to above and all throw-away systems.

Magnesium oxide scrubbing	Regenerable process/wet absorption by magnesium oxide slurry; fly ash removed prior to or after scrubbing; magnesium oxide regenerated by calcining with carbon; SO_2 by-product can be converted to sulfuric acid or sulfur.	90 percent SO_2 removal/particulates removal as required by prescrubber.	Similar to lime/limestone but oil-fired boilers will not require particulate control upstream of scrubber.	May be more reliable than lime/limestone process; no known waste disposal problems; regeneration facility need not be located at utility.	Cost of regeneration; marketing of sulfur products; reheat.
Wellman-Lord	Regenerable process/sodium base scrubbing with sulfite to produce duce bisulfite; regeneration in an evaporative crystalizer. Sulfate formed either purged or removed by selective crystallization.	>90 percent SO_2 removal; particulate removal as above by prescrubber.	As above.	More reliable than lime/limestone system based on Japanese experience; simplicity of unit operations in regenerator; waste disposal problems reduced.	Some bleed of solution to remove undesirable reaction products a source of water pollution, otherwise as above.

Source: U.S. Environmental Protection Agency Publication 450/2-75-007

TABLE 14-9. Control Alternatives for SO_2

Control Technique	Sulfur Removal Efficiency (3% S base), %	Energy Penalty, %	Incremental Cost, mills/kWh	Timing
Nonregenerable (lime/ limestone) scrubbing (new plants)	90	3–7	3.0	1975–77
Nonregenerable scrubbing (existing plants)	90	3–7	4.4–6.1	1975–77
Regenerable scrubbing (new plants)	80–95	3–7	3.4–7.0	1980
Physical coal cleaning	30–50	3–10	0.6–1.6	1974–80
Coal liquefaction (solvent refined cleaning)	75–85	20	10	1985
Low Btu gas	80–98	10–40	0–16	1980–85 1980
Pipeline quality	98+	40	NA	1980 1985
Fluidized bed	90+	Minimal	0–3	1985
Refuse combustion	20	Minimal	Minimal	1985–90
Low sulfur western coals	75–85	Minimal	0–5.3	1980
SIP compliance coal (Eastern)	33–70	Minimal	1.0–3.0	1975
Desulfurized residual oil 0.3%	85	–	2.0	1980–85

Source: U.S. Environmental Protection Agency Publication 450/2-75-007

Annual Capacity, 10^6 tons coal	Likely Applicability	Other Environmental Effects	Comments
230 (1930)	All new utilities and large industrial boilers	Waste disposal problems	Operation load factor 80%, 1000-MW power plant
60 (1980)	Boilers $\leq$ 100 MW $\leq$ 20 yr age	Waste disposal problems	Operating load factor 60%, 300-MW power plant
–	New utilities and large industrial boilers	Reduced waste disposal problems	Limited data on utilization and costs
–	All coal combustors	Land, air, waste disposal at cleaning site	Cleaning costs $1.50/ton
110	Existing and mine mouth power plants, industrial boilers, small point sources	Sulfur and other emissions partially concentrated at central plants; increased mining necessary; unidentified air, water, and solid waste problems	Economics of scale possible
13 200	New and large existing power plants		Energy penalty reduced by combined cycle application. Capital intensive
29 50–100	Small and area sources		Minimal energy penalty if used to replace electric heat
150	New power plants and industrial boilers	Substantial reductions in NO_x emissions possible	Energy efficiency can be greater than conventional power plants
50	Coal mix, 10–20% refuse	Benefits due to solid waste reduction	
250	Western and fringe eastern boilers	Increased particulate control necessary	3% energy penalty for transportation to eastern areas
95	Eastern existing boilers		Cost for 1% S or less compared to 3% S coal
NA	Existing oil burners	Also removes V, other trace elements	

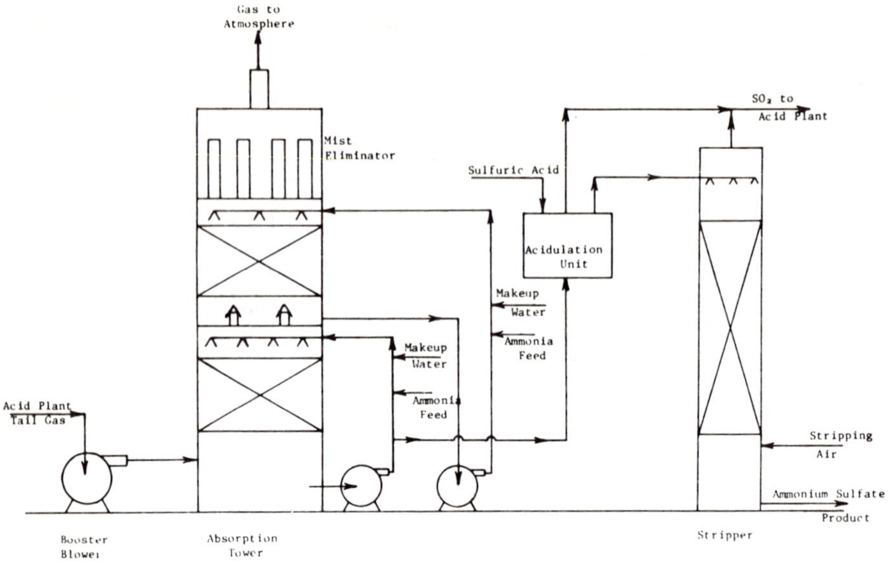

Fig. 14-22. Davy powergas SO_2 process. ***Source:*** **Davy Powergas.**

alumina, having special porosity which will selectively pick up certain gaseous compounds by adsorption. Molecular sieve units are currently being used on sulfuric plants for sulfur dioxide removal, nitric acid plants for NO_X removal, and caustic chlorine plants for mercury removal.

The sulfur dioxide removal system is accomplished in a fixed bed adsorption system, and the acid plant adsorber tower overhead is compressed, cooled, and passed through a high efficiency mist eliminator. The tail gas then passes through the molecular sieve bed for selective adsorption of the sulfur dioxide. Completely automatic regeneration is accomplished by heat. The required regeneration gas is obtained from either acid dried air from the plant or ambient air. After thermal stripping of the adsorbent bed, the sulfur dioxide-rich gas used for regeneration is recycled to the basic acid plant.

A typical system is the Union Carbide Purasiv S system. The Purasiv S system is capable of reducing the sulfur dioxide emission level to a guaranteed average of less than 100 ppm by volume, while recycling the recovered sulfur dioxide back to the acid plant for an additional 2 to 3% of acid production. The first commercial unit was started in 1973 at Coulton Chemical Corporation, Argon, Ohio. This unit treats a tail gas from a 200 ton/day acid plant which processes a mixture of spent acid and refinery hydrogen sulfide off-gas. The unit is designed to remove and recover up to 3500 ppm of sulfur dioxide along with small amounts of sulfur trioxide, sulfuric acid vapor, and sulfuric acid mist. The tail gas also contains 7-1/2% carbon dioxide and 10 ppm of NO_X. The maximum design tail gas flow rate for the unit is 10,500 SCFM at a sulfur dioxide concentration of 2800 ppm.

The air which is used for regeneration in this system is also dried by a molecular sieve so that the unit contains two sulfur dioxide adsorbers and two air driers, each of which is operating on a 4 hr on and 4 hr off regeneration cycle.

The unit recovers 99% of the sulfur dioxide present in the tail gas stream and has provided an average daily effluent concentration in the range of 15 to 25 ppm under a wide range of feed conditions. This is equivalent to a daily emission rate of 0.4 lb of sulfur dioxide per ton of acid.

14-6 GASEOUS EMISSION CONTROL, GENERAL

To this point, we have discussed the removal of particulate materials from various air and gas streams, as well as the removal of sulfur dioxide, which is a prime pollutant. However, there are literally thousands of different types of waste gases that can be considered pollutants which may need to be removed before a stack gas can be acceptable to the atmosphere. Some of these are unacceptable because they are toxic or represent health hazards, and others because of an aesthetic problem having to do with their odors. In either case, the control methods are generally quite similar.

The major control techniques can be divided into the following: dilution and dispersion through the use of tall stacks and additional clean air; chemical reaction which is primarily used for odor counteraction by means of other chemicals; incineration, which will destroy any organic materials present in the waste gas stream, either directly or through catalytic methods; adsorption on the surface of activated carbon or silica gel with subsequent regeneration; wet scrubbing which uses adsorption and possible oxidation with potassium permanganate, hypochlorites, or chlorine dioxide; or ozonation, which makes use of the highly oxidizing nature of ozone.

14-6.1 Dilution

Any new industrial facility must be judged on the basis of the pollutants emitted from the plant in relation to the regulatory performance standards and the possible degredation of air quality in the area. The impact of these emissions may be predicted by a mathematical diffusion model. Inputs to the model are an inventory of the emission sources, meteorological data, and a map of the impact area showing the location of emission sources and receptor points. The meteorological data are in the form of an array covering wind direction, wind velocities, and atmospheric stability conditions for at least one year. Good sources of such records are the National Weather Records Center in Asheville,

North Carolina, and local weather stations, and environmental meteorological support units.

It is possible to use a tall stack or dispersing system to ensure that the predicted ground level concentration from the source does not exceed the standards for nondegradation and also that the predicted level concentration when added to the background air quality levels for the area in question does not exeeed the secondary air quality standards.

The predicted contributions of suspended particulate, sulfur dioxide, or other pollutants in the ambient air in the vicinity of the new facility can be plotted on maps as lines of equal concentration called isopleths. These lines provide a basis for the judgment of the impact of these pollutants on the surrounding area; the values can then be compared to ambient air quality standards. In order to determine whether a stack or similar type of dispersion apparatus is going to be suitable, there are a number of steps necessary. They are as follows:

1. A definition of the study area which includes the physical size, the selection of a coordinate system, and the plotting of both the source and the receptor coordinates.
2. The collection of source data which includes the stack height, diameter, velocity and temperature, the emission rates, and the physical location in the coordinates.
3. A collection of meteorological array which is the mixing depth, the ambient temperature, the ambient pressure, and the wind data, including speed, direction, and stability.
4. A collection of sampling station data which gives the arithmetic mean of the pollutant levels and the background information.
5. The selection of a mode of operation which determines whether the pollutant concentration isopleths are to be plotted, how the calibration is to be handled, and what pollutants are to be considered.
6. An estimation of concentration of source emissions on the surrounding area using a diffusion model.
7. An estimation and evaluation of the impact which includes adding the background air quality levels to the predicted impact and comparing new air quality levels with standards to determine their acceptability.

There are certain meteorological parameters and process emission information which are necessary for the estimation of down-wind concentrations from an air pollution source. Diffusion models based upon mathematical equations are being used to estimate the effects on ambient air quality in a region of specific industrial operation. The equation generally used for making these estimates is the Gaussian Diffusion Equation. If the plume has Gaussian distribution in both the horizontal and vertical dimensions with standard deviations of plume concentrations of distribution of vertical and horizontal of σ_z and σ_y respectively, the mean wind speed affecting the plume is $\bar{U}$, the height of the emission is H, the crosswind distance from the plume center line is σ_y, the source of strength is Q, and the total reflection of the plume takes place at the earth's surface. Then, the calculation at ground level

$$C = \frac{Q}{\pi \bar{U} \sigma_y \sigma_z} \text{ exponent} \left[-\frac{1}{2} \left(\frac{H}{\sigma_z} \right)^2 \right] \text{ exponent} \left[-\frac{1}{2} \left(\frac{Y}{\sigma_y} \right)^2 \right] \quad \text{(Eq. 14-1)}$$

where

C = μg/meter^3
Q = grams/sec
$\bar{U}$ = meters/sec
σ_y = meters
σ_z = meters
H = meters
Y = meters

The values of the σ_z and σ_y must be in terms of downwind distance X. The computation of concentration C is therefore made for specific downwind distances. Also C is the total dosage in gm-sec/meters^3 of curies-sec/meters^3 for the same distance for the passage of the entire plume, if the total release is Q.

This equation is valid at diffusion if the downwind direction can be neglected and if the duration of release is equal to or greater than the travel time from the source to the location of interest. The downwind diffusion can be safely assumed to be negligible.

One of the inputs from the diffusion equation is the effective stack height, which is a combination of the physical stack height and the plume rise due to the velocity and temperature of the stack gases. The Holland Stack Rise equation is as follows:

$$\Delta H = \frac{V_s d}{U} \left(1.5 + 2.68 \times 10^{-3}\, p \frac{\Delta T}{T_s} d \right) \quad \text{(Eq. 14-2)}$$

where

ΔH = Plume rise above the stack, meters
V_s = Stack velocity meters/sec
d = I.D. of stack, meters
U = Wind speed, meters/sec
p = Atmospheric pressure, millibars
T_s = Stack gas temperature
ΔT = Difference between stack gas temperature and ambient temperature

There are many different diffusion and stack rise equations which may be used in an analysis of this type. Models may consider topographic conditions, different time averaging periods, and other parameters, but the intent here is to present the methodology of environmental assessment rather than the pros and cons of the various diffusion models.

There are several major considerations in the use of a diffusion dilution or dispersion type of device. The first is the general area in which it is going to be used. If the area is highly industrialized with many other stacks, it may be in your favor because state implementation plans are designed to lower air quality levels. Thus, over a given period of time, your impact added to a high background condition may be under the air quality standards, although you may have been marginal to start with. Also, the present nondegradation clause in the air quality amendments to the Clean Air Act may require essentially 100% control of many sources, if this is strictly enforced. Consequently, diffusion does not look like a good method of disposing of waste gases in a concentrated area.

Secondly, the weather and meteorological conditions of the particular area are all-important. Areas in which atmospheric inversions occur at regular intervals will probably be very poor locations unless a stack is tall enough to protrude through the normal height of the inversion layer. Normal wind direction will also be a major factor since, if there are large population areas downwind from the stack, this will usually cause problems.

Dilution of stack gases before entering the stack is usually not permitted by air pollution regulation in most states. For example, incinerator flue gases could be easily diluted for cooling and dispersion, but the effluent volume is usually corrected to 12% carbon dioxide in the flue gas which means that there is essentially no dilution permitted. Even if dilution is used, this correction negates any advantages of dilution in terms of sampling both particulate and gaseous pollutants in the stack.

14-6.2 Odor Counteraction

There are many industrial gaseous effluents which are not necessarily toxic but which are objectionable because of odor, sometimes to the point of being nauseating. One of the possibilities for handling the odor, especially if the gas is not toxic and does not cause any nuisance other than odor, is an odor counteractant method. Wastewater treatment plants are not usually designed to smell like perfume despite modern treatment methods. Many plants will always cause complaints when the wind is in a certain direction and, while this odor is objectionable, it is not necessarily dangerous. The concentration of organic materials in the air is well below established minimums, and except for the aesthetic aspect of the odor, it does not affect the ecology of the neighborhood.

To use any of the other methods of gaseous waste control, it would be necessary to capture all the gases coming off the wastewater treatment plant and either incinerate them or adsorb them on some adsorbing medium. These plants are very large in physical area and enclosures are impractical (except for those based on high-purity oxygen) so that odor dissemination into the air often is a way of life. Some municipalities have used odor counteraction to solve the problem.

Strong odors tend to mask weaker ones. If a chemical with a strong odor is injected into a waste gas having a weak odor, the olfactory nerve, the odor recipient, might perceive only the stronger one; this is known as "odor masking." Certain odors in appropriate relative concentrations could also tend to destroy the odorous properties of both thus diminishing the intensity of each; this is called "odor counteraction or neutralization."

Since odors are due to gases, the material used as an odor counteractant must also be in a gaseous or vapor form, and this is done by injecting the odor counteractant in a liquid atomized form into the air through a special nozzle designed to provide a fine spray. This is usually done at ground level; the odor-counteractant chemical is pumped from a storage tank or drum through the atomizing nozzle under pressure. It could also, however, be applied at the top of a stack so that ground level application is not the only possibility. Every odor has some type of particular counteractant (there are several companies currently providing odor counteracting chemicals to industry). This is not a total solution to the problem and it is certainly not recommended if there is any toxicity associated with the gaseous effluent, but there are many applications of this type of odor control for bodies of water, landfills, ovens, storage tanks, and other places where flumes cannot easily be captured and ducted to an incinerator or some other form of treatment. The counteractant should never be a material that desensitizes the olfactory nerve because without the sense of smell, it would be impossible to identify a toxic emission that might be released in the same area.

14-6.3 Thermal Incineration

The methods for "control" of gaseous effluents which have been discussed to this point have not involved true control methods. Dispersion and odor control by odor counteraction are strictly methods of making the gaseous effluent more acceptable to the atmosphere. However, there are several methods for controlling gaseous emissions which do destroy organic content and render the gaseous effluent suitable for final disposal to the atmosphere. The first of these is the method of thermal incineration. Thermal incineration involves oxidation of organic content in the gaseous effluent to the point where all of the material vented to the atmosphere is carbon dioxide, water vapor, nitrogen, and various inorganic compounds depending on elemental composition of the organic incinerated.

The term often used in thermal incineration systems is "after burner." The afterburner is used to destroy gases coming from any combustion operation and is essentially a secondary step which completes the combustion operation. In the process of waste copper wire reclaimation, insulation is burned off in a furnace or retort. The combustion in the first stage usually is not complete because it is impossible to get enough air through the bundle of waste copper to complete combustion of the insulation. Therefore, the stack from this furnace will contain combustible materials which should not be released to the atmosphere. By placing an auxiliary burner in the stack to reheat the waste gas to a higher temperature and by adding additional air or oxygen, an afterburning is achieved which gives acceptable products of combustion. Essentially what has been done is to take a poor combustion operation and improve it by the addition of air to complete what really might have been done in the first place. Thermal incineration may take on the aspect of afterburning, but it is not necessarily required if good primary burning is accomplished.

The afterburner should be differentiated from fume incineration even though they are both forms of thermal incineration. Afterburning is necessitated by poor combustion in the process preceding the afterburner, but fume incineration is a primary burning operation and is installed following a process which produces a combustible or partially combustible mixture of hydrocarbons and air. Fume incineration, or thermal incineration, finds its application in a variety of processes, such as curing or drying ovens which volatilize a solvent, or in chemical processes which result in a hydrocarbon emission, or a petroleum refining process which releases light hydrocarbons.

Baking and curing operations which release oils or solvents to the atmosphere in various quantities mixed with air, or even food processing operations where the hydrocarbons may not be toxic but produce a definite odor, need to be treated in some way to make their effluent acceptable to the atmosphere.

It is absolutely necessary when a hydrocarbon is being burned that a stoichemetric or more than a stoichemetric amount of oxygen be provided in order to get good combustion. Therefore, to insure good combustion with a rich effluent, there should always be at least enough oxygen in the mixture to combine stoichiometrically with the hydrocarbons present to make the reaction go to completion.

14-6.3.1 Basic Considerations

In mixtures of combustible materials and oxygen or air, there are several critical combinations. One is known as the lower explosive limit and one is known as the higher explosive limit. Theoretically, between these two limits there is a combustible or explosive mixture, and these explosive limits are related to the hydrocarbon or the mixture of hydrocarbons in the air or other combinations of carrier gases containing oxygen. In most fume incineration applications, we are concerned with these limits with respect to air. It should be recognized that oxygen rather than air will tend to widen these limits, and inert gases such as carbon dioxide and nitrogen would tend to narrow them.

The lower explosive limit, often known as the lean limit, is the minimum hydrocarbon content in air which will support combustion once ignited. The higher explosive limit, or rich limit, is the limit above which there is not enough oxygen present to effect a combustion reaction at normal ignition temperatures; in other words, there is too much hydrocarbon and not enough oxygen. Between these two limits, however, there are a variety of mixtures which will burn upon ignition and, without controlled conditions might explode. Flammability limits are important in the handling of all wastes, especially solvent air mixtures. For additional information, the U.S. Bureau of Mines has published a paperback volume entitled, *Limits of Flammability of Gases and Vapors,* Bulletin No. 503.

For most applications involving the thermal incineration of fumes, insuring agencies, such as FM and FIA, require that the gases passing into a thermal incinerator be less than 25% of the lower explosive limit. This is done so that ignition of the fumes cannot occur in the ductwork to the incinerator nor can flame from the incinerator propogate back to the source of the fumes.

Fume incineration is basically the consideration of three conditions: time, temperature, and turbulence. The time is the residence time in the incinerator; the temperature is the temperature at which the reaction must be carried out; and the turbulence is the degree of mixing between air and/or combustion products from a preheat burner which is required to raise the mixture to a temperature where it will oxidize. Any chemical reaction takes a given period of time. It may never go to 100% completion as in the case with most combustion reactions, but the degree of incompleteness is insignificant if the residence time is sufficient. Residence time is measured as that period in which the reactants are held at a given temperature after being mixed in some fashion. The normal time for a combustion reaction for a hydrocarbon fuel varies from 1/10 of a second to several seconds, depending upon the temperature and the degree of mixing.

The operating temperature of the reaction is that temperature which will promote completion of the reaction between the reactants in the desired time period. If the temperature is increased, the reaction is accelerated. For example, a well-mixed system may be able to complete a given reaction in 3/10 of a second at 1800°F, but it may take 3 sec to complete the same reaction at 1300°F.

Turbulence is the degree of mixing between the reactants. In a combustion reaction, this is usually the mixing between air and some hydrocarbon. If the oxygen in the air does not come in contact with the hydrocarbon to be burned, then the hydrocarbon cannot burn; consequently, longer time periods are needed for the reaction or higher temperatures are needed for the reaction if good turbulence is not present.

There are basically three types of systems where thermal incineration can be employed. The first is a waste gas or fume that will sustain combustion without auxiliary fuel but with the addition of air to the waste gas. In this type of incineration system, the gas is fuel rich, and the waste gas may be fed directly into a combuster, such as a flare burner as may be seen in a refinery or chemical plant. The composition of combustibles in the waste gas is above the upper explosive limit so the material is nonexplosive until it is mixed with air. With this type of waste gas there usually is no concern of flashback because there is not enough air present to propogate a flame.

The second type of waste gas is one that will sustain combustion without the addition of auxiliary fuel and without the addition of air for combustion. Such a fume is a combustible mixture and is extremely dangerous because it is in the combustible range. Its combustibility lies above the lower explosive limit and below the upper explosive limit; therefore, a flame can propogate from the point of ignition back through the pipe or duct carrying the waste gas. Safeguards to prevent this must be applied in any system handling this type of waste.

The third type of waste gas or fume, and the one most normally handled in industrial thermal incineration processes, is one that will not maintain combustion without the addition of auxiliary fuel and contains enough air for combustion of its hydrocarbon content.

14-6.3.2 Equipment

Thermal incineration of waste gases is usually accomplished in one of two types of systems. If the gas is rich enough to burn by itself, then it may be flared either in an elevated flare or a ground flare. This merely requires some means of ignition and the gas will burn in atmospheric air, Figure 14-23. If it is a Type 2 waste gas within the combustible limits, then it will be extremely dangerous to burn and usually should be diluted below the lower explosive limit before ignition is accomplished.

The Type 3 waste gas, which is most commonly found in industry, may be handled in a thermal incinerator using an auxiliary fuel burner to bring the hydrocarbon-air mixture to a point several hundred degrees above the autoignition temperature of the hydrocarbon present in the waste stream. At this point, rapid oxidation will occur and the products of combustion will be clean carbon dioxide, nitrogen, and water vapor, with perhaps a small trace of residual

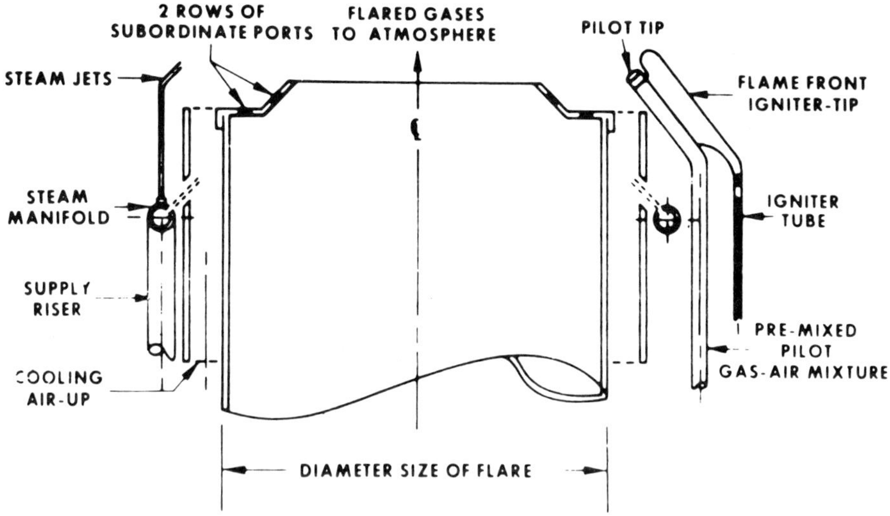

Fig. 14–23. Flare burner.

hydrocarbons (and inorganic compounds), depending upon the operating temperature.

Most solvents burn at temperatures in the 1300 to 1500°F range; however, it should be recognized that there will be some residual carbon monoxide at these temperatures since carbon monoxide will never completely oxidize to carbon dioxide until a temperature of about 1800°F is achieved. Several different types of thermal incinerators are shown in Figures 14–24 and 14–25.

14-6.4 Catalytic Oxidation

An alternative to straight thermal oxidation of gaseous wastes is the use of catalytic oxidation. The catalyst does nothing but accelerate the rate of reaction; it often allows a gaseous waste to burn at temperatures somewhat below the temperature at which it would burn in a thermal oxidizing unit. The catalytic system usually employs a noble metal catalyst coating on a ceramic base. This ceramic base may be a ceramic coated wire, pellets, or a honeycomb, onto which the noble metal catalyst has been applied, usually in solution form, so that when dried a certain residual of noble metal remains. It is not necessary in every case to use a noble metal catalyst, but for most hydrocarbon incineration applications, platinum, palladium or a mixture of platinum and palladium is used.

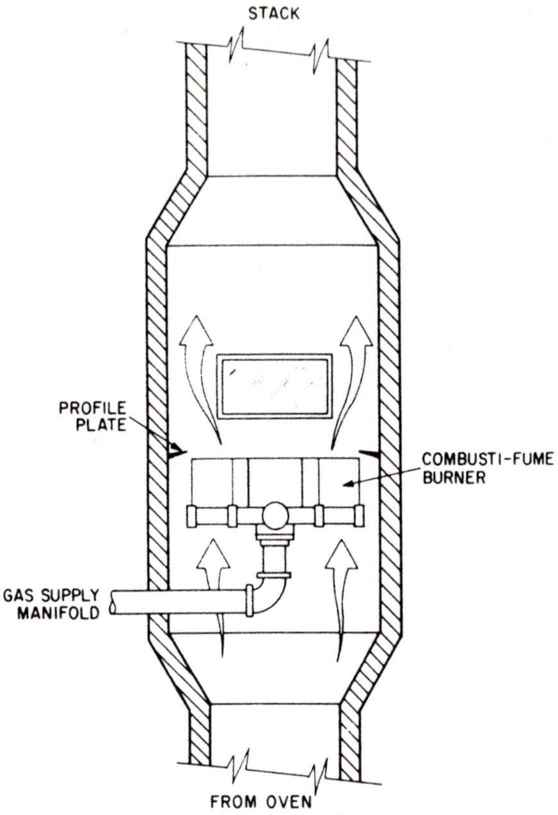

Fig. 14-24. Duct-type fume incinerator. ***Source: Air Pollut. & Ind.*** **Ref. 14-12. (*Courtesy:* Maxon Corp.).**

In the catalytic incinerator, the waste gas is heated by means of an auxiliary fuel burner to some predetermined temperature, passed over the catalyst where a definite temperature rise will occur and then out to atmosphere, if not through some heat recovery device. The hydrocarbon in the air burns on the surface of the catalyst in a localized combustion process. The catalytic type incinerator usually saves fuel because it achieves a temperature rise on the surface of the catalyst and requires a lower preheat temperature than the thermal type incinerator. However, the amount of heat released by the oxidation of the hydrocarbon should be exactly the same. The advantage of the catalyst type system is lower fuel consumption, but the catalyst in itself is more expensive and, consequently, the incinerator usually costs more. The catalyst is also subject to fouling by any particulate material and may be attacked by a variety of

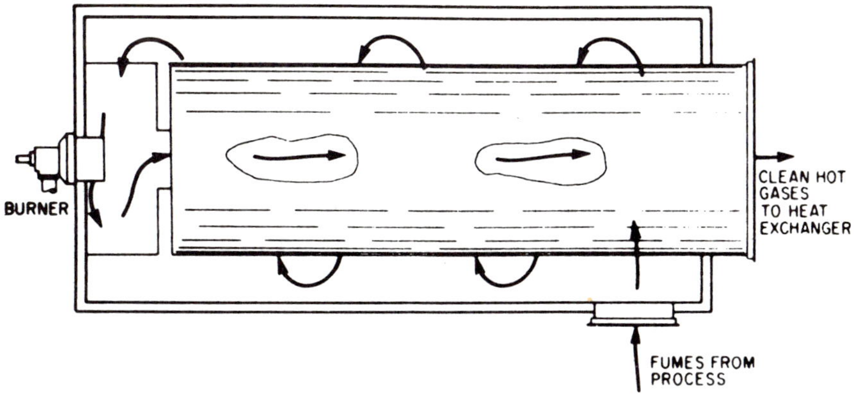

Fig. 14-25. Thermal fume incinerator. ***Source: Air Pollut. & Ind.*** **Ref. 14-12.**

chemical constituents in the waste gas, such as vanadium or other heavy metals, sulfur, and chlorine. All of these materials poison the catalyst to a varying degree and render it inactive.

The choice between catalytic and thermal incineration is usually an economic one and should be considered carefully. There have been dozens of economic comparisons made between catalytic and thermal gaseous waste incineration systems but no specific conclusions can be drawn without analyzing each case in detail. As a general statement, thermal incineration is less costly initially and represents a slightly higher operating cost. The catalytic system is usually applicable only for very lean hydrocarbon fumes since most catalysts will not survive at temperatures above 1500 to 1600°F.

14-6.5 Adsorption

Another possibility for the treatment of gaseous waste effluents is the process of adsorption. Generally, this involves the use of activated carbon. Vapor-phase activated carbon adsorption utilizes the very high surface area of activated carbon to pick up the hydrocarbon constituents in various waste gases. Figure 14-26 shows a typical two-adsorber system using activated carbon which is regenerated with steam. This type of system is used not only to prevent air pollution but to recover valuable solvents. In a typical organic solvent recovery plant, the contaminated air passes through an air filter to remove particulates and possibly a preheating coil to preheat the mixture and then through a bed of activated charcoal. The solvent-laden air will pass through this bed until the bed is eventually saturated or comes to a point called "break-through." The contaminated gas is then switched to the second bed, while the first bed is isolated by valves and regenerated by back-flushing with low pressure steam.

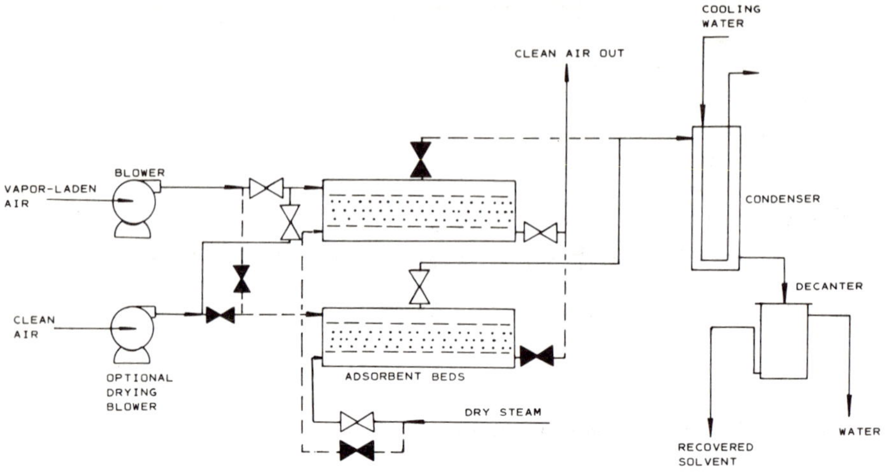

Fig. 14–26. Two bed carbon adsorber.

The combination of the steam and organic vapor then passes through a water-cooled condenser and if the solvent is immiscible in water, the solvent can be decanted from the solvent layer and the remaining water is discharged to the sewer. If, on the other hand, the solvent is miscible with water, then the solution must be distilled to recover the organic material.

Figure 14–27 shows a type of system which is regenerated with air in which a fume incinerator is used to burn the solvent removed from the regenerated activated charcoal bed. This type of system, unlike the steam system, does not permit the recovery of the solvent.

Adsorption systems are generally most economical when handling a reasonably concentrated fume stream, i.e., one which may be above the upper explosive limit. Organic compounds with a molecular weight over 45 or a boiling point over 0°C are potential candidates for such systems; the adsorption capacity of activated carbon tends to increase with the increasing concentration. The adsorption capacity of activated carbon also varies inversely with temperature, so it is necessary to maintain lower temperatures on the adsorbing cycle and higher temperatures on the regeneration cycle.

Equipment for activated carbon systems generally utilize large vessels with deep horizontal beds of activated charcoal. The incoming gaseous waste passes into the bottom of the vessel and up through the bed, exiting at the top. When the first bed becomes saturated then the valves switch and the bed is regenerated while the second bed begins the adsorption cycle. Three bed systems are often used so that one bed may be cooled after regeneration by high temperature steam. This merely increases the bed's capacity for the next adsorption cycle.

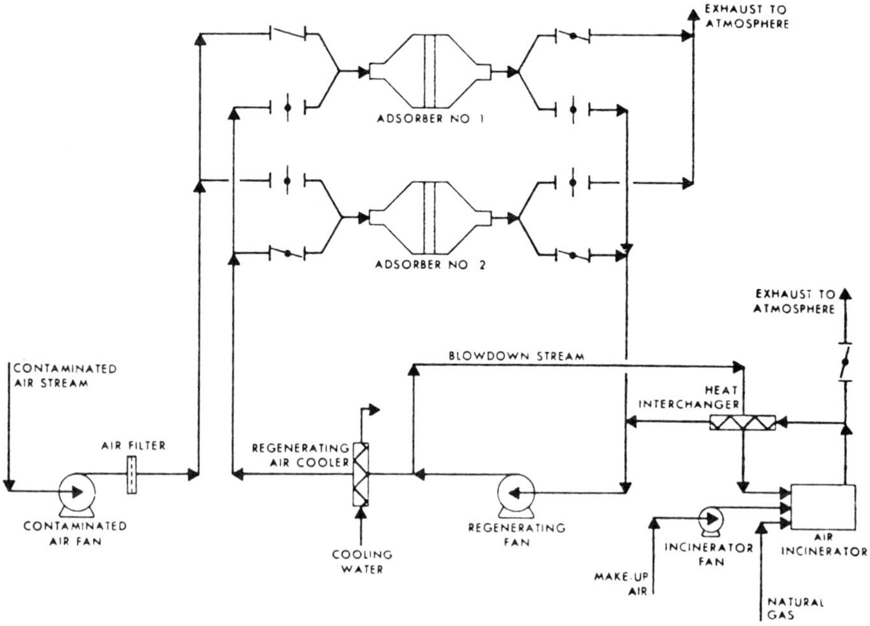

Fig. 14-27. Air regenerated carbon adsorption system.

Generally, activated carbon adsorption systems are expensive, but they are quite efficient and will remove more than 99% of all the hydrocarbon in the waste gas. They are, however, most economically efficient when they recover a high priced solvent for reuse in the plant operation.

There are other types of adsorption systems that use a number of vertical thin beds of fine carbon and others that use a rotating carbon bed arrangement which is alternatively adsorbing and desorbing, but functionally all work in the same manner.

14-6.6 Absorption and Chemical Oxidation

Waste gases that contain materials which are soluble in various liquids are suitable for absorption techniques. Such systems usually involve the scrubbing or absorption of some inorganic material, such as hydrogen chloride or sulfur trioxide, rather than a hydrocarbon material, but, depending upon the scrubbing medium used, various types of hydrocarbon materials could also be absorbed in such a system.

Absorption devices vary so widely that it would be impossible to discuss all the possibilities here. However, in most cases, they take the form of a packed tower. The tower may be packed with ceramic, metallic or plastic elements such as saddles, rings, or beads, or wire packing, or virtually any packing material which will provide a large surface area for absorption to take place. Since absorption is essentially a surface phenomenon, it requires excellent contact between the scrubbing liquid and the material to be absorbed.

For example, incineration might be used as the first step in disposing of a waste gas from a vinyl chloride monomer plant. Such waste contains small quantities of vinyl chloride and other chlorinated hydrocarbons. Incineration will convert these chlorinated materials into hydrogen chloride gas, water vapor, and carbon dioxide. If these materials are then contacted with water in a packed tower or similar type of scrubbing device, the hydrogen chloride is absorbed by the water and forms weak hydrochloric acid. With this type of system, it is possible by recirculation to make hydrochloric acid up to about 18% strength; at this point, we reach the azeotrope of hydrogen chloride and water where it is impossible to go to higher concentrations without using other methods.

Sulfur dioxide and sulfur trioxide may be absorbed and reacted in caustic, soda ash, or lime solutions as described earlier in this chapter. This is a combination of absorption and chemical reaction.

The equipment used for absorption applications consist of packed columns, plate columns, sparging systems, spray towers, and high energy scrubbers. Packed columns employ high liquid and gas rates, whereas plate columns are used for the absorption of gases and vapor when low liquid rates are desirable. Sparging systems are used for high gas rates which must be distributed throughout a large liquid volume, such as the carbonation of municipal water supplies with carbon dioxide. Spargers are not commonly used for waste gas absorption systems because they are only a single stage absorber, but they can be used where the waste gas or vapor is highly soluble. Spray towers are more often used for humidification of waste gases or for the reduction of temperature. Spray towers use large quantities of water in the form of fine droplets and absorption of the soluble vapor or gas occurs within the chamber. This is a less efficient method than plate or packed columns. High energy scrubbers (Venturis) are not generally used for gas absorption because of the higher energy requirements.

Sorption is also possible in the dry form, such as described previously for the sieve process; many of the earlier sulfur dioxide control processes used a dry sorption system. The dolomite injection process developed some years ago injected dry dolomite into a boiler burning a fossil fuel. The limestone reacted with the sulfur dioxide in the flue gases directly in the boiler forming calcium sulfate or gypsum which was removed by subsequent wet scrubbing.

The Cat-Ox process developed by Monsanto passes the flue gases through a vanadium pentoxide catalyst after initial particulate removal to absorb sulfur dioxide which is then used to produce sulfuric acid. The Reinluft process absorbs sulfur dioxide on a char formed by calcining coal at about 1100°F, and the alkalyzed aluminum process developed by the U.S. Bureau of Mines absorbs sulfur dioxide on a bed of sodium aluminate. The absorber in this case is a fluidized bed. The alkalyzed alumina reacts with the sulfur dioxide in the flue gases converting it to a sulfate which is regenerated by heating to a temperature of about 700°F in the fluid bed using a reducing gas obtained by reforming natural gas or a suitable hydrocarbon feed stock. The product of this process is hydrogen sulfide which can then be converted by the Claus process to elemental sulfur.

Permanganate and Chlorine Oxidation. Strong oxidizing agents, such as potassium permanganate and chlorine or sodium hypochlorite or calcium hypochlorite, can be used to oxidize small quantities of hydrocarbon materials, especially where odor problems exist. These materials in relatively dilute solution can be used to scrub waste gases containing small quantities of hydrocarbons and oxidize them in the scrubber so that there is no residual hydrocarbon in the stack. The major problem with the use of chlorine, hypochlorite, or permanganate is corrosivity of the solution.

In the case of permanganate, this can be overcome by dissolving the potassium permanganate in a mildly alkaline solution where it will produce an insoluble precipitate of manganese dioxide and no permanganic acid. Permanganate oxidation is expensive, and the use of chlorine or chlorine dioxide or various hypochlorite solutions is usually less expensive but not as oxygen efficient as the permanganate oxidation. It should be remembered that these chemicals are only used for very low concentrations of hydrocarbons (generally not over several hundred parts per million) because they are too expensive to use for high concentrations.

14-6.7 Condensation

There is the possibility of using condensation as a means of removing gaseous pollutants, although this is usually only practical where the vapor present has a high dewpoint and recovery of the condensed vapor is simple and profitable. For example, a solvent vapor coming from a vat or mixing tank will usually contain very little air and possibly some inert gases if the inert gas has been used to blanket the tank. By dropping the temperature below the dewpoint of this solvent-air mixture, either by cooling with chilled water or a refrigerant, most of the solvent can be recovered and reused. Condensation is usually applicable to high concentrations of organics or water vapor in the effluent. Condensation can take place in shell and tube type condensers or air cooled condensers or it can be handled in a direct contact type condenser; however, the direct contact condenser is only applicable to materials which will be discarded rather than saved.

REFERENCES

14-1. Report of the Secretary of Health, Education and Welfare to the U.S. Congress, "National Emissions Standards Study," U.S. Government Printing Office, March 1970.

14-2. Environmental Protection Agency, "The National Air Monitoring Program: Air Quality and Emission Trends," Vol. **1**, pp. 1-6 ff, 1973.

14-3. "Control Techniques for Particulate Air Pollutants," AP-51, U.S. Department of HEW, Public Health Service, January 1969.

14-4. Stumph, Terry L. and Duprey, Robert L., "Trends in Air Pollution Control Regulations," NAPCA, June 1968.

14-5. *Journal of the Air Pollution Control Association,* from a paper presented at the May 18, 1973, meeting in Baltimore of the South Atlantic States Section of APCA outlining the reasoning which prompted the Industrial Gas Cleaning Institute Product Division to prepare this consensus data. The paper was presented by Eugene P. Stastny, Koppers Co., Inc., 1973, Vice President IGCI.

14-6. Faith, W. L., "An Evaluation of Fugitive Dust Regulations," presented at the APCA Meeting, Denver, Colorado, June 9-13, 1974.

14-7. "Compilation of Air Pollutant Emission Factors," 2nd Ed., AP-42, USEPA, April 1973.

14-8. *Stack Sampling News,* Technomic Publishing Co., Inc., Vol. **3**, No. 2, August, 1975.

14-9. "Interlaboratory Cooperative Study of Air Quality Measurement Methods," DS 55-56, ASTM, June 1974.

14-10. "Abatement of Particulate Emissions from Stationary Sources," reproduced with permission from the National Academy of Sciences. A report of the National Academy of Engineering. Published by the National Academy of Science.

14-11. *Stack Sampling News,* Technomic Publishing Co., Inc., Vol. **3**, No. 4, October 1975.

14-12. "Air Pollution and Industry," R. D. Ross, ed., Van Nostrand Reinhold Co., Table 8-1, pp. 342-343, 1972.

14-13. Cross, Frank L., Jr., "Handbook on Air Pollution Control," Technomic Publishing Co., Inc., Westport, Connecticut.

14-14. Friedman, Leonard J., "Ammonia Scrubbing of Sulfuric Acid Tail Gas—Recent Davy Powergas Experience."

14-15. "Technology–Air Pollution Control," *Chem. Eng.,* Deskbook Issue, June 21, 1971.

14-16. Weir, A., Jr., "Clearing the Opacity Issue," *Environ. Sci. Technol.,* Vol. **11**, No. 6, June 1977.

14-17. Easton, E. B. and O'Donnell, F. J., "The Clean Air Act Amendments of 1977," *J. Air Pollut. Control Assoc.,* Vol. **27**, No. 10, October 1977.

14-18. Bethea, R. M., "Control of Industrial Air Pollution," *Pollut. Eng.,* 1977 Yearbook Issue.

14-19. Harrison, E. B., "How Did We Get Where We Are?", *Hydrocarbon Processing,* October 1977.

14-20. Doerschlag, C. and Miczek, G., "How to Choose a Cyclone Dust Collector," *Chem. Eng.,* February 14, 1977.

14-21. Editorial, "EPA Sets Its Sights on Nixing CPI NO_X Emissions," *Chem. Eng.,* February 14, 1977.

14-22. Stern, A. C., "Prevention of Significant Deterioration," *J. Air Pollut. Control Assoc.,* Vol. **27**, No. 5, May 1977.

14-23. Cheremisinoff, P. and Young, R., *Air Pollution Control and Design Handbook,* Marcel Dekker, New York 1977.

General Appendix

Metric Conversion Tables

Recommended Units				
Description	Unit	Symbol	Comments	Customary* Equivalents
Length	meter	m	Basic SI unit	39.37 in. = 3.28 ft = 1.09 yd
	kilometer	km		0.62 mi
	millimeter	mm		0.03937 in.
	centimeter	cm		0.3937 in.
	micrometer	μm or μ		3.937×10^{-5} in. = 10^4 Å
Area	square meter	m^2		10.76 sq ft = 1.196 sq yd
	square kilometer	km^2		0.3861 sq mi = 247 acres
	square centimeter	cm^2		0.155 sq in.
	square millimeter	mm^2		0.00155 sq in.
	hectare	ha	The hectare (10,000 m^2) is a recognized multiple unit and will remain in international use.	2.471 acres
Volume	cubic meter	m^3		35.314 cu ft = 1.3079 cu yd
	cubic centimeter	cm^3		0.061 cu in.
	liter	l	The liter is now recognized as the special name for the cubic decimeter	1.057 qt = 0.264 gal = 0.81×10^{-4} acre-ft
Mass	kilogram	kg	Basic SI unit	2.205 lb
	gram	g		0.035 oz = 15.43 gr
	milligram	mg		0.01543 gr
	tonne	t	1 tonne = 1,000 kg	0.984 ton (long) = 1.1023 ton (short)
Time	second	s	Basic SI Unit	
	day	day	Neither the day nor the year is an SI unit but both are important.	
	year	yr or a		
Force	newton	N	The newton is that force that produces an acceleration of 1 m/s^2 in a mass of 1 kg.	0.22481 lb (weight) = 7.2 poundals
Velocity linear	meter per second	m/s		3.28 fps

Recommended Units				
Description	Unit	Symbol	Comments	Customary* Equivalents
	millimeter per second	mm/s		0.00328 fps
	kilometers per second	km/s		2,230 mph
angular	radians per second	rad/s		9.549 rpm
Flow (volumetric)	cubic meter per second	m^3/s	Commonly called the cumec	15,850 gpm = 2,120 cfm
	liter per second	l/s		15.85 gpm
Viscosity (dynamic)	millipascal second	mPa · s		1 centipoise
Pressure	newton per square meter	N/m^2	The newton is not yet well-known as the unit of force and kgf/cm^2 will clearly be used for sometime. In this field the hydraulic head expressed in meters is an acceptable alternative.	0.000145 psi
	kilonewton per square meter	kN/m^2		0.145 psi
	kilogram (force) per square centimeter	kgf/cm^2		14.223 psi
	pascal	Pa		0.000145 psi
	kilopascal	kPa		0.145 psi
Temperature	degree Kelvin	K	Basic SI unit	$\frac{5F}{9} - 17.77$
	degree Celsius	C	The Kelvin and Celsius degrees are identical. The use of the Celsius scale is recommended as it is the former centigrade scale.	
Work, energy, quantity of heat	joule	J	1 joule = 1 N-m	2.778×10^{-7} kw hr = 3.725×10^{-7} hp-hr = 0.73756 ft-lb = 9.48×10^{-4} Btu
	kilojoule	kJ		2.778×10^{-4} kw-hr
Power	watt	W	1 watt = 1 J/s	44.25 ft-lb/min
	kilowatt	kW		1.341 hp
	joule per second	J/s		3.412 Btu/hr

Metric Conversion Tables (Continued)

Application of Units				
Description	**Unit**	**Symbol**	**Comments**	**Customary* Equivalents**
Precipitation, run-off, evaporation	millimeter	mm	For meteorological purposes it may be convenient to measure precipitation in terms of mass/unit area (kg/m^2). 1 mm of rain = 1 kg/sq m	
Flow	Cubic meter per second	m^3/s		35.31 cfs
	liter per second	l/s		15.85 gpm
Discharges or abstractions, yields	cubic meter per day	m^3/day	1 l/s = 86.4 $m^3/$day	0.1835 gpm
	cubic meter per year	m^3/yr		264.2 gal/yr
Density	kilogram per cubic meter	kg/m^3	The density of water under standard conditions is 1,000 kg/m^3 or 1,000 g/l	0.0624 lb/cu ft = e kg/m^3 62.4 lb/cu ft = density of water
Concentration	milligram per liter	mg/l		1 ppm
BOD loading	kilogram per cubic meter per day	kg/m^3 day		0.0624 lb/cu-ft day
Hydraulic load per unit area; e.g. filtration rates	cubic meter per square meter per day	m^3/m^2 day	If this is converted to a velocity, it should be expressed in mm/s (1 mm/s = 86.4 m^3/m^2 day).	3.28 cu ft/sq ft day
Hydraulic load per unit volume; e.g. biological filters, lagoons	cubic meter per cubic meter per day	m^3/m^3 day		
Optical units	lumen per square meter	$lumen/m^2$		0.093 ft candle/sq ft

*Miles are U.S. statute, qt and gal are U.S. liquid, and oz and lb are avoirdupois.
Source: *Upgrading Lagoons,* EPA Technology Transfer Publication, EPA-625/4-73-001b, U.S. Environmental Protection Agency, Washington, D.C.

Index

Index